T0206106

STELLAR EXPLOSIONS
Hydrodynamics and Nucleosynthesis

Series in Astronomy and Astrophysics

The Series in Astronomy and Astrophysics includes books on all aspects of theoretical and experimental astronomy and astrophysics. Books in the series range in level from textbooks and handbooks to more advanced expositions of current research.

Series Editors:
M Birkinshaw, University of Bristol, UK
J Silk, University of Oxford, UK
G Fuller, University of Manchester, UK

Series in Astronomy and Astrophysics

STELLAR EXPLOSIONS
Hydrodynamics and Nucleosynthesis

Jordi José

Department of Physics, Technical University of Catalonia (UPC), Barcelona

CRC Press
Taylor & Francis Group
Boca Raton London New York

CRC Press is an imprint of the
Taylor & Francis Group, an **informa** business

CRC Press
Taylor & Francis Group
6000 Broken Sound Parkway NW, Suite 300
Boca Raton, FL 33487-2742

First issued in paperback 2020

© 2016 by Taylor & Francis Group, LLC
CRC Press is an imprint of Taylor & Francis Group, an Informa business

No claim to original U.S. Government works

Version Date: 20151110

ISBN 13: 978-0-367-57515-1 (pbk)
ISBN 13: 978-1-4398-5306-1 (hbk)

Visit the Taylor & Francis Web site at
http://www.taylorandfrancis.com

and the CRC Press Web site at
http://www.crcpress.com

In loving memory of my parents,
Tomàs José [1926–2012] and
Rosa Pont [1927–2014],
wherever their atoms have gone.

Contents

Foreword

Stars and their explosions evoke a profound feeling of awe. Most elements, including the carbon that life is based on, the oxygen in the air we breathe, the calcium in our bones, the iron in our blood cells, and the iodine in our thyroids, were synthesized in stars and were expelled by their violent explosions. The Solar System formed from matter enriched by stellar ejecta. Therefore, the story of our origin is closely connected to the complex interplay of physical processes in these stars, spanning from theoretical astrophysics to observational astronomy, from cosmochemistry to nuclear physics. Stars and their explosions also tell a story of paramount interest to human thought: the formation of white dwarfs, neutron stars, and black holes, and the expansion of the universe. Surprisingly, very few books for students and young scientists have been written so far on this fascinating subject.

Stellar Explosions: Hydrodynamics and Nucleosynthesis by Dr. Jordi José explores this story in all its beauty and complexity. Every chapter abounds with enthusiasm. The book accomplishes the difficult task of presenting a staggering amount of information in a clear, complete, and coherent manner. It begins with a chapter on stellar hydrodynamics, and then continues with chapters on nuclear physics aspects, thermonuclear reactions, cosmochemistry, and presolar grains. Subsequent chapters are devoted to specific explosive events: novae, thermonuclear (type Ia) supernovae, X-ray bursts, and, finally, core-collapse supernovae. Each chapter gradually introduces the reader to key ideas and concepts that are necessary to grasp the complex interplay between different topics. With its captivating discussions and explanations, readers from a wide range of educational levels will be able to comprehend the book.

Dr. José is an accomplished expert on stellar astrophysics. Over the past quarter century, he has published numerous original research articles on the broad topics covered in this book. The reader will be treated to an accessible and rigorous discussion commingled with a highly engaging style.

Science will continue to uncover the unimaginable, as long as young scientists immerse their minds into the world of astrophysics. This book is destined to ignite the flame of curiosity and excitement in the reader, and will no doubt play a major role in motivating the next generation of stellar astrophysicists. I am confident that this masterful account will become a cornerstone for stellar astrophysics education.

Christian Iliadis
Professor of Physics and Astronomy
University of North Carolina at Chapel Hill

Preface

Stars are nuclear furnaces in which the chemical abundance pattern of the universe is continuously reshaped. Much has been achieved since the early, naked-eye observations of stars carried out by ancient astronomers. In fact, our current understanding of the physics of the stars and of their role as nucleosynthesis factories owes much to cross-fertilization of different, somehow disconnected fields. Access to space, for instance, has opened new windows to study the cosmos from a novel panchromatic perspective. Indeed, since the last decades, UV, X- and γ-ray space-borne observatories have been used simultaneously to ground-based optical and radiotelescopes to analyze the behavior of stars at different wavelengths. In parallel to the elemental stellar abundances inferred spectroscopically, cosmochemists are now providing astonishingly precise isotopic abundance ratios in micron-sized presolar grains extracted from meteorites. Encapsulated in those grains, there is pristine information about the suite of nuclear processes that took place in their stellar progenitors, which translate into huge isotopic anomalies with respect to bulk solar system material. The dawn of supercomputing has also provided astrophysicists with the appropriate arena in which complex physical phenomena that require a truly multidimensional approach can now be tackled (e.g., convective transport, mixing, and flame propagation in stars). Last but not least, nuclear physicists have developed new experimental techniques to measure nuclear interactions at or close to stellar energies, at the so-called *Gamow window*, thus reducing the burden (and also the risks) of extrapolation from laboratory energies down to stellar energies. Understanding of the progress achieved so far in nuclear astrophysics and of its current challenges requires, therefore, some background on observational astronomy, cosmochemistry, computational astrophysics, and experimental (and theoretical) nuclear physics. Providing such a multidisciplinary background in a coherent way has been the driving force for this book.

Frequently, young PhD students and postdocs attend international conferences and workshops early in their scientific careers. The two most relevant, periodic forums in this field, "Nuclei in the Cosmos" and "Nuclear Physics in Astrophysics," combine technical talks in the different areas mentioned above. In this framework, the young (and sometimes the not so young) researchers, whose basic training is predominantly either astrophysics or nuclear physics, but rarely both, may get easily lost, overwhelmed by a number of questions: *How are stellar abundances determined? What are corundum grains? Which features distinguish a 1D hydro code and a 1-zone code? What is the difference between a deflagration and a detonation, and in which context are those modes of flame propagation expected to occur in a star? How are nuclear reactions actually measured in the lab and how do the bombarding energies adopted compare to stellar energies? Why and when can underground facilities be useful to perform such measurements?* This book attempts to answer some of these questions, providing basic background on computational astrophysics, cosmochemistry, and nuclear physics along the first three chapters. The basic concepts addressed get intertwined in all their full splendor in subsequent chapters, when applied to the realm of stellar explosions (i.e., classical and recurrent novae, type I and II supernovae, X-ray bursts and superbursts, and stellar mergers), proving how multidisciplinarity has been instrumental in our understanding of nucleosynthesis in stars (and of stellar explosions, in particular).

Each chapter finishes with two summary boxes, with an overview of achievements and facts, together with a list of current problems and remaining questions. They represent somewhat the *greatest hits and the B-sides* of each topic (those readers old enough to remember music's vinyl era will likely understand the comment). The list of mysteries and challenges, by no means exhaustive, highlights a number of questions that deserve further attention, and likely new ideas and approaches that may be undertaken and explored in future PhD theses or research projects. Supplementary material, including color plots, animations, and a solutions' manual for the exercises proposed, is available at an accompanying webpage, http://fisica.upc.edu/ca/users/jjose/CRC-Downloads.

The book contains as well two appendices, aimed at further extending the description and development of a hydrodynamic code. They include a source code specifically developed to illustrate the art of numerical simulation applied to a simple test case: the free-fall collapse of a homogeneous sphere. The code is written in Fortran. This may sound terribly old-fashioned as other programming languages (e.g., C++, Python...) are gaining momentum, and may eventually dominate this field in the nearby future. The fact is that many astrophysicists (self-)learned Fortran as the main programming language at a very young age (in some sense, astrophysicists are "Fortran-native speakers"). As such, many existing codes, inherited from generation to generation, are actually written in this language and, therefore, even the newest generations of researchers also get accidentally exposed to Fortran. Nevertheless, a version of the free-fall collapse code in C will be posted soon at the accompanying webpage mentioned above, for those who may consider themselves "Fortran-challenged."

Every single book has a story behind it, and the one that you are holding in your hands (or reading on a tablet or computer screen) is not an exception. The project traces back to a meeting of the American Physical Society that took place in Washington, D.C. in 2010. I was cordially invited by Dr. Christian Iliadis to give a talk on X-ray bursts, in a session specifically devoted to nucleosynthesis and stellar explosions. The title of my talk caught the eye of Dr. John Navas, senior editor at Taylor & Francis/CRC Press, who was also attending the event. I had the chance to meet Dr. Navas in Washington and discuss his proposal to write a book on nucleosynthesis. I thought that a book simultaneously addressing computational astrophysics, cosmochemistry, and nuclear physics at an introductory level, was missing and, therefore, accepted the challenge. A few months later, I submitted a full proposal, focused on stellar explosions, which was positively reviewed by four anonymous referees. The project, therefore, got a green light...

Writing a book is a truly titanic goal. It usually starts as a smooth marathon that ends up frantically, like in a 400-m sprint (400-m hurdles, to be more precise). Indeed, all sorts of unexpected things interfered with the preparation of this manuscript, sometimes dramatically. I ended up sympathizing with Sisyphus, king of Ephyra in Greek mythology, known for his chronic frustration, doomed with an unreachable goal that he was compelled to pursue over and over. Against all odds, and after what seems like an eternity, the *doom* is finally over. But not without a toll: Even though, in its current final form, the manuscript exceeds by more than 150 pages its expected length, many interesting aspects had to be painfully left aside and did not make it into the book. This includes a more in-depth description of certain aspects of observational astronomy (spectroscopy, photometry), atomic physics (equation of state, opacities), and likely a more detailed account of the physics of core-collapse supernovae and stellar compact mergers.

The manuscript has survived a number of editorial changes at Taylor & Francis/CRC Press. Francesca McGowan replaced John Navas as the editor in charge of this project, while Rachel Holt (John Navas's editorial assistant) was subsequently replaced by Sarah Gelson and, later on, by Sarfraz Khan, as Francesca McGowan's assistants. Kathryn Everett has, however, survived as project coordinator through the entire project. I owe them all a debt of gratitude for continuous support and assistance. I would also like to extend acknowledgments to those who have been closely working with me during the last stages of production, soon after completion of the full manuscript. In particular, Robin Lloyd-Starkes (project editor) and Alex Edwards (editorial assistant). Special acknowledgments are also devoted to Shashi Kumar, the LaTex expert at CRC Press, who solved all my increasingly difficult questions during preparation of the camera-ready version of the manuscript. On the graphic design front, Josep Escarp is wholeheartedly acknowledged for his commitment and care in preparing all figures and plots displayed in this book, frequently facing the challenge of a poor-quality original. Finally, Dr. Jordi Casanova, a former PhD student of mine (and despite of this, a true friend), unselfishly devoted quite some time to provide the figures and movies of the hydro test cases presented in Chapter 1.

Some people claim that physicians bury their mistakes, lawyers send them to jail, while physicists get them published... In an attempt to minimize such a threat, a number of distinguished colleagues agreed to review each of the individual chapters of this book. Their contributions have dramatically improved the manuscript, and I'm extremely grateful to them for such an unselfish investment of time to this project (any potential typo or error leftover is, of course, my entire fault). In particular, I would like to thank Sachiko Amari, Alan C. Calder, Alain Coc, Roland Diehl, Alessandro Ederoclite, Domingo García-Senz, Margarita Hernanz, Christian Iliadis, Jean in't Zand, Jordi Isern, Karl-Ludwig Kratz, John Lattanzio, Manuel Linares, Richard Longland, Maria Lugaro, Uli Ott, Anuj Parikh, Glòria Sala, Hendrik Schatz, Michael Shara, Ivo Seitenzahl, Steven N. Shore, Francis X. Timmes, Helena Uthas, and the late Ernst Zinner. I would also like to extend my gratitude to Joana Figueira and David Martin, members of my research group at UPC, who also scrutinized the text with sharp eyes, providing valuable feedback.

A book constitutes also a wonderful opportunity to look backward in time and gain perspective. In this regard, I'm deeply indebted to Ramon Canal and Jordi Isern, for reinforcing my interest (passion, actually) in stellar explosions, and for countless scientific discussions and enlightment. To Margarita Hernanz, for continuous support, encouragement, advice, and a long-lasting, fruitful collaboration that has spanned more than a quarter of a century and will continue through the years to come. Along the past 25 years I also had the chance to interact with many colleagues. I feel terribly indebted to many of them. Even though they may have not directly contributed to this book, their continuous efforts to shed light into scientific darkness, shared with colleagues in countless meetings, conferences, and workshops, have been vital in the often tortuous but breathtaking path to scientific discovery. In fact, the type of interactions that I experienced with the whole nuclear astrophysics community bears a clear resemblance with the contents of this book, since they range from strong to weak interactions, and even occasionally, scattering or breakup. To all of them, my deepest appreciation.

And last but by no means least, I feel terribly indebted to my wife, Pilar, and to my son, Arnau, for so many, too many stolen moments invested in the last years in writing this book.[1]

Jordi José
Professor of Applied Physics
Barcelona

---

[1] The author would be glad to receive feedback from the readers, whether positive or negative. They may contact him by email at `jordi.jose@upc.edu`.

First, my fear; then, my curtsy; last, my speech.
My fear is, your displeasure; my curtsy, my duty;
and my speech, to beg your pardons. If you look for a good speech now,
you undo me; for what I have to say is of mine own making; and what
indeed I should have say will, I doubt, prove mine own marring. But to
the purpose, and so to the venture.

William Shakespeare, *Henry IV*, Part II (1597)

Symbol Description

a   Acceleration (cm s^{-2}).

A   Mass number.

B(Z,N)   Binding energy of a nucleus (MeV).

c   Speed of light in vacuum, 2.998×10^{10} cm s^{-1}.

dv Volume element (cm^3).

E   Internal energy per unit mass (erg g^{-1}).

F   Force (dyn).

G   Gravitational constant, 6.67×10^{-11} $\text{dyn cm}^2 \text{ g}^{-2}$.

h   Planck constant, 6.626×10^{-27} erg s.

k   Boltzmann constant, 1.381×10^{-16} erg K^{-1}.

L   Luminosity (erg s^{-1}).

L_{nuc} Nuclear luminosity (erg s^{-1}).

$L_\odot$ Solar luminosity, 3.846×10^{33} erg s^{-1}.

m(r)   Mass enclosed in a sphere of radius r (g).

M_{sun} Absolute magnitude of the Sun, 4.76^{mag}.

$M_\odot$ Solar mass, 1.989×10^{33} g.

n   Number density (cm^{-3}).

N_{Av} Avogadro number, 6.022×10^{23} mol^{-1}.

p   Linear momentum (kg m s^{-1}).

P   Pressure (dyn cm^{-2}).

q   Artificial viscosity (dyn cm^{-2}).

Q   Heat (erg).

Q   Reaction Q-value (MeV).

r   Radial distance from the stellar center (cm).

$R_\odot$ Solar radius, 6.96×10^{10} cm.

S(E)   Astrophysical S-factor (eV b)

t   Time (s).

$T_{1/2}$ Half-life (s).

u   Atomic mass unit, 1.66054×10^{-24} g ($931.494 \text{ MeV}/c^2$).

u   Velocity (cm s^{-1}).

V   Specific volume ($\text{cm}^3 \text{ g}^{-1}$).

W   Work (erg).

X   Mass fraction.

Y   Mole fraction (mol g^{-1}).

z   Redshift.

Z   Metallicity.

Z   Atomic number.

ε Nuclear energy generation rate ($\text{erg g}^{-1} \text{ s}^{-1}$).

λ Decay constant (s^{-1}).

ρ Mass density (g cm^{-3}).

σ Stephan–Boltzmann constant, 5.671×10^{-5} $\text{erg cm}^{-2} \text{ K}^{-4} \text{ s}^{-1}$.

σ Cross-section (b).

τ Mean lifetime (s).

1

Computational Hydrodynamics

"For many problems in the theory of the stellar interior the speed of numerical integrations by hand is entirely sufficient. A person can usually accomplish more than twenty integration steps per day for a set of differential equations [...] Thus for a typical single integration consisting of, say, forty steps less than two days are needed. Correspondingly, if, for example, a set of models is to be determined and if these models are to be constructed of a one-parameter family starting from the surface and a one-parameter family starting from the core, and if each of these two families can be represented with sufficient accuracy by, say, six individual integrations, then the entire numerical work for this fairly typical case can be accomplished by one person in one month. However, if extensive evolutionary model sequences including a variety of physical complications are to be derived, then numerical integrations by hand may become prohibitive and the advantage of large electronic machines will be incontestable."

Martin Schwarzschild, *Structure and Evolution of the Stars* (1958)

The big picture that emerged from the theory of stellar evolution owes substantial credit to computational astrophysics. Certainly, a number of observational breakthroughs have helped to pave the road by shedding light onto long-standing enigmas of the physics of stars. But only through numerical simulation have astrophysicists witnessed the full lifecycle of a star, from birth to death, spanning millions/billions of years of evolution, or have experienced the extreme conditions that characterize stellar interiors during their titanic explosions. Computational astrophysics has experienced a dramatic progress in recent years. The computational capabilities now available (Figure 1.1) were just a dream a few decades ago. Indeed, the concept of numerical models *integrated by hand*, as stressed in the quote from Martin Schwarzschild, can hardly be taken seriously anymore, in the era of state-of-the-art, multidimensional simulations.

Stars can be described by the set of conservation equations of fluid dynamics (i.e, mass, energy, and momentum conservation), together with supplementary relations that account for energy production (e.g., nuclear reactions), energy transport (by radiation, conduction, or convection), plus a suitable equation of state that describes the thermodynamics of the fluid. The evolution of continuous, Newtonian fluids[1] can be described by the Navier–Stokes equations, a set of nonlinear, partial differential equations with no known analytical solu-

---

[1] Fluids are often classified according to different properties. For instance, we distinguish between compressible and incompressible fluids, depending whether they undergo significant changes in density as they flow. Even though all real fluids are compressible, to some extent, liquids can be approximated as incompressible fluids; gases, instead, are highly compressible. Viscous forces acting on a fluid can be large or small compared to inertial forces. This yields to the distinction between viscous and inviscid fluids. Inviscid flows, for instance, are characterized by a *Reynolds number* $R_e \gg 1$ (i.e., ratio of inertial to viscous forces defined as $R_e = \rho u L / \mu$, where ρ is the density, u the velocity, L a characteristic length, and μ the dynamic viscosity of the fluid). Fluids are called Newtonian (non-Newtonian) when the viscous stresses arising from their flow are (are not) linearly proportional to the local strain rate, or rate of variation of its deformation over time. Finally, we can distinguish between laminar and turbulent flows, depending on the way the fluid flows (in parallel vs. random, superposed layers). Flows characterized by $R_e > 5000$ are typically—but not necessarily!—turbulent, while those at low Reynolds numbers usually remain laminar. In the case of a flow moving through a straight pipe with a circular cross-section, transition between laminar and turbulent flows occurs at a Reynolds number of ~ 2000.

FIGURE 1.1

The MareNostrum supercomputer, located inside an ancient chapel at the Technical University of Catalonia (UPC, Barcelona). In its current configuration, MareNostrum has 48,896 Intel SandyBridge processors in 3056 nodes, and 84 Xeon Phi 5110P in 42 nodes, with a peak performance of 1.1 petaflops. A color version of this picture is available at `http://fisica.upc.edu/ca/users/jjose/CRC-Downloads`. Image courtesy of Barcelona Supercomputing Center (www.bsc.es).

tion. Unless a number of simplifications are introduced, Navier–Stokes equations require a numerical approach. To this end, they have to be rewritten in a suitable, computer-friendly form (e.g., finite differences, finite elements, particles) that does not always bear full resemblance to the original equations. In this framework, numerical solutions have to be taken with caution and must be considered only as approximate solutions. Indeed, the robustness of the solution has to be checked against any control parameter or simplification undertaken (e.g., from time-step constraints to the size of the computational domain, or the resolution adopted). From a wider perspective, numerical codes should undergo extensive verification and validation, testing their performance in the framework of physical problems with a well-established solution. Unfortunately, test problems with analytical solution are frequently so simple that only a subset of modules or subroutines, rather than the full performance of the code, are actually checked. A thorough comparison of results obtained with similar codes turns out to be a reliable alternative, although it is hard to overcome the natural reluctancy of competing teams in sharing information to this end. Intuition on the behavior of the expected solution may also help (particularly when intuition proves right, which is not always the case!). Laboratory experiments offer an appealing but challenging alternative for validation of numerical codes. However, the framework in which such experiments can be actually conducted is often far from the conditions achieved in the stellar interiors. Besides, one can easily figure out the problems faced by experiments aimed at validating models of thermonuclear explosions, not only because of the extreme physical conditions required, but also because of the classified nature of such tests.

The arsenal of mathematical methods used in astrophysics encompasses many aspects of linear algebra and calculus (in particular, the solution of ordinary and partial differential equations). This is a very rich field, certainly too broad for an in-depth analysis in an introductory book on stellar explosions. Interested readers can find a wealth of information on these topics in many specialized publications [209,227,637,1448,1500,1812]. This chapter, instead, specifically focuses on the realm of stellar evolution codes (with particular emphasis on Lagrangian hydrodynamic formulations), reviewing the basic equations of stellar physics, the numerical techniques frequently adopted, and a suite of specific hydrodynamic concepts often encountered at different stages during the evolution of stars [209,279,988,1083,1634].

1.1 To Grid or Not to Grid: A Primer on Hydrocodes

Fluids are substances (i.e., liquids, gases, plasmas, and even solids with large plasticity) that deform under the influence of shear stresses. As such, they are ubiquitous in the universe on all scales, which in turn makes fluid dynamics one of the widest research areas, with interests spanning from weather predictions to airplane design, or stellar evolution. Even though the physical conditions describing a star or, say, a terrestrial cloud (e.g., temperature, density, pressure ...) differ by orders of magnitude, the numerical tools used in their analyses share many common features and rely on a similar set of coupled, partial differential equations, which are not usually amenable to analytical solutions, except under highly restrictive conditions. Numerical solutions are actually obtained for a finite, discrete number of points of the computational domain (either a volume, a surface, or a line), which often, but not always, relies on a computational mesh or grid, representing a subdivision of the domain into cells or elements. These computational domains can be mapped by means of structured and unstructured (or even hybrid) meshes. Structured (or regular) meshes are families of gridlines, topologically equivalent to Cartesian grids. They are usually applied to simple computational domains, since complex problems may result in unwanted grid distortions and consequently may lead to a computational failure. Unstructured meshes instead are more versatile and better suited for arbitrary domains. However, they are more difficult to implement. All numerical methods that rely on a mesh, whether structured or not, are usually referred to as *grid-* or *mesh-based* methods. In contrast, *gridless* or *meshfree* methods directly rely on the simulated object, avoiding mesh tangling and distortion, which are frequently encountered in traditional mesh-based methods in the presence, for instance, of nonlinear behavior, discontinuities, or singularities.

Two different prescriptions are frequently adopted in the analysis of fluid flows (see Figures 1.2 and 1.3). In the first one, a computational grid is attached to the fluid, which yields what is commonly termed as Lagrangian (or comoving) representation. A second approach relies on a reference frame that is fixed in space, allowing material to flow through the grid. This yields a Eulerian representation[2].

---

[2]Some codes combine the best of both worlds into what is known as an Arbitrary Lagrangian–Eulerian (ALE) approach. See Hirt, Amsden, and Cook [808] for the pioneering paper on ALE methods.

FIGURE 1.2
A diving sequence is followed by means of a Lagrangian grid attached to the diver. Severe grid distortion and tangling results. Obviously, the diver will never leave the computational domain. A color version of this picture is available at `http://fisica.upc.edu/ca/users/jjose/CRC-Downloads`. Diver sequence by Jennifer Williams; reproduced with permission. Artwork by h2o creative communications Ltd.

Let's consider a scalar variable T (e.g., temperature) that depends upon position, r, and time, t. In the Eulerian approach, the total rate of change of T is simply given by $\partial T/\partial t$, evaluated at a fixed location, r. In the Lagrangian approach, however, the total rate of

FIGURE 1.3
Same as Fig. 1.2, but for a Eulerian grid fixed in space. No grid distortion results, although the diver, who moves across the grid, may leave the computational domain. A color version of this picture is available at `http://fisica.upc.edu/ca/users/jjose/CRC-Downloads`.

change of variable T (also known as material or substantial time derivative) is expressed as:

$$\frac{DT}{Dt} = \frac{\partial T}{\partial t} + \frac{\partial T}{\partial x}\frac{dx}{dt} + \frac{\partial T}{\partial y}\frac{dy}{dt} + \frac{\partial T}{\partial z}\frac{dz}{dt} \tag{1.1}$$

Interpreting the time derivatives of the space coordinates x, y, z as components of a (Eulerian) velocity vector, the material time derivative can be written as:

$$\frac{DT}{Dt} = \frac{\partial T}{\partial t} + \frac{\partial T}{\partial x}v_x + \frac{\partial T}{\partial y}v_y + \frac{\partial T}{\partial y}v_z \tag{1.2}$$

or in a more compact form,

$$\frac{DT}{Dt} = \frac{\partial T}{\partial t} + \vec{v} \cdot \nabla T, \tag{1.3}$$

with ∇T being the gradient of variable T. Whereas DT/Dt represents the time derivative of T for a moving fluid element (Lagrangian frame), $\partial T/\partial t$ is the local time derivative of T at a fixed point (Eulerian frame), and $\vec{v} \cdot \nabla T$ represents an advectional change (also known as *convective* velocity) driven by the fluid displacement.

By construction, the independent variables used in Lagrangian formulation are mass, m, and time, t, whereas spatial coordinates, r, and time, t, are better suited for Eulerian formulations. Transformation between the two coordinates m and r is handled through application of the chain's rule. For spherical coordinates,

$$\frac{\partial}{\partial m} = \frac{\partial}{\partial r}\frac{\partial r}{\partial m} = \frac{1}{4\pi r^2 \rho}\frac{\partial}{\partial r}. \tag{1.4}$$

Both Lagrangian and Eulerian approaches present pros and cons, and no obvious choice can be made *a priori*. For non-turbulent fluids, Lagrangian formulation is probably the simplest choice, since the grid is directly attached to the fluid, and therefore boundary conditions are easy to implement. Moreover, mass is conserved by default, since no material is transferred between adjacent cells. However, since the computational grid moves with the fluid, severe mesh distortion may result. In the Eulerian approach, the computational grid is fixed and the fluid may flow through the mesh. Boundary conditions are therefore harder to implement, since fluid boundaries do not necessarily coincide with the edges of the computational domain. An advantage with respect to a Lagrangian frame is the lack of distortion, since a Eulerian grid is fixed in space. Nevertheless, Eulerian grids are usually larger than their Lagrangian counterparts, as there is some risk that the fluid may leave the computational domain (particularly in rapid fluid flows like those characterizing stellar explosions).

To date, detailed nucleosynthesis simulations coupling hydrodynamics and large nuclear reaction networks have mostly been perfomed with Lagrangian, 1D codes[3], since Lagrangian frames do not induce artificial mixing between adjacent cells. In 2D and 3D, turbulent flows may cause large distortions of a Lagrangian grid, and, accordingly, a Eulerian approach is frequently adopted.

Different discretization strategies can be implemented to transform the original set of coupled, partial differential equations into a system of algebraic equations that could be solved numerically (that is, the mathematical model describing the system in a continuous way is discretized in space and transformed into a discrete model with a finite number of degrees of freedom). Finite differences are a form of discretization frequently adopted in stellar evolution codes[4]. They are simple and effective, although their applicability is restricted to structured meshes. In this technique, variables are assigned to individual grid points (either

---

[3] According to Bill Rider (Sandia National Labs), the first hydrodynamic calculation appeared in a report from Los Alamos National Laboratory on June 20, 1944. The leading author was Hans Bethe, while the calculation leader was Richard Feynman. The work is still classified. The first codes were 1D and Lagrangian. Eulerian codes were first developed for weather prediction and used in weapons labs already in the mid-1950s.

[4] Other techniques, widely used in physics and engineering, include finite element methods, finite volume methods, spectral methods (which are based on global transformations of the system of equations—often

located at the edges, corners, or centers of each computational cell; see Section 1.4.1), and the corresponding derivatives are evaluated as differences of variables in neighboring cells.

Let's consider, for instance, a given function $F(x,t)$. For a fixed time, t, F can be approximated by means of a Taylor series expansion, in the form

$$F(x_o + \Delta x) = F(x_o) + \left(\frac{\partial F}{\partial x}\right)_{x_o} \Delta x + \left(\frac{\partial^2 F}{\partial x^2}\right)_{x_o} \Delta x^2 + O(\Delta x^3). \tag{1.5}$$

Truncation of the Taylor polynomial to first order yields

$$\left(\frac{\partial F}{\partial x}\right) = \frac{F(x_o + \Delta x) - F(x_o)}{\Delta x}, \tag{1.6}$$

which corresponds to the first-order space derivative of F in a forward-difference approximation (that is, between x_o and the *next* grid point located at $x_o + \Delta x$). A similar expression, the backward-difference approximation, results from a Taylor expansion of F between x_o and $x_o - \Delta x$:

$$\left(\frac{\partial F}{\partial x}\right) = \frac{F(x_o) - F(x_o - \Delta x)}{\Delta x}, \tag{1.7}$$

which, like the forward-difference expression, is first-order accurate. Any of these finite-difference expressions can be used in principle to approximate space derivatives and therefore turn out to be essential in the numerical analysis of differential equations. However, in the process of replacing differential equations by algebraic equations based on finite differences, unexpected surprises may arise. Truncation errors, for instance, can lead to unacceptable solutions when many integration steps are required and first-order, finite-difference expressions may no longer be appropriate. Often, more complex numerical schemes are required to handle differential equations efficiently[5]. This reflects the intricacies and difficulties in the art of numerical modeling. As an example, combining both forward- and backward-Taylor expansions of F up to third order, one gets:

$$\left(\frac{\partial F}{\partial x}\right) = \frac{F(x_o + \Delta x) - F(x_o - \Delta x)}{2\Delta x}, \tag{1.8}$$

---

relying on Fourier transforms—into a sum of functions that must satisfy the original differential equations [864]), and meshfree methods (like the *smoothed-particle hydrodynamics*, or *SPH* method; see Section 1.5).

Finite elements and finite volumes are particularly suited for arbitrary meshes (unstructured grids, in particular). Finite elements operate by splitting the domain of the problem into a number of subdomains, called *finite elements*, each represented by a set of equations [250,912]. But in contrast to the direct (*strong-form*) approach adopted in finite difference techniques (also in SPH), by which the partial differential equations are directly discretized and solved, finite element methods rely on an indirect or *weak-form* approach, in which the original equations are transformed, often in integral form, on the basis of trial solutions and weighting functions. The collection of local equations is then reassembled into a global system of equations through a transformation of coordinates.

Finite-volume methods [507, 1855] are gaining popularity as a suitable form of discretization in multidimensional codes. They are somewhat similar to finite-difference and finite-element methods, since they rely on the discretization of a system of partial differential equations onto a discrete set of points of a grid-based geometry. To this end, the computational domain is divided into a set of discrete, nonoverlapping discretization cells or *control volumes*, with physical variables assigned to the centroid of each control volume. The set of partial differential equations is then integrated over each control volume, resulting in balance equations that are subsequently discretized. The cornerstone of finite-volume methods is the discretization of the fluxes at the boundaries of each control volume. This discretization technique guarantees conservation of physical magnitudes for any control volume as well as for the overall computational domain. Note that discretization is again performed on local balance equations rather than on the original partial differential equations (as in finite-differences). For more information on these and other methods used in computational fluid dynamics, the reader is referred to LeVeque et al. [1083] and Toro [1812].

[5] See, e.g., Fryxell, Müller, and Arnett [566], Benz [153], Gershenfeld [637], or Bodenheimer, Laughlin, Rózyczka, and Yorke [209] for a detailed account on the most extensively used numerical schemes (e.g., Lax–Wendroff, Crank–Nicolson, DuFort–Frankel, Gauss–Seidel, Jacobi, alternating direction implicit (ADI) methods).

which is second-order accurate, while being as simple to implement as any of the former, first-order expressions.

1.1.1 From One-Zone Models to Multidimensional Codes

The specific literature on stellar explosions and their associated nucleosynthesis is sometimes confusing with regard to the numerical tools undertaken in the study of such events. In fact, it is extremely important to clarify the framework in which such simulations have been performed to properly assess the applicability and generality of the conclusions reached.

One-zone models have extensively been used in nucleosynthesis studies of classical novae and X-ray bursts (see Chapters 4 and 6). They rely on the time evolution of the temperature, T, and density, ρ, of a selected single layer. Such thermodynamic quantities are calculated by means of semianalytical models, or correspond to T-ρ profiles directly extracted from hydrodynamic simulations. This approach, while representing an extreme oversimplification of the physical conditions governing a star, has been widely used to overcome the time limitations that arise when large nuclear reaction networks are coupled to computationally intensive numerical codes. More recently, one-zone models have also been used to estimate the impact of nuclear uncertainties on the final yields, in a number of astrophysical sites (e.g., classical novae [880], or type I X-ray bursts [1385, 1386]), requiring thousands of test calculations that would be prohibitive with standard hydrodynamic codes. A somewhat related approach involves postprocessing, multizone calculations based on a suite of T-ρ profiles extracted from hydrodynamic simulations at different locations of the computational domain. This has extensively been used, for instance, in supernova nucleosynthesis calculations (see Chapters 5 and 7), as well as in reaction-rate sensitivity studies [1388].

The state-of-the-art modeling of stellar explosions and nucleosynthesis relies, however, on *hydrodynamic models*. Until the 1990s, most of the available hydro codes were 1D, that is, assumed spherical symmetry. This simplifying hypothesis excludes an entire sequence of events in the modeling of stellar explosions, for instance, the way in which an ignition front develops and propagates. In fact, spherical symmetry demands that ignition simultaneously occurs along a spherical shell, while it most likely initiates at one, maybe several, spots. Moreover, 1D models face severe limitations in the treatment of convective mixing, a key mechanism for energy transport in stars that can only accurately be modeled in three dimensions (see Section 1.2.4.3).

Multidimensional models are extremely time consuming and often require the use of parallelization techniques to distribute the overwhelming computational load among different processors. There has been a large number of multidimensional simulations of supernova explosions to date. In sharp contrast, multidimensional models of novae or X-ray bursts have been not only scarce, but limited to somewhat small computational domains (i.e., a box containing a fraction of the overall star) and to reduced nuclear reaction networks that include only a handful of isotopes to approximately account for the energetics of the explosion. Hence, while appropriate to tackle key phenomena that truly require a three-dimensional approach (e.g., convection and mixing, turbulence, shear), multidimensional models still need to rely on postprocessing techniques for detailed nucleosynthesis studies that require huge nuclear reaction networks.

1.2   Equations of Stellar Structure

Stars are massive spheres of plasma held together by gravity. Indeed, gravity is the driving force behind stellar evolution, leading the way toward thermonuclear fusion through matter compresion (see Chapter 2). The structure of a star is stabilized by the internal (thermal) pressure, which prevents the otherwise inevitable collapse. Indeed, when these two forces— gravity and pressure—are not in balance, the star destabilizes, and a suite of different explosive events may result (see Chapters 4–7). Stars are also characterized by an internal energy source, often powered by thermonuclear processes—although occasionally driven by gravitational contraction. A fraction of this energy, transported through the star either by radiation, convection, or conduction, is ultimately emitted into space. In this framework, a star can be described by a set of conservation equations (mass, momentum, and energy), energy sources, and transport mechanisms:

- Mass conservation

- Momentum conservation

- Energy conservation

- Energy transport

These four basic equations, supplemented by a suitable equation of state, an account of the main processes contributing to stellar opacity (a measure of how opaque—transparent— the plasma is to the radiation field), and to nuclear energy generation and nucleosynthesis, constitute the building blocks of a stellar evolution code[6].

The simplest, commonly used stellar codes rely on spherical symmetry, and on the assumption of local thermodynamic equilibrium (LTE), under which temperature is assumed to be identical both for matter and radiation (photons). This approximation often turns out to be appropriate, since the photon mean free path, or average distance covered between successive collisions, is much smaller than the length over which the temperature varies substantially. In the following subsections, we will describe the basic equations of stellar structure under these assumptions and will later reconsider their validity.

1.2.1   Mass Conservation: The Continuity Equation

Let's consider a tiny element of mass dm within a star, small enough to guarantee that its temperature, density, and chemical composition are approximately uniform.

In spherical symmetry, dm corresponds to a thin shell comprised between radii r and $r + dr$ (Figure 1.4), hence occupying a volume $dv = 4\pi r^2 dr$. In terms of the density, ρ,

$$dm = \rho\, dv = \rho\, 4\pi r^2 dr\,. \tag{1.9}$$

Under the assumption of spherical symmetry, the physical properties that characterize a star depend only on the radial distance from its center, r. Unfortunately, it is often the case that the radius of a star varies dramatically by orders of magnitude during the course of its evolution. Hence, relying on the mass enclosed in a sphere of radius r (hereafter, $m(r)$)

---

[6]Two types of stellar evolution codes are usually mentioned in the literature: hydrostatic and hydrodynamic. Essentially, they rely on the same set of equations, but hydrodynamic codes allow for an acceleration term, du/dt, in the momentum conservation equation, resulting from the inbalance between the gravitational force and the pressure gradient. When $du/dt = 0$, as in hydrostatic codes, the momentum conservation equation simply reduces to the hydrostatic equilibrium equation.

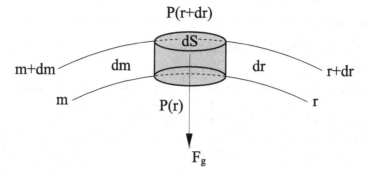

FIGURE 1.4
Schematic of the relevant forces acting on a mass element, dm, located at a distance r from the center of a star.

rather than on the radial distance itself, turns out to be more useful. Note that $m(r)$ acts exactly as a pointlike mass located at the stellar center exerting an inward gravitational acceleration equal to $Gm(r)/r^2$.

Rearranging terms in Equation 1.9, one obtains:

$$V \equiv \frac{1}{\rho} = 4\pi r^2 \frac{\partial r}{\partial m} = \frac{4}{3}\pi \frac{\partial r^3}{\partial m}, \qquad (1.10)$$

where V is known as the specific volume. Equation 1.10 is usually referred to as the *mass conservation* or *continuity* equation.

1.2.2 Energy Conservation

The first law of thermodynamics, an expression of the principle of energy conservation, states that the internal energy of a system can be modified by heat exchange and work. If δQ is the amount of heat absorbed (> 0) or emitted (< 0) by a mass element dm, and δW the work done on dm over a time δt, the change in internal energy per unit mass, δE, can be expressed as:

$$\delta E \, dm = \delta Q + \delta W. \qquad (1.11)$$

The work done on the system (contraction) is taken as $-Pdv$, whereas the work done by the system during expansion is Pdv. Therefore, one can write:

$$\delta W = -P \, \delta(dv) = -P \, \delta\left(\frac{dv}{dm} dm\right) = -P \, \delta\left(\frac{1}{\rho}\right) dm, \qquad (1.12)$$

where mass conservation has been assumed. Note that a compression ($\delta(dv) < 0$) requires an addition of energy to the mass element, whereas an expansion ($\delta(dv) > 0$) is achieved at the expense of the element's own energy.

Two sources of heat for a mass element dm can be considered:

- Nuclear energy, often expressed in terms of an energy generation rate, ε, or energy released per unit mass and unit time.

- The net balance of energy fluxes entering/leaving the mass element, expressed in terms of the luminosity, L, or energy flux per unit time.

Hence, for a spherically symmetric thin shell (Figure 1.4), we have:

$$\delta Q = \varepsilon \, dm \, \delta t + L(m) \, \delta t - L(m + dm) \, \delta t \,. \tag{1.13}$$

Using a first-order Taylor expansion of $L(m + dm)$, in the form

$$L(m + dm) = L(m) + \left(\frac{\partial L}{\partial m} \right) dm, \tag{1.14}$$

Equation 1.13 can be rewritten as:

$$\delta Q = \left[\varepsilon - \left(\frac{\partial L}{\partial m} \right) \right] dm \, \delta t \,. \tag{1.15}$$

Plugging Equations 1.12 and 1.15 into Equation 1.11, one gets

$$\delta E \, dm = \delta Q + \delta W = \left[\varepsilon - \left(\frac{\partial L}{\partial m} \right) \right] dm \, \delta t - P \delta \left(\frac{1}{\rho} \right) dm \,. \tag{1.16}$$

Eliminating dm,

$$\delta E = \left[\varepsilon - \left(\frac{\partial L}{\partial m} \right) \right] \delta t - P \, \delta V \tag{1.17}$$

and rearranging terms, one finally gets the *energy conservation equation*

$$\frac{\partial E}{\partial t} = \varepsilon - \frac{\partial L}{\partial m} - P \frac{\partial V}{\partial t} \,. \tag{1.18}$$

Note that under thermal equilibrium conditions (i.e., $\frac{\partial}{\partial t} = 0$), Equation 1.18 reduces to

$$\varepsilon = \frac{\partial L}{\partial m} \,. \tag{1.19}$$

The previous expression can be integrated over the entire star, in the form

$$\int_0^M \varepsilon \, dm = \int_0^M dL \,, \tag{1.20}$$

where $\int_0^M dL$ corresponds to the overall luminosity of the star, L. Defining $\int_0^M \varepsilon \, dm$ as a *nuclear luminosity*, L_{nuc}, or energy released by nuclear processes per unit time, one finally gets

$$L = L_{\mathrm{nuc}} \,. \tag{1.21}$$

1.2.3 Momentum Conservation

Newton's second law states that the net force acting on a body (of constant mass) leads to its acceleration, in the form $a = F/m$. More generally, the net force applied to a body is equal to the time derivative of its momentum, $F = dp/dt$, with $p = mv$.

Let's now consider a small cylindrical volume element within a star, dv, with a cross-sectional area dS, and an axis of length dr, radially comprised between r and $r + dr$. The mass contained in this volume element can be expressed as:

$$dm = \rho \, dv = \rho \, dr \, dS \,. \tag{1.22}$$

As shown in Figure 1.5, the forces acting on the volume (mass) element are:

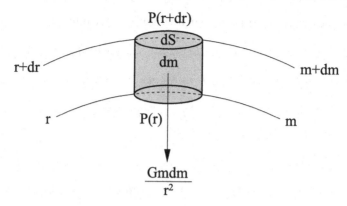

FIGURE 1.5

Net forces acting on a volume (mass) element within a star.

- The gravitational force, $-G\,m\,dm/r^2$, exerted onto a mass element dm by the mass m enclosed in a sphere of radius r, as described by Gauss's law for gravity.

- Forces associated with the pressure gradient along the surrounding gas.

In this context, the equation of motion for the mass element dm can be written as:

$$dF = dm\,a \rightarrow -Gm\frac{dm}{r^2} + P(r)\,dS - P(r+dr)\,dS = \frac{\partial^2 r}{\partial t^2}\,dm. \tag{1.23}$$

Note that pressure forces acting on both vertical sides of the cylinder (Figure 1.5) cancel mutually in the framework of spherical symmetry.

Equation 1.23 can be rearranged by using a first-order Taylor expansion of $P(r+dr)$ in the form:

$$P(r+dr) = P(r) + \left(\frac{\partial P}{\partial r}\right)dr, \tag{1.24}$$

which leads to

$$\frac{\partial^2 r}{\partial t^2}\,dm = -Gm\frac{dm}{r^2} - \left(\frac{\partial P}{\partial r}\right)dr\,dS = -Gm\frac{dm}{r^2} - \left(\frac{\partial P}{\partial r}\right)\frac{dm}{\rho}. \tag{1.25}$$

Dividing all terms by dm,

$$\frac{\partial^2 r}{\partial t^2} = -G\frac{m}{r^2} - \left(\frac{\partial P}{\partial r}\right)\frac{1}{\rho}, \tag{1.26}$$

replacing the second-order space derivative by the first-order derivative of the velocity, u, and applying the chain's rule to transform $\frac{\partial P}{\partial r}$ into $\frac{\partial P}{\partial m}$ as follows,

$$\frac{\partial P}{\partial r} = \frac{\partial P}{\partial m}\frac{\partial m}{\partial r} = \rho\,4\pi r^2\frac{\partial P}{\partial m}, \tag{1.27}$$

one finally obtains the *momentum conservation* equation:

$$\frac{\partial u}{\partial t} = -G\frac{m}{r^2} - 4\pi r^2\left(\frac{\partial P}{\partial m}\right). \tag{1.28}$$

If the acceleration term becomes negligible, Equation 1.28 reduces to the *hydrostatic equilibrium* equation:

$$\frac{\partial P}{\partial m} = -\frac{Gm}{4\pi r^4}. \tag{1.29}$$

It is worth noting that since the right-hand side of this equation is always negative, P must decrease outward.

1.2.4 Energy Transport

The energy emitted by a star through its surface frequently originates from deep inside reservoirs. Radiation, conduction, and convection are the mechanisms by which heat, and sometimes blobs of matter, can be carried throughout the star, aided by a temperature gradient between the central, hot layers (occasionally, the envelope base, like in novae or X-ray bursts) and the outer, cooler regions.

1.2.4.1 Radiation

Radiative transfer is a mechanism of energy transport by means of electromagnetic radiation—that is, photons. In their voyage through the dense stellar plasma, photons are affected by absorption, emission, and scattering processes, resulting from the tight interaction between radiation and matter.

Analytic solutions to the differential equation governing radiative transfer only exist for simple cases, because of the complexity posed by the multiple interactions faced by photons and the fact that radiation is emitted from the source in various directions and at different energies—or, equivalently, frequencies.

The distance travelled by a photon between two consecutive interactions is known as the *mean free path*, l_{phot}, and is given by:

$$l_{phot} = \frac{1}{\kappa \rho}, \tag{1.30}$$

where κ is the mean absorption coefficient or opacity (a measure of the cross-section or probability of interaction per unit mass and averaged over frequency).

For a star like the Sun, the mean free path of a photon, based on average values for the density and opacity (i.e., $\rho \sim 1$ g cm^{-3}, $\kappa \sim 1$ cm^2 g^{-1}), is about ~ 1 cm (i.e., the Sun is nearly opaque to radiation). This is very short[7], compared with the typical length scale at which structural changes occur within the star. Since the temperature gradient between the center and the surface of the Sun is $dT/dr \sim 10^{-4}$ K cm^{-1}, the radiation field is expected to be highly isotropic and close to thermal equilibrium. In such conditions, the radiation field can be approximated by a blackbody through Planck's intensity function[8] or spectral radiance, which quantifies the power emitted by a blackbody at temperature T per unit surface area and unit solid angle at a frequency ν:

$$B_\nu(T) = \frac{2h\nu^3}{c^2} \frac{1}{e^{h\nu/kT} - 1}, \tag{1.31}$$

where h is the Planck constant, c is the speed of light in vacuum, and k is the Boltzmann constant. In thermodynamic equilibrium, the spectral radiance is related to the (spectral) energy density, $U_\nu(T)$, and to the radiation pressure, $P_{rad,\nu}$, as

$$U_\nu(T) = \frac{4\pi}{c} B_\nu(T) \tag{1.32}$$

$$P_{rad,\nu} = \frac{4\pi}{3c} B_\nu(T). \tag{1.33}$$

---

[7]Close to the center, the mean free path of a photon is even shorter because of the large local density, while it becomes longer in the outer regions (e.g., at the Sun's photosphere) following the drop in density.

[8]Note that while the radiation field is described in terms of a constant temperature, T, a net flux of radiation resulting from a nonzero (macroscopic) temperature gradient throughout the star is simultaneously assumed. This is commonly referred to as local thermodynamic equilibrium.

FIGURE 1.6
Radiation emitted by human bodies at infrared wavelengths ($T = 61\,°\text{F} = 16\,°\text{C}$). In the image, light areas are hotter than dark regions. A color version of this image is available at `http://fisica.upc.edu/ca/users/jjose/CRC-Downloads`.

Integration of the spectral radiance over the entire frequency range yields

$$B(T) = \int_0^\infty B_\nu(T)d\nu = \frac{2k^4\pi^4}{15h^3c^2}T^4 = \frac{\sigma}{\pi}T^4, \tag{1.34}$$

where σ is the Stefan–Boltzmann constant. Note than σT^4 is the power radiated by a blackbody per unit surface area across all wavelengths (Stefan's law). Maximum emission, corresponding to a maximum of the spectral radiation function, peaks at a specific wavelength (or frequency) that depends only on the blackbody temperature, as described by Wien's displacement law:

$$\lambda_{\text{max}}T = 0.2898\,\text{cm K}. \tag{1.35}$$

Hence, while the Sun, characterized by a (photospheric) blackbody temperature of about 5780 K, has a peak emission at about 500 nm (i.e., visible light), mammals like humans emit predominantly in the infrared ($T \sim 310$ K $\to \lambda_{\text{max}} \sim 9\,\mu$m; see Figure 1.6).

Since the mean free path of a photon in the optically thick regions of a star is very small compared with its characteristic length—the radius R—, radiative transport can be approximated as a diffusion process[9]. Accordingly, the flux of particles per unit area and unit time, $\vec{j}$, between two points characterized by different particle densities, n, can be expressed as [988]

$$\vec{j} = -D\nabla n, \tag{1.36}$$

where D is a diffusion coefficient determined by the average speed of the particles and their mean free path. In spherical symmetry, the radiative energy flux, F, has only a radial component. Hence, averaging over all possible wavelengths, and replacing n by the radiation energy density, $U = aT^4$, yields:

$$F = -D_{\text{rad}}\frac{\partial T}{\partial r}, \tag{1.37}$$

---

[9]A more detailed description of radiative transfer can be found, e.g., in references [224, 296, 1242, 1641].

where $D_{\rm rad}$ is the radiative diffusion coefficient defined as

$$D_{\rm rad} = \frac{4ac}{3} \frac{T^3}{\overline{\kappa}\rho} \tag{1.38}$$

and $\overline{\kappa}$ is the average opacity integrated over the whole range of wavelengths, frequently taken as the Rosseland mean opacity,

$$\frac{1}{\overline{\kappa}} \equiv \frac{\int_0^\infty \frac{1}{\kappa_\nu} \frac{\partial B_\nu}{\partial T} d\nu}{\int_0^\infty \frac{\partial B_\nu}{\partial T} d\nu} \ . \tag{1.39}$$

Note that $\overline{\kappa}$ corresponds to the harmonic mean of κ_ν, with the weighting function $\partial B(T)/\partial T$. Finally, from the relationship between flux and luminosity,

$$L_{\rm rad} = 4\pi r^2 F \ , \tag{1.40}$$

one gets

$$L_{\rm rad} = -4\pi r^2 \frac{4ac}{3} \frac{T^3}{\overline{\kappa}\rho} \frac{\partial T}{\partial r} \ , \tag{1.41}$$

or, equivalently, an expression for the radiative temperature gradient in the diffusion approximation

$$\frac{\partial T}{\partial r} = -\frac{3}{4ac} \frac{\overline{\kappa}\rho}{T^3} \frac{L_{\rm rad}}{4\pi r^2} \ . \tag{1.42}$$

1.2.4.2 Conduction

In high-density environments, like in the interior of white dwarfs, energy can be transferred through frequent collisions during the random thermal motion of the particles. If the plasma is nondegenerate or weakly degenerate, each particle has an average energy $E = \frac{3}{2}kT$, which corresponds to a velocity

$$v = \sqrt{\frac{3kT}{m}} \ . \tag{1.43}$$

Accordingly, electrons move faster than ions (by the square root of the mass ratio between ions and electrons), and, therefore, electrons constitute the most relevant source of heat conduction[10].

By analogy with radiative transport, the electron thermal conductivity can be expressed in terms of an energy flux, in the form of a diffusion equation:

$$F_{\rm cond} = -D_{\rm cond} \frac{dT}{dr} \ , \tag{1.44}$$

with $D_{\rm cond}$ being a thermal diffusion coefficient that depends on the conductive opacity, $\kappa_{\rm cond}$,

$$D_{\rm cond} = \frac{4acT^3}{3\kappa_{\rm cond}\rho} \ . \tag{1.45}$$

Globally, the energy flux carried by radiative and conductive heat transport can be expressed as

$$F_{\rm rad+cond} = -\frac{4ac}{3} \frac{T^3}{\kappa_{\rm ef}\rho} \frac{dT}{dr} \ , \tag{1.46}$$

where

$$\frac{1}{\kappa_{\rm ef}} = \frac{1}{\kappa_{\rm rad}} + \frac{1}{\kappa_{\rm cond}} \ . \tag{1.47}$$

---

[10]Typical densities and temperatures of main sequence stars yield mean free paths for electrons several orders of magnitude smaller than for photons. Moreover, electrons travel at velocities much smaller than the speed of light. This translates into smaller electron diffusion coefficients, compared with those for photons.

1.2.4.3 Convection

Convective transport relies on macroscopic mass elements (usually called *bubbles, blobs,* or *convective elements*) exchanged between hotter and cooler layers. Such convective mass elements will ultimately dissolve in the surroundings, delivering their excess heat. The description of convective heat transfer is only possible through a phenomenological approach, due to inherent uncertainties and complexity. The most widely used prescription in stellar evolution codes to date is the *mixing-length theory*, developed by Ludwig Pradtl [1443], who sketched a simple description of convection in analogy to molecular heat transfer[11]. This theory relies on the definition of the *mixing-length*, l_m, a free parameter that measures the distance travelled by a convective blob before dissolving in the surroundings. Despite its simplicity, the mixing-length theory yields a reasonable qualitative description of convective heat transfer. However, it is important to stress that the theory assumes hydrostatic equilibrium conditions, and their implementation in the modeling of stellar explosions has to be taken with caution. Unfortunately, no self-consistent prescriptions of convective transport exist, in 1D, for dynamic conditions and theoreticians have resigned themselves to rely on the mixing-length theory for their stellar evolution—and explosion!—calculations[12].

Arguments based on the buoyancy forces acting on moving convective elements, under the assumption of pressure equilibrium between the rising bubbles and their surroundings, yield a criterion for convective instability. That is, convection settles whenever

$$\left| \frac{dT}{dr} \right| > \left| \frac{dT}{dr} \right|_{\text{ad}} \tag{1.48}$$

or, in terms of $\nabla \equiv d\ln T / d\ln P$,

$$\nabla > \nabla_{\text{ad}} , \tag{1.49}$$

where ∇ is the *actual* gradient at a given grid point, and

$$\nabla_{\text{ad}} \equiv \left| \frac{d\ln T}{d\ln P} \right|_{\text{ad}} = -\frac{P\left(\frac{\partial P}{\partial T}\right)_{\text{V}}}{C_{\text{p}} \left(\frac{\partial P}{\partial V}\right)_{\text{T}}} \tag{1.50}$$

is the *adiabatic* gradient. C_{p} is the heat capacity at constant pressure, related to the heat capacity at constant volume, C_{v}, through

$$C_{\text{p}} = C_{\text{v}} - T \frac{\left(\frac{\partial P}{\partial T}\right)_{\text{V}}^2}{\left(\frac{\partial P}{\partial V}\right)_{\text{T}}} = \left(\frac{\partial U}{\partial T}\right)_{\text{V}} - T \frac{\left(\frac{\partial P}{\partial T}\right)_{\text{V}}^2}{\left(\frac{\partial P}{\partial V}\right)_{\text{T}}} . \tag{1.51}$$

Equation 1.49 is known as the *Schwarzschild criterion* for convection. A related expression, the *Ledoux criterion*, is used for chemically inhomogeneous regions:

$$\nabla > \nabla_{\text{ad}} + \frac{\phi}{\psi} \left(\frac{\partial \ln \mu}{\partial \ln P} \right)_{\rho, \text{T}} , \tag{1.52}$$

where

$$\phi = \left(\frac{\partial \ln \rho}{\ln \mu} \right)_{\text{P,T}} \tag{1.53}$$

---

[11] See also Böhm-Vitense [214] and Cox and Giuli [384] for early detailed descriptions of the mixing-length theory. The classic Cox and Giuli book has been recently revised and extended by A. Weiss, W. Hillebrandt, H.-C. Thomas, and H. Ritter [1919].

[12] The rise of multidimensional simulations has allowed ways to overcome this limitation through the implementation of prescriptions based on 3D models of convection [70, 71, 1222, 1251].

$$\psi = -\left(\frac{\partial \ln \rho}{\ln T}\right)_{P,\mu}.$$ (1.54)

In the mixing-length theory, the myriad convective bubbles, characterized by a suite of different sizes, shapes, velocities, and lifetimes, are somewhat *averaged*, so that all bubbles located at a given distance r from the stellar center have the same physical properties.

A rising bubble is characterized by a lower density than the surroundings. Because of the assumption of pressure equilibrium, a lower density implies a higher temperature, and therefore, a rising bubble implies a thermal energy transport upwards. In mixing length theory, hot bubbles rise, traveling a distance given by the mixing length before dissolving in the surroundings, while cool bubbles sink about the same distance before losing their identity. Following Cox and Giuli [384], the relation between the temperature excess and the density deficit between a bubble and its surroundings is given by

$$\Delta\rho \simeq -Q\frac{\rho}{T}\Delta T,$$ (1.55)

where

$$-Q = \left(\frac{\partial \ln \rho}{\partial \ln \mu}\right)_{P,T}\left(\frac{\partial \ln \mu}{\partial \ln T}\right)_{P} + \left(\frac{\partial \ln \rho}{\partial \ln T}\right)_{\mu,P},$$ (1.56)

which reduces to

$$Q = -\left(\frac{\partial \ln \rho}{\partial \ln T}\right)_{\mu,P}$$ (1.57)

in chemically homogeneous regions. Due to the density contrast, a buoyancy force per unit volume acts on any rising bubble, $f_b = -g\Delta\rho$. The velocity of this moving element can be estimated from the corresponding equation of motion

$$\frac{d^2 r}{dt^2} = -g\frac{\Delta\rho}{\rho}.$$ (1.58)

Assuming that the initial velocity of the bubble at r is zero, and the density contrast increases linerly with r, from r to $r + \Delta r$, the work done by the buoyancy force per unit volume acting on the bubble along a distance Δr is given by

$$W(\Delta r) = \int_o^{\Delta r} f_b(r)dr = -\frac{1}{2}g\Delta\rho.\Delta r$$ (1.59)

Accordingly, the average work done along a distance equal to the mixing length is given by

$$\overline{W} = \frac{1}{4}W(l_m) = -\frac{1}{8}g\Delta\rho\, l_m,$$ (1.60)

where the factor $1/4$ is adopted in order to match Böhm-Vitense's work [214]. Assuming that one half of the average work translates into kinetic energy of the rising bubble, one gets

$$\frac{1}{2}\overline{W} = \overline{\frac{1}{2}\rho v^2} = \frac{1}{2}\rho.\overline{v}^2$$ (1.61)

And, since $v = 0$ at r,

$$\overline{v}^2 = -\frac{1}{8}g\frac{\Delta\rho}{\rho}l_m = \frac{1}{8}gQ\frac{\Delta T}{T}l_m.$$ (1.62)

The temperature excess, ΔT, is often expressed in terms of the *pressure scale height*, H_P,

$$H_P \equiv -\frac{dr}{d\ln P} = -P\frac{dr}{dP},$$ (1.63)

a measure of the characteristic length of the radial variation of P, which under hydrostatic equilibrium conditions is simply given by

$$H_P = \frac{P}{\rho g}. \tag{1.64}$$

All in all, the temperature excess can be written as

$$\Delta T = T \frac{l_m}{H_P}(\nabla - \nabla'), \tag{1.65}$$

with $\nabla \equiv d\ln T/d\ln P$ being the average temperature gradient with respect to pressure of all the matter at a given location r, and $\nabla' \equiv d\ln T'/d\ln P$ the gradient of the rising (sinking) bubble, often approximated as [384, 1037]

$$\nabla' = \nabla_{ad}. \tag{1.66}$$

Equations 1.62, 1.65, and 1.66 yield the expression for the (local) convective velocity,

$$\bar{v} = \frac{1}{2^{3/2}} \frac{l_m}{r} \left[-\frac{Gm}{H_P}(\nabla - \nabla_{ad}) \left(\frac{\partial \ln \rho}{\partial \ln T} \right)_{P,\mu} \right]^{1/2}. \tag{1.67}$$

The convective energy flux can be obtained from the product of the thermal energy per gram carried by the rising bubble (which moves at constant pressure), $C_p \Delta T$, and the mass flux, $\frac{1}{2}\rho\bar{v}$. The factor $1/2$ results from the assumption that upward and downward flows are identical, and, therefore, only one half of the matter moves upwards at any time:

$$F_{conv} = \frac{1}{2}\rho\bar{v}C_p\Delta T. \tag{1.68}$$

Finally, plugging Equations 1.65 and 1.67 into Equation 1.68, and recalling that $L = 4\pi r^2 F$, one gets an expression for the (local) convective luminosity:

$$L_{conv} = \pi \sqrt{-\frac{G}{2}\left(\frac{\partial \ln \rho}{\partial \ln T}\right)} H_P^{-3/2} l_m^2 C_p \frac{rTm^{1/2}}{V}(\nabla - \nabla_{ad})^{3/2}. \tag{1.69}$$

Equation 1.69, which appears explicitly in the energy transport equation whenever superadiabatic temperature gradients are established within the stellar plasma, depends on the mixing-length, l_m, which is taken as a free parameter. Different prescriptions are adopted for l_m on the basis of plausibility arguments. Frequently, l_m is taken as a multiple of the pressure scale height, in the form

$$l_m = \alpha H_P, \tag{1.70}$$

with adopted values for α in the range 1–3. There are reasons to favor $l_m \sim H_P$ (that is, $\alpha \sim 1$) [384]: On one hand, convective cells much smaller than one scale height in size would not efficiently carry energy away, first because of their small size (radiative energy losses will be important), and second because of the small distance travelled before dissolving. Therefore, only large convective bubbles play a relevant role in convective energy transport. On the other hand, convective elements cannot retain their identities when traveling through distances larger than a few scale heights. Empirical determinations of α around ~ 1.7–1.9 have been inferred by fitting stellar evolution models to the observed properties of the Sun [1181]. However, it is unclear how to extrapolate this value of α to other types of stars, particularly under explosive conditions[13].

---

[13] Another limitation of the mixing-length theory is its inability to account for convective overshoot, that is, the displacement of fluid elements beyond the boundaries of the convective regions. In sharp contrast to mixing-length formulation, it is expected that convective mixing would not stop abruptly at the edge of a convective region. Indeed, evidence of convective overshoot has been observed in the Sun in the form of granulation at the surface.

It is finally worth noting that many phases of the evolution of stars are characterized by steady conditions, and, therefore, by large timescales. In such conditions, it is justified to assume a time-independent prescription for convective transport. However, when a star proceeds in a nearly dynamical timescale (i.e., during explosive stages or collapses; see Section 1.3.1), the assumption of a time-independent convection does not hold, and one should rely on a time-dependent formalism (see, e.g., reference [1955]).

1.3 A Touch of Hydrodynamics

1.3.1 Timescales

Some characteristic timescales have proved particularly useful in unveiling the physical mechanisms that drive stellar explosions and the conditions in which these take place. In this section, a number of relevant timescales employed within this book are reviewed.

The equation of momentum conservation (Equation 1.28) establishes that if the pressure gradient imbalances the gravitational force, an acceleration results. Let's assume, for convenience, that the pressure term just vanishes. Equation 1.28 then becomes

$$\frac{\partial u}{\partial t} = -G\frac{m}{r^2}, \tag{1.71}$$

or, in terms of the surface gravity, g,

$$\frac{\partial^2 r}{\partial t^2} = g. \tag{1.72}$$

The resulting equation merely describes the unavoidable free-fall collapse of the star, which would take place with a characteristic timescale τ_{ff} that can be inferred by setting $\partial^2 r/\partial t^2 \sim R/\tau_{ff}^2$, with R being the radius of the star. All in all, we get

$$\frac{R}{\tau_{ff}^2} \sim g \to \tau_{ff} \sim \sqrt{\frac{R}{g}}, \tag{1.73}$$

which in terms of the average density of the star, $\bar{\rho} = 3M/4\pi R^3$, becomes

$$\tau_{ff} \sim \frac{1}{\sqrt{G\bar{\rho}}}. \tag{1.74}$$

Similar dynamical timescales, within factors of the order of unity, can be derived[14]. For instance, the free-fall (or escape velocity) in a gravitational field, $v_{esc} = \sqrt{2GM/R}$, yields

$$\tau_{dyn} \sim \frac{R}{v_{esc}} = \sqrt{\frac{R^3}{2GM}} \sim \frac{1}{\sqrt{G\bar{\rho}}}. \tag{1.75}$$

---

[14] Another frequently used estimate for τ_{ff} can be obtained by considering a spherical shell initially at a distance r_o from the center of a star. The free-fall speed of a point in the shell during collapse, for an arbitrary distance to the center, r, can be inferred from energy conservation:

$$\frac{1}{2}u^2 = \frac{1}{2}\left(\frac{dr}{dt}\right)^2 = \left(-\frac{Gm(r_0)}{r_0}\right) - \left(-\frac{Gm(r_0)}{r}\right).$$

Defining $\tau_{ff} \equiv \int_0^{r_0}(1/u)dr$, and replacing u by a suitable expression from the energy conservation equation, one gets

$$\tau_{ff} = \sqrt{\frac{3\pi}{32\bar{\rho}G}}.$$

Note that such dynamical timescales provide an estimate of the time required by a star to recover hydrostatic equilibrium before a collapse or an explosion results. For a star like the Sun, $\tau_{\mathrm{dyn}} \sim 1000$ s.

A crude estimate of the characteristic timescale for a stellar explosion can actually be inferred by, instead, setting the gravitational force to zero in the momentum conservation equation (Equation 1.28):

$$\frac{\partial^2 r}{\partial t^2} = -4\pi r^2 \left(\frac{\partial P}{\partial m} \right) = \frac{1}{\rho} \left(\frac{\partial P}{\partial r} \right). \tag{1.76}$$

Imposing, as before, $\partial^2 r / \partial t^2 \sim R/\tau_{\mathrm{expl}}^2$ and $\partial P / \partial r \sim \bar{P}/R$, one gets

$$\frac{R}{\tau_{\mathrm{expl}}^2} \sim \frac{1}{\bar{\rho}} \frac{\bar{P}}{R} \rightarrow \tau_{\mathrm{expl}} \sim R \sqrt{\frac{\bar{\rho}}{\bar{P}}}, \tag{1.77}$$

where $\bar{P}$ and $\bar{\rho}$ are averaged over the whole star. Since the speed of sound of a wave propagating through an elastic medium is given by

$$c_{\mathrm{s}} = \sqrt{\left(\frac{\partial P}{\partial \rho} \right)}, \tag{1.78}$$

τ_{expl} is of the order of R/c_{s}, which implies that the time required by an explosion to propagate from the center to the surface of a star is of the order of the sound-crossing time.

A suitable definition for the nuclear timescale is given by

$$\tau_{\mathrm{nuc}} \sim \frac{c_{\mathrm{p}} T}{\varepsilon}, \tag{1.79}$$

where c_{p} is the specific heat and ε the nuclear energy generation rate.

The relevance of convective transport, particularly for mixing of chemical species, can be evaluated by means of the convective turnover time, or ratio of the size of the convective region, l_{conv}, over the characteristic velocity of the rising blobs, v_{conv} (see Section 1.2.4.3):

$$\tau_{\mathrm{conv}} \sim \frac{l_{\mathrm{conv}}}{v_{\mathrm{conv}}}. \tag{1.80}$$

Another useful timescale, particularly for accretion-driven stellar explosions, is the accretion timescale, defined in terms of the mass-accretion rate, $\dot{M}$, and the overall accreted mass, ΔM_{acc}:

$$\tau_{\mathrm{acc}} \sim \frac{\Delta M_{\mathrm{acc}}}{\dot{M}}. \tag{1.81}$$

Usually, mass-accretion is relevant whenever $\tau_{\mathrm{acc}} \leq \tau_{\mathrm{nuc}}$.

1.3.2 Instabilities

The astrophysical plasmas encountered in the study of stellar explosions frequently undergo shocks, turbulence, chaos, and a number of phenomena that drive the formation of a suite of different instabilities.

Mixing at the core-envelope interface during classical nova outbursts is likely driven by the onset of Kelvin–Helmholtz instabilities (see Chapter 4). Named after Lord Kelvin and Hermann von Helmholtz, these instabilities are driven by velocity differences across the interface between two fluids (or by a velocity shear in a single fluid). Kelvin–Helmholtz

(a)

(b)

FIGURE 1.7
Characteristic Kelvin–Helmholtz instabilities in stormy clouds spotted at (a) the Birmingham–Shuttlesworth International Airport (USA) in december 2011, and (b) around Sagunt (Valencia, Spain). Color versions of these images are available at `http://fisica.upc.edu/ca/users/jjose/CRC-Downloads`. Credits: (a) WBMA–TV, Birmingham (Alabama, USA) and (b) Minerva Gracia; reproduced with permission.

instabilities develop into characteristic ripples that are often seen in terrestrial fluids, such as clouds (see Figure 1.7) or ocean/sea waves, and have even been spotted in Saturn's atmosphere and in the solar corona.

In sharp contrast, Rayleigh–Taylor instabilities require density gradients. Named after

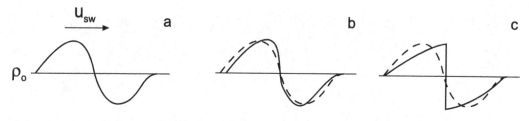

FIGURE 1.8
Formation of a discontinuity in a pressure wave propagating through a uniform fluid.

Lord Rayleigh and Geoffrey I. Taylor, these instabilities appear in every fluid, or at the interface between two fluids, characterized by a density inversion, in particular when a lighter fluid pushes the heavier one. Therefore, the driving mechanism for Rayleigh–Taylor instabilities is buoyancy, not shear (unlike Kelvin–Helmholtz instabilities). The characteristic fingering accompanying Rayleigh–Taylor instabilities has been observed in terrestrial plasmas (e.g., in plasma fusion reactors; see also Section 1.7.2). They are also present in astrophysical plasmas, like in the Crab nebula, a famous supernova remnant (see Chapters 5 and 7), when accelerated gas from the innermost regions collides with a denser gas shell. Strong Rayleigh–Taylor instabilities have also been discussed in the context of SN 1987A [69,1269] (see Chapter 7). Magnetically modulated Rayleigh–Taylor instabilities have also been spotted in the solar corona, in the form of upflow plumes, when a relatively dense solar prominence overlies a lower desity plasma bubble [158].

   A variation of Rayleigh–Taylor instabilities, known as Richtmyer–Meshkov instabilities, occur when a shock wave impinges perpendicularly upon an interface that separates two fluids with different densities [251, 1230, 1499]. Richtmyer–Meshkov instabilities develop, for instance, during implosion in inertial confinement fusion experiments and have been linked to the initiation of the deflagration to detonation transition in type Ia supernova models [1351, 1785] (see Chapter 5). The interested reader can find a detailed overview on fluid instabilities in Shore [1634].

1.3.3   Shock Waves and the Physics of Combustion

The nonlinear nature of the Navier–Stokes equations of fluid dynamics gives rise to a number of peculiar patterns, in the form of contact discontinuities, shock fronts, and rarefaction waves. See Section 1.7.1.1 for an example of a hydrodynamic problem in which these three types of discontinuities are actually encountered.

   Discontinuities arise, for instance, when a large-amplitude pressure wave propagates through a uniform fluid [224]. Two distinct regions can be identified in the pressure pulse: A compression region, characterized by $\rho > \rho_o$, and a rarefaction region, where $\rho < \rho_o$ (with ρ_o being the density of the unperturbed fluid; see Figure 1.8). As the pulse moves forward, and because the speed of sound increases as the density increases, the compression region begins to move faster than the rarefaction region. Eventually, a discontinuity develops[15]. Special numerical techniques must be implemented to handle such discontinuities, since they cannot be adequately represented by a grid.

   Contact discontinuities are surfaces that separate regions of different density and temperature, while maintaining pressure constant on both sides. Let's consider the velocity

---

[15]A similar outcome is expected for a pulse that propagates outward across a star, through layers of decreasing density. Note that if the maximum pressure is not much larger than that of the unperturbed fluid, the wave will move at approximately the speed of sound.

vector describing a fluid flow, $\vec{u} = (u_{\mathrm{n}}, u_{\mathrm{t}})$, where u_{n} and u_{t} represent the normal and tangential components of $\vec{u}$, with respect to the surface of the discontinuity. When u_{t} also varies substantially across the discontinuity, this is referred to as a slip discontinuity. In contrast, both shocks and rarefaction waves result from abrupt changes in pressure, with shock fronts associated to compression episodes and rarefaction waves accompanying fluid expansions. Shock fronts, in particular, separate regions characterized by different pressure, density, temperature, and velocity (normal component), which may vary abrupty across the discontinuity. Shock fronts frequently occur, for instance, in the context of stellar explosions.

While small-amplitude perturbations propagate throughout a fluid adiabatically, at the local speed of sound, c_{s}, strong perturbations can propagate supersonically. Here, a discontinuity between the regions behind and ahead of the disturbance will develop. Both regions would become locally disconnected. Indeed, the yet unperturbed fluid would not be able to react to the imminent arrival of the supersonic shock (through expansion, for instance), since such information is propagated at the speed of sound.

Let's assume that a strong shock wave, propagating at a supersonic speed u_{sw}, separates two regions of a fluid. The undisturbed region ahead of the shock wave, initially at rest ($u_1' = 0$), is characterized by a pressure P_1, density ρ_1, and temperature T_1. The postshocked region moves at speed u_2', with a pressure, density, and temperature given by P_2, ρ_2, and T_2. In the reference frame comoving with the shock wave at speed u_{sw} (see Figure 1.9), the undisturbed fluid flows toward the discontinuity at speed $u_1 = u_{\mathrm{sw}}$, while the postshocked fluid left behind moves at speed $u_2 = u_{\mathrm{sw}} - u_2'$. Note that the speed of the shock front[16] must fulfill $u_{\mathrm{sw}} > c_{\mathrm{s}}$, otherwise the discontinuity would smoothen and eventually disappear.

The relation between the physical variables in the unperturbed and postshocked regions of the fluid across a planar shock wave (i.e., a shock perpendicular to the flow) is given by the following set of conservation equations, known as *Rankine–Hugoniot*, *junction*, or *jump conditions* [854, 855, 1468]:

- Mass flux conservation

$$\rho_1 \, u_1 = \rho_2 \, u_2 \tag{1.82}$$

- Energy flux conservation

$$u_1 \left(\frac{1}{2} \rho_1 u_1^2 + E_1 \rho_1 + P_1 \right) = u_2 \left(\frac{1}{2} \rho_2 u_2^2 + E_2 \rho_2 + P_2 \right), \tag{1.83}$$

where E is the internal energy per unit mass.

- Momentum flux conservation

$$P_1 + \rho_1 \, u_1^2 = P_2 + \rho_2 \, u_2^2 \tag{1.84}$$

1.3.3.1 Deflagrations vs. Detonations

The passage of a shock wave may raise the temperature of the plasma to ignition values. The associated combustion front can propagate either subsonically (deflagration) or supersonically (detonation).

In a detonation, combustion proceeds on a timescale shorter than the characteristic hydrodynamic timescale. The corresponding burning front propagates simultaneously but slightly behind the shock (since the material needs first to be heated up to the characteristic burning temperature), incinerating the material lying right behind the shock front. Both

---

[16]It is also worth noting that the fluid dissipates part of its kinetic energy into heat upon crossing the shock. Therefore, an entropy jump is also encountered across the shock.

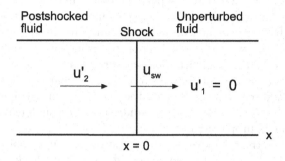

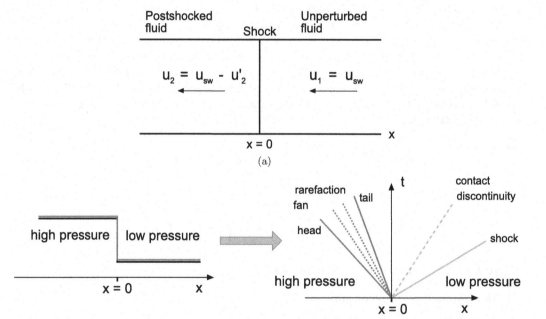

FIGURE 1.9

(Top) The passage of a shock wave as seen from the laboratory and from a reference frame comoving with the shock at speed u_{sw}. (Bottom) Diagram depicting the different types of discontinuities driven by a sharp pressure jump in a shock tube (see Section 1.7.1.1). A color version of this picture is available at http://fisica.upc.edu/ca/users/jjose/CRC-Downloads.

the detonation and the combustion fronts propagate at the same supersonic speed with respect to the unperturbed fluid.

A deflagration, instead, propagates by means of heat (and mass) exchange between preshocked and postshocked material. Here, the shock wave itself is unable to initiate the incineration of the plasma, and therefore both the (subsonic) burning front and the shock front propagate independently.

The evolution of a fluid from an initial state 1 to a postincineration state 2 is determined by a number of characteristic curves [1105, 1940] and is schematically shown on a $P - V$ plane in Figure 1.10(a). Let's define $J \equiv \rho_1 u_1 = \rho_2 u_2$ as the constant mass flux across the shock. Plugging this expression into the momentum flux conservation equation (Equation 1.84) yields

$$P_1 + Ju_1 = P_2 + Ju_2, \qquad (1.85)$$

which, after rearranging terms, can be written as:

$$J^2 = \frac{P_2 - P_1}{u_1 - u_2}. \qquad (1.86)$$

The energy flux conservation equation can be rewritten in the form

$$u_1 \left(\frac{1}{2} u_1^2 \rho_1 + H_1 \rho_1 \right) = u_2 \left(\frac{1}{2} u_2^2 \rho_2 + H_2 \rho_2 \right), \qquad (1.87)$$

where H is the enthalpy of the fluid per unit mass ($H = E + PV$), and $V = 1/\rho$ the specific volume. It follows that

$$H_1 - H_2 = \frac{1}{2}(P_1 - P_2)(V_1 + V_2). \qquad (1.88)$$

Equation 1.88 defines the *Hugoniot curve* or *shock adiabat*.

The combination of mass and momentum flux equations, in turn, yields

$$P_1 + \frac{J^2}{\rho_1} = P_2 + \frac{J^2}{\rho_2}. \qquad (1.89)$$

Note that, in terms of the specific volume, V, this corresponds to a straight line on the $P - V$ plane, $P + J^2 V$, known as the *Rayleigh line*. Equation 1.89 can be conveniently rewritten as

$$P_2 = P_1 + u_1^2 \left[\rho_1 - \frac{\rho_1^2}{\rho_2} \right] \equiv \alpha + \beta u_1^2 V_2, \qquad (1.90)$$

which shows that the slope of the Rayleigh line is proportional to the square of the detonation velocity, u_1. Therefore, one can think of a series of Rayleigh lines resulting for different values of u_1. In the simplest theory of combustion developed by D. L. Chapman and J. C. E. Jouguet [313, 934, 935], the final state of the postshocked fluid (that is, the solution to the flux conservation equations) is determined by the intersection between the Rayleigh line and the Hugoniot curve (see reference [522] for details).

The upper left branch of the Hugoniot curve in Figure 1.10(b) identifies the possible postshocked states of the fluid after the passage of a detonation front, which precompresses the fluid prior to incineration. On a $P - V$ diagram, this corresponds to a trajectory given by the Rayleigh line from the initial (V_1, P_1) state up to the so-called Von Neumann state[17] (V_N, P_N) on the shock adiabat H_o (or Hugoniot curve corresponding to zero energy released by nuclear reactions, as calculated by Equation 1.88). The Hugoniot curve shifts, however,

---

[17]The Von Neumann state corresponds to the high-pressure point on the unreacted Hugoniot curve, H_o, reached after compression of the fluid from the initial, unperturbed state along the Rayleigh line (see Figure 1.10).

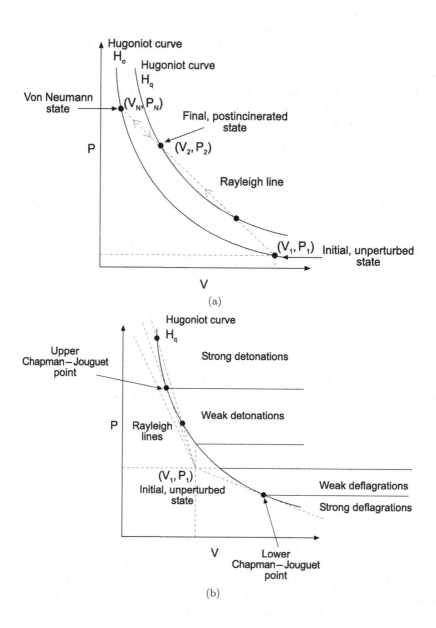

FIGURE 1.10
(Top) The loci of all possible final states of a fluid after the passage of a combustion front. (Bottom) Final states reached through weak/strong deflagrations and detonations. Color versions of these plots are available at `http://fisica.upc.edu/ca/users/jjose/CRC-Downloads`.

toward the upper right corner of the $P-V$ diagram as nuclear fusion sets in. The final curve, H_q, corresponding to complete incineration, is determined by the addition of the nuclear energy released per unit mass, q_{nuc}, to the overall energy flux conservation Equation 1.84. As incineration goes to completion, the fluid moves from (V_N, P_N) down the Rayleigh line up to a point (V_2, P_2) on the Hugoniot curve H_q.

Depending on the specific value of u_1, the Rayleigh line may actually intersect the Hugoniot curve at two distinct locations, may be tangent to the Hugoniot curve, or may not intersect at all (the latter corresponding to an unphysical case that does not satisfy the whole set of conservation equations). The single intersection point obtained when the Rayleigh line is tangent to the Hugoniot curve is known as the (upper) Chapman–Jouguet point, CJ (see Figure 1.10). This CJ point corresponds to a minimum of the detonation velocity, and splits the upper Hugoniot curve into regions of strong (I) and weak detonations (II). Chapman–Jouguet detonations (corresponding to the tangential solution) are characterized by postincinerated products flowing at the speed of sound relative to the shock front. In contrast, the flow moves supersonically in the case of weak detonations, and subsonically for strong detonations. This means that Chapman–Jouguet and weak detonations are self-sustained, since any pressure wave generated in the incinerated products cannot catch up with the detonation front and attenuate it. Strong detonations, being potentially hampered by rarefaction waves, are not self-propagating and require the support of a moving piston[18] (see, for instance, reference [522]).

With regard to the lower right branch of the Hugoniot curve (Figure 1.10), it corresponds to deflagration fronts. Here, the fluid evolves from the initial state (V_1, P_1) directly onto a point on the Hugoniot curve H_q through a Rayleigh line. As in the detonation case, the Rayleigh line that intersects the Hugoniot curve in just one point, the lower Chapman–Jouguet point, separates the regions of weak (III) and strong deflagrations (IV). The lower CJ point represents in this case a maximum of the deflagration velocity, and as in the previous detonation case, corresponds to postincinerated material propagating at the speed of sound, in the reference frame of the combustion wave. Note that while the postincinerated material is characterized by subsonic speeds in the weak deflagration region, strong (supersonic) deflagrations actually correspond to unphysical solutions.

The interested reader can find more information on the physics of shock waves in the reference text books by Landau and Lifshitz [1048] and Zeldovich and Raizer [2018]. See also Bowers and Deeming [224], Shore [1634], Bodenheimer et al. [209], or Longair [1145] for thorough descriptions of shocks and supersonic flows in astrophysical environments.

1.4 Grid-Based Methods: The Realm of Finite Differences

Most of the numerical codes used in astrophysical hydrodynamics are based on finite difference approximations to the set of partial differential equations. Such schemes, as discussed in Section 1.1, rely on a computational grid or mesh[19]. The simplest hydrodynamic codes

---

[18]Even though three possible detonation states have been identified in the previous analysis, one must carefully take into account the structure of the detonation front to determine whether such states are actually feasible. Indeed, Y. B. Zeldovich [2019], J. von Neumann [1868], and W. Döring [459] independently came up in the 1940s with a simple model for the structure of the detonation (known as ZND theory), in which the leading front is a shock wave that travels at the CJ velocity through the undisturbed fluid. Thus, in the framework of the ZND theory, weak detonations cannot occur. However, Fickett and Davis [522] proved that weak detonations may result if some of the restrictions imposed by the ZND theory are relaxed. It is finally worth noting that weak detonations do actually occur but are very seldom observed.

[19]Meshfree codes, such as SPH, are discussed in Section 1.5.

assume spherical symmetry (1D) and are frequently implemented in Lagrangian formulation. This, unfortunately, is not always feasible in 2D and 3D, since Lagrangian meshes often suffer severe distortion and tangling in the presence of nonlinear phenomena, discontinuities, or singularities. Therefore, while the vast majority of 1D codes rely on Lagrangian grids, multidimensional codes are frequently Eulerian.

This section focuses on the main features of 1D, Lagrangian hydrodynamic codes[20]. They rely on the following set of coupled mechanical and thermal equations that describe the structure and time evolution of a star (see Section 1.2):

- Conservation of mass

$$V = \frac{4}{3}\pi \frac{\partial r^3}{\partial m} . \tag{1.91}$$

- Conservation of momentum

$$\frac{\partial u}{\partial t} + 4\pi r^2 \frac{\partial (P+q)}{\partial m} = -G\frac{m}{r^2} , \tag{1.92}$$

where q is an artificial viscosity[21] term.

- Conservation of energy

$$\frac{\partial E}{\partial t} = \varepsilon - \frac{\partial L}{\partial m} - (P+q)\frac{\partial V}{\partial t} , \tag{1.93}$$

where ε is the overall energy generation rate by nuclear reactions and energy losses by neutrinos ($\varepsilon = \varepsilon_{nuc} - \varepsilon_\nu$).

- Energy transport (by radiation and convection)

$$L = -256\sigma\pi^2 r^4 \frac{T^3}{3\kappa}\frac{\partial T}{\partial m} + L_{\text{conv}} , \tag{1.94}$$

where L_{conv} the convective luminosity (Section 1.2.4.3).

- Lagrangian velocity

$$\frac{\partial r}{\partial t} = u . \tag{1.95}$$

The set of partial differential equations is linked to the constitutive equations:

- Pressure and internal energy: $P = P(T, \rho, X)$, $E = E(T, \rho, X)$

- Opacity: $\kappa = \kappa(T, \rho, X)$

- Nuclear energy generation and neutrino losses: $\varepsilon = \varepsilon(T, \rho, X)$,

where X represents the mass fractions of the chemical species present in the stellar plasma (see Chapter 2).

---

[20]For convenience, we will mostly rely on the choices adopted in building up the 1D, Lagrangian, implicit, hydrodynamic code SHIVA [919], extensively used in the modeling of classical nova explosions and type I X-ray bursts.

[21]The first hydrodynamic simulations of thermonuclear explosions (e.g., the implosion bomb model developed in 1943–1994) failed in the presence of shocks. Shock waves are modeled as discontinuities that are difficult to handle numerically unless certain tricks are implemented. One such tricks relies on the inclusion of an unphysical term, known as artificial viscosity, to smear out discontinuities, thickening the narrow shock transition zones (which extend over a few mean-free paths) to widths that could be resolved computationally. Pioneering work on the use of artificial viscosity was performed by R. D. Richtmyer in 1948 (even if the author never used that term but the words *ficticius* or *mock* to describe the concept), remaining classified until 1993. His first report, LA–671, is actually available at http://fas.org/sgp/othergov/doe/lanl/index1.html. A paper on this subject, coauthored by von Neumann and Richtmyer, was published in 1950 [1870].

1.4.1 Equations of Stellar Structure in Finite Differences

Solutions to the system of partial differential equations, Equations 1.92–1.95, are obtained for a finite, discrete number of grid points that define the computational domain. To this end, the domain is subdivided into N individual cells (also called elements or shells). In a 1D framework, this corresponds to segments that define N concentric shells and the corresponding $N + 1$ intershells. Hereafter, each intershell will be labeled with a subscript i, ranging from 1 (at the star's center or innermost shell) to $N + 1$ (stellar surface).

The goal of any stellar hydrocode is to determine the time evolution of a set of physical variables (e.g., luminosity, L; radius, r; velocity, u; temperature, T; and specific volume, V) at each shell. Such variables must be assigned to specific grid points, frequently involving cell-centered locations or cell edges. The choice, however, is somewhat arbitrary and certainly not unique, as a thorough comparison between existing codes would prove. A convenient choice in 1D assigns variables L, r, and u to each of the $N + 1$ intershells, while V and T are regarded as shell-centered variables (i.e., assigned to a mass-coordinate given by the geometric mean $m_{i+1/2} = \sqrt{m_{i+1}\, m_i}$). Note that in a Lagrangian formulation, the most convenient independent variables are the mass interior to the ith-intershell, m_i, and time, t^n.

In this chapter, a superscript will be used to indicate whether a given variable is evaluated at current time, t^{n+1}, or at previous time, t^n. Accordingly, note that the time-step is simply given by $\Delta t \equiv t^{n+1} - t^n$. Following Potter [1441], partial differential equations involving time-derivatives, in the form

$$\frac{\partial y}{\partial t} = Ly, \qquad (1.96)$$

where L represents an operator, can be discretized in a general way as

$$y^{n+1} = y^n + Ly^n(1 - \beta)\Delta t + Ly^{n+1}\beta\Delta t, \qquad (1.97)$$

where y^n and y^{n+1} are the values of function y at times t^n and t^{n+1}, respectively, while β is an interpolation parameter ($0 \leq \beta \leq 1$) that defines different numerical schemes. The choice $\beta = 0$ leads to an *explicit* determination of y^{n+1} from the previous, known value y^n. This condition defines the so-called *explicit* methods. Instead, the choice $\beta \neq 0$ defines *implicit* methods. In general, explicit schemes are easier to implement than implicit schemes. However, explicit schemes are only stable if the time-step is limited by the *Courant–Friedrichs–Levy condition* (see Problem P5 at the end of this chapter), that prevents any disturbance traveling at the sonic speed from traversing more than one numerical cell, thus leading to unphysical results [1500]. On the other hand, implicit schemes allow larger time-steps than explicit schemes, with no precondition on the time-step, but they require an iterative procedure to solve the system at each step. $\beta = 0.5$ is the only choice that ensures second-order accuracy in the time-derivates, resulting in neither artificial damping nor unphysical amplification [1037, 1268].

The system of differential equations needs to be supplemented by a suitable set of boundary conditions at the edges of the computational grid. A convenient choice for a 1D, Lagrangian code (see, e.g., reference [1037]) is:

- Innermost shell: $m = m_1 \Rightarrow u = 0, L = L_1, r = r_1$. When the innermost shell corresponds to the star's center, $m_1 = 0 \Rightarrow u_1 = r_1 = L_1 = 0$.

- Surface: $m = M_{tot} \Rightarrow T^4 = \frac{3}{4}T_{\text{eff}}^4\left(\frac{2}{3} + \tau\right), P = P_{\text{rad}}$, where τ is the optical depth, T_{eff} the effective temperature[22], and P_{rad} the radiation pressure.

---

[22]The effective temperature of a star is the temperature of a blackbody with the same luminosity of the star per surface area.

Prior to discretizing the set of differential equations into an algebraic system of equations in finite differences, one has to consider whether rescaling of certain variables is required to avoid potential numerical problems. A suitable choice, for instance, includes:

- $m_i \rightarrow Q_i = 1 - m_i/M_{tot}$

- $r_i \rightarrow R_i = \ln r_i$

- $V_{i+1/2} \rightarrow W_{i+1/2} = \ln V_{i+1/2}$

- $T_{i+1/2} \rightarrow Z_{i+1/2} = \ln T_{i+1/2}$

- $L_i \rightarrow B_i = L_i/L_\odot$,

where M_{tot} is the overall mass of the star and $L_\odot$ is the solar luminosity.

All in all, the system of equations of stellar structure in finite difference form[23] can be written as:

- Conservation of mass

$$\frac{4\pi}{3M_{tot}} \frac{(r_i^3 - r_{i-1}^3)^{n+1}}{Q_i - Q_{i-1}} = -V_{i-1/2}^{n+1} \tag{1.98}$$

- Conservation of momentum

$$\frac{u_i^{n+1} - u_i^n}{\Delta t} = (1-\beta)\mathbf{F}_i^n + \beta\mathbf{F}_i^{n+1} + \frac{4\pi}{M_{tot}} \frac{1}{Q_{i+1/2} - Q_{i-1/2}}$$

$$\left((1-\beta)(r_i^2)^n \left[q_{i+1/2}^{n+1/2} \frac{V_{i+1/2}^{n+1}}{V_{i+1/2}^n} - q_{i-1/2}^{n+1/2} \frac{V_{i-1/2}^{n+1}}{V_{i-1/2}^n} \right] + \beta(r_i^2)^{n+1}(q_{i+1/2}^{n+1/2} - q_{i-1/2}^{n+1/2}) \right) , \tag{1.99}$$

where

$$\mathbf{F}_i^n = 4\pi \frac{(r_i^2)^n}{M_{tot}} \frac{P_{i+1/2}^n - P_{i-1/2}^n}{Q_{i+1/2} - Q_{i-1/2}} - G\frac{m_i}{(r_i^2)^n}$$

- Conservation of energy

$$\frac{E_{i-1/2}^{n+1} - E_{i-1/2}^n}{\Delta t} = \varepsilon_{i-1/2}^n + (1-\beta)\mathbf{G}_{i-1/2}^n + \beta\mathbf{G}_{i-1/2}^{n+1}$$

$$-q_{i-1/2}^{n+1/2}V_{i-1/2}^{n+1} \frac{W_{i-1/2}^{n+1} - W_{i-1/2}^n}{\Delta t} , \tag{1.100}$$

where

$$\mathbf{G}_{i-1/2}^{n+1} = \frac{L_\odot}{M_{tot}} \frac{B_i^{n+1} - B_{i-1}^{n+1}}{Q_i - Q_{i-1}} - P_{i-1/2}^{n+1}V_{i-1/2}^{n+1} \frac{W_{i-1/2}^{n+1} - W_{i-1/2}^n}{\Delta t}$$

---

[23]Both convective energy transport and nuclear energy generation are implemented explicitly in the finite difference scheme outlined here (see also reference [1037]). Other fully implicit formulations are obviously possible.

- Energy transport (by radiation and convection[24])

$$B_i^{n+1} = 256\sigma\pi^2(r_i^4)^{n+1}\frac{(T_i^4)^{n+1}}{3L_\odot M_{\text{tot}}\kappa_i^{n+1}}\frac{Z_{i+1/2}^{n+1} - Z_{i-1/2}^{n+1}}{Q_{i+1/2} - Q_{i-1/2}} - \mathbf{H}_i^n\frac{r_i^{n+1}T_i^{n+1}}{V_i^{n+1}}, \qquad (1.101)$$

where

$$\mathbf{H}_i^n = \begin{cases} 0, & \text{if } \nabla_i^n \le \nabla_{\text{ad},i}^n \\ \frac{\pi}{L_\odot}\sqrt{G/2}H_{\text{p},i}^{-3/2}l_{\text{m},i}^2 m^{1/2}c_{\text{p},i}\sqrt{\left(\frac{\partial W}{\partial Z}\right)}(\nabla_i - \nabla_{\text{ad},i})^{3/2}, & \text{if } \nabla_i^n > \nabla_{\text{ad},i}^n \end{cases}$$

- Lagrangian velocity

$$\frac{R_i^{n+1} - R_i^n}{\Delta t} = (1-\beta)\frac{u_i^n}{r_i^n} + \beta\frac{u_i^{n+1}}{r_i^{n+1}} \qquad (1.102)$$

Special care has to be devoted to the implementation of the energy transport equation at the surface, taking into account the corresponding boundary conditions. Following Kutter and Sparks [1037], when convection is neglected between grid points $N + 1/2$ and $N + 1$, the energy transport equation becomes

$$B_{\text{N}+1}^{n+1} = 16\sigma\pi(r_{\text{N}+1}^2)^{n+1}\frac{(T_{\text{N}+1/2}^4)^{n+1}}{3L_\odot}\left[-\frac{M_{\text{tot}}}{4\pi}(Q_{\text{N}+1}-\right.$$

$$\left. Q_{\text{N}+1/2})\frac{\kappa_{\text{N}+1/2}^{n+1}}{2\sqrt{(r_{\text{N}+1}^3 r_{\text{N}})^{n+1}}} + \frac{2}{3}\right]^{-1}. \qquad (1.103)$$

It is worth noting that an artificial viscosity term, q, has been included in the conservation of momentum (Equation 1.99) to handle shock waves. Following von Neumann and Richtmyer [1870], q can be expressed as the divergence of the velocity, which ensures that artificial viscosity remains negligible in the absence of shocks. A simple procedure switches on artificial viscosity whenever any mass-shell gets compressed (i.e., $\rho_{i+1/2}^{n+1} > \rho_{i+1/2}^n$):

$$q_{i+1/2}^{n+1/2} = \begin{cases} 0, & \text{if } \rho_{i+1/2}^{n+1} \le \rho_{i+1/2}^n \\ q_0\frac{(r_{i+1}^{n+1} - r_i^{n+1})^2(W_{i+1/2}^{n+1} - W_{i+1/2}^n)^2}{V_{i+1/2}^{n+1}(\Delta t^{n+1/2})^2}, & \text{if } \rho_{i+1/2}^{n+1} > \rho_{i+1/2}^n \end{cases}, \qquad (1.104)$$

where q_0 is a parameter of the order of unity [1037].

The algebraic system of finite difference equations in implicit formulation described above is solved through the application of an iterative procedure (a Newton–Raphson method; see Section 1.6 for a detailed account), until a given accuracy criterion is satisfied. Due to the inclusion of artificial viscosity, the overall conservation of energy[25] is carefully checked [332]:

$$\frac{\partial}{\partial t}\left[\frac{u^2}{2} - G\frac{m}{r} + E\right] + \frac{\partial L}{\partial m} + \frac{\partial}{\partial m}\left[4\pi r^2 u(P+q)\right] = \varepsilon_{\text{nuc}} - \varepsilon_\nu \qquad \text{erg g}^{-1}\text{s}^{-1} \qquad (1.105)$$

---

[24]For simplicity, only the time-independent prescription for convection is given in the expression for the energy transport. Interested readers are referred to Wood [1955] for a suitable time-dependent scheme.

[25]In most of the simulations performed with the 1D code SHIVA, energy is conserved to within a few percent.

1.4.2 Nuclear Reaction Networks

Even though the energetics of a stellar explosion can often be approximated by a handful of nuclides and nuclear processes, state-of-the-art nucleosynthesis studies require hundreds of isotopes linked through hundreds or thousands of nuclear interactions. Their implementation in a stellar evolution hydro code may cause a severe slowdown in its performance, becoming the truly time-limiting factor, to the point that nucleosynthesis calculations are often not viable without the aid of high-power supercomputers, unless postprocessing techniques are employed. Special care, therefore, has to be devoted to the numerical techniques used to handle nuclear reaction networks, in order to allow larger step-sizes in full hydrodynamical models or to increase the performance in postprocessing calculations.

The differential equations that dictate the time evolution of the abundances in a stellar plasma frequently contain terms that may lead to rapidly varying solutions. For a particular nuclide i, and in the absence of diffusion of chemical species, they can be written as:

$$\frac{dY_i}{dt} = \dot{R}_i, \tag{1.106}$$

where $Y_i = X_i/M_i$ is the mole fraction of species i, or ratio between mass-fraction abundance, X_i, and mass, M_i (see Chapter 2 for suitable definitions). $\dot{R}_i$ is the overall reaction rate, that is, a balance of all reactions producing and destroying species i. Stiffness actually requires extremely small step-sizes to avoid potential numerical failures driven by severe instabilities, particularly in the context of explicit schemes. A number of implicit and semiimplicit techniques have been implemented over the last decades to overcome those problems. Some of these methods [73, 1826, 1873, 1969], particularly Wagoner's two-step linearization technique [1873], have become the standard in nucleosynthesis studies for nearly 40 years, and in spite of their limitations, are still widely used.

1.4.2.1 Wagoner's Method

Wagoner's two-step linearization technique [1873] is a semiimplicit, second-order Runge–Kutta method that can be viewed as an extension of the first numerical schemes employed in the study of type Ia supernova nucleosynthesis [62, 74, 1826] (see Chapter 5).

Let Y^n be the vector that contains the mole fractions of all species included in the network at time t^n (i.e., $Y_1^n, Y_2^n, ...Y_i^n...Y_N^n$). In Wagoner's method, the change in mole fractions after one timestep, $t^{n+1} = t^n + \Delta t$, is estimated as

$$Y^{n+1} = Y^n + \frac{1}{2}\Delta t \left[\frac{dY}{dt}(Y^n, t^n) + \frac{dY}{dt}(\tilde{Y}^{n+1}, t^{n+1}) \right], \tag{1.107}$$

where

$$\tilde{Y}^{n+1} \equiv Y^n + \Delta t \frac{dY}{dt}(Y^n, t^n) \tag{1.108}$$

is actually determined from the Jacobian matrix $\mathbf{J} = \partial(dY^n/dt)/\partial Y$, through the equation

$$[\mathbf{I} - \Delta t\,\mathbf{J}]\tilde{Y}^{n+1} = Y^n, \tag{1.109}$$

where $\mathbf{I}$ is the unity matrix. All in all, mole fractions at time t^{n+1} are determined by averaging the time-derivates of Y evaluated at times t^n and estimated for t^{n+1}.

The determination of $\tilde{Y}^{n+1}$ is actually the most computationally challenging step in Wagoner's method. The Jacobian matrix $\mathbf{J}$, containing all possible forms of nuclear interaction between the N species of the network, can actually be quite large[26]. It is, however, worth

---

[26] A typical nova nucleosynthesis study is characterized by a Jacobian containing 100×100 species, about 400×400 species in the case of type Ia supernovae, and 600×600 species for X-ray bursts.

noting that nuclear interactions for a given species are mostly limited to proton, neutron, electron, or alpha captures, plus beta disintegrations (that is, involve only light particles). Indeed, reactions between heavier species are mostly prevented by the large Coulomb barriers at the temperatures of interest (exceptions include $^{12}C + ^{12}C$, or $^{12}C + ^{16}O$, for type Ia supernovae; see Chapter 2). Therefore, the matrix of interactions exhibits a characteristic sparse pattern, with large portions essentially containing zero elements, which could be efficiently handled by means of standard, optimized routines (see references [1148, 1804]).

While the best asset of Wagoner's method is its ease of implementation, it lacks a proper way to consistently infer the next time-step. Often, criteria based on the largest abundance variations from the previous step,

$$\Delta t' \sim \Delta t \left[\frac{Y^{n+1}}{Y^{n+1} - Y^n} \right]_{min}, \tag{1.110}$$

are implemented to this end, while mass conservation within a given tolerance (i.e., $\sum_{i=1}^{N} X_i \equiv 1$) has to be satisfied to guarantee accuracy in the procedure.

1.4.2.2 Bader–Deuflhard's and Gear's Methods

Bader–Deuflhard's method [107] is a semiimplicit technique based on a generalization of the Bulirsch–Stoer algorithm (an application of Richardson extrapolation technique) for solving stiff systems of ordinary differential equations [1448]. The method relies on the following implicit form of the so-called midpoint rule, applied to a first-order differential equation $dY/dt = f(Y)$ (see Equation 1.106):

$$Y^{n+1} - Y^{n-1} = 2\Delta t\, f\left(\frac{Y^{n+1} + Y^{n-1}}{2} \right). \tag{1.111}$$

Linearization of the right-hand side of the equation about $f(Y^n)$ yields the semiimplicit midpoint rule:

$$\left[1 - \Delta t \frac{\partial f}{\partial Y} \right] Y^{n+1} = \left[1 + \Delta t \frac{\partial f}{\partial Y} \right] Y^{n-1} + 2\Delta t \left[f(Y^n) - \frac{\partial f}{\partial Y} Y^n \right]. \tag{1.112}$$

But rather than solving this equation for an arbitrary large timestep, $\Delta \tau$, Bader–Deuflhard's strategy adopts m substeps, each of length $\Delta t = \Delta \tau / m$. The method assumes that the solution corresponding to a large step $\Delta \tau$ is a function of the number of substeps, which can be proved by solving the equation for a suitable range of values of m. Once this function is found, the solution can be extrapolated to an infinite number of substeps, thus yielding converged abundances. Bader and Deuflhard inferred the values of m that provide best convergence. See references [1148, 1448, 1804] for details.

Bader–Deuflhard's technique requires that the system of equations is solved a large number of times, even if convergence is reached promptly. Therefore, the method requires large steps to offset their computational cost [1804]. In sharp contrast to Wagoner's technique, in Bader–Deuflhard's method a new timestep can be directly inferred from the extrapolation function truncation error for the desired accuracy.

Another useful method in the analysis of stiff differential equations (although much harder to implement) is Gear's backward differentiation technique [278, 619]. Gear's method relies on information available on previous steps to infer the evolution of a system. Two steps are used to determine the time evolution of the system from t^n to t^{n+1}: a predictor step and a corrector step[27]. Aside from the possibility of directly inferring the new timestep, the main advantage of Gear's method relies on the relatively large step-sizes achievable while maintaining a moderate computational load.

---

[27]The reader is referred to Longland et al. [1148] for details on the implementation of Gear's method.

The efficiency of Wagoner's, Bader–Deuflhard's, and Gear's methods has been assessed through a test suite of reaction networks and postprocessing profiles from a variety of stellar nucleosynthesis sites by Richard Longland and collaborators [1148] (see also references [1804] for a detailed comparison between Wagoner's and Bader–Deuflhard's schemes). According to this study, both Bader–Deuflhard's and Gear's methods exhibit dramatic improvements in both speed and accuracy with respect to Wagoner's method.

1.5 Gridless Methods: Smoothed-Particle Hydrodynamics

The traditional, Lagrangian grid-based methods described in previous sections face potentially important shortcomings in 2D and 3D formulations. Indeed, the presence of shear or turbulence, for instance, may result in severe grid deformation and tangling, likely causing a computational failure. On the other hand, multidimensional Eulerian methods are harder to implement and have also a limited applicability, since they cannot handle expanding fluids properly—i.e., material may leave the computational domain. Alternative methods, specifically suited for astrophysics, have been developed to overcome the drawbacks of grid-based formulations.

This section focuses on SPH, a method that drastically eliminates grid distorsions by removing the grid itself. In essence, SPH codes are closely related to *N-body methods*, where the traditional computational cells of grid-based methods are replaced by particles of a finite length. However, SPH codes need to include pressure terms to characterize stellar plasmas in most astrophysical applications[28]. Pioneering implementations of Lagrangian SPH codes were developed by Lucy [1156] and Gingold and Monaghan [646], and were subsequently applied to a wide range of astrophysical scenarios[29]. Those directly related with the contents of this book include type Ia supernovae [154, 245, 246, 613, 614], type II supernovae [770–772, 1278], and mergers of stellar binary systems (involving white dwarfs, neutron stars, and black holes; see Rosswog [1533] and references therein)[30].

1.5.1 Briefing on SPH Methods: Weighted Sums, Kernels, and Smoothing Lengths

In SPH, the physical properties of a fluid element are locally reconstructed by interpolating the properties of the neighboring particles. The *smoothed* value of a space-dependent physical variable, $f(\vec{r})$, is therefore defined as

$$\langle f(\vec{r}) \rangle = \int f(\vec{r}') W(\vec{r} - \vec{r}', h) d\vec{r}' , \qquad (1.113)$$

where $\vec{r}$ is the location of particle i, $W(\vec{r}, h)$ is an interpolating (or weight) function known as *kernel*, and h is the *smoothing length*, a parameter that somehow defines the *size* of the

---

[28]See, e.g., references [11, 209] for an in-depth comparison between SPH and grid-based methods.

[29]See also references [647, 1256, 1260].

[30]SPH has also been used in several movies that feature fluid simulations. This includes Gollum's fall into the lava, in *The Lord of the Rings: The Return of the King* (2003), and several scenes in *Superman Returns* (2006), to quote a few examples.

kernel[31]. Note that for $h \to 0$, $W(\vec{r})$ becomes a delta function [1257],

$$\lim_{h \to 0} W(\vec{r} - \vec{r}', h) = \delta(\vec{r} - \vec{r}') . \tag{1.115}$$

Two conditions are imposed to any kernel: First, it must be normalized,

$$\int W(\vec{r}, h) d\vec{r} \equiv 1, \quad \forall h \tag{1.116}$$

and second, it must correspond to an even function,

$$W(\vec{r}, h) = W(-\vec{r}, h) . \tag{1.117}$$

In numerical applications, integrals are replaced by discrete summations, such that Equation 1.113 becomes

$$\langle f(\vec{r}) \rangle = \sum_j m_j \frac{f(\vec{r_j})}{\rho(\vec{r_j})} W(\vec{r} - \vec{r_j}, h) , \tag{1.118}$$

where $m_j / \rho(\vec{r_j})$ represents the discretized version of the volume element $d\vec{r}'$. Knowing the location and mass of the N particles of a system, one can determine the local density as

$$\langle \rho(\vec{r}) \rangle = \int \rho(\vec{r}') W(|\vec{r} - \vec{r}'|, h) d\vec{r}' = \sum_j m_j W(\vec{r} - \vec{r_j}, h) . \tag{1.119}$$

The above definitions clearly stress the pivotal role played by the kernel in SPH simulations. However, the choice of the kernel is not unique, and different functions have been proposed in the literature, including the following expressions, second-order accurate in h:

- Gaussian [646]

$$W(r, h) = \frac{1}{\pi^{3/2} h^3} \exp(-w^2) \tag{1.120}$$

- Exponential [1953]

$$W(r, h) = \frac{1}{8\pi h^3} \exp(-w) \tag{1.121}$$

- Cubic spline [1261]

$$W(r, h) = \frac{1}{\pi h^3} \begin{cases} 1 - 3w^2/2 + 3w^3/4, & 0 \le w \le 1 \\ (2 - w)^3 / 4, & 1 \le w \le 2 \\ 0, & \text{otherwise} , \end{cases} \tag{1.122}$$

---

[31] In this brief introduction to SPH methods, the smoothing length is considered constant, for simplicity. Certain practical situations, however, require a variable smoothing length to guarantee, for instance, a relatively constant number of neighbors throughout the computation. To this end, Hernquist and Katz [789] have proposed the simple prescription,

$$h_i^{t+\Delta t} = \frac{h_i^t}{2} \left[1 + (2^\eta - 1) \frac{n_n^0}{n_n^i} \right]^{1/\eta} , \tag{1.114}$$

where h_i^t and $h_i^{t+\Delta t}$ are the values of the smoothing length for particle i evaluated at times t and $t + \Delta t$, n_n^0 and n_n^i are the desired and actual number of neighbors for particle i, and η a parameter that controls the speed of the adjustments required to achieve $n_n^i \simeq n_n^0$. See also Price and Monaghan [1454] for another formalism based on the inclusion of a correction term to the gravitational force related to the gradient of the smoothing length.

where $w \equiv r/h$ and $r \equiv |\vec{r} - \vec{r_j}|$.

The cubic spline kernel has an important advantatge with respect to other formalisms: It is defined on a *compact support*, that is, it has nonzero values only within a limited domain. This limits the number of neighbors that effectively interact with a given particle, thus reducing the computational load[32]. A more general cubic spline kernel, of the form

$$W(r,h) = \frac{M}{\pi h^3} \begin{cases} 1 + aw + bw^2 + cw^3, & 0 \le w \le 1 \\ d(2-w)^3, & 1 \le w \le 2 \\ 0, & \text{otherwise}, \end{cases} \tag{1.123}$$

has been introduced by Domingo García-Senz and collaborators [613]. Note that the choice $M = 1$, $a = 0$, $b = -3/2$, $c = 3/4$, and $d = 1/4$ reproduces Monaghan and Lattanzio's cubic spline kernel.

A similar procedure can be applied to calculate derivatives of an arbitrary function:

$$\langle \nabla_r f(\vec{r}) \rangle = \sum_j \nabla_r \left(\frac{m_j}{\rho(\vec{r_j})} f(\vec{r_j}) \right) W(\vec{r} - \vec{r_j}, h) + \frac{m_j}{\rho(\vec{r_j})} f(\vec{r_j}) \nabla_r W(\vec{r} - \vec{r_j}, h) =$$

$$\sum_j \frac{m_j}{\rho(\vec{r_j})} f(\vec{r_j}) \nabla_r W(\vec{r} - \vec{r_j}, h), \tag{1.124}$$

1.5.2 SPH Equations

SPH relies on the same set of conservation equations described for grid-based methods. Following, e.g., Hernquist and Katz [789] and Benz [153], to which the reader is referred for details, the SPH version of the main conservation equations describing an astrophysical plasma can be written as:

- Mass conservation

$$\langle \rho_i \rangle = \sum_j m_j W(\vec{r_i} - \vec{r_j}, h), \tag{1.125}$$

where ρ_i is the density of particle i (i.e., $\rho(\vec{r_i})$).

- Energy conservation

$$\frac{dE_i}{dt} = \frac{P_i}{\rho_i^2} \sum_j m_j (\vec{u_i} - \vec{u_j}) \nabla_i W(|\vec{r_i} - \vec{r_j}|, h) + \frac{1}{2} \sum_j m_j \prod_{ij} (\vec{u_i} - \vec{u_j}) \nabla_i W(|\vec{r_i} - \vec{r_j}|, h), \tag{1.126}$$

where E_i, u_i, and P_i are the internal energy, the velocity, and the pressure of particle i, and ∇W is the derivative of the kernel. The term $\prod_{ij}$ corresponds to the artificial viscosity (see Section 1.4), needed to handle shocks properly, and for which several prescriptions have been proposed. A widely used recipe is the one advised by Gingold and Monaghan [647], which relies on a local estimate of the divergence of the velocity:

$$\mu_{ij} = \frac{h(\vec{u_i} - \vec{u_j})(\vec{r_i} - \vec{r_j})}{|\vec{r_i} - \vec{r_j}|^2 + \epsilon h^2}, \tag{1.127}$$

---

[32]Other kernels with higher-order accuracy (i.e., h^4) have also been proposed, but their applicability is challenged by a number of effects. See Benz [153] for a detailed discussion.

where ϵh^2 is added to avoid divergence for small values of $|\vec{r_i} - \vec{r_j}|$. This translates into a pressure term that is added to the momentum conservation equation in the form

$$\prod_{ij} = \frac{M}{\pi h^3} \begin{cases} \frac{-\alpha c_{ij}\mu_{ij} + \beta \mu_{ij}^2}{\rho_{ij}}, & \text{if}(\vec{u_i} - \vec{u_j})(\vec{r_i} - \vec{r_j}) \leq 0 \\ 0, & \text{otherwise} , \end{cases} \tag{1.128}$$

where $c_{ij} \equiv (c_i + c_j)/2$ and $\rho_{ij} \equiv (\rho_i + \rho_j)/2$ are the average speed of sound and density, respectively.

- Momentum conservation

$$\frac{d\vec{u_i}}{dt} = -\sum_j \left(\frac{P_i}{\rho_i^2} + \frac{P_j}{\rho_j^2} + \prod_{ij} \right) \nabla W(\vec{r_i} - \vec{r_j}, h) - \nabla \Phi_i , \tag{1.129}$$

where $\nabla \Phi_i$ corresponds to the gravitational force per unit mass acting on particle i, which can be obtained by application of Poisson's equation as

$$\nabla \Phi_i = G \sum_j 4\pi \frac{\vec{r_{ij}}}{|\vec{r_{ij}}|^3} \int_0^{|\vec{r_{ij}}|} W(x, h)x^2 dx . \tag{1.130}$$

The above expression does not vanish unless the distance between particles i and j is infinite. Moreover, if the summation is not done properly, the method will translate into a computational cost proportional to N^2, where N is the number of particles. A possible way out involves the use of a *hierarchical tree*, which allows a fast track of the neighboring particles and reduces the computational cost to $N \log N$. To this end, a widely used prescription in the Barnes–Hut algorithm [130], first implemented in SPH codes by Hernquist and Katz [789], to which the reader is referred for further details. Integration and time evolution of the system is finally handled by means of a simple algorithm (e.g., Runge–Kutta, predictor-corrector). Further details on SPH methods can be found in references [1120, 1258, 1259, 1531, 1533, 1688, 1689].

1.6   Building a 1D Hydrodynamic Code

In the absence of magnetic fields, a nonrotating stellar plasma can be described in spherical symmetry by the system of nonlinear, coupled differential equations discussed in Section 1.2. The system cannot be solved analytically, and numerical techniques need to be implemented.

This section outlines the basic strategies undertaken in the design of a stellar evolution code[33] in a simplified framework. To this end, a 1D, spherically symmetric, Lagrangian, implicit, hydrodynamic code, aimed at characterizing the free-fall collapse of a homogeneous sphere, will be built from scratch[34].

1.6.1   Differential Equations for the Free-Fall Collapse Problem

Let's consider a homogeneous, spherical plasma in hydrostatic equilibrium (i.e., the gravitational pull is balanced by a pressure gradient). If forces due to pressure gradients are

---

[33]See Table 1.1 for examples of hydro codes used in the study of stellar explosions.

[34]The computer program, `freefall.f`, written in Fortran, is actually listed in Appendix B and can also be downloaded from `http://fisica.upc.edu/ca/users/jjose/CRC-Downloads`.

TABLE 1.1

Examples of Hydro Codes Used in the Study of Stellar Explosions

Code	Stellar* explosion	Main properties and reference papers
AGILE	SNII, XRB	1D, implicit/explicit, Lagrangian, general relativistic [533, 1109]
AxisSPH	WD+WD mergers	2D (axisymmetric) SPH, explicit, Lagrangian; Parallelized [616]
DJEHUTY	CN, SNIa, SNII	3D, explicit, ALE; Parallelized [143, 421]
FLASH	CN, XRB, SNIa	3D, explicit, Eulerian; Parallelized [567]
FRANEC	SNII	1D, fully implicit, Lagrangian, with rotation [324, 1110]
GADGET2	CN	3D SPH, explicit, Lagrangian; Parallelized [1687]
KEPLER	CN, XRB, SNIa, GRB, SNII	1D, implicit, Lagrangian with rotation; Parallelized [1475, 1911]
MESA	CN, XRB, SNIa	1D, implicit/explicit, Lagrangian, with rotation; Parallelized [1396, 1397]
NOVA	CN, XRB, SNIa	1D, implicit/explicit, Lagrangian [1037, 1706]
PROMETHEUS	CN, SNIa SNII	3D, explicit, Eulerian; Parallelized [566, 985]
SHIVA	CN, XRB, SNIa	1D, implicit/explicit, Lagrangian, with rotation; Parallelized [919, 927]
SPH	WD+WD(NS) mergers	3D SPH, explicit, Lagrangian [97, 1151]
SPHYNX	SNIa, SNII	3D SPH, explicit, Lagrangian; Parallelized [613]
TYCHO	SNIa, SNII	1D, implicit/explicit, Lagrangian [67, 2007]
VULCAN	CN, SNIa, SNII	2D, implicit/explicit, ALE; Parallelized [649, 1130]

*Acronyms: Classical novae (CN); type Ia supernovae (SNIa); type II supernovae (SNII); X-ray bursts (XRB); γ-ray bursts (GRB); white dwarfs (WD); neutron stars (NS); smoothed-particle hydrodynamics (SPH); arbitrary Lagrangian–Eulerian (ALE).

neglected (i.e., $P = 0$), the dynamics of the system is governed only by gravity, and a free-fall collapse will ensue. Since no shock is expected during the collapse, artificial viscosity can be removed ($q = 0$). In such conditions, the collapse can be described by the following set of differential equations:

- Conservation of mass

$$\frac{1}{\rho} = \frac{4}{3}\pi\frac{\partial r^3}{\partial m}$$ (1.131)

- Conservation of momentum ($P = q = 0$)

$$\frac{\partial u}{\partial t} = -G\frac{m}{r^2}$$ (1.132)

- Lagrangian velocity

$$\frac{\partial r}{\partial t} = u$$ (1.133)

1.6.2 Variable Assignment

Solutions to the set of differential equations described in Section 1.6.1 will be obtained for a discrete number of points within the sphere. To this end, a suitable computational grid is built by dividing the sphere into N concentric layers or shells, with equal or different masses and sizes (for simplicity, we will assume hereafter that all concentric shells have the same mass).

A decision has to be made with regard to variable assignment to specific grid points. Let's assume, for instance, that radii and velocities are assigned to each intershell while densities are instead evaluated at midpoints, on the basis of symmetry arguments inspired by the form of the mass conservation equation (see Figure 1.11). For convenience, the mass enclosed in each spherical shell, or *interior mass*, m_i, is also assigned to each intershell. Note, however, that this choice is somewhat arbitrary and alternative criteria for variable assignment could be adopted.

1.6.3 Discretization

The set of partial differential equations reviewed in Section 1.6.1 is subsequently replaced by the following system in finite difference form:

- Conservation of mass

$$\frac{1}{(\rho_{i+1/2})^{n+1}} = \frac{4}{3}\pi\frac{(r_{i+1}^3)^{n+1} - (r_i^3)^{n+1}}{m_{i+1} - m_i}$$ (1.134)

- Conservation of momentum

$$\frac{u_{i+1}^{n+1} - u_{i+1}^n}{\Delta t} = (1-\beta)\left(\frac{-Gm_{i+1}}{r_{i+1}^2}\right)^n + \beta\left(\frac{-Gm_{i+1}}{r_{i+1}^2}\right)^{n+1}$$ (1.135)

- Lagrangian velocity

$$\frac{r_{i+1}^{n+1} - r_{i+1}^n}{\Delta t} = (1-\beta)u_{i+1}^n + \beta u_{i+1}^{n+1},$$ (1.136)

where superscripts n and $n+1$ denote variables evaluated at times t^n and $t^{n+1} = t^n + \Delta t$. β is the interpolation parameter introduced in Equation 1.96 that characterizes explicit ($\beta = 0$) and implicit ($\beta \neq 0$) schemes.

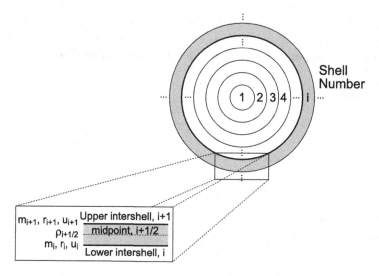

FIGURE 1.11

Shell structure and variable assignment in the 1D free-fall collapse problem. Note that masses, radii, and velocities are assigned to the intershells, while densities are instead assigned to shell midpoints.

1.6.4 Initial Models, Boundary Conditions, and Scaling

If the N concentric shells contain the same mass, $\Delta m \equiv M_{tot}/N$, the interior mass variable m_i is simply given by:

$$m_{i+1} = i \, \Delta m \qquad (i = 1, N) \, . \tag{1.137}$$

Note that $m_{N+1} \equiv M_{tot}$, by construction, with $M_{tot} = \frac{4}{3}\pi R_o^3 \rho_o$ for a homogeneous sphere of initial radius R_o and density ρ_o.

If the computational domain extends all the way from the center of the sphere to its surface[35], the radius and interior mass variable at the first intershell trivially become $r_1 = m_1 = 0$. Moreover, for a homogenous sphere, $\rho_{i+1/2} \equiv \rho_o$ $(i = 1, N)$, by definition, and assuming an initial static configuration, $u_i = 0$ $(i = 1, N + 1)$ at $t = 0$.

Once boundary conditions are applied to the innermost shell, mass conservation equation becomes

$$r_2 = \left(\frac{3m_2}{4\pi\rho_{3/2}} \right)^{1/3} \tag{1.138}$$

for $r_1 = m_1 = 0$, while the radii of the subsequent intershells can be obtained through

$$r_{i+1} = \left(\frac{3(m_{i+1} - m_i)}{4\pi\rho_{i+1/2}} + r_i^3 \right)^{1/3} =$$

$$\left(\frac{3\Delta m}{4\pi\rho_{i+1/2}} + r_i^3 \right)^{1/3} \qquad (i = 2, N) \, . \tag{1.139}$$

---

[35] In some applications, the computational domain does not necessarily cover the whole physical system. Localized, stellar thermonuclear explosions, for instance, can be numerically handled by restricting to smaller computational domains, therefore increasing the resolution at the regions of interest while speeding up the calculations. In such cases, the innermost shell is not located at the center of the star (i.e., $m_1 \neq 0$), but placed at a suitable depth where no physical changes are actually expected. Accordingly, the inner boundary conditions are simply given by the initial values of the radius and velocity at m_1, which are assumed to remain constant throughout the calculation.

Finally, rescaling of certain physical variables will be used to avoid numerical problems. In particular, the interior mass m_i will be rewritten as $Q_i = 1 - m_i/M_{tot}$, while $V_{i+1/2} = 1/\rho_{i+1+2}$ and r_i will be replaced by their natural logarithms, $R_i = \ln(r_i)$, $W_{i+1/2} = \ln(V_{i+1/2})$, thus limiting their ranges of variation.

1.6.5 Henyey Method

Several methods have been proposed for solving the stellar structure equations with fixed boundary conditions. They are often classified into two broad categories: shooting[36] and relaxation methods.

The most extensively used in stellar evolution are a class of relaxation methods generically known as Newton–Raphson. They rely on the linearization of the set of finite difference equations. The methodology adopted in a generalized Newton–Raphson method (sometimes called Henyey method[37]) [764,987] will be the subject of an in-depth analysis in this section.

1.6.5.1 Equations for the Innermost Shell

The finite difference equations for the innermost shell of the sphere, with the corresponding boundary conditions, can be written in compact form as a function C that depends on a number of unknowns. Let's start with the mass conservation equation: for $m_1 = r_1 = u_1 = 0$, it can be written as

$$C^1 = \frac{1}{(\rho_{3/2})^{n+1}} - \frac{4}{3}\pi\frac{(r_2^3)^{n+1}}{m_2} = C^1(r_2, \rho_{3/2}) = 0. \tag{1.140}$$

Momentum conservation and the equation for the Lagrangian velocity can, in turn, be expressed as:

$$C^2 = \frac{u_2^{n+1} - u_2^n}{\Delta t} - (1-\beta)\left(\frac{-Gm_2}{r_2^2}\right)^n - \beta\left(\frac{-Gm_2}{r_2^2}\right)^{n+1} =$$

$$C^2(u_2, r_2) = 0, \tag{1.141}$$

$$C^3 = \frac{r_2^{n+1} - r_2^n}{\Delta t} - (1-\beta)u_2^n - \beta u_2^{n+1} = C^3(u_2, r_2) = 0. \tag{1.142}$$

Globally, this set of equations can be written as a function of just 3 unknowns, u_2, r_2, and $\rho_{3/2}$, such that

$$C^j = C^j(\rho_{3/2}, u_2, r_2) = 0 \qquad (j = 1, 3). \tag{1.143}$$

1.6.5.2 Equations for the Intermediate Shells

The same procedure is then applied to the $N - 2$ intermediate shells ($i = 2, N - 1$), in the form:

$$F_i^1 = \frac{1}{(\rho_{i+1/2})^{n+1}} - \frac{4}{3}\pi\frac{(r_{i+1}^3)^{n+1} - (r_i^3)^{n+1}}{m_{i+1} - m_i} = F_i^1(r_{i+1}, r_i, \rho_{i+1/2}) = 0, \tag{1.144}$$

---

[36]Shooting methods, also known as fitting or matching point methods, rely on a double integration of the stellar structure equations, one starting from the center outward, the other from the surface inward. In general, both integrations do not match, and variations of the initial guess values adopted at both edges are required. While useful for constructing equilibrium configurations, it turned out that shooting methods are in general unsuitable for complex stellar models (see discussion in reference [988]).

[37]See Appendix A for an extension of the Henyey method to arbitrary hydrodynamic problems. Further generalizations of the Henyey method for an arbitrary set of differential equations can be found, e.g., in references [1313, 1948].

$$F_i^2 = \frac{u_{i+1}^{n+1} - u_{i+1}^n}{\Delta t} - (1-\beta)\left(\frac{-Gm_{i+1}}{r_{i+1}^2}\right)^n - \beta\left(\frac{-Gm_{i+1}}{r_{i+1}^2}\right)^{n+1} =$$

$$F_i^2(u_{i+1}, r_{i+1}) = 0\,, \tag{1.145}$$

$$F_i^3 = \frac{r_{i+1}^{n+1} - r_{i+1}^n}{\Delta t} - (1-\beta)u_{i+1}^n - \beta u_{i+1}^{n+1} = F_i^3(u_{i+1}, r_{i+1}) = 0\,, \tag{1.146}$$

or globally,

$$F_i^j = F_i^j(\rho_{i+1/2}, r_i, r_{i+1}, u_{i+1}) = 0 \qquad (i = 2, N-1; j = 1, 3)\,. \tag{1.147}$$

1.6.5.3   Equations for the Outermost Shell

Finally, for the outermost shell ($i = N$), we have

$$S^1 = \frac{1}{(\rho_{N+1/2})^{n+1}} - \frac{4}{3}\pi\frac{(r_{N+1}^3)^{n+1} - (r_N^3)^{n+1}}{m_{N+1} - m_N} = S^1(r_{N+1}, r_N, \rho_{N+1/2}) = 0\,, \tag{1.148}$$

$$S^2 = \frac{u_{N+1}^{n+1} - u_{N+1}^n}{\Delta t} - (1-\beta)\left(\frac{-Gm_{N+1}}{r_{N+1}^2}\right)^n - \beta\left(\frac{-Gm_{N+1}}{r_{N+1}^2}\right)^{n+1} =$$

$$S^2(u_{N+1}, r_{N+1}) = 0\,, \tag{1.149}$$

$$S^3 = \frac{r_{N+1}^{n+1} - r_{N+1}^n}{\Delta t} - (1-\beta)u_{N+1}^n - \beta u_{N+1}^{n+1} = S^3(u_{N+1}, r_{N+1}) = 0\,, \tag{1.150}$$

which can be written as:

$$S^j = S^j(\rho_{N+1/2}, r_N, r_{N+1}, u_{N+1}) = 0 \qquad (j = 1, 3)\,. \tag{1.151}$$

1.6.6   Linearization

Let x^0 be a vector containing the *exact* values of the physical variables of the problem, r, ρ, and u, at $t^0 = 0$ (i.e., initial model), or, in general, at a given time, t^n. Let x^1 be the corresponding vector after one step, $t^1 = t^0 + \Delta t$. For small enough values of the timestep, Δt, all physical variables would have scarcely varied from their values at t^0 (i.e., $x^1 \sim x^0$). Therefore, let's consider, as a first approximate guess, that $x^1 \equiv x^0$. In general, such a choice would not yield the exact values of the variables at t^1, so that $C^j(x^1) \neq 0$, $F_i^j(x^1) \neq 0$, and $S^j(x^1) \neq 0$. Nevertheless, since $x^1 \sim x^0$, one can think of a set of corrections, δx, that added to the first guess values, $x^1 = x^0 + \delta x$, will actually satisfy $C^j(x^1) = 0$, $F_i^j(x^1) = 0$, and $S^j(x^1) = 0$. For small corrections, the whole set of structure equations can be written in the form

$$C^j(x^1) = C^j(x^0) + \delta C^j = 0$$

$$F_i^j(x^1) = F_i^j(x^0) + \delta F_i^j = 0$$

$$S^j(x^1) = S^j(x^0) + \delta S^j = 0\,, \tag{1.152}$$

which correspond to the following linearized system of equations,

$$C^j + \frac{\partial C^j}{\partial \rho_{3/2}}\delta\rho_{3/2} + \frac{\partial C^j}{\partial u_2}\delta u_2 + \frac{\partial C^j}{\partial r_2}\delta r_2 = 0 \tag{1.153}$$

$$F_i^j + \frac{\partial F_i^j}{\partial \rho_{i+1/2}}\delta\rho_{i+1/2} + \frac{\partial F_i^j}{\partial r_i}\delta r_i + \frac{\partial F_i^j}{\partial r_{i+1}}\delta r_{i+1} + \frac{\partial F_i^j}{\partial u_{i+1}}\delta u_{i+1} = 0$$

$$(j = 1, 3; i = 2, N - 1) \tag{1.154}$$

$$S^j + \frac{\partial S^j}{\partial \rho_{N+1/2}} \delta \rho_{N+1/2} + \frac{\partial S^j}{\partial r_N} \delta r_N + \frac{\partial S^j}{\partial r_{N+1}} \delta r_{N+1} + \frac{\partial S^j}{\partial u_{N+1}} \delta u_{N+1} = 0$$

$$(j = 1, 3), \tag{1.155}$$

where all partial derivatives can be determined analytically. For instance, for the innermost shell (see Section 1.6.5.1), we have:

$$\frac{\partial C^1}{\partial \rho_{3/2}} = -\frac{1}{\rho_{3/2}^2} \tag{1.156}$$

$$\frac{\partial C^1}{\partial u_2} = 0 \tag{1.157}$$

$$\frac{\partial C^1}{\partial r_2} = -4\pi \frac{r_2^2}{m_2} \tag{1.158}$$

$$\frac{\partial C^2}{\partial \rho_{3/2}} = 0 \tag{1.159}$$

$$\frac{\partial C^2}{\partial u_2} = \frac{1}{\Delta t} \tag{1.160}$$

$$\frac{\partial C^2}{\partial r_2} = -\beta \frac{Gm_2}{r_2^2} \tag{1.161}$$

$$\frac{\partial C^3}{\partial \rho_{3/2}} = 0 \tag{1.162}$$

$$\frac{\partial C^3}{\partial u_2} = -\beta \tag{1.163}$$

$$\frac{\partial C^3}{\partial r_2} = \frac{1}{\Delta t}. \tag{1.164}$$

Note that all variables evaluated at previous time (i.e., t^0, or in general, at an arbitrary time t^n) are fixed, therefore nonzero derivatives involve only variables evaluated at the current time, t^1 (t^{n+1}, in general).

This system of linearized equations can also be written in matrix form, $\mathbf{A} \cdot \mathbf{B} = \mathbf{C}$ (see Equation 1.165). In principle, one would be tempted to obtain the matrix containing the set of correction values δ from $\mathbf{B} = \mathbf{A}^{-1} \cdot \mathbf{C}$. Since typically 1D computational domains are discretized in $N \sim 100 - 1000$ shells, for the free-fall collapse problem with 3 unknowns per shell, matrix $\mathbf{A}$ has a characteristic size $3N \times 3N$, that is, contains a number of elements ranging from 300×300 to 3000×3000. Such large matrices are, in general, not easy to handle numerically. Note, however, that the linearized system of equations for the innermost shell corresponds to

$$C^j + \frac{\partial C^j}{\partial \rho_{3/2}} \delta \rho_{3/2} + \frac{\partial C^j}{\partial u_2} \delta u_2 + \frac{\partial C^j}{\partial r_2} \delta r_2 = 0, \tag{1.166}$$

$$
\begin{pmatrix}
-C^1 \\
-C^2 \\
-C^3 \\
-F_2^1 \\
-F_2^2 \\
-F_2^3 \\
-F_3^1 \\
-F_3^2 \\
-F_3^3 \\
\vdots \\
-S^1 \\
-S^2 \\
-S^3
\end{pmatrix}
=
\begin{pmatrix}
\dfrac{\partial C^1}{\partial \rho_{3/2}} & \dfrac{\partial C^1}{\partial r_2} & 0 & 0 & 0 & 0 & 0 & 0 & 0 & \cdots & 0 & 0 & 0 \\[2mm]
0 & \dfrac{\partial C^2}{\partial r_2} & \dfrac{\partial C^2}{\partial u_2} & 0 & 0 & 0 & 0 & 0 & 0 & \cdots & 0 & 0 & 0 \\[2mm]
0 & \dfrac{\partial C^3}{\partial r_2} & \dfrac{\partial C^3}{\partial u_2} & 0 & 0 & 0 & 0 & 0 & 0 & \cdots & 0 & 0 & 0 \\[2mm]
0 & \dfrac{\partial F_2^1}{\partial r_2} & 0 & \dfrac{\partial F_2^1}{\partial \rho_{5/2}} & \dfrac{\partial F_2^1}{\partial r_3} & 0 & 0 & 0 & 0 & \cdots & 0 & 0 & 0 \\[2mm]
0 & 0 & 0 & 0 & \dfrac{\partial F_2^2}{\partial r_3} & \dfrac{\partial F_2^2}{\partial u_3} & 0 & 0 & 0 & \cdots & 0 & 0 & 0 \\[2mm]
0 & 0 & 0 & 0 & \dfrac{\partial F_2^3}{\partial r_3} & \dfrac{\partial F_2^3}{\partial u_3} & 0 & 0 & 0 & \cdots & 0 & 0 & 0 \\[2mm]
0 & 0 & 0 & 0 & \dfrac{\partial F_3^1}{\partial r_3} & 0 & \dfrac{\partial F_3^1}{\partial \rho_{7/2}} & \dfrac{\partial F_3^1}{\partial r_4} & 0 & \cdots & 0 & 0 & 0 \\[2mm]
0 & 0 & 0 & 0 & 0 & 0 & 0 & \dfrac{\partial F_3^2}{\partial r_4} & \dfrac{\partial F_3^2}{\partial u_4} & \cdots & 0 & 0 & 0 \\[2mm]
0 & 0 & 0 & 0 & 0 & 0 & 0 & \dfrac{\partial F_3^3}{\partial r_4} & \dfrac{\partial F_3^3}{\partial u_4} & \cdots & 0 & 0 & 0 \\[2mm]
\vdots & \vdots & \vdots & \vdots & \vdots & \vdots & \vdots & \vdots & \vdots & \ddots & \vdots & \vdots & \vdots \\[2mm]
0 & 0 & 0 & 0 & 0 & 0 & 0 & 0 & 0 & \cdots & \dfrac{\partial S^1}{\partial \rho_{N+1/2}} & \dfrac{\partial S^1}{\partial r_{N+1}} & 0 \\[2mm]
0 & 0 & 0 & 0 & 0 & 0 & 0 & 0 & 0 & \cdots & 0 & \dfrac{\partial S^2}{\partial r_{N+1}} & \dfrac{\partial S^2}{\partial u_{N+1}} \\[2mm]
0 & 0 & 0 & 0 & 0 & 0 & 0 & 0 & 0 & \cdots & 0 & \dfrac{\partial S^3}{\partial r_{N+1}} & \dfrac{\partial S^3}{\partial u_{N+1}}
\end{pmatrix}
\begin{pmatrix}
\delta \rho_{3/2} \\
\delta r_2 \\
\delta u_2 \\
\delta \rho_{5/2} \\
\delta r_3 \\
\delta u_3 \\
\delta \rho_{7/2} \\
\delta r_4 \\
\delta u_4 \\
\vdots \\
\delta \rho_{N+1/2} \\
\delta r_{N+1} \\
\delta u_{N+1}
\end{pmatrix}
\tag{1.165}
$$

which, after moving the C^j term to the right-hand side of the equation, can be written in the following simple matrix form:

$$
\begin{pmatrix}
\dfrac{\partial C^1}{\partial \rho_{3/2}} & \dfrac{\partial C^1}{\partial u_2} & \dfrac{\partial C^1}{\partial r_2} \\[2mm]
\dfrac{\partial C^2}{\partial \rho_{3/2}} & \dfrac{\partial C^2}{\partial u_2} & \dfrac{\partial C^2}{\partial r_2} \\[2mm]
\dfrac{\partial C^3}{\partial \rho_{3/2}} & \dfrac{\partial C^3}{\partial u_2} & \dfrac{\partial C^3}{\partial r_2}
\end{pmatrix}
\begin{pmatrix}
\delta \rho_{3/2} \\[1mm] \delta u_2 \\[1mm] \delta r_2
\end{pmatrix}
=
\begin{pmatrix}
-C^1 \\ -C^2 \\ -C^3
\end{pmatrix},
\tag{1.167}
$$

Since the system contains 3 equations and just 3 unknowns (i.e., $\delta \rho_{3/2}$, δu_2, and δr_2), it can be solved in a straightforward way.

Now, let's scrutinize the linearized system of equations for the $N-2$ intermediate shells,

$$
F_i^j + \frac{\partial F_i^j}{\partial \rho_{i+1/2}} \delta \rho_{i+1/2} + \frac{\partial F_i^j}{\partial r_i} \delta r_i + \frac{\partial F_i^j}{\partial r_{i+1}} \delta r_{i+1} + \frac{\partial F_i^j}{\partial u_{i+1}} \delta u_{i+1} = 0
$$

$$
(j = 1, 3; i = 2, N-1),
\tag{1.168}
$$

which as well can be written in matrix form:

$$
\begin{pmatrix}
\dfrac{\partial F_i^1}{\partial \rho_{i+1/2}} & \dfrac{\partial F_i^1}{\partial u_{i+1}} & \dfrac{\partial F_i^1}{\partial r_{i+1}} \\[2mm]
\dfrac{\partial F_i^2}{\partial \rho_{i+1/2}} & \dfrac{\partial F_i^2}{\partial u_{i+1}} & \dfrac{\partial F_i^2}{\partial r_{i+1}} \\[2mm]
\dfrac{\partial F_i^3}{\partial \rho_{i+1/2}} & \dfrac{\partial F_i^3}{\partial u_{i+1}} & \dfrac{\partial F_i^3}{\partial r_{i+1}}
\end{pmatrix}
\begin{pmatrix}
\delta \rho_{i+1/2} \\[1mm] \delta u_{i+1} \\[1mm] \delta r_{i+1}
\end{pmatrix}
=
\begin{pmatrix}
-F_i^1 - \dfrac{\partial F_i^1}{\partial r_i} \delta r_i \\[2mm]
-F_i^2 - \dfrac{\partial F_i^2}{\partial r_i} \delta r_i \\[2mm]
-F_i^3 - \dfrac{\partial F_i^3}{\partial r_i} \delta r_i
\end{pmatrix}.
\tag{1.169}
$$

For $i = 2$, δr_2 is no longer an unknown, since such correction was already obtained in the analysis of the innermost shell. The same applies to δr_i for an arbitrary intermediate shell, i. For convenience, δr_i can be simply moved to the right-hand side of the equations. As for the innermost shell, the system contains 3 equations ($j = 1, 3$) and just 3 unknowns (i.e., $\delta \rho_{i+1/2}$, δu_{i+1}, and δr_{i+1}) and can again be easily solved. The procedure described above for the first intermediate shell can be extended to the additional intermediate shells, as well as to the surface layer. Therefore, the solution of the system that characterizes the free-fall collapse problem involves N matrices containing 3×3 elements, rather than dealing with a huge $3N \times 3N$ matrix[38].

The freshly determined δ corrections are then added to the first guess values x^o, such that $x^1 = x^o + \delta x$. The process is then iterated: Functions and derivatives are recalculated with the improved values x^1, and a new set of δ corrections is obtained. When small enough corrections result, for all variables at all shells, satisfying $\delta X / X < \epsilon$ (with ϵ being a pre-determined accuracy parameter), the iteration process is halted, and the system is advanced one step, $t^2 = t^1 + \Delta t$.

1.6.7 Theory vs. Simulation

The procedure described above (Sections 1.6.2–1.6.6) has been specifically designed for the simulation of the free-fall collapse of a homogeneous sphere. The problem has an analytical

---

[38]Unfortunately, the usual situation faced in computational hydrodynamics is not that simple, and one has to deal unavoidably with the decomposition of the overall matrix, for which special techniques have been developed (see Appendix A).

solution [366] that can be obtained inserting the definition of velocity in the momentum conservation equation (with $P = 0$, and no artificial viscosity), while assuming that the homogeneous sphere is initially at rest ($u = 0$ cm s^{-1}, at $t = t_o$). All in all, the corresponding solution can be written as:

$$\left(8\pi G \frac{\rho_o}{3}\right)^{1/2} (t - t_o) = \left(1 - \frac{R}{R_o}\right)^{1/2} \left(\frac{R}{R_o}\right)^{1/2} + \arcsin\left(1 - \frac{R}{R_o}\right)^{1/2}, \quad (1.170)$$

where ρ_o and R_o are the initial density and radius of the sphere, and G is the gravitational constant.

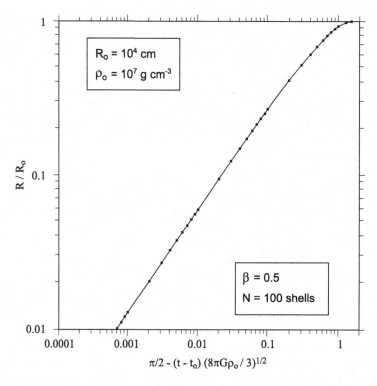

FIGURE 1.12
Simulation of the free-fall collapse of a homogeneous sphere. Note the agreement between the analytical solution (solid line) and the numerical results (points). Simulation performed with the code `freefall.f` (see Appendix B), with $\beta = 0.5$.

Results of a simulation for an initial density $\rho_o = 10^7$ g cm^{-3} and radius $R_o = 10^4$ cm, modeled as 100 equal mass, numerical shells, with the code `freefall.f` (see Appendix B) are shown in Figures 1.12 and 1.13. Note the excellent agreement between the numerical results and the analytical solution, even when the sphere has already shrunk in size by two orders of magnitude.

In about 0.6 s, the density of the sphere has already increased by six orders of magnitude (Figure 1.13) but, as theoretically predicted, homogeneity is maintained throughout the sphere (with a relative error $\leq 10^{-6}$%). This defines a homologous collapse, with a constant temperature and a uniform density across the sphere. Note also that since $\rho \propto R^{-3}$, a plot of the density versus radius has a constant slope in a log ρ–log R plane.

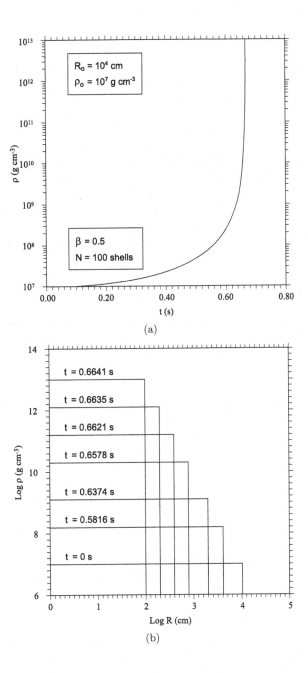

FIGURE 1.13
(a) Evolution of the central density as a function of time and (b) density profile versus radius, for the free-fall collapse problem.

1.7   Code Validation and Verification

Hydrocodes are computer-friendly models of nature, in which the standard suite of partial differential equations (often representing conservation laws) are commonly replaced by simpler, approximate expressions. As such, the way in which a hydrocode describes a physical phenomenon can only be regarded as approximate. The degree in which a numerical solution represents the true solution of a problem is a major concern for theorists and often implies substantial tests to validate and verify each code's capabilities and their performance.

Verification and validation are fundamental aspects in code development [996, 1338, 1505, 1790]. While a verification assessment is aimed at proving whether the computational implementation of the conceptual model is correct, validation basically determines whether a simulation agrees with nature [283]. As nicely summarized by P. Roache [1505], verification tests whether a code "solves the equations right", while validation checks whether a code is "solving the right equations". Remarkably, verification and validation are necessary but not sufficient strategies to check the performance of a code, and one should not forget that through such tests, one can never prove whether a code works correctly, only that a code fails or does not work properly.

Verification tests often rely on a suite of problems with analytical solution. The performance of the code is therefore checked by means of a thorough comparison between the analytical and the numerical solutions. Unfortunately, test cases with analytical solution are frequently so simple and rare that only a subset of code modules or subroutines are actually checked. Code-to-code comparison turns out to be an appealing alternative, although it is not always easy to overcome the natural reluctancy of competing teams in showing "the kitchen", so to speak[39].

On the other hand, validation involves confirmation of the numerical results through laboratory experiments. Often, validation problems are much more complex than verification tests, and therefore are frequently devoid of analytical solution. Validation tests are used to assess resolution issues and the level of refinement required to reproduce the relevant physics of a problem. Such tests may be hampered by a number of experimental issues, such as limited experimental resolution, or not well-known material properties. On the computational side, main difficulties arise from the impossibility to resolve all the relevant length- and timescales of the problem, which make a direct comparison with experiments sometimes hard to interpret. And frequently, stellar codes lack fundamental physics inputs to reproduce a real experiment in terrestrial conditions (e.g., thermal diffusion, appropriate equation of state, viscosity).

All in all, verification and validation can be regarded as important but challenging stages in the art of scientific computing and, in particular, in the modeling of stellar explosions.

1.7.1   Verification Tests

The free-fall collapse problem of a homogeneous sphere, discussed in Section 1.6.7, is just one of myriad classical tests frequently used in the verification of hydrodynamic codes. Other examples, computed with the multidimensional code FLASH, are described in the following subsections.

---

[39]While a common practice in multidimensional hydrodynamics, comparison between stellar evolution codes is much more scarce. An initiative in this regard was pushed by the AGB modeling community back in the 1990s, in the framework of the early Torino workshops, in an effort to share and compare the performance of the different codes used in their simulations.

1.7.1.1 Sod's Shock Tube Test

Sod's shock tube problem, a standard Riemann problem in Newtonian fluid dynamics, has become the quintessential benchmark test for hydro codes. Named after Gary Sod [1675], it tests the ability of a numerical algorithm to handle discontinuities and shocks (in particular, whether the Rankine–Hugoniot conditions are satisfied; see Section 1.3.3). An asset of this test relies on the possibility of comparing results with an analytical solution when a simple equation of state[40] is implemented.

The test consists of two fluids, initially at rest, with different densities and pressures, separated by a planar interface. When the interface is removed ($t = 0$), a characteristic fluid pattern that can be experimentally validated emerges [1724]. Five distinct regions are identified (Figure 1.14): the undisturbed left fluid (region I), the expanding left fluid (II), the decompressed left fluid (III), the compressing right fluid (IV), and the undisturbed right fluid (V).

Even though the analytical solution of Sod's shock tube problem is rather complex, a number of relevant features can be easily addressed[41]. The flow in regions III and IV is characterized by a constant density. Such regions are actually separated by a *contact discontinuity*, that is, a dividing line that separates two fluids with different entropy but identical pressure and velocity (see Figures 1.14 and 1.15). Therefore, $u_{III} = u_{IV}$ and $P_{III} = P_{IV}$. The contact discontinuity propagates with the fluid, at $u_{cont} = u_{III}$, and hence, its location is determined by $x_{cont} = u_{cont} t + x_0$, with x_0 being the initial position of the planar interface. Region II, the only nonsteady region in the solution, is instead characterized by an x-dependent density profile, imposed by the propagation of an expansion wave that moves to the left (i.e., a *rarefaction fan*; see Figure 1.14).

Regions IV and V are in turn separated by a shock wave that propagates to the right. Since region V is unperturbed, $u_V = 0$. The speed at which the shock wave propagates can be directly evaluated from mass conservation,

$$u_{shock} = u_{IV} \frac{\rho_{IV}}{\rho_{IV} - \rho_V}, \tag{1.171}$$

while the location of the shock is simply given by $x_{shock} = u_{shock} t + x_0$.

Following Bodenheimer et al. [209], the velocity in region IV is given by

$$u_{IV} = (P_{IV} - P_V)\sqrt{\frac{1 - \omega^2}{\rho_V(P_{IV} + \omega^2 P_V)}}, \tag{1.172}$$

while the relation between u and P in region II can be written as

$$u_{II} = (P_I^\psi - P_{II}^\psi)\sqrt{\frac{P_I^{1/\gamma}(1 - \omega^4)}{\rho_I \omega^4}}, \tag{1.173}$$

where $\psi = (\gamma - 1)/2\gamma$ (see reference [741] for a full derivation of Equations 1.172 and 1.173).

On a $P - u$ plane, Equations 1.172 and 1.173 represent the loci of all the possible states of the fluid in regions II to IV. Both curves intersect at a point, where the portion of the fluid expanded in regions II and III smoothly matches the shocked flow in region IV. Therefore,

---

[40] In this section, a polytropic equation of state, $P = K\rho^\gamma = K\rho^{(n+1)/n}$, is adopted, with K the polytropic constant, and n the polytropic index.

[41] See references [741, 1149] for a detailed derivation. A Fortran code aimed at testing Sod's shock tube problem, developed by Bruce Fryxell and Frank Timmes, is available at http://cococubed.asu.edu/code_pages/exact_riemann.shtml.

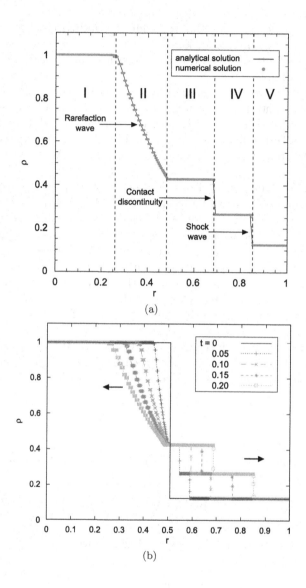

FIGURE 1.14

(a) Density profile at $t = 0.2$, in Sod's shock tube problem, as computed with the FLASH code. The plot displays the five distinct regions predicted theoretically, as well as the formation of a rarefaction wave, a shock wave, and a contact discontinuity (see text for details). (b) Density profiles at times $t = 0$, 0.05, 0.1, 0.15, and 0.2 units, in Sod's shock tube problem. Color plots and a movie portraying the time-evolution of the density in the shock tube test are available at http://fisica.upc.edu/ca/users/jjose/CRC-Downloads.

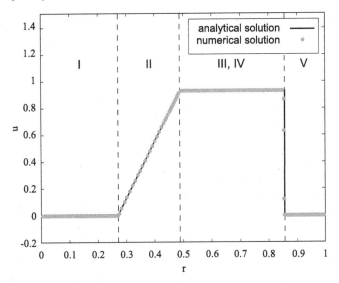

FIGURE 1.15
Comparison between the numerical velocity profile at t = 0.2 units and the analytical solution in Sod's shock tube problem. A color version of this plot is available at http://fisica.upc.edu/ca/users/jjose/CRC-Downloads.

since $u_{III} = u_{IV}$ and $P_{III} = P_{IV}$, P_{III} can be determined numerically by imposing $u_{II} = u_{IV}$:

$$(P_{III} - P_V)\sqrt{\frac{1 - \omega^2}{\rho_V(P_{III} + \omega^2 P_V)}} = (P_I^\psi - P_{III}^\psi)\sqrt{\frac{P_I^{1/\gamma}(1 - \omega^4)}{\rho_I \omega^4}}. \tag{1.174}$$

Once P_{III} (P_{IV}) is found, u_{III} (u_{IV}) can be determined from Equation 1.172.

The density and pressure ratios across the shock (i.e., at the boundary between regions IV and V) are given by the Rankine–Hugoniot conditions:

$$\frac{\rho_{IV}}{\rho_V} = \frac{P_{IV} + \omega^2 P_V}{P_V + \omega^2 P_{IV}}, \tag{1.175}$$

with $\omega^2 = (\gamma - 1)/(\gamma + 1)$. This yields the value of ρ_{IV}.

Regions I, II, and III have not been perturbed by the shock wave, and therefore are characterized by the same equation of state. Accordingly,

$$\rho_{III} = \rho_I \left(\frac{P_{III}}{P_I}\right)^{1/\gamma}, \tag{1.176}$$

which yields ρ_{III}.

Finally, the physical conditions in the x-dependent region II are given by (see reference [741]):

$$u_{II}(x, t) = (1 - \omega^2)\left(\frac{x - x_o}{t} + c_{s,I}\right) \tag{1.177}$$

and

$$\rho_{II}(x, t) = \left(\frac{\rho_I^\gamma}{\gamma P_I}\left[u_{II}(x, t) - \frac{x - x_o}{t}\right]^2\right)^{1/\gamma - 1}, \tag{1.178}$$

where $c_{s,I} = \sqrt{\gamma P_I / \rho_I}$ is the speed of sound in region I. Note that the location of the head and tail of the rarefaction wave as a function of time are simply given by:

$$x_{\text{head}}(t) = x_{\text{o}} - c_{s,I}\, t \tag{1.179}$$

and

$$x_{\text{tail}}(t) = x_{\text{o}} + (u_{III} - c_{s,III})\, t, \tag{1.180}$$

where $c_{s,III}$ is the speed of sound in region III.

The time evolution of the system, solved numerically, is depicted in Figures 1.14 and 1.15. Calculations have been performed with the multidimensional code FLASH. To this end, two ideal fluids, characterized by a polytropic equation of state, $P = K\rho^{\gamma}$, with an adiabatic index corresponding to a diatomic molecular gas, $\gamma = 7/5$, are placed inside of a 1 unit-long tube, and separated by a planar interface located at $x_{\text{o}} = 0.5$ units. The initial conditions of the fluids are $\rho_I = 1$ and $P_I = 1$ units at the left-hand side of the interface, and $\rho_V = 0.125$ and $P_V = 0.1$ at the right-hand side. Once the interface is removed, the presence of a sharp discontinuity induces a shock that propagates to the right, into the undisturbed, low-density and low-pressure fluid. Simultaneously, a contact discontinuity drifts to the right, while a rarefaction wave propagates to the left.

The simulation, performed with a minimum resolution of 0.00056 units, shows excellent agreement with the analytical solution. In particular, it keeps the fluid discontinuities sharp, particularly when special techniques (e.g., adaptive mesh refinement[42]) are implemented to invest more resolution in the regions of the computational grid, where physical variables exhibit the largest variations.

It is worth noting that Sod's shock tube test[43] (and any test, in general) only checks a few specific features of a code (e.g., its ability to handle shocks), and, therefore, only a handful of subroutines are effectively tested and verified.

1.7.1.2   Emery's Wind Tunnel Test

Another problem aimed at testing the performance of a hydrodynamic code is the *wind tunnel test with a step*, first described by Ashley Emery [495]. Specifically, the test checks the code's ability to handle multiple shock interactions in a multidimensional framework, particularly with irregular computational boundaries. The problem, however, has no analytical solution, and the performance of a specific code has to be tested by comparison with previously reported simulations of the same problem (e.g., references [567, 1958]).

Figure 1.16 displays the response of the FLASH code to Emery's wind tunnel test. The simulation relies on a 2-dimensional, rectangular domain, with a dimensionless area of 3 × 1, containing a 0.2 units high and 2.4 units wide step. The resolution adopted is 0.0016. A horizontal wind, with a dimensionless velocity of 3, is imposed on the left-hand wall. The fluid embedded in the domain is set to a pressure $P_{\text{o}} = 1$ and a density $\rho_{\text{o}} = 1.4$ ($\gamma = 1.4$), such that the local speed of sound yields a value of

$$c_s = \sqrt{\frac{\gamma P_{\text{o}}}{\rho_{\text{o}}}} = 1, \tag{1.181}$$

and hence, the wind corresponds to a supersonic, Mach 3 flow.

---

[42]See references [156, 157, 282, 1141, 1174] for an introduction to adaptive mesh refinement techniques in computational hydrodynamics.

[43]A variety of Sod's test, the *strong shock tube problem*, relies on larger pressure and density contrasts between both fluids (e.g, $P_I/P_V = 100$ and $\rho_I/\rho_V = 10$, a factor of 10 larger than in the standard Sod's test) [567, 741, 742]. This test is used to check the performance of a code in handling much stronger discontinuities and narrower density peaks, and hence becomes a much more demanding test than Sod's.

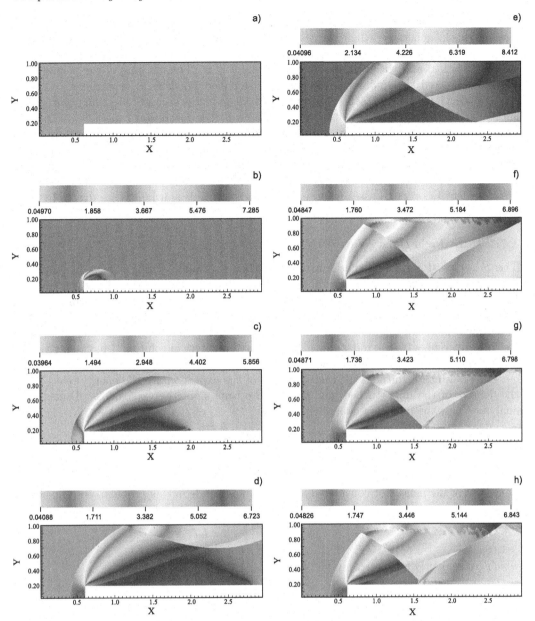

FIGURE 1.16
Evolution of the density field in Emery's wind tunnel test, at times 0 (panel a), 0.1 (b), 0.5 (c), 0.8 (d), 1.3 (e), 2.6 (f), 3.3 (g), and 4 (h). See text for details. Color plots and a movie portraying the time-evolution of the density in Emery's wind tunnel test are available at http://fisica.upc.edu/ca/users/jjose/CRC-Downloads.

A shock wave develops in front of the step ($t \sim 0.1$), curving progressively around the step's corner and growing in size until it eventually hits the upper boundary of the computational domain ($t \sim 0.7$), where reflecting boundary conditions have been imposed. Indeed, the step's corner acts as a "rarefaction fan", connecting the unperturbed gas located above the step and the shocked gas in front of it [567]. A numerical boundary layer is formed all along the surface of the step, which translates into a small overexpansion of the gas and a weak shock, generated when the gas falls back onto the step. The first reflected wave experiences additional reflections when colliding, back and forth, with the surface of the step (at $t = 1.2$) and with the upper boundary ($t = 3.3$). The merging of the primary and the first reflected shock waves gives rise to a so-called *Mach stem* at about $t = 1.8$. At $t = 3$, a second, spurious Mach stem seems to develop at the second reflection point, on the surface of the step, but this feature dissapears in higher resolution runs [567]. The shear zone behind the upper Mach stem gives rise to Kelvin–Helmholtz instabilities, clearly visible in the fluid at $t > 2$, likely originated by the amplification of numerical errors produced at the shock intersection. This stresses the need of an in-depth analysis when interpreting numerical results [567]. Indeed, simulations should be ideally repeated at different levels of refinement in order to asses whether a specific feature is spurious (e.g., driven by noise amplification) or real [1958].

1.7.1.3   Sedov's Blast Wave Problem

The *Sedov blast wave test* checks a code's ability to handle strong shocks in nonplanar symmetry. The test follows the evolution of a cylindrical or spherical blast wave driven by the deposition of an instantaneous amount of energy in a small area of an otherwise homogeneous medium. The problem attracted the interest of several physicists during World War II, when nuclear weapons became feasible. L. I. Sedov and J. von Neumann [1602,1869] (see also reference [1048]), independently derived an analytical solution[44] to the problem of a blast wave propagating into a cold gas.

The relevant magnitudes for the dynamics of the explosion are the energy released, E; the density of the unperturbed gas, ρ_o; and the time since the explosion, t. Dimensional analysis yields the following dependences:

$$[E] = ML^2T^{-2} \tag{1.182}$$

$$[\rho] = ML^{-3}. \tag{1.183}$$

To get rid of the mass, not considered a relevant magnitude in this problem, one can simply write

$$\left[\frac{E}{\rho}\right] = L^5T^{-2} \rightarrow \left[\frac{E}{\rho}\right]^{1/5} = LT^{-2/5}, \tag{1.184}$$

which conveniently multiplied by $t^{2/5}$ yields

$$R(t) = k\left(\frac{E}{\rho}\right)^{1/5} t^{2/5}. \tag{1.185}$$

Equation 1.185 defines the characteristic length-scale of the expanding blast wave or *fireball*[45], with k a normalization constant of the order of unity.

The three partial differential equations governing the evolution of the system (i.e., mass, momentum, and energy conservation) can be transformed into a set of ordinary differential

---

[44]The backbone of the Sedov blast wave theory, the similarity relations, was also derived by G. I. Taylor in the 1940s [1783, 1784]. Using a series of snapshots of the Trinity explosion, Taylor tested his scaling hypothesis and calculated the energy of the blast (see Problem P8 at the end of this chapter).

[45]Arguments based on the energetics of the explosion yield the same scaling law: The mass of swept-up gas after a time t, when the fireball has expanded to a size R, is given by $m(t) \sim \rho_o R^3$. The postshocked material velocity, immediately behind the shock, is of the order of the speed of the shock front, $u \sim R/t$. Therefore, the kinetic energy of the swept-up gas can be estimated as $\sim \rho_o R^3 u^2 \sim \rho_o R^5/t^2$. A similar expression can be obtained for the internal energy of the postshocked fluid. All in all, a scaling law $R = k(E/\rho_o)^{1/5}t^{2/5}$ can be inferred.

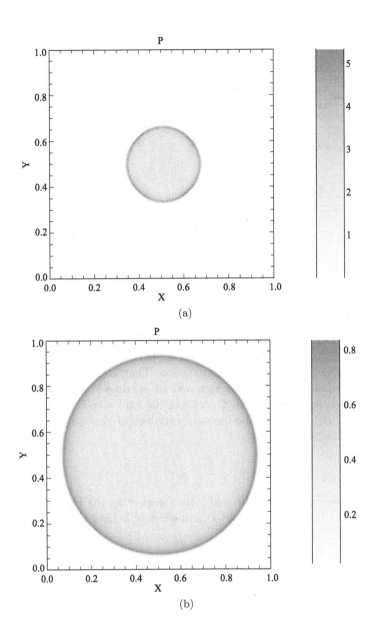

FIGURE 1.17

Snapshots of the pressure at (a) t = 0.03 and (b) 0.21 units, in Sedov's blast wave problem, showing that the spherical symmetry of the blast wave is preserved as the detonation sweeps the computational domain. Color plots and a movie portraying the time-evolution of the pressure in Sedov's blast wave test are available at http://fisica.upc.edu/ca/users/jjose/CRC-Downloads.

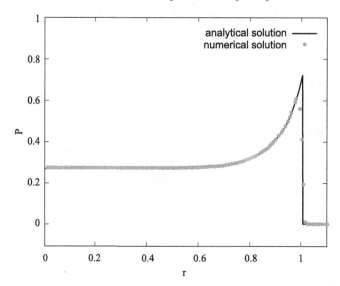

FIGURE 1.18
Comparison between the analytical pressure and the numerical value at t $=$ 0.27
units, in Sedov's blast wave problem. A color version of this plot is available at
`http://fisica.upc.edu/ca/users/jjose/CRC-Downloads`.

equations with analytical solution by considering that the time-evolution of the sphere is
independent of the details of the initial expansion. That is, the shape of the radial profiles
of pressure, density, and velocity are assumed to be time-independent (i.e., a similarity
solution). Once the characteristic length-scale of the explosion is identified with the radius
of the blast wave, $R(t)$, the thermodynamic variables of the system can be expressed as
functions of the dimensionless variable

$$\xi = \frac{r}{R(t)}. \tag{1.186}$$

Assuming strong shock jump conditions[46], thus neglecting the preshock pressure, one can
get the postshocked values of u, ρ, and P immediately behind the shock (i.e., at $\xi = 1$):

$$u = \frac{2}{\gamma + 1} u_{\text{sw}} \tag{1.187}$$

$$\rho = \rho_{\text{o}} \frac{\gamma + 1}{\gamma - 1} \tag{1.188}$$

$$P = \rho_{\text{o}} u_{\text{sw}}^2 \frac{2}{\gamma + 1}, \tag{1.189}$$

where ρ_{o} is the immediate preshock density and $u_{\text{sw}} \equiv dR(t)/dt \sim (2/5)(E/\rho_{\text{o}})^{1/5} t^{-3/5}$ is
the speed of the shock wave.

    The Sedov's blast wave problem has also been tested with the FLASH code, assuming
an ideal-gas fluid (with $\gamma = 1.4$) at rest in a square, two-dimensional computational domain
of 1×1. The resolution adopted was 0.0039 in both coordinates. The initial dimensionless
density was set to 1, while the adopted dimensionless pressure was 10^{-5}. A dimensionless

---

[46]This assumption limits the applicability of the formulae derived up to the time when the shock moves
at $\sim$ Mach 1, and therefore, the hypothesis of a strong shock no longer holds.

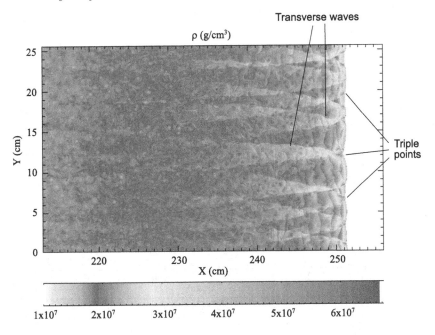

FIGURE 1.19
Snapshot of the density field at t = 185 ns, in the cellular problem. Incident shock waves, triple points, and transverse waves can be identified, with a postshock structure extending about 20 cm behind the front. A color plot and a movie portraying the time-evolution of the density in the cellular problem are available at http://fisica.upc.edu/ca/users/jjose/CRC-Downloads.

energy $E = 0.85$ was deposited in a small region of radius $\delta r \sim 0.014$, centered in the domain. Following Fryxell and collaborators [567], the pressure inside this tiny volume is given by

$$\delta P = \frac{3(\gamma - 1)\epsilon}{(\nu + 1)\pi \delta r^{\nu}}, \tag{1.190}$$

with $\nu = 3$ for a spherical geometry. Snapshots of the evolution of the system, at $t = 0.03$ and $t = 0.21$ units, are shown in Figure 1.17.

The simulation is stopped at $t = 0.21$, when the explosion hits the boundaries of the computational domain. A comparison between the numerical results and the analytical solution (Figure 1.18) reveals that the simulation qualitatively reproduces the analytical solution, but the shock discontinuity is not well resolved. Indeed, the pressure peak, for instance, is reduced by 13% with respect to the analytical solution. The same pattern applies to the density peak, which exhibits a reduction by 20%. The velocity peak, instead, is just 4% below the analytical value. Better agreement between the numerical results and the analytical solution can be obtained with higher resolution runs.

1.7.1.4 Cellular Test

Another test, specifically suited for stellar explosion codes—type Ia supernovae, in particular—is the *cellular problem*. The test explores the performance of a code in handling detonations under extreme physical conditions [216, 601, 1809], allowing the analysis of detonation structures in a multidimensional framework, which lie beyond current detonation theory [1940].

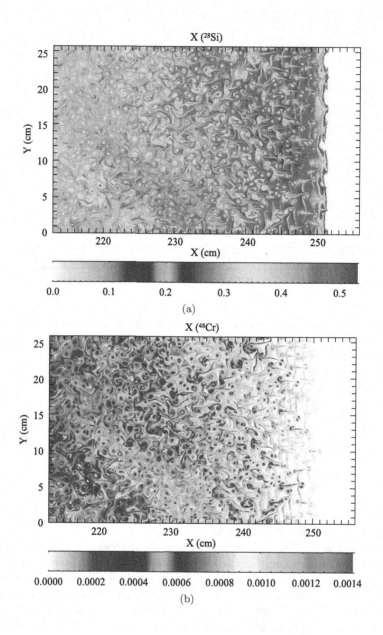

FIGURE 1.20

Distribution of ^{28}Si and ^{48}Cr in mass fractions at t = 185 ns in the cellular problem, showing the onset of Kelvin–Helmholtz instabilities. Note that while ^{28}Si increases from left to right, the abundance of ^{48}Cr decreases from left to right. Color plots and movies portraying the time-evolution of the abundances of ^{28}Si and ^{48}Cr in the cellular problem are available at http://fisica.upc.edu/ca/users/jjose/CRC-Downloads.

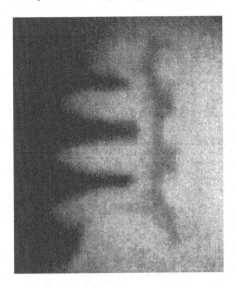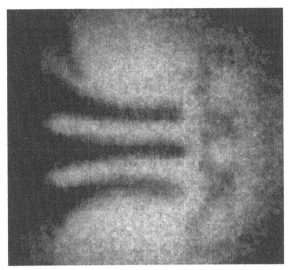

FIGURE 1.21
X-ray radiographs taken at 39.9 ns (left panel) and 66.0 ns (right panel) in a laser-driven shock experiment. The shock propagates through layers of decreasing density, made of Cu, polyimide plastic (with an embedded strip of a brominated hydrocarbon), and a carbonized resorcinol formaldehyde (CRF) foam, driving the development and growth of Richtmyer–Meshkov instabilities (see text for details). Figure from A. Calder, B. Fryxell, T. Plewa, and collaborators [282], reproduced with permission.

The test consists of a pure ^{12}C plasma, at $T_o = 2 \times 10^8$ K and $\rho_o = 10^7$ g cm^{-3}, initially at rest. The fluid is encapsulated in a tiny two-dimensional, computational domain of 256×25.6 cm^2. The evolution of the system has been followed with the FLASH code, as in previous tests, with a resolution of 0.025 cm.

The region $x < 25.6$ cm is perturbed by artificially increasing the initial temperature to 4.4×10^9 K and the density to 4.2×10^7 g cm^{-3}. The velocity of this postshocked material is fixed at 2.9×10^8 cm s^{-1}. As a result, a detonation moving forward to the right is born. When t = 45 ns, density perturbations show up, transforming the initial, planar detonation into a complex cellular structure, with incident shocks moving forward, triple points[47], and transverse shock waves (generated at the triple points) moving backward. The interaction between backward and forward shock waves yields, in turn, the formation of new triple points (see Figure 1.19).

The rich nucleosynthesis accompanying this complex detonation structure is characterized by a fast ^{12}C consumption, accompanied by the synthesis of a number of intermediate-mass species. Figure 1.20 depicts the distribution of two species, ^{28}Si and ^{48}Cr, at $t = 185$ ns, when Kelvin–Helmholtz hydrodynamic instabilities are clearly appreciated. All in all, the test has proved the FLASH code performance in handling complex hydrodynamical structures, such as Kelvin–Helmholtz instabilities, under challenging physical conditions that require extremely small timesteps (i.e., $\Delta t \sim 9 \times 10^{-12}$ s).

---

[47]Triple points appear in steady and unsteady flows when shocks interfere in the presence of a wall. At a triple point three shocks intersect: the incident shock, a reflected shock, and a Mach stem [761].

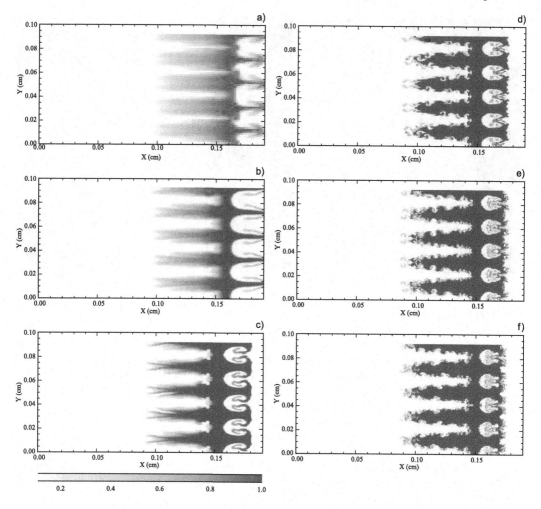

FIGURE 1.22
Plots based on laser-driven shock simulations depicting the onset of Richtmyer–Meshkov instabilities in a CH abundance map at 66.0 ns. Spatial resolution increases from panels a to f. Figure from A. Calder, B. Fryxell, T. Plewa, and collaborators [282], reproduced with permission.

1.7.2   Validation Tests

One of the active research areas in the emerging field of laboratory astrophysics is the analysis of astrophysical hydrodynamic problems through laboratory experiments. Astrophysical objects are obviously characterized by much larger scales than laboratory systems. Therefore, any laboratory experiment aimed at addressing a specific astrophysical phenomenon must be able to be extrapolated with the aid of a good scaling model. In this regard, numerical simulations seem better suited than experiments to tackle astrophysical plasmas, but they face the impossibility to reach the extremely small scales required by microphysics.

A number of experiments have been conducted in the lab to address specific phenomena produced in astrophysical plasmas during supernova explosions (see references [466,1486] for reviews). Among others: unmagnetized expansion experiments with the Helios laser system at Los Alamos National Laboratory, aimed at characterizing the dynamics of supernova remnants [220,1773]; blast-wave deceleration experiments at the KI-1 facility at Novosibirsk (Russia), aimed at simulating the deceleration of the blast wave in supernova remnants [53, 93,1352]; blast-wave instability experiments conducted at the Naval Research Laboratory, designed to study the formation of instabilities driven by reverse shocks in young supernova remnants [690]; contact-surface hydrodynamic experiments aimed at addressing the local, hydrodynamic evolution of supernova remnants at the Nova laser facility of the Lawrence Livermore National Laboratory [467,468]; fuel burning experiments aimed at understanding the physics of deflagration-to-detonation transitions in type Ia supernovae (see Chapter 5) [981,982,1626]; and interface instability experiments with the Nova laser, aimed at testing the role played by several hydrodynamic instabilities during core-collapse supernovae (see Chapter 7) [942,943,1483–1485].

Hereafter, we will briefly focus on some experiments conducted with high-energy-density lasers, aimed at testing the performance of the FLASH code, as an example of the difficulties faced in validation assessments. Figure 1.21 shows two X-ray radiographs taken at 39.9 ns and 66.0 ns in an experiment with laser-driven shocks conducted at the Omega laser facility (Laboratory for Laser Energetics, University of Rochester, New York). Shock waves propagate through a multilayered target of decreasing density, made of Cu, polyimide plastic (with an embedded strip of a brominated hydrocarbon), and a carbonized resorcinol formaldehyde (CRF) foam, generating Rayleigh–Taylor and Richtmyer–Meshkov instabilities (see Section 1.3.2) [283,447,753,944]. The long, dark fingers of the images are actually spikes of expanding Cu, while the horizontal band to the right of the spikes corresponds to the brominated plastic showing how the instabilities grow at the plastic-foam interface. The validation experiment revealed an observed pattern qualitatively in agreement with the results of the numerical simulations performed with the FLASH code and reproduced in Figure 1.22. Unfortunately, a thorough comparison between simulation and experiment is frequently hampered by the experimental diagnostic resolution [1917]. Imaging of the evolution of the driven instabilities through face-on or side-on X-ray radiography turns out to be challenging. Moreover, while the experiment confirmed that the code under scrutiny can actually reproduce the main observational properties of the flow and the resolvable morphology, higher-resolution runs revealed a degradation in the agreement between simulations and experiments. Indeed, the increase of resolution translates into the development of small-scale structures that affect the long-term evolution of larger structures, such as the spikes. This likely reveals the lack of key physics modules that, while irrelevant in the astrophysical scenarios of interest, turn out to be essential for a perfect match with terrestrial experiments. This stresses the need (as well as the challenge) to find out the appropriate experiments that can succesfully lead to the validation of a numerical code under the same physical conditions in which it will be used.

Box I. The Computational Hydrodynamics "Hall of Fame"

A selection of facts on numerical modeling and hydrodynamic codes

1. Stellar codes rely on a set of conservation equations (mass, momentum, and energy), energy sources, and energy transport mechanisms (radiation, conduction, and convection).

2. The most extensively used form of discretization in stellar evolution codes of the original set of coupled, partial differential equations into a system of algebraic equations that could be solved numerically are finite differences. Other methods, such as finite elements or finite volumes, are frequently used in other areas of computational fluid dynamics.

3. Different numerical tools have been used to date in the modeling of stellar explosions and their associated nucleosynthesis, ranging from one-zone to multidimensional models. State-of-the-art nucleosynthesis in novae and X-ray bursts relies mostly on 1D, spherically symmetric, hydrodynamic models, while nucleoynthesis in (types Ia and II) supernovae and stellar mergers requires a postprocessing approach based on temperature and density versus time profiles.

4. Traditional Lagrangian methods, with a computational grid attached to the fluid, constitute the simplest choice for building a hydrodynamic code, in the context of nonturbulent fluids. Eulerian methods, with a grid fixed in space, provide a good alternative to Lagrangian methods in the framework of turbulent fluids, because of the lack of grid distortion. While the vast majority of 1D hydro codes are Lagrangian, grid-based multidimensional codes mostly rely on an Eulerian formulation.

5. Gridless or meshfree methods, such as smoothed-particle hydrodynamics, provide a suitable alternative to avoid mesh tangling and distortion, frequently encountered in traditional grid-based methods in the presence of nonlinear behavior, discontinuities, or singularities.

6. The dawn of supercomputing has provided theoreticians with the appropriate arena in which truly multidimensional processes can be tested. The last decades had witnessed a number of multidimensional models of stellar explosions, with increasing levels of resolution.

7. The degree in which a numerical solution represents the true solution of a problem requires efforts aimed at verifying and validating each code's capabilities. A suite of classical tests (e.g., Sod's shock tube, Sedov's blast wave) is frequently used in the verification stage of hydrodynamic codes. Laboratory experiments (e.g., laser-driven shocks) offer an appealing but challenging path for validation.

Box II. Mysteries, Unsolved Problems, and Challenges

- 1D, stellar evolution codes face severe limitations in the treatment of convection, a key mechanism for energy transport in stars. Convection is a truly multidimensional process that is, however, implemented through phenomenological approaches—frequently, the mixing-length theory—due to inherent uncertainties and complexity. New prescriptions based on multidimensional models are needed for further implementation in 1D codes, particularly in the framework of stellar explosions, such as novae or X-ray bursts.

- Some stellar evolution codes are still constructed, for simplicity, as nonrotating and nonmagnetic. In contrast, stars obviously rotate and a number of relevant magnetic effects frequently accompany stellar plasmas. Efforts to include rotation and magnetic fields in current stellar hydro codes are highly advisable.

- A number of implicit and semiimplicit schemes have been proposed to tackle the stiff differential equations that govern the time evolution of the abundances in a stellar plasma. While in the past most hydro codes relied on Wagoner's two-step linearization technique, other schemes like Bader–Deuflhard's or Gear's backward differentiation proved far more accurate and fast. Since nuclear reactions often constitute the time-limiting factor, at least in most 1D codes, it would be advisable to rely on faster reaction network integration schemes.

- Most multidimensional codes implemented to date are explicit, with a time-step limited by the Courant–Friedrichs–Levy condition. Fully implicit multidimensional codes are needed to address full cycles of stellar explosions, from the long, quiet accretion stages—unaffordable with explicit schemes—to the truly dynamic stages of the event.

- ALE codes are still not widely used in astrophysical hydrodynamics. They reduce the shortcomings of purely Lagrangian or Eulerian codes by combining the best features of both formulations. Multidimensional ALE codes are particularly well suited to simultaneously address mixing and convection (which result in undesired mesh tangling and distortion in Lagrangian schemes), as well as the dynamic explosion and ejection stages (which cannot be easily followed with an Eulerian scheme).

- Nucleosynthesis accompanying stellar explosions mostly relies on 1D hydrodynamic models or on postprocessing techniques. To date, multidimensional models frequently implement reduced nuclear reaction networks, containing only a handful of isotopes to approximately account for the energetics of the explosion.

- Special efforts must be devoted to develop and implement new ways to tackle the vast range of relevant length scales that characterize most stellar explosions, which are currently impossible to resolve. In this regard, the advances made in type Ia supernova modeling (e.g., level-set techniques [1354, 1518, 1525]) should also be applied to models of classical novae and X-ray bursts.

1.8   Exercises

P1. A strong shock wave, propagating at a supersonic speed, separates two regions of a fluid. Prove that a fluid characterized by a polytropic equation of state, $P = K\rho^\gamma$, satisfies the following relations:

$$\frac{P_2}{P_1} = \frac{2\gamma Ma_1^2 - (\gamma - 1)}{\gamma + 1}$$

$$\frac{\rho_2}{\rho_1} = \frac{u_2}{u_1} = \frac{\gamma + 1}{(\gamma - 1) + 2/Ma_1^2},$$

where subscript 1 refers to the unperturbed fluid, subscript 2 to the postshocked fluid, and $Ma_1 = u_1/c_{s,1}$ is the Mach number in the unperturbed fluid.

Show also that in the limit of very strong shocks, $Ma_1 \gg 1$:

$$\frac{P_2}{P_1} = \frac{2\gamma Ma_1^2}{\gamma + 1}$$

$$\frac{\rho_2}{\rho_1} = \frac{\gamma + 1}{\gamma - 1}.$$

P2. Derive the analytical solution of the free-fall collapse problem (i.e., Equation 1.170).

P3. Use program `freefall.f` (Appendix B) and compute the free-fall collapse of a homogeneous sphere, for the same initial conditions described in Section 1.6.7 but different choices of β (i.e., 0, 0.25, 0.5, 0.75, 1). Compare the results obtained with the corresponding analytical solution.

P4. A variable y(t) varies according to the equation

$$\frac{dy}{dt} = Cy, \tag{1.191}$$

where C is a constant.

a) Solve the differential equation exactly. Determine y for $t \to \infty$ and $C < 0$.

b) Solve the differential equation numerically, using Euler's method (the simplest numerical scheme, corresponding to a first-order Taylor expansion):

$$y(t + \Delta t) = y(t) + \frac{dy}{dt}\Delta t. \tag{1.192}$$

c) Compare qualitatively the exact and the numerical solution for $C > 0$, $-1 < \Delta t\, C < 0$, and $-2 < \Delta t\, C < -1$.

d) Solve the differential equation using a second-order Runge–Kutta scheme:

$$y(t + \Delta t) = y(t) + \Delta t\, F\left[t + \frac{\Delta t}{2}, y(t) + \frac{\Delta t}{2}F(t, y(t))\right], \tag{1.193}$$

with $F \equiv dy/dt$.

P5. A simple method to assess the stability of a numerical scheme (that is, a way to determine the robustness of a scheme when small perturbations are introduced in an equation) was derived by John von Neumann at Los Alamos National Lab around 1944 (see reference [153]). Let's consider the (hyperbolic) partial differential equation:

$$\frac{\partial u}{\partial t} = -v\frac{\partial u}{\partial x} \tag{1.194}$$

whose solution has the form $u(x,t) = f(x-vt)$. A possible numerical scheme to approximate the equation (assuming Δx = constant and Δt = constant) is given by

$$\frac{u_j^{n+1} - u_j^n}{\Delta t} = -v\frac{u_{j+1}^n - u_{j-1}^n}{2\Delta x}, \tag{1.195}$$

where superscripts n and $n+1$ refer to times t^n and $t^{n+1} = t^n + \Delta t$, and subscripts $j-1$, j, and $j+1$ correspond to grid points $x_{j-1} = x_j - \Delta x$, x_j, and $x_{j+1} = x_j + \Delta x$.

a) Prove that the scheme is first-order accurate in time while second-order accurate in space.

The scheme can also be written as:

$$u_j^{n+1} = u_j^n - \frac{v\Delta t}{2\Delta x}\left(u_{j+1}^n - u_{j-1}^n\right). \tag{1.196}$$

In von Neumann's stability analysis, the solution is expanded in a finite Fourier series, in the form

$$u_j^n = A^n e^{ikjx}. \tag{1.197}$$

Plugging this expression into the numerical scheme (Equation 1.196) yields

$$A^{n+1}e^{ikjx} = A^n e^{ikjx} - \frac{v\Delta t}{2\Delta x}\left(A^n e^{ik(j+1)x} - A^n e^{ik(j-1)x}\right). \tag{1.198}$$

b) Show that the amplification factor from t^n to t^{n+1} is given by

$$A = 1 - i\frac{v\Delta t}{\Delta x}\sin(kx) \tag{1.199}$$

and prove that whenever $\mid A \mid > 1$ the scheme is unstable.

c) Consider a similar scheme in which u_j^n is replaced by an average between neighbors (i.e., Lax method):

$$u_j^{n+1} = \frac{u_{j+1}^n - u_{j-1}^n}{2} - \frac{v\Delta t}{2\Delta x}\left(u_{j+1}^n - u_{j-1}^n\right). \tag{1.200}$$

Show that the amplification factor is now given by

$$A = \cos(kx) - i\frac{v\Delta t}{\Delta x}\sin(kx) \tag{1.201}$$

and prove that $\mid A \mid \leq 1$ requires

$$\frac{\mid v \mid \Delta t}{\Delta x} \leq 1, \tag{1.202}$$

which is known as the Courant–Friedrichs–Levy (CFL) stability condition (i.e., the velocity at which information propagates within the algorithm, $\Delta x/\Delta t$, must be faster than the velocity of the solution, v).

P6. Consider the 1D diffusion equation (an example of parabolic equation), with a constant diffusion coefficient, D:

$$\frac{\partial u}{\partial t} = \frac{\partial}{\partial x}\left(D\frac{\partial u}{\partial x}\right) = D\frac{\partial^2 u}{\partial x^2}. \tag{1.203}$$

a) Determine the performance of the following numerical scheme through von Neumann's stability analysis

$$\frac{u_j^{n+1} - u_j^n}{\Delta t} = D\frac{u_{j+1}^n - 2u_j^n + u_{j-1}^n}{(\Delta x)^2} \tag{1.204}$$

and prove that the method is stable only for small step sizes, Δx.

b) Repeat the analysis for the alternative scheme (i.e., Crank–Nicolson)

$$\frac{u_j^{n+1} - u_j^n}{\Delta t} = \frac{D}{2}\frac{(u_{j+1}^{n+1} - 2u_j^{n+1} + u_{j-1}^{n+1}) + (u_{j+1}^n - 2u_j^n + u_{j-1}^n)}{(\Delta x)^2} \tag{1.205}$$

and determine the corresponding stability condition.

P7. The surface blackbody temperature of the Sun is 5780 K. Use Stefan's law to determine the Sun's rest mass lost per second and the fraction lost each century to radiation by the Sun.

P8. Consider the snapshots of the test explosion of the world's first atomic bomb (Trinity site, near Alamogordo, New Mexico, July 16, 1945) shown in Figure 1.23. Mesure the radius of the spherical shock wave or *fireball*, for each shot.

a) Plot the radius, r(m), as a function of time, t(s), in a logarithmic scale (i.e., 5/2 log r vs. log t) and prove that the data points can be fitted by a straight line.

b) Assume that the time-evolution of the fireball follows a Sedov–Taylor explosion until $t \sim 100$ ms, and consider a density for the nonperturbed air surrounding the explosion of $\rho = 1.25$ g cm^{-3}. Estimate the energy released by the explosion in kilotons [1784]. Compare the result with the on-site estimate performed by Enrico Fermi, ~ 10 kilotons, on the basis of the displacement of small pieces of paper, before, during, and after the passage of the blast wave, at about 10 miles from the explosion site[48].

c) Consider the expansion of the Crab nebula, a supernova remnant (see Chapters 5 and 7). Rewrite Equation 1.185 in the convenient form

$$R(t) = 2.3\,\mathrm{pc}\left(\frac{E}{10^{51}\,\mathrm{erg}}\right)^{1/5}\left(\frac{\rho_o}{10^{-24}\,\mathrm{g\,cm^{-3}}}\right)^{-1/5}\left(\frac{t}{100\,\mathrm{yr}}\right)^{2/5}, \tag{1.206}$$

and assume that the energy released in a supernova explosion is about 10^{51} erg, the density of the interstellar medium swept by the supernova blast has a mean density of 10^{-24} g cm^{-3}, and the size of the Crab nebula is ~ 3 pc. Estimate the age of the supernova remnant and its expansion velocity.

---

[48]For a detailed acount, see *The Day the Sun Rose Twice* (1984) by F. M. Szasz [1757].

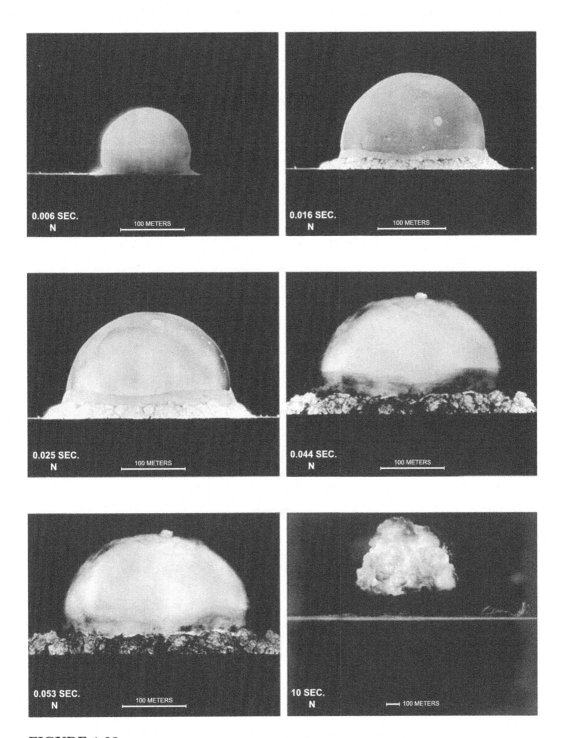

FIGURE 1.23

Snapshots of the test explosion of the world's first atomic bomb (Trinity site, near Alamogordo, New Mexico, July 16, 1945). Image courtesy of Los Alamos National Laboratory.

2

Nuclear Physics

"—[...] Radium is an element that is breaking up and flying to pieces. But perhaps all elements are doing that at less perceptible rates. Uranium certainly is; thorium—the stuff of this incandescent gas mantle—certainly is; actinium. I feel that we are but beginning the list. And we know now that the atom, that once we thought hard and impenetrable, and indivisible and final—and lifeless—is really a reservoir of immense energy. [...]

Carolinum, which belonged to the ß-Group of Hyslop's so-called 'suspended degenerator' elements, once its degenerative process has been induced, continued a furious radiation of energy, and nothing could arrest it. Of all Hyslop's artificial elements, Carolinum was the most heavily stored with energy and the most dangerous to make and handle. [...] What the earlier twentieth–century chemists called its half period was seventeen days."

H.G. Wells, *The World Set Free* (1914)

The universe has dramatically evolved from the chemically poor ashes synthesized a few minutes after the Big Bang during primordial nucleosynthesis. Indeed, the main observable legacy of the Big Bang, aside from the cosmic microwave background, was in the form of relic ^{1}H (75% by mass), ^{4}He (25%), and traces of deuterium (d) and ^{3}He (at the level of $\sim 10^{-5}$%), and ^{7}Li ($\sim 10^{-7}$%; see Figure 2.1(a)). Hydrogen and helium are still the most abundant species in today's universe[1], accounting for ~ 98% of its mass. But the chemical composition of the human body (Figure 2.1(b)) or of the Solar System itself (Figure 2.2) is far more rich in nuclear species and reveals a complex history.

Most of the ordinary (visible) matter in the universe, from a terrestrial rock to a giant star, is composed of protons and neutrons, baryonic particles that can assemble in a suite of different nuclear configurations or *nuclides*. Each nuclide is represented in the form $^A_Z X$, where Z is the number of protons or *atomic number*, A the number of *nucleons* (i.e., Z protons plus N neutrons) or *mass number*, and X represents the symbol assigned to each species. Nuclides characterized by the same atomic number but different number of neutrons (and therefore, different mass number) are called *isotopes*. There are 82 elements with stable isotopes[2], all the way from hydrogen (H; $Z = 1$) to lead (Pb; $Z = 82$), except for technetium (Tc; $Z = 43$) and promethium (Pm; $Z = 61$). Several dozen elements with $Z > 82$ have only unstable isotopes, naturally abundant on Earth or artificially produced in nuclear physics facilities[3]. But if the early universe was essentially made of hydrogen and helium, while devoid of heavier species[4], as observations of very distant (and therefore, very old) cosmic objects reveal, where did the rest of the elements come from? The explanation traces back to the 19th century, at the time when renowned physicists like J. R. Mayer, J.

---

[1]According to the prolific American writer Harlan Ellison, "*The two most common elements in the universe are hydrogen and stupidity.*"

[2]Throughout this book, unless otherwise stated, species with a half-life $T_{1/2} > 10^9$ yr are considered *stable* for practical purposes.

[3]The heaviest nucleus discovered to date, temporarily known as "ununoctium" or ^{294}UUo, has 118 protons, 176 neutrons, and a half-life of 0.89 ms [1344].

[4]In astronomy, the chemical composition is traditionally expressed in terms of hydrogen (X), helium (Y), and *metals* (Z). Therefore, the term *metallicity* refers to the chemical content in everything except hydrogen and helium.

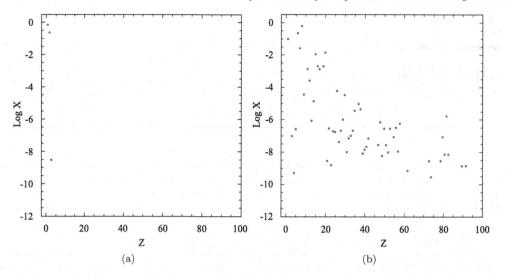

(a)   (b)

FIGURE 2.1

(a) Elemental composition of the primitive universe after Big Bang nucleosynthesis, in mass fractions, X, vs. atomic number, Z. (b) Elemental composition of the human body based on data from J. Emnsley [496]. Color versions of these plots are available at http://fisica.upc.edu/ca/users/jjose/CRC-Downloads.

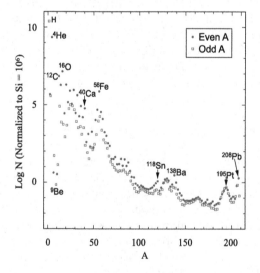

FIGURE 2.2

Solar System isotopic abundances by number, N, vs. mass number, A. Abundances are normalized to silicon in the figure ($N(\mathrm{Si}) = 10^6$), and are based on meteoritic samples as well as on values inferred from the solar photosphere (see Chapter 3). Values taken from Lodders [1135]. A color version of this figure is available at http://fisica.upc.edu/ca/users/jjose/CRC-Downloads.

J. Waterson, H. von Helmholtz, or W. Thomson (Lord Kelvin) were struggling in search for a feasible energy source, capable of explaining the long-lasting luminosity of the Sun. Early explanations suggested that the energy emitted by the Sun originates from the efficient conversion of gravitational potential energy into heat. The estimated lifetime of the Sun, according to Kelvin's meteor theory[5], yielded about 20–40 Myr, an age clearly at odds with estimates based on geological records.

Clues on the nature of the energy source that powers the Sun[6] were revealed shortly after the serendipitous discovery of radioactivity by A. H. Becquerel in 1896. An early series of precise atomic mass measurements conducted by F. W. Aston proved that four hydrogen nuclei are heavier than a single helium nucleus [85]. Nuclear physics escalated to the forefront of modern astrophysics when Arthur S. Eddington, in his 1920 speech to the British Association for the Advancement of Science, suggested that Aston's results indicate that the Sun could shine through nuclear fusion of hydrogen to helium [482]. A number of experimental and theoretical breakthroughs followed, including pioneering studies on Coulomb barrier penetration by G. Gamow (1928) and R. W. Gurney and E. U. Condon (1929), which led R. Atkinson and F. Houtermans to conclude that quantum tunneling is a key mechanism for energy generation in stars through nuclear fusion [87].

Pioneering work on the different modes for hydrogen fusion, that is, proton-proton (pp) chains [86, 177] and CNO cycles [173, 1871] (see Section 2.4.1), paved the road for the first self-consistent studies of element production in stars, the so-called *nucleosynthesis theory*, conducted by F. Hoyle [842, 843]. Equally influential was the compilation of reliable Solar System abundances reported by H. Suess and H. Urey [1739]. The distribution of abundances, dominated by hydrogen and helium, exhibits a complex pattern characterized by the presence of several maxima for both odd- and even-A species[7] (Figure 2.2). Maxima for $A \geq 56$ were soon attributed to a number of nuclear physics effects (e.g., the existence of tightly bound nuclei, such as ^{56}Fe, or the role played by closed shell configurations with *magic numbers* at 50, 82, and 126 nucleons[8], to quote a few examples).

Stars appeared indeed as active nuclear furnaces where most of the cosmic elements were being cooked, but observational evidence of their active role as nucleosynthesis sites was yet missing. The detection of technetium in the spectra of several giant stars by P. W. Merrill [1229] provided smoking-gun evidence to this conjecture. Technetium is the lightest element with no stable isotopes. Since its longest-lived isotope has a rather short half-life, $T_{1/2}(^{98}\text{Tc}) \sim 4.2$ Myr, compared with the age of the universe, its detection proved that nucleosynthesis is still ongoing in the cosmos. Further evidence has been provided by the detection of other radioactive species throughout the Galaxy, such as ^{26}Al or ^{44}Ti (see Chapters 4, 5, and 7).

Two seminal papers that provided the theoretical basis for the origin of the chemical species were published in 1957—almost exactly a century after Darwin's treatise on the origin of biological species (1859)—by E. M. Burbidge, G. R. Burbidge, W. A. Fowler, and

---

[5]Kelvin suggested that the energy emitted by the Sun was provided by meteors and asteroids that were falling into it at a regular rate.

[6]In 1919, H. N. Russell [1547], using simple physical arguments, inferred that the rate of the unknown process that supplies energy to stars must increase with increasing stellar temperature. He also concluded that such dependence of the energy production on temperature would make stars remain stable for long periods of time.

[7]Nuclei with an even number of nucleons are in general more abundant than odd-A nuclei because of *pairing effects*, which favor the stability of species by assembling pairs of identical nucleons (i.e., protons and protons, neutrons and neutrons, but not neutrons and protons) [170].

[8]Other tightly bound configurations are also obtained for 2, 8, 20, and 28 nucleons. Nuclides characterized by both magic neutron and proton numbers are known as *double magic*.

F. Hoyle [268], and independently by Cameron[9]. Our modern conception of the nuclear processes operating in stars is deeply rooted in and owes substantial credit to both papers.

A thorough description of all the nuclear physics inputs used in the study of stars, with full derivations of the relevant equations, is beyond the scope of this book. This chapter, instead, is aimed at providing background and a comprehensive guide to the major nuclear processes at work during stellar explosions. There are several monographs on the subject, such as the texts by Clayton [344], Rolfs and Rodney [1513], Arnett [67], Boyd [228], and, most notably, Iliadis [878], to which the reader is referred for details.

2.1 Nuclear Prelude: Abundances, Masses, and Binding Energies

Let's assume that a chunk of material (e.g., a terrestrial rock, an interstellar gas cloud, a compact star), of mass M and volume V, contains different chemical species. The *mass fraction* of species i in this volume, X_i, is given by

$$X_i = \frac{N_i m_i}{M},$$ (2.1)

where N_i is the number of nuclei of species i present in the sample and m_i is the mass (in grams) of each nucleus of species i. Rewritting the overall mass M in terms of the density, $M = \rho V$, yields

$$X_i = \frac{N_i m_i}{\rho V} = \frac{n_i m_i}{\rho} \rightarrow X_i = \frac{n_i M_i}{\rho N_{Av}},$$ (2.2)

where $n_i = N_i/V$ is the *number density* of species i (in cm^{-3}), and $N_{Av} \equiv M_i/m_i$ is the Avogadro number, or number of nuclei of species i that yield a mass M_i. By definition, $\sum_i X_i \equiv 1$.

A related quantity is the *mole fraction*, Y_i, defined as

$$Y_i = X_i/M_i$$ (2.3)

and commonly expressed in mol g^{-1}.

The number density can be written in terms of the mole fraction in the form

$$n_i = \rho N_{Av} Y_i.$$ (2.4)

Note that $\sum_i Y_i \neq 1$.

The time evolution of the number density (or the overall number of nuclei) of an unstable species is frequently expressed as:

$$\frac{dn_i}{dt} = -\lambda n_i$$ (2.5)

where λ is the probability of decay of species i per unit time, or *decay constant*. Integration of Equation 2.5 yields

$$n_i(t) = n_{i,o} e^{-\lambda t},$$ (2.6)

where $n_{i,o}$ is the initial number density of the species i at $t = 0$.

The amount of time required for a sample to decrease its initial number density to one-half is known as the *half-life*, $T_{1/2}$. Since, by definition, $n_o/2 \equiv n_o e^{-\lambda T_{1/2}}$, this implies

$$T_{1/2} = \frac{\ln 2}{\lambda}.$$ (2.7)

---

[9]Cameron's foundational paper has been recently edited and retyped [288].

TABLE 2.1

Binding Energy Per Nucleon* for Some
Representative Nuclear Species

Nucleus	A	B(Z,N)/A (MeV)
^{1}H	1	0
d (^{2}H)	2	1.112283
α ^{4}He	4	7.073915
^{12}C	12	7.680144
^{16}O	16	7.976206
^{18}F	18	7.631638
^{22}Na	22	7.915667
^{26}Al	26	8.149764
^{40}Ca	40	8.551303
^{44}Ti	44	8.533520
^{56}Fe	56	8.790342
^{58}Fe	58	8.792239
^{62}Ni	62	8.794546
^{64}Ge	64	8.528820
^{235}U	235	7.590906

*Values adapted from the 2012 atomic mass evaluation by Wang, Audi, Wapstra, and collaborators [1899].

A related quantity is the *mean lifetime*, τ, which accounts for a decrease of the initial number density by a factor $1/e$. Hence,

$$\tau = \frac{1}{\lambda} = \frac{T_{1/2}}{\ln 2}. \tag{2.8}$$

If a species can decay through different channels, the total decay probability is given by $\lambda = \sum_i \lambda_i$, or $1/\tau = \sum_i 1/\tau_i$.

Naively, the mass of a nucleus, M_{nuc}, can be thought as the sum of the masses of the individual nucleons, $M_{\text{nuc}} \sim Zm_{\text{p}} + Nm_{\text{n}}$, where m_{p} and m_{n} are the masses of the proton and the neutron, respectively. However, as pointed out by Aston (1920), the mass of a nucleus is actually smaller than the overall mass of its individual constituents because of the energy invested in assembling the nucleus. The so-called *nuclear mass defect*, ΔM, is therefore given by

$$\Delta M = Zm_{\text{p}} + Nm_{\text{n}} - M_{\text{nuc}}, \tag{2.9}$$

which corresponds to an energy $B(Z,N) = \Delta Mc^2$, known as the *binding energy of the nucleus*, or energy required to split a nucleus into its individual constituent nucleons,

$$B(Z,N) = (Zm_{\text{p}} + Nm_{\text{n}} - M_{\text{nuc}})c^2. \tag{2.10}$$

The binding energy increases with the mass number as more energy is required to stick all nucleons together. In short, the binding energy of a nucleus results from nuclear attraction (through the strong nuclear force) minus the contribution of electrostatic repulsion. Most nuclides are characterized by a binding energy per nucleon, $B(Z,N)/A$, around 7 or 8 MeV (Table 2.1). Interestingly, when plotted against the mass number A (Figure 2.3), the $B(Z,N)/A$ curve shows a maximum (which corresponds to the most tightly bound nucleus) at ^{62}Ni (8.794546 ± 0.000008 MeV), followed by ^{58}Fe (8.792239 ± 0.000008 MeV) and ^{56}Fe (8.790342 ± 0.000008 MeV) [1899].

Nuclear processes lead to energy generation whenever the binding energy per nucleon of

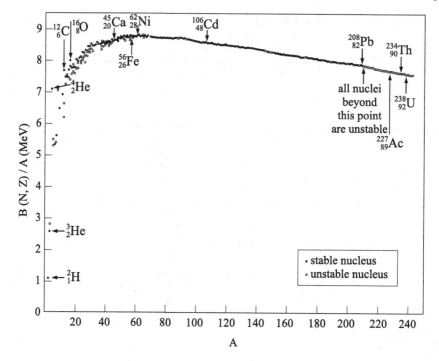

FIGURE 2.3

Binding energy per nucleon, $B(Z, N)/A$, plotted against mass number, A, for some representative nuclides.

the final products exceeds that of the interacting species. Therefore, nuclear energy can be released by fusion of nuclei lighter than Fe, or by fission of nuclei heavier than Fe. This is usually indicated by the *reaction Q-value*. In a nuclear interaction of the form $a + b \rightarrow c + d$, where a and b are the interacting particles (e.g, nuclei, protons, neutrons, photons) and c and d the reaction products[10], the *Q-value* is defined as:

$$Q \equiv (m_a + m_b - m_c - m_d)c^2 \,, \tag{2.11}$$

where m_i is the rest mass of species i. Equivalently, from total relativistic energy conservation,

$$Q = K_c + K_d - K_a - K_b \,, \tag{2.12}$$

where K_i is the kinetic energy of species i. Reactions with $Q > 0$ are called exothermic, while $Q < 0$ defines endothermic processes.

    An important drawback in the determination of reaction Q-values is the difficulty associated with nuclear mass measurements. Direct measurements are indeed challenging because of the presence of electrons. Instead, atomic masses, M_{at}, can actually be measured with high precision[11]. This is why atomic rather than nuclear masses are the quantities usually

---

[10]When particles a and b are identical to the reaction products c and d, the interaction is merely a scattering process. Otherwise, it defines a nuclear reaction. If species a is a photon, the interaction is known as a photodisintegration (see Section 2.3.4), while if species c is a photon, the process is known as a radiative capture reaction. Nuclear reactions are frequently written in the compact form $a(b, c)d$, where a represents the target and b the beam or projectile.

[11]Mass measurements have been traditionally conducted through indirect techniques (i.e., several reactions and decays), that yield Q-values or energy differences, and through direct methods (mass spectrometry), that rely on time-of-flight or cyclotron-frequency measurements. The highest precision in direct mass

tabulated in mass evaluations (see, e.g., reference [1899]). The relation between atomic and nuclear masses (for a neutral atom) is given by:

$$M_{at} = M_{nuc} + Zm_e - B_e(Z), \tag{2.13}$$

where m_e is the electron mass and $B_e(Z)$ is the electron binding energy in an atom. A related quantity is the *atomic mass excess*, which in energy units is given by

$$\Delta = (M_{at} - Au)c^2, \tag{2.14}$$

where A is the mass number and u is the atomic mass unit, defined as 1/12th of the mass of a ^{12}C atom. The reaction Q-value can be expressed in terms of the mass excess as

$$Q = \Delta_a + \Delta_b - \Delta_c - \Delta_d. \tag{2.15}$$

2.2 Nuclear Fusion

Fusion requires two (or more) interacting particles to approach closely enough, within the short range of the (attractive) strong nuclear force, $\lesssim 10^{-15}$ m, to form a new nucleus with $A = A_1 + A_2$. Charged interacting particles need to overcome the existing Coulomb barrier, of the form[12]

$$V(r) = \frac{Z_1 Z_2 e^2}{r}, \tag{2.16}$$

where Z_1 and Z_2 are the charges of projectile and target nuclei and r is their separation.

The simplest representation of this interaction is in the form of an attractive square well plus a repulsive square barrier, as depicted in Figure 2.4. Classically, an incoming particle with total energy $E < V = Z_1 Z_2 e^2 / (R_n + R_c)$ cannot penetrate the barrier. In fact, in a stellar plasma at temperature T, the height of the barrier is usually larger than the corresponding thermal energy of the particles, kT. But quantum mechanics allows the incoming particle to penetrate the barrier through tunneling.

2.2.1 Barrier Penetration

A simple, illustrative analysis of quantum tunneling can be performed in a 1D framework. In quantum mechanics, a particle can be described in terms of a wave function, $\psi(r)$, that corresponds to the solution of the time-independent Schrödinger equation for the characteristic potential of the problem, $V(r)$,

$$\left[-\frac{\hbar^2 \nabla^2}{2m} + V(r) \right] \psi(r) = E\psi(r), \tag{2.17}$$

where $m \equiv m_1 m_2 / (m_1 + m_2)$ is the reduced mass of the system. The probability of barrier penetration or *transmission coefficient*, T, can be directly obtained from the wave function by application of Born's rule [684].

---

measurements, ranging from 0.1 eV/c^2 (light atoms) to 20 eV/c^2 (heavy atoms) [1277], has been achieved with storage rings and Penning traps, on the basis of determination of the frequency with which a specific ion revolves in a magnetic field, which in turn depends on the m/q ratio of the ion. See references [197,1164,1165] for details.

[12]Note that the electrostatic constant or Coulomb's constant, $k \equiv 1/4\pi\epsilon_o$ is taken as 1 in cgs units.

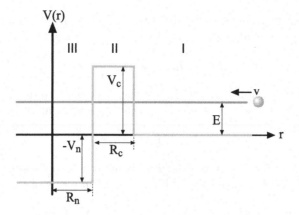

FIGURE 2.4

A square-well-plus-square-barrier potential, representing the simplest model for charged-particle-induced nuclear reactions. In the scheme, a particle is incident from the right, with total energy E. The particle first faces a repulsive barrier, characterized by a height V_c and a width R_c, followed by an attractive square well that has a depth $-V_n$ and a width R_n.

Solutions to the Schrödinger equation at each of the three regions shown in Figure 2.4 are well known and can be found in any basic quantum mechanics monograph (see, e.g., reference [684]). In general, the wave function in regions where the total energy of the incident particle, E, satisfies $E > V$, can be described by means of two complex exponentials, one corresponding to a wave that propagates to the right ($e^{ikx-\omega t}$) and another to the left ($e^{-ikx-\omega t}$). In region II, where $E < V$, the wave function is instead given by two real exponentials. All in all, the wave function (the eigenfunction, actually) in each of the three regions can be written, for the 1D case, as

$$\psi_{\mathrm{I}} = A e^{ik_{\mathrm{I}}x} + B e^{-ik_{\mathrm{I}}x}, \qquad\qquad x \geq R_{\mathrm{n}} + R_{\mathrm{c}} \qquad (2.18)$$

$$\psi_{\mathrm{II}} = C e^{k_{\mathrm{II}}x} + D e^{-k_{\mathrm{II}}x}, \qquad\qquad R_{\mathrm{n}} \leq x \leq R_{\mathrm{n}} + R_{\mathrm{c}} \qquad (2.19)$$

$$\psi_{\mathrm{III}} = F e^{ik_{\mathrm{III}}x} + G e^{-ik_{\mathrm{III}}x}, \qquad\qquad x \leq R_{\mathrm{n}}, \qquad (2.20)$$

where $k_{\mathrm{I}} = \sqrt{2mE}/\hbar$, $k_{\mathrm{II}} = \sqrt{2m(V_c - E)}/\hbar$, and $k_{\mathrm{III}} = \sqrt{2m(E + V_n)}/\hbar$ are the corresponding wave numbers of the incoming particle in each region.

The system of equations, characterized by six unknown coefficients, A, B, C, D, F, and G, can be solved by imposing $A \equiv 0$ (assuming that the particle is incident from the right in region I) and continuity of the wave function and its first-order spatial derivative at the boundaries,

$$(\psi_{\mathrm{I}})_{x=R_{\mathrm{n}}+R_{\mathrm{c}}} = (\psi_{\mathrm{II}})_{x=R_{\mathrm{n}}+R_{\mathrm{c}}} \qquad (2.21)$$

$$(\psi_{\mathrm{III}})_{x=R_{\mathrm{n}}} = (\psi_{\mathrm{II}})_{x=R_{\mathrm{n}}} \qquad (2.22)$$

$$\left(\frac{d\psi_{\mathrm{I}}}{dx}\right)_{x=R_{\mathrm{n}}+R_{\mathrm{c}}} = \left(\frac{d\psi_{\mathrm{II}}}{dx}\right)_{x=R_{\mathrm{n}}+R_{\mathrm{c}}} \qquad (2.23)$$

$$\left(\frac{d\psi_{\mathrm{III}}}{dx}\right)_{x=R_{\mathrm{n}}} = \left(\frac{d\psi_{\mathrm{II}}}{dx}\right)_{x=R_{\mathrm{n}}}. \qquad (2.24)$$

The above system of equations allows us to express four unknown coefficients in terms of the amplitude of the incoming particle in region I, $\mid B \mid^2$. The resulting transmission coefficient through the barrier, defined as the ratio of transmitted over incident particle fluxes, has the

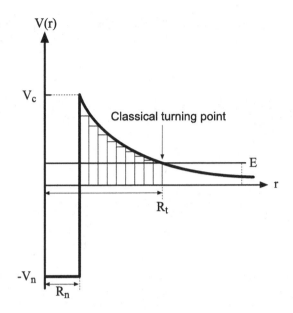

FIGURE 2.5
A square-well-plus-Coulomb potential, representing a more realistic model for a charged-particle nuclear reaction.

form

$$T = \frac{k_{\text{III}} \, | \, G \, |^2}{k_{\text{I}} \, | \, B \, |^2} \tag{2.25}$$

and is given by [878]

$$\frac{1}{T} = \frac{1}{\sqrt{E(E+V_{\text{n}})}} \left[\left(2E + V_{\text{n}} + 2\sqrt{E(E+V_{\text{n}})}\right) + \left(E + V_{\text{n}} + V_{\text{c}} + \right. \right.$$
$$\left. \left. \frac{E(E+V_{\text{n}})}{V_{\text{c}} - E}\right) \sinh^2\left(\sqrt{\frac{2m}{\hbar^2}(V_{\text{c}} - E)}R_{\text{c}}\right) \right], \tag{2.26}$$

which, for the limit of low bombarding energy (or thick barrier), $k_{\text{II}}R_{\text{c}} \gg 1$, reduces to:

$$T \approx \exp\left[\frac{-2}{\hbar}\sqrt{2m(V_{\text{c}} - E)}R_{\text{c}}\right]. \tag{2.27}$$

The study of the square-well-plus-square-barrier outlined above can be easily extended to the more relevant case of a Coulomb barrier. An estimate of the overall transmission coefficient can be obtained by approximating the Coulomb barrier by a set of square, narrow barriers of width dr and increasing height, as the incoming particle progressively approaches the target. This is illustrated in Figure 2.5. In this framework, for a sufficiently large number of narrow barriers, and in the limit of low energy, Equation 2.27 becomes

$$T = T_1 T_2 T_3 ... T_{\text{N}} \approx \exp\left[\frac{-2}{\hbar}\int_{R_{\text{n}}}^{R_{\text{t}}} \sqrt{2m(V(r) - E)}\,dr\right], \tag{2.28}$$

where R_{t} is the location of the classical turning point, defined as $E = V(R_{\text{t}})$. For a Coulomb potential of the form $V(r) = Z_1 Z_2 e^2 / r$, Equation 2.28 can be approximated by [878]

$$T \approx \exp\left(-\frac{2}{\hbar}\sqrt{\frac{2m}{E}} Z_1 Z_2 e^2 \left[\frac{\pi}{2} - 2\sqrt{\frac{E}{V_c}} + \frac{1}{3}\left(\frac{E}{V_c}\right)^{3/2}\right]\right). \tag{2.29}$$

The first term in Equation 2.29 dominates the transmission coefficient, for the case in which the incoming projectile has a small energy compared with the height of the barrier,

$$T \approx \exp\left[-\frac{2\pi}{\hbar}\sqrt{\frac{m}{2E}} Z_1 Z_2 e^2\right] \equiv e^{-2\pi\eta}, \tag{2.30}$$

where η is the *Sommerfeld parameter*,

$$2\pi\eta = 0.98951 Z_1 Z_2 \sqrt{\frac{m(u)}{E(\text{MeV})}}. \tag{2.31}$$

The term $e^{-2\pi\eta}$ is known as the *Gamow factor*, and as discussed in Section 2.2.2, is related to a rather useful quantity, the astrophysical S-factor.

2.2.2   Fusion Cross-Sections

The probability of occurrence of a specific nuclear interaction is quantified by means of its *cross-section*, σ. Classically, the probability that a projectile will hit a target increases as their respective *cross-sectional* areas increase. In this framework, the cross-section is defined as $\sigma = \pi(R_p + R_t)^2$, where R_p and R_t are the radii of projectile and target, respectively, and as discussed below, it has units of area. In the case of nuclear reactions, which are governed by quantum mechanics, the cross-section relies instead on the *reduced de Broglie wavelength* of the system,

$$\lambdabar = \frac{m_p + m_t}{m_t} \frac{\hbar}{\sqrt{2m_p E_{\text{lab}}}}, \tag{2.32}$$

where m_p and m_t are the masses of projectile and target, and E_{lab} is the laboratory bombarding energy of the projectile. All in all,

$$\sigma = \pi\lambdabar^2 \propto \frac{1}{E_{\text{lab}}}. \tag{2.33}$$

Let's consider now that N_i projectiles of species i are emitted as a beam of cross-sectional area S_i. The beam is focused onto a target, characterized by N_j particles of type j within the beam. Let's also assume that the total number of interactions per unit time, N_{int}/t, equals the overall emission rate of interaction products, N_e/t. The cross-section σ of the interaction is defined as the number of interactions per unit time, N_{int}/t, divided by the number of incoming projectiles i per unit area and time, $N_i/(S_i t)$, and by the number of (nonoverlapping) target nuclei j within the beam, N_j,

$$\sigma = \frac{N_{\text{int}}/t}{[N_i/(S_i t)]N_j} = \frac{N_e/(N_j t)}{N_i/(S_i t)}. \tag{2.34}$$

Note that the cross-section has units of an area. In nuclear physics, the adopted cross-section unit is the barn (1 b = 10^{-24} cm^2), which approximately corresponds to the cross-sectional area of an uranium nucleus, $\sim \pi R^2 \sim \pi(R_o A^{1/3})^2$, with $R_o = 1.3 \times 10^{-13}$ cm and $A = 92$ [1016].

Sufficiently large cross-sections can be experimentally determined by measuring the rate of interaction products emitted, N_e/t, for a given projectile current density, $N_i/(S_i t)$, and number of target nuclei within the beam, N_j. To this end, detectors are usually located at a given angle with respect to the direction of the incoming projectiles. In this regard, a more useful experimental quantity is the *differential cross-section*, $d\sigma/d\Omega$, defined as

$$\frac{d\sigma}{d\Omega} = \frac{N_e'/(N_j t)}{N_i/(S_i t)} \frac{1}{d\Omega},\tag{2.35}$$

where N_e' is the number of interaction products emitted at an angle θ into a solid angle $d\Omega$. Integration of the differential cross-section over all possible solid angles yields

$$\sigma = \int \frac{d\sigma}{d\Omega}\, d\Omega.\tag{2.36}$$

Cross-sections for different interactions can differ by many orders of magnitude, for the same bombarding energies. For instance, the slowest in the suite of reactions that occur during hydrogen burning (see Section 2.4.1), $p + p \rightarrow d + e^+ + \nu$, has a cross-section of 8×10^{-24} b at a center-of-mass energy of 0.5 MeV (i.e., laboratory proton bombarding energy of 1 MeV; see Problem 4 at the end of this chapter). In sharp contrast, the reaction $^{16}O + p \rightarrow {}^{17}F + \gamma$ has a cross-section of 10^{-7} b at the same energy[13], $E_{c.m.} = 0.5$ MeV. It is worth noting that, as anticipated by the energy dependence of the transmission coefficient (Equation 2.27), the cross-section drops dramatically at low energies. This reflects the lower probability of barrier penetration for decreasing energies of the incoming particles. For instance, the cross-section of the $p + p$ reaction decreases to 6×10^{-25} b at $E_{c.m.} = 50$ keV and to 4×10^{-27} b at $E_{c.m.} = 5$ keV, which basically corresponds to the energy at which this reaction occurs in the Sun. Let's imagine that an experiment is aimed at determining the nuclear cross-section of the $p + p$ reaction at solar energies. To this end, an intense 1 mA beam of protons is used to hit a dense hydrogen target containing 10^{20} protons cm^{-2}. The expected count rate at 5 keV would be about 1 interaction in 13 Myr! This shows the challenges (and impossibilities) faced by experimentalists in the determination of small nuclear cross-sections, which unavoidably have to be estimated on theoretical grounds (see, e.g., reference [10] for examples of solar fusion cross-sections).

Cross-section curves are characterized by a suite of possible shapes when plotted against bombarding energy. Some nuclear interactions, such as $d + p \rightarrow {}^3He + \gamma$ (Figure 2.6(a)), or the above-mentioned reactions $p + p$ and $^{16}O + p$, exhibit smooth, almost structureless shapes. On the other hand, reactions like $^{24}Mg + p \rightarrow {}^{25}Al + \gamma$ or $^{12}C + p \rightarrow {}^{13}N + \gamma$ (Figure 2.6(b)) show the presence of one or multiple spikes. Smooth cross-section curves can be naturally explained in terms of the energy dependence of the transmission coefficient (Equation 2.27), obtained for a 1D approximation of the square-well-plus-square-barrier model. The characterization of the spikes, however, requires a careful analysis of the full radial wave function in a 3D framework (see Iliadis [878] for a detailed account). In this context, the spikes are identified with resonances resulting from favorable wave function matching conditions at the nuclear boundary, $r = R_n$, obtained for specific values of the nuclear potential depth, V_n. Accordingly, nuclear interactions are classified as *nonresonant*—characterized by smooth, weakly energy-dependent cross-sections—and *resonant*—with strongly varying cross-sections around specific energies.

2.2.2.1 Resonances

Many experimental cross-sections reveal, in general, complex patterns with multiple resonances superimposed onto an otherwise smooth curve that decreases as the energy of the

---

[13]Such variation in the corresponding cross-sections illustrates the effect of the different forces that govern nuclear interactions, with σ(strong force) [e.g., (p, α)] $\gg \sigma$(electromagnetic force) [e.g., (p, γ)] $\gg \sigma$(weak force) [e.g., as in p(p, $e^+\nu$)d. See Section 2.4.1.1].

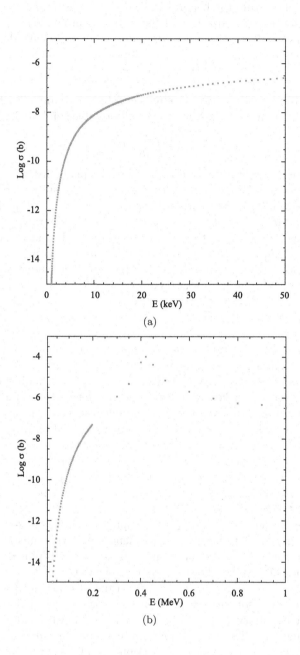

FIGURE 2.6

(a) Experimental cross-section of the reaction d(p, γ)^{3}He. Data courtesy of S. Goriely, based on the NACRE II compilation [1996]. See `http://www.astro.ulb.ac.be/nacreii/` for details. (b) Same as Figure 2.6(a), for the reaction ^{12}C(p, γ)^{13}N. Color versions of these plots are available at `http://fisica.upc.edu/ca/users/jjose/CRC-Downloads`.

incident projectile decreases. While some resonances are narrow[14] and isolated, others are actually broad, with long tails, such that they may even quantum-mechanically interfere with other neighboring resonances. Although some observed resonances can be described by means of single-particle potentials, like the one adopted in the analysis of nonresonant nuclear processes, the interactions occurring inside the nucleus between multiple nucleons cannot, in general, be appropriately described in this framework. A many-body approach was first introduced by Niels Bohr who, in the framework of neutron-nucleus interactions [215], envisaged nuclear reactions as two-stage processes: In the first stage, the projectile gets absorbed by the target nuclei, which leads to the formation of a quasi-stationary state in a compound nucleus; this is followed, in the second stage, by the decay of the compound state by means of one or several channels. In this scenario, resonances correspond to states in the compound nucleus formed by the interaction of many nucleons.

The simplest analytical treatment restricts to the case of narrow, isolated resonances[15]. This occurs at an energy of the incident projectile that matches the energy of an excited state in the compound nucleus, E_r. This is frequently described by the Breit–Wigner formula [196, 878]:

$$\sigma_{\mathrm{BW}}(E) = \frac{\lambda^2 \omega}{4\pi} \frac{\Gamma_i \Gamma_o}{(E_r - E)^2 + \Gamma^2/4}, \tag{2.37}$$

where λ is the de Broglie wavelength; ω is a statistical factor containing total angular momentum (spin),

$$\omega = \frac{2J + 1}{(2J_a + 1)(2J_b + 1)}, \tag{2.38}$$

with J_a and J_b the total spins of the interacting particles a and b, and J the spin of the resonance state in the compound nucleus; E_r is the resonance energy; Γ_i and Γ_o are the energy-dependent partial widths of the entrance and exit channels (i.e., the probability of formation and decay[16] of the state in the compound nucleus into the desired channel), and Γ is the total width given by the sum of all partial widths. The formula applies to both charged-particles and neutron projectiles.

Partial widths play a critical role in nuclear physics. In short, they describe the probability, in energy units, per unit time of the formation or decay of a compound state from a particular channel. In terms of Heisenberg's principle,

$$\Gamma_{i,o} \, \tau_{i,o} \sim \hbar, \tag{2.39}$$

where $\tau_{i,o}$ is the characteristic formation or decay time of the compound state. In the simplest case of only one possible channel for formation or decay of the compound state, the partial width is given by [877, 878]

$$\Gamma_c = 2\frac{\hbar^2}{mR^2} P_c C^2 S \theta_{\mathrm{pc}}^2, \tag{2.40}$$

where $C^2 S$ is the *spectroscopic factor*, or probability that the nucleons will arrange according to the final state configuration, θ_{pc}^2 is the *dimensionless reduced single-particle width*, or

---

[14]Several definitions of a narrow resonance can be found in the literature. A widely used criterion requires $\Gamma \ll E_r$. Alternatively, a resonance is considered narrow if the partial widths are approximately constant over the entire resonance width.

[15]A useful concept in the characterization of a nucleus is the *level density*, a measure of the number of levels per unit energy at a certain excitation energy. In other words it quantifies the different ways in which individual nucleons can be placed, for an excitation energy in the range from E to $E + dE$. A low-level density in the compound nucleus during a nuclear interaction, for instance, is a requirement for the presence of isolated (nonoverlapping) resonances in the cross-section.

[16]Note that the decay may refer to a particle or to a γ-ray.

probability that a proton will appear at the nuclear boundary, and P_c is the *penetration factor*, or probability of penetration of the incident particle through the Coulomb—for the case of charged-particle interactions—and centrifugal barriers[17]. P_c and θ^2_{pc} can be determined numerically [129,877], whereas the spectroscopic factor, C^2S, a nuclear structure quantity, can sometimes be estimated through the nuclear shell-model or experimentally determined using indirect methods.

While nuclear cross-sections are preferentially determined through experiments, overcoming the limitations that arise from the limited knowledge of the precise nuclear potential, no data are available for, e.g., very small cross-sections, or reactions that involve short-lived radioactive species. In this framework, the interest of the Breit–Wigner formula becomes crystal clear, as the cross-section can only be inferred theoretically.

2.3   Nuclear Interactions

In stars, particles describe random motions with a translational kinetic energy of thermal origin. Accordingly, nuclear processes in stellar plasmas are referred to as *thermonuclear reactions*[18].

Let's consider a generic interaction in the form $a + b \rightarrow c + d$, in which a and b (i.e., targets and projectiles) have rest mass (see Section 2.3.4 for photon-induced interactions). The rate at which species a and b interact per second and per unit volume, r_{ab}, in a plasma containing N_a and N_b particles depends on the velocity of the interacting species through the cross-section $\sigma(v)$, in the form (see Equation 2.34):

$$r_{ab} = \frac{N_{int}}{tV} = \frac{\sigma N_a N_b}{S_b tV}. \tag{2.41}$$

Rearranging terms,

$$r_{ab} = \sigma \frac{N_a}{V} \frac{N_b}{S_b t}, \tag{2.42}$$

and bearing in mind that in a time t, all projectiles within a volume $S_b v t$ would hit the target, one can write

$$r_{ab} = \sigma \frac{N_a}{V} \frac{N_b v}{V}. \tag{2.43}$$

Equation 2.43 is usually expressed in terms of the number densities, n_a and n_b, in the form

$$r_{ab} = n_a n_b v \sigma(v). \tag{2.44}$$

Note that v actually corresponds to the relative velocity between projectiles and targets. In a stellar plasma, however, this velocity is not unique, since particles are characterized by a certain distribution of velocities. This is described by a probability function, $P(v)$, such that $P(v)dv$ represents the probability that the relative velocity ranges between v and $v + dv$, and satisfies:

$$\int_0^\infty P(v)dv \equiv 1. \tag{2.45}$$

---

[17]Centrifugal barriers correspond to the nonradial solutions of the full 3D Schrödinger equation for the nuclear potential [878].

[18]At sufficiently high densities, as in the neutron star interior, density-induced interactions known as *pycnonuclear reactions* can also occur even at low temperature [285].

Accordingly, Equation 2.44 is generalized in the form

$$r_{ab} = n_a n_b \int_0^\infty v P(v) \sigma(v) dv \equiv n_a n_b \langle \sigma v \rangle_{ab}. \tag{2.46}$$

The quantity $\langle \sigma v \rangle_{ab}$ corresponds to the *reaction rate per particle pair*. Note, however, that many reaction rate compilations[19] actually rely on the related quantity $N_{Av} \langle \sigma v \rangle_{ab}$ (in cm^3 mol^{-1} s^{-1}).

In many astrophysical scenarios, stellar plasmas consist of nonrelativistic and nondegenerate particles whose velocities can be described by means of a Maxwell–Boltzmann distribution,

$$P(v) dv = \left(\frac{m}{\sqrt{2\pi kT}} \right)^{3/2} e^{-mv^2/(2kT)} 4\pi v^2 dv, \tag{2.47}$$

where k is the Boltzmann constant and m is the reduced mass of the projectile-target system. For convenience, Equation 2.47 is often expressed in terms of the energy, $E = 1/2 mv^2$, in the form

$$P(v) dv = P(E) dE = \frac{2}{\sqrt{\pi}} \frac{1}{(kT)^{3/2}} \sqrt{E} e^{-E/kT} dE. \tag{2.48}$$

All in all, the reaction rate per particle pair can finally be written as

$$\langle \sigma v \rangle_{ab} = \left(\frac{8}{\pi m} \right)^{1/2} \frac{1}{(kT)^{3/2}} \int_0^\infty E\sigma(E) e^{-E/kT} dE. \tag{2.49}$$

It is worth noting that $\langle \sigma v \rangle_{ab}$ provides an estimate of the characteristic timescale of a nuclear reaction. The rate of destruction of species a due to nuclear interactions with species b can be written as

$$\left(\frac{dn_a}{dt} \right)_b = -r_{ab} = -n_a n_b \langle \sigma v \rangle_{ab}. \tag{2.50}$$

Combining Equations 2.5 and 2.50 yields to

$$\frac{1}{\tau_b(a)} = -n_b \langle \sigma v \rangle_{ab}. \tag{2.51}$$

If destruction occurs by more than one channel, the mean lifetime of species a is given by

$$\frac{1}{\tau(a)} = \sum_i \frac{1}{\tau_i(a)}. \tag{2.52}$$

In general, reaction rates can always be obtained from numerical integration of Equation 2.49 (provided that $\sigma(E)$ is known!), but some particular cases of interest, where simplifications apply, will be addressed in the following subsections.

2.3.1 Nonresonant Charged-Particle Reactions

Cross-section data of nonresonant reactions, such as d(p, γ)^{3}He (Figure 2.6(a)), exhibit a dramatic decrease at low energies due to quantum tunneling, as reflected in the energy-dependence of the tranmission coefficient through the Coulomb barrier (Equation 2.30),

---

[19]See, e.g., references [50, 304, 882, 1996].

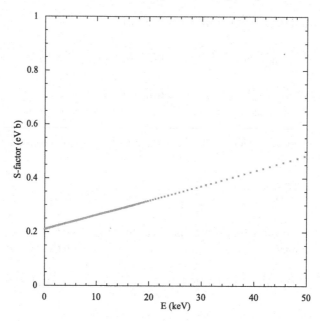

FIGURE 2.7

Astrophysical S-factor for the reaction d(p, γ)^{3}He. Compare the low-energy dependence of the data with the large variation exhibited by the raw cross-section displayed in Figure 2.6(a). Data courtesy of S. Goriely, based on the NACRE II compilation [1996]. See `http://www.astro.ulb.ac.be/nacreii/` for details. A color version of this plot is available at `http://fisica.upc.edu/ca/users/jjose/CRC-Downloads`.

$$\sigma(E) \propto e^{-2\pi\eta}.$$

On the other hand, nuclear cross-sections satisfy (Equation 2.33)

$$\sigma(E) \propto \pi\lambdabar^2 \propto \frac{1}{E}.$$

A convenient quantity, the *astrophysical S-factor*, $S(E)$, has extensively been used in the field to remove the energy dependence of the Coulomb barrier penetration from the cross-section, in the form

$$S(E) \equiv E e^{2\pi\eta}\sigma(E). \tag{2.53}$$

As shown in Figure 2.7 for the reaction d(p, γ)^{3}He, S(E) exhibits a much lower energy dependence than σ(E), and is therefore better suited for representing the effects of nuclear structure on the cross-section[20] (particularly, as will be discussed later, when extrapolation of experimental values to low energies is required).

In terms of the S-factor, the rate of a nonresonant nuclear reaction can be written as

$$\langle \sigma v \rangle_{ab} = \left(\frac{8}{\pi m} \right)^{1/2} \frac{1}{(kT)^{3/2}} \int_0^\infty S(E) e^{-2\pi\eta} e^{-E/kT} dE. \tag{2.54}$$

Many nonresonant reactions are characterized by a nearly constant S-factor, regardless of

---

[20]The S-factor has become so widely used in nuclear astrophysics—also for resonant reactions—that it has almost replaced cross-sections in the literature.

the energy. Under this assumption, $S(E) \equiv S_0 = \text{const}$, the reaction rate is fully determined by the product of two exponentials: The Gamow factor, $e^{-2\pi\eta}$, and the Boltzmann factor, $e^{-E/kT}$. As shown in Figure 2.8(a), this product defines the likely energy range, known as the Gamow peak [602], at which a specific nuclear reaction actually occurs, for a given temperature, T (see, e.g., Figure 2.8(b)). The maximum of the Gamow peak, E_0, can be calculated from the first-order derivative of the product, and is given by

$$E_0 = \left[\left(\frac{\pi}{\hbar} \right)^2 (Z_a Z_b e^2)^2 \frac{m}{2} (kT)^2 \right]^{1/3}. \tag{2.55}$$

As shown by Equation 2.55, E_0 increases with increasing projectile or target charge. For instance, the maximum of the Gamow peak for the p + p reaction occurs at $E_0 = 5.88$ keV, for $T = 15$ GK (a value corresponding to the central temperature of the Sun). At the same temperature, ^{14}N + p occurs at $E_0 = 26.5$ keV, ^{12}C + α at $E_0 = 56.1$ keV, and ^{16}O + ^{16}O at $E_0 = 237$ keV.

There is no simple analytical solution for the width of the Gamow peak. A frequently adopted approach is to approximate the shape of the peak by a Gaussian of the same height, centered on E_0. The full width of the Gaussian, Δ, at E_0/e is then determined by imposing that both curves have the same curvature (i.e., second derivatives) at E_0, which corresponds to

$$\Delta = \frac{4}{\sqrt{3}} \sqrt{E_0 kT}. \tag{2.56}$$

Therefore, we conclude that thermonuclear reactions occur mainly over an energy window ranging from $E_0 - \Delta/2$ and $E_0 + \Delta/2$, except in the presence of narrow resonances (see, however, reference [1290]).

Replacement of the Gamow peak by a Gaussian in Equation 2.54 leads to the following expression for the reaction rate

$$\langle \sigma v \rangle_{ab} = \left(\frac{8}{\pi m} \right)^{1/2} \frac{1}{(kT)^{3/2}} S_0 e^{-3E_0/kT} \int_0^\infty \exp\left[-\left(\frac{E - E_0}{\Delta/2} \right) \right] dE, \tag{2.57}$$

and extending the integral all the way from $-\infty$ to ∞, one finally obtains[21]

$$\langle \sigma v \rangle_{ab} = \left(\frac{2}{m} \right)^{1/2} \frac{\Delta}{(kT)^{3/2}} S_0 e^{-3E_0/kT}. \tag{2.58}$$

It is finally worth noting that, for increasing charges of projectiles and targets, the Gamow peak becomes broader, while the area under the Gamow peak—which determines the total reaction rate—decreases dramatically. This result stresses that, for a given temperature, nuclear energy is driven by reactions with the smallest Coulomb barriers.

2.3.2 Resonant Charged-Particle Reactions

The presence of resonances can fully dominate a nuclear cross-section. As discussed in Section 2.2.2.1, isolated and narrow resonances are frequently described by means of the Breit–Wigner formula (Equation 2.37):

$$\sigma_{BW}(E) = \frac{\lambda^2 \omega}{4\pi} \frac{\Gamma_i \Gamma_o}{(E_r - E)^2 + \Gamma^2/4},$$

---

[21]More accurate expressions that take into account the often weak energy-dependence of the S-factor, plus the error introduced by the replacement of the asymmetric Gamow peak by a symmetric Gaussian, can be found, e.g., in references [546, 878].

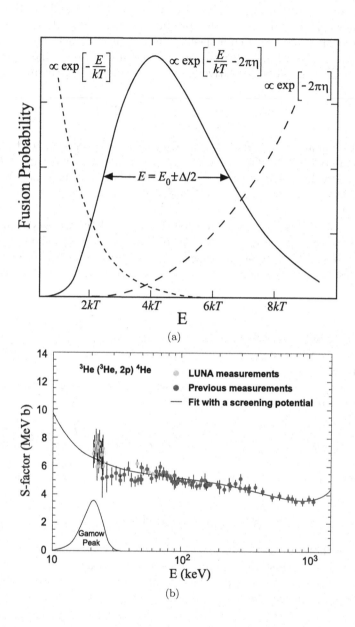

FIGURE 2.8

(a) Convolution of the Maxwell–Boltzmann energy distribution and the transmission probability through the Coulomb barrier showing the formation of the Gamow peak at energies in the range $E_0 \pm \Delta/2$. The size of the Gamow peak has been magnified in the figure, for the sake of clarity. (b) Experimental cross-section of the reaction $^3\text{He}(^3\text{He}, 2p)^4\text{He}$, showing the location of the Gamow Peak at the solar central temperature [10]. Figure from Junker et al. [937], reproduced with permission.

with

$$\omega = \frac{2J + 1}{(2J_a + 1)(2J_b + 1)}. \tag{2.59}$$

An expression for the reaction rate between particles a and b, in the case of a single, narrow resonance, can be obtained combining the Breit–Wigner formula for the cross-section and Equation 2.49:

$$\langle \sigma v \rangle_{ab} = \left(\frac{8}{\pi m} \right)^{1/2} \frac{1}{(kT)^{3/2}} \int_0^\infty E\sigma(E)e^{-E/kT} dE =$$

$$\frac{\sqrt{2\pi}\hbar^2}{(mkT)^{3/2}} \omega \int_0^\infty \frac{\Gamma_i \Gamma_o}{(E_r - E)^2 + \Gamma^2/4} e^{-E/kT} dE. \tag{2.60}$$

In the context of narrow resonances, both the Boltzmann factor $\exp(-E/kT)$ and the partial widths $\Gamma_{i,o}$ are assumed to be approximately constant over the total width of the resonance, and therefore can be replaced by their value at $E = E_r$. This yields

$$\langle \sigma v \rangle_{ab} = \left(\frac{2\pi}{mkT} \right)^{3/2} \hbar^2 e^{-E_r/kT} \omega\gamma, \tag{2.61}$$

where $\gamma \equiv \Gamma_i \Gamma_o/\Gamma$. Note that, at $E = E_r$, $\omega\gamma$ is proportional to the product of the maximum cross-section and the total width,

$$\sigma \Gamma = \frac{\lambda^2}{\pi}\omega\gamma, \tag{2.62}$$

which constitutes a measure of the contribution of a resonance to the overall reaction rate. Accordingly, $\omega\gamma$ (with units of energy) is referred to as the *resonance strength*.

Equation 2.61 can be extended to the case of several narrow and isolated resonances, in the form

$$\langle \sigma v \rangle_{ab} = \left(\frac{2\pi}{mkT} \right)^{3/2} \hbar^2 \sum_i e^{-E_{r,i}/kT}(\omega\gamma)_i, \tag{2.63}$$

where the different contributions to the overall reaction rate are simply summed incoherently. Equations 2.61 and 2.63 reveal that the contribution to the overall reaction rate for narrow resonances does not depend on the shape of the cross-section curve. Moreover, the fact that resonance energies enter exponentially in the reaction rate formula requires a precise determination of their values to achieve an accurate knowledge of the rate.

It is also worth noting that not all resonances equally contribute to the overall reaction rate. Although it is frequently stated in the literature that resonances within the Gamow peak dominate the contribution to the reaction rate in charged-particle interactions, this is not always the case. For instance, for resonance energies above $\simeq 0.5$ MeV, the rate is often governed by low-lying resonances, because of the dominant effect of the Boltzmann factor, $e^{-E/kT}$.

Cross-section curves frequently reveal the presence of broad resonances[22] whose tails may even interfere, and for which the energy-dependence of their partial widths cannot be ignored (e.g., $^{18}F(p, \alpha)$). Although several approximate formulae have been obtained and published in the literature, it is safer to include the contribution of such broad resonances by means of the Breit–Wigner formula, with energy-dependent partial widths, and then

---

[22]Occasionally, an excited state in the compound nucleus may be characterized by an energy E_r that is actually smaller than the Q-value of the corresponding decay. Such a state is known as a subthreshold resonance. The high-energy tail of a broad subthreshold resonance may have a dramatic effect on the overall reaction rate. Examples of subthreshold resonances can be found in, e.g., $^{14}N(p, \gamma)^{15}O$ ($E_{c.m.} = -22$ keV) and $^{20}Ne(p, \gamma)^{21}Na$ ($E_{c.m.} = -7$ keV).

evaluate the integral in the reaction rate expression numerically. In the presence of multiple resonances, the reaction rate is evaluated in the framework of statistical models, such as Hauser–Feshbach[23] [740].

All in all, the overall reaction rate of a specific nuclear interaction is, in general, determined by many different terms that include (narrow and broad) resonant and nonresonant contributions as well as subthreshold resonances and, if applicable, interferences. Once the total reaction rate has been determined, the nuclear energy released per reaction, or *energy generation rate*, can be calculated as

$$\varepsilon = r_{ab}Q/\rho \ (\text{erg g}^{-1}\text{s}^{-1}),\tag{2.64}$$

where ρ is the density of the plasma.

2.3.3 Electron Screening

In a stellar plasma, interacting nuclei cannot be regarded as isolated systems, since they are surrounded by free electrons and neighboring nuclei. The structure of the Coulomb potential of the nuclei is certainly affected by the presence of such distributions of charged particles, and therefore has an impact on the overall reaction rate. Free electrons actually *shield* the nucleus in such a way that the incoming projectile, b, feels a reduced Coulomb barrier when approaching the target, a, compared with the corresponding value for a fully isolated, bare nucleus[24]. This enhances the reaction rate, which has to be multiplied by a *screening factor*, f, that depends on the thermodynamic conditions of the plasma through specific regimes (i.e., weak, intermediate, or strong screening).

Following Salpeter [1558], the screened potential seen by the incoming projectile can be written as

$$V'(r) \sim \frac{Z_a Z_b e^2}{r} e^{-r/R_D},\tag{2.65}$$

where R_D is the Debye–Hückel radius, a measure of the size of the charged cloud that surrounds each nucleus,

$$R_D = \sqrt{\frac{kT}{4\pi e^2 \rho N_{Av}\zeta}}.\tag{2.66}$$

The parameter ζ represents the root mean square charge average,

$$\zeta = \sqrt{\sum_i Z_i^2 X_i/A_i + \theta_e \sum_i Z_i X_i/A_i},\tag{2.67}$$

where θ_e is the electron degeneracy factor, which quantifies the degree of participation of

---

[23] Many nuclear reactions of interest proceed through compound systems characterized by a large number of level densities. Under these circumstances (i.e., ≥ 10 levels [1477]), statistical methods can provide a reliable description of the nuclear interaction. The physical framework that leads to formation and decay of compound nuclei was developed by Ewing and Weisskopf [1921], on the basis of a formalism that assumed energy, mass, and charge conservation. The theory was later on extended by Hauser and Feshbach [740], who included angular momentum and parity dependences. A simple derivation of the Hauser–Feshbach model can be obtained from an energy average over isolated Breit–Wigner resonances [445]. This derivation turns out to be correct only at low energies, where the resonances do not overlap. Derivations for the more general case with overlapping resonances are also available [959]. The Hauser–Feshbach technique has been widely used in the modeling of stellar explosions, such as type Ia supernovae or X-ray bursts (see Chapters 5 and 6).

[24] Electron screening has also been observed in low-energy laboratory measurements of nuclear reactions, in the presence of metals. There are, however, important discrepancies between the experimental data obtained by different groups. See, e.g., references [392, 1858] for recent reports on this subject.

electrons in the screening as a function of degeneracy (see reference [983]). Values of ζ are of order unity.

Weak screening, the most frequent regime in stellar plasmas, occurs when the Coulomb potential energy of the projectile-target system, E_{Coul}, is smaller than the thermal energy of the interacting particles, kT. This takes place when the Debye–Hückel radius is much larger than the average distance between neighboring nuclei. In this regime, the screening factor, f, is given by

$$f = \exp\left[\frac{Z_a Z_b e^2}{R_D kT}\right] \simeq \exp\left[0.188 Z_a Z_b \sqrt{\frac{\rho\zeta}{T_6^3}}\right], \tag{2.68}$$

where T_6 is the temperature expressed in MK. At moderate densities, screening factors in the weak screening regime often translate into a limited enhancement of the corresponding reaction rates. For instance, for typical thermodynamic conditions achieved during nova outbursts (see Chapter 4), $T \sim 10^8$ K and $\rho \sim 10^3$ g cm^{-3}, the screening factor for the reaction ^{12}C(p, γ) is of the order of $f \sim 1.1$.

As the density of the stellar plasma increases, screening effects become progressively more important. In fact, when $E_{\text{Coul}} \gg kT$, the weak screening approximation no longer holds. Analytic fits for strong screening conditions in dense stellar plasmas have been obtained, in the liquid phase, by Itoh et al. [896]. Other analytic fits for the different screening regimes can also be found, e.g., in references [436, 674].

2.3.4 Photodisintegrations

Let's now consider the case in which the incoming projectile is a photon, $\gamma + d \rightarrow a + b$, an interaction known as a photodisintegration. Since photons move at the speed of light, c, the reaction rate can, in principle, be written as (Equation 2.44)

$$r_{\gamma d} = n_d n_\gamma c\, \sigma(E_\gamma). \tag{2.69}$$

However, since the number density of photons depends on the energy, the reaction rate must be properly expressed as

$$r_{\gamma d} = c\, n_d \int_0^\infty n_\gamma(E_\gamma)\sigma(E_\gamma) dE_\gamma, \tag{2.70}$$

from which the corresponding decay constant, $\lambda_\gamma(d)$, or probability of interaction per nucleus and per unit time, can be obtained:

$$\lambda_\gamma(d) = r_{\gamma d}/n_d = c \int_0^\infty n_\gamma(E_\gamma)\sigma(E_\gamma) dE_\gamma. \tag{2.71}$$

Moreover, the energy density of a radiation field containing photons of frequencies ranging between ν and $\nu + d\nu$, at a given temperature T, is determined from Planck's law (Section 1.2.4) as

$$U_\nu(T)d\nu = \frac{8\pi h \nu^3}{c^3} \frac{1}{e^{h\nu/kT} - 1} d\nu. \tag{2.72}$$

Expressing the frequency in terms of the energy of the photon through the relation $E_\gamma = h\nu$ yields

$$U(E_\gamma)dE_\gamma = \frac{8\pi}{(hc)^3} \frac{E_\gamma^3}{e^{E_\gamma/kT} - 1} dE_\gamma, \tag{2.73}$$

and recalling that $n_\gamma(E_\gamma)dE_\gamma \equiv U(E_\gamma)/E_\gamma\, dE_\gamma$, one finally obtains

$$\lambda_\gamma(d) = \frac{8\pi}{h^3 c^2} \int_0^\infty \frac{E_\gamma^2}{e^{E_\gamma/kT} - 1} \sigma(E_\gamma)dE_\gamma. \tag{2.74}$$

Most photodisintegration cross-sections have not been directly measured, since they can be derived from the inverse reaction[25] $c + d \to a + \gamma$. According to the *reciprocity theorem*, which has been experimentally verified (see, e.g., reference [1872]), reactions between particles with spin satisfy [196]

$$k^2(2J_a + 1)(2J_b + 1)\sigma_{ab} = k'^2(2J_c + 1)(2J_d + 1)\sigma_{\gamma d}\,, \tag{2.75}$$

where k and k' are the corresponding wave numbers, J_i is the spin of particle i, and $(2J_i + 1)$ is the total number of states of orientation available for particle i [878, 1232]. In the case of a photon, $(2J_\gamma + 1) = 2$. Therefore,

$$\frac{\sigma_{\gamma d}}{\sigma_{ab}} = \frac{(2J_a + 1)(2J_b + 1)}{2(2J_d + 1)} \frac{2m_{ab}c^2 \, E_{ab}}{E_\gamma^2}. \tag{2.76}$$

Note that since most capture reactions are characterized by $Q > 0$, the corresponding inverse photodisintegrations are obviously endothermic ($Q < 0$).

2.3.5　Neutron-Induced Reactions

Free neutrons have a short half-life, ~ 10 min [159, 331]. Accordingly, only those released through nuclear interactions within this timescale can be present in a stellar plasma. In such environments, neutrons quickly thermalize through mutual interactions, resulting in a Maxwell–Boltzmann velocity distribution. At thermal energies $E \sim kT$, neutron-induced reactions are dominated by (s-wave) neutrons with $l = 0$ angular momentum (i.e., no centrifugal barriers). Interactions with s-wave neutrons are characterized by a cross-section with a simple dependence on energy, given by [344, 878, 1513]:

$$\sigma \sim \frac{1}{\sqrt{E}} \sim \frac{1}{v}\,, \tag{2.77}$$

or equivalently,

$$\sigma v = \text{const} \equiv v_T \sigma_T\,, \tag{2.78}$$

where $v_T = (2kT/m)^{1/2}$ is the most likely velocity at thermal energies (with $m \equiv m_a m_b/(m_a + m_b)$ being the reduced mass of the target-projectile system), and σ_T is the cross-section for thermalized neutrons (corresponding to a velocity v_T; see discussion in Clayton [344]).

Such characteristic cross-sections involving low-energy neutrons translate into reaction rates of the form

$$N_{Av}\langle \sigma v \rangle = N_{Av} \int_0^\infty vP(v)\sigma dv = \text{const.} \tag{2.79}$$

While a single measurement of the cross-section in the thermal energy range would suffice to infer the corresponding reaction rate, several measurements at different energies are usually done to verify whether $\sigma \propto 1/v$. The presence of resonances, for instance, will lead to deviations from such simple velocity dependence (see discussion in reference [878]).

With increasing neutron energy, contribution from $l > 0$ terms (p-waves) are no longer

---

[25] Inverse reactions associated to low Q-value reactions also play an important role in nucleosynthesis.

negligible, and cross-sections that differ from the $1/v$ law (i.e., $\langle \sigma v \rangle \neq$ const) have to be considered. Suitable expressions corresponding to an expansion of $\langle \sigma v \rangle$ in a Taylor series around $E = 0$ in terms of the velocity (or $\sqrt{E}$) can be found, e.g., in references [344, 878, 1513]. Resonances in neutron-induced reactions are handled following the same formalism described for charged-particle interactions, with the effective energy range around the Gamow peak replaced by that corresponding to $E \sim kT$.

2.3.6   Weak Interactions: Electron Captures and β-Decays

Weak interactions, such as electron captures[26], β-decays (leading to electron/positron emission), and neutrino interactions, are part of the rich variety of nuclear processes that occur inside the stars. In radioactive species, the β-decay channels can favorably compete with charged-particle interactions (mediated by the strong nuclear force) as the main destruction mechanisms. Electron captures play also a key role in the collapse of the ONe-rich cores of intermediate-mass stars ($\sim 8 - 10$ M$_\odot$), and in the presupernova evolution of massive stars, where they determine the *neutron excess* of the stellar plasma, or number of excess neutrons per nucleon, defined as $\eta \equiv \sum_i (N_i - Z_i) X_i / M_i$. N_i, Z_i, X_i, and M_i denote the number of neutrons and protons, the mass fraction, and the atomic mass (in atomic mass units) of species i, respectively[27]. Moreover, the interaction between myriad neutrinos emitted in massive stars with the dense stellar plasma plays a key role in the dynamics of type II supernovae (see Chapter 7 for details).

Any species $^A_Z X_N$ undergoing an electron capture or a β-decay mutates into another element, characterized by different atomic and neutron numbers, Z and N, but equal mass number, A, in the form:

$$^A_Z X_N \rightarrow \, ^A_{Z+1} X'_{N-1} + e^- + \bar{\nu}, \qquad\qquad \beta^- - \text{decay} \qquad (2.80)$$

$$^A_Z X_N \rightarrow \, ^A_{Z-1} X'_{N+1} + e^+ + \nu, \qquad\qquad \beta^+ - \text{decay} \qquad (2.81)$$

$$^A_Z X_N + e^- \rightarrow \, ^A_{Z-1} X'_{N+1} + \nu, \qquad\qquad \text{electron capture.} \qquad (2.82)$$

The light particles released (i.e., electrons $[e^-]$, positrons $[e^+]$, and electron neutrinos and antineutrinos $[\nu, \bar{\nu}]$) are all leptons, and as such are not affected by the strong nuclear force. With respect to the energy released, Q, it is shared between the different reaction products. For instance, the overall energy released in a β^+-decay[28] is carried away as kinetic energy by the positron and the electron neutrino, with different fractions, $Q \equiv K_e + E_\nu$. Electron captures, however, release only one lepton, which, consequently, is monoenergetic, $Q \equiv E_\nu$.

Unless the density of the stellar plasma is very large (i.e., $\rho \geq 10^{11}$ g cm^{-3}), the suite of weakly-interacting neutrinos emitted leaves the star mostly unaffected (see Chapter 7 for details). Accordingly, the energy carried away by these neutrinos has to be subtracted from the overall energy content of the star. Estimates of the mean energy losses carried away by neutrinos can be found, for instance, in Fowler, Caughlan, and Zimmerman [546]:

$$E_\nu(\beta) \sim \frac{m_e c^2}{2} x \left(1 - \frac{1}{x^2}\right) \left(1 - \frac{1}{x} - \frac{1}{9x^2}\right), \qquad (2.83)$$

---

[26] At stellar temperatures above $\gtrsim 1$ GK, a large number of photons have energies that exceed the threshold for pair production. Even though positrons quickly annihilate with free electrons in such environments, the pair production rate may become so large that both electron and positron captures by nuclei must be taken into account.

[27] The neutron excess parameter is related to the electron mole fraction, Y_e, by $\eta = 1 - 2Y_e$. For example, a plasma consisting only of nuclei with equal number of protons and neutrons has $\eta = 0$ or $Y_e = 0.5$.

[28] Occasionally, a β-decay may also populate unstable levels in the daughter nucleus. Such levels decay through a suite of possible channels, including emission of light particles (i.e., protons, neutrons, or α-particles), referred to as β-delayed particle decays.

where $x = (1 + Q_\beta/m_e c^2)$ and Q_β is the energy released in a β-decay.

In a stellar plasma, excited states in the parent nucleus become thermally populated, and subsequently undergo energetically-allowed β-decay transitions to the ground or excited states in the daughter nucleus. All in all, the number of possible transitions between daughter and parent nuclei is far more rich in a hot stellar plasma than in the lab[29]. The most relevant transitions in weak interaction processes are known as Gamow–Teller and Fermi transitions[30]. Calculations of different weak interactions for intermediate-mass elements, on the basis of Fermi and Gamow–Teller transitions, have been reported by Fuller, Fowler, and Newman (for $A = 21 - 60$) [576–578], Oda et al. ($A = 17 - 39$) [1341], and Langanke and Martínez-Pinedo ($A = 45 - 65$) [1050, 1051].

Following Oda et al. [1341] and Langanke and Martínez-Pinedo [1050, 1051], in the framework of the general formalism for stellar weak interactions via large-scale shell model calculations, the rate of transition λ_{ij}^{ψ} (in s^{-1}) from the ith state of a parent nucleus into a jth state of the daughter nucleus, can be expressed as

$$\lambda_{ij}^{\psi} = \frac{\ln 2}{K} B_{ij} \Phi_{ij}^{\psi}, \tag{2.84}$$

where ψ is merely a label that specifies the interaction channel (i.e, electron and positron captures, β^+ and β^- decays).

The first term in Equation 2.84 exclusively contains fundamental constants, and can be expressed as

$$\frac{\ln 2}{K} = \frac{G_F^2 V_{ud}^2 g_V^2 m_e^5 c^4}{2\pi^3 \hbar^7}, \tag{2.85}$$

where G_F is the Fermi coupling constant (a measure of the strength of the interaction), V_{ud} is an element in the Cabibbo–Kobayashi–Maskawa matrix of the standard model of particle physics, and $g_V = 1$ is the weak vector coupling constant [159]. Estimates based on Fermi transitions yield a value of $K = 6146 \pm 6$ s [1813].

B_{ij} yields the probability of the nuclear transition, for which Fermi and Gamow–Teller contributions are distinguished, $B_{ij} = B_{ij}(F) + B_{ij}(GT)$. Both $B_{ij}(F)$ and $B_{ij}(GT)$ terms depend upon several nuclear properties (i.e., isospin, coupling constants) whose explicit dependencies can be found, e.g., in Langanke and Martínez-Pinedo [1050, 1051].

Finally, the term Φ_{ij}^{ψ} corresponds to a phase space integral (see again Langanke and Martínez-Pinedo [1050, 1051] or Oda et al. [1341] for suitable expressions for each of the different processes listed in Equations 2.80–2.82), which depends on the total mass and momentum of the emitted particle, the overall energy available in the decay, and the corresponding distribution function[31].

The electron released in a β^--decay feels an attractive Coulomb force near the nucleus

---

[29] β-decay transitions can even occur in a stellar plasma from excited states in the daughter nucleus to the ground or to excited states in the parent nucleus. In this context, the terms *parent* and *daughter* lose somehow their original meaning.

[30] Nuclei can decay through a number of possible transitions. Allowed Gamow–Teller transitions obey a number of selection rules for certain nuclear properties, such as nuclear spins ($\Delta J = |J_p - J_d| = 0$ or 1, where J_p and J_d are the spins of the parent and daughter nuclei, respectively) and parities. A transition characterized by $\Delta J = 1$ implies that angular momentum is carried away by the released electron and antineutrino requiring that their spins are actually parallel. In sharp contrast, conservation of angular momentum during Fermi transitions implies that the spin of the electron and the antineutrino are antiparallel.

[31] At stellar conditions, electrons are well described by a Fermi–Dirac distribution, in the form

$$S_e = \frac{1}{\exp[(E_e - U_e)/kT] + 1}, \tag{2.86}$$

where E_e is the total energy of the electron, and U_e is the chemical potential. Positrons are also described by a Fermi–Dirac distribution, with $U_p = -U_e$.

(repulsive for a positron in a β^+-decay). Accordingly, a correction factor that depends on the charge of the daughter nucleus and the energy (or momentum) of the emitted electron (or positron), known as *Fermi function*[32], is applied to the corresponding phase space integral [1050, 1051, 1341].

At the characteristic physical conditions of stellar plasmas, nuclei frequently occupy excited states with a probability given by

$$P_i = \frac{(2J_i + 1)\exp(-E_i/kT)}{\sum_l (2J_l + 1)\exp(-E_l/kT)}, \tag{2.87}$$

where E_i and J_i are the excitation energy and angular momentum of state i. Note that the sum extends over all possible states, l. The total rate of transition[33] through process ψ in the parent nucleus is therefore given by

$$\lambda^\psi = \sum_{ij} P_i \lambda_{ij}^\psi = \frac{\ln 2}{K} \sum_i \frac{(2J_i + 1)e^{-E_i/kT}}{G(Z,A,T)} \sum_j B_{ij} \Phi_{ij}^\psi, \tag{2.88}$$

where again i and j refer to states in the parent and daughter nuclei, and $G(Z,A,T) \equiv \sum_l \exp(-E_l/kT)$ is known as the partition function of the parent nucleus.

It is finally worth noting that in the laboratory, parent nuclei usually remain at ground states. The corresponding decay rates, λ, can be obtained from the transition probabilities for all the allowed decay branches. Such rates are both temperature- and density-independent. In sharp contrast, stellar rates depend on multiple states of the parent nuclei, which are thermally populated. Therefore, stellar decay rates are both temperature- and density-dependent, and in general, differ from the corresponding laboratory values, sometimes dramatically[34]. Note, however, that the ratio of stellar to laboratory decay rates is approximately equal to the ratio of the electron densities[35] at the nucleus for the stellar and laboratory environments, and therefore stellar rates can be estimated from known laboratory decay rates [878].

2.4 Stellar Evolution in a Nutshell: H- to Si-Burning

Stars form by gravitational contraction in huge, interstellar molecular clouds, typically composed of hydrogen, helium, and traces of other species. If the mean kinetic energy of a cloud is smaller than half of its overall gravitational potential energy, gravity dominates over gas pressure and the cloud begins to contract[36]. This condition points toward massive

---

[32]See reference [673] for an extended tabulation of Fermi functions for $Z = 6 - 95$ and transition energies ranging from 0.01 MeV to 10 MeV.

[33]Alternative expressions for transition rates rely on *ft-values*, which can be experimentally obtained from measurements of half-lifes and energies of the emitted electrons or positrons (see http://www.nndc.bnl.gov/logft/ for a database of log ft-values, and Sigh et al. [1660] for a review on this subject).

[34]This, in turn, affects the corresponding half-life of a radioactive species, $T_{1/2} \equiv \ln 2/\lambda$ (see Equation 2.8). For instance, radioactive ^{26}Al (ground state) decays into ^{26}Mg with a half-life of $T_{1/2} \equiv \ln 2/\lambda = 7.2 \times 10^5$ yr at laboratory conditions. This corresponds to the stellar value for $T \leq 50$ MK, but decreases as the temperature of the plasma increases: 3.6×10^5 yr at 100 MK, 5.5 days at 300 MK, and just 17 min at 1 GK (values from Fuller et al. [577, 578]).

[35]The electron density of a stellar plasma, n_e, is given by $n_e = N_{Av}\rho(1 - \eta)/2$.

[36]According to the virial theorem, the gravitational potential energy of a bound system in hydrostatic equilibrium, $\overline{V}$, and its pressure, P, satisfy the condition $\overline{V} + 3\int_0^M (P/\rho)dm = 0$. This relation results from

TABLE 2.2

Nuclear Burning Stages for a 20 $M_\odot$ Star, Based on
Models by Woosley et al. [1977]

Burning stage	T (GK)	ρ (g cm^{-3})	Main products	Timescale
H	0.037	4.5	He	8.1 Myr
He	0.19	970	C, O	1.2 Myr
C	0.87	1.7×10^5	O, Ne	2800 yr
Ne	1.6	3.1×10^6	O, Mg	600 yr
O	2.0	5.6×10^6	Si, S	1.3 yr
Si	3.3	4.3×10^7	Fe, Ni	12 d

and cold clouds as the likely *stellar nurseries*. The critical mass required for a cloud to undergo collapse, known as the *Jeans mass* [908], depends on its specific temperature and density, but typically[37] ranges between 1000–10,000 $M_\odot$.

During the early stages of the collapse, the gas is optically thin (i.e., transparent to radiation). Accordingly, the gravitational potential energy released is mostly radiated away, and the cloud remains almost isothermal. The existing pressure gradients are too small to have an effect, and, therefore, the collapse proceeds basically in free-fall conditions (see Section 1.6.7). Pioneering models of these early contraction stages in the life of stars were developed in the 1950–1960s by Levée [1081], Henyey et al. [766], Hayashi [744], and Ezer and Cameron [510]. The simulations assumed a *homologous collapse*, with a constant gas temperature and a uniform density across the proto-star. Density contrasts are, however, expected within interstellar clouds, with regions characterized by slightly larger densities than their surroundings. Those regions become seeds for stellar formation as contraction of the cloud initiates and the existing density gradients get progressively amplified by effect of the gravitational force. Soon, structures that somewhat resemble stars begin to appear, with inner, higher-density layers—i.e., stellar proto-cores—surrounded by lower-density material.

As shown by Iben [869] and Ezer and Cameron [509], who pioneered the first systematic studies of nonhomologous gravitational contraction[38] for a wide range of stellar masses, such density differences translate into a faster contraction of the stellar proto-cores than the surrounding material. As the gravitational collapse proceeds, the stellar proto-core progressively becomes optically thick[39]. Radiation gets trapped and the temperature in the innermost layers of the proto-star begins to increase. Accordingly, the collapse evolves from isothermal to adiabatic conditions. The temperature rise is accompanied by a pressure increase. The reinforced pressure gradient substantially slows down the collapse in the central regions of the proto-star, which nearly achieve hydrostatic equilibrium. The surrounding layers, still in free-fall, proceed basically unaffected by the structural changes that have occurred below, and at some point, collide with the slowly contracting proto-core, generating a shock wave.

The continuous infall of material progressively heats the stellar proto-core. At about $T \sim 2000$ K, dissociation of molecular hydrogen, followed by ionization of the very abundant hydrogen and helium atoms, occurs in those inner layers. Since part of the thermal energy of the system is invested to this end, temperature and pressure drop, and, consequently,

---

the integration of Equation 1.26 for $\partial^2 r / \partial t^2 = 0$, after multiplication by $4\pi R^3$ (see, e.g., reference [988]). For an ideal gas, the virial theorem can be expressed as $2\overline{K} + \overline{V} = 0$, where $\overline{K}$ is the mean kinetic energy of the system. Therefore, systems characterized by $2\overline{K} > \overline{V}$ expand, while those with $2\overline{K} < \overline{V}$ contract.

[37] See, e.g., references [296, 872, 1633] for a full derivation of the Jeans mass.

[38] See also references [208, 210, 508, 745, 748, 1053].

[39] For a 1 $M_\odot$ proto-star, this occurs at a density of $\sim 10^{-13}$ g cm^{-3} [1053].

the collapse of the stellar proto-core reaccelerates. A second shock wave forms when the shrinking proto-core reestablishes a suitable pressure gradient, slows down, and is hit again by the flow of infalling material. This continues until the protostar achieves hydrostatic equilibrium, after ~ 0.01–100 Myr of gravitational contraction [869]. Remarkably, while planets reach equilibrium configurations at relatively low central temperatures (e.g., 2400 K for Mercury, 36,000 K for Jupiter), stellar core temperatures are high enough to ignite thermonuclear reactions. Therefore, and in sharp contrast to planets, stars have an internal energy source.

Thermonuclear reactions initiate short fusion episodes during the gravitational contraction of a star, when the central temperature reaches $T \sim 10^6$ K. With plenty of hydrogen and helium present in the stellar plasma, one may expect dominant nuclear interactions of the type p + p $\to$ ^{2}He + γ, or p + ^{4}He $\to$ ^{5}Li + γ. However, both reactions are inhibited by the lack of bound ^{2}He and ^{5}Li configurations. Instead, the main nuclear activity at these early stages involves proton-capture reactions by other light nuclei, such as d, Li, Be, and B [206, 207, 1559]. These reactions are characterized by larger cross-sections than the alternative processes by which hydrogen burning actually occurs (see Section 2.4.1). However, they involve extremely underabundant species, and, therefore, such nuclear fuels get rapidly consumed.

The first true burning stage[40] in the life of a star is *hydrogen burning*. Hydrogen has the lowest Coulomb potential, and, therefore, hydrogen burning requires the lowest temperatures ($\sim 10^7$ K) of all stellar burning stages[41]. During central H-burning, stars occupy a characteristic band, known as *main sequence*, on plots of absolute magnitude (or luminosity) versus effective temperature (see Chapter 1) known as *Hertzsprung–Russell* (H–R) *diagrams*. Since hydrogen is the main component of newly born stars, and because its fusion proceeds through a slow chain of reactions, H-burning constitutes the longest stage in the life of a star (see Table 2.2).

When hydrogen is consumed near the center, the star departs from hydrostatic equilibrium and therefore undergoes structural changes (Figure 2.9). The central regions recontract under the influence of gravity, while the H-rich layers that surround the core ignite. This causes the star to leave the main sequence as it gradually grows in size and becomes a red giant (reaching 100–1000 R$_\odot$). Meanwhile, the temperature in the contracting core steadily increases, to a point where helium, a species with the lowest Coulomb barrier after hydrogen, may fuse, provided that the core achieves the required temperature ($\sim 10^8$ K). The energy released through helium burning stabilizes the star against further contraction.

Stars may undergo several burning regimes, following a complex pattern of subsequent contraction and equilibrium stages. This strongly depends on the initial mass and composition of the star (see Figure 2.10 for an overview of solar composition stars). However, the specific mass cuts between the different regimes are not well established and significant differences can be found in the literature. Roughly speaking, stars up to ~ 0.5 M$_\odot$ only burn hydrogen; stars up to ~ 7–8 M$_\odot$ undergo H-burning, followed by He-burning; stars up to ~ 10 M$_\odot$ experience three different burning stages: H-, He-, and C-burning. Once nuclear burning comes to completion, low- and intermediate-mass stars evolve into compact, inert bodies of planetary size known as *white dwarfs*[42]. Note, however, that white dwarfs can be potentially revitalized through mass transfer episodes if the progenitor stars are not isolated but form a close binary system[43]. The actual decoupling between core and envelope,

---

[40]In astrophysics, "burning" traditionally refers to thermonuclear fusion rather than chemical combustion, as the term may actually suggest.

[41]The minimum mass required to power hydrogen burning is about 0.08 M$_\odot$. Note that reaction rates down to 10^7 K are needed for stellar evolution.

[42]Indeed, the Sun, after H- and He-burning, will evolve into a CO-rich white dwarf of about 0.6 M$_\odot$ [1555].

[43]Note that the mass cuts given in this section rely on single stellar evolution. Different values result from

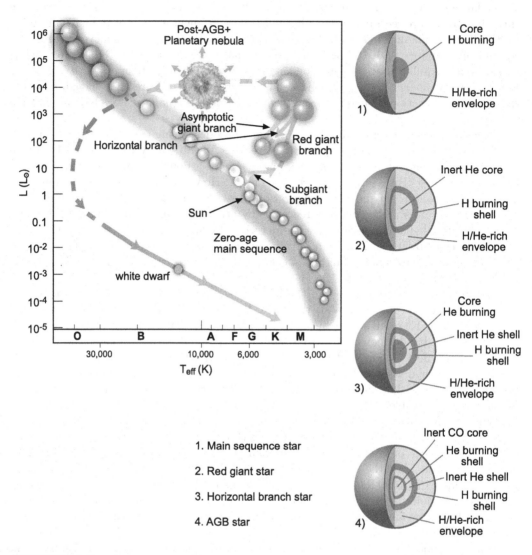

FIGURE 2.9

Evolution of the Sun in an H–R diagram, from core H-burning to the white dwarf stage. The life of any star begins with core H-fusion (sphere 1), while occupying a characteristic location on the diagram: the main sequence. In a star like the Sun, H will be depleted near the center after ~ 10 Gyr. Following recontraction, H will ignite in the shells that surround the He-rich core, leading to a dramatic increase in size, approximately reaching the orbit of the Earth. Its path along the H–R diagram will subsequently follow the subgiant and red giant branches, while the Sun becomes a giant star (2). The temperature rise at the center during recontraction will eventually lead to core He-ignition. Its evolution will continue through the horizontal branch (3). After 1 Gyr, helium will be depleted at the center and double H- and He-shell burning will set in. The Sun, by then a bright star (4), will climb the asymptotic giant branch (AGB). The late stages will be characterized by a gentle ejection of the outer layers at $v \sim 50 - 100$ km s^{-1}, giving rise to a planetary nebula, while the inner, CO-rich 0.6 $M_\odot$ of the Sun will smoothly evolve into a stable, white dwarf configuration. A color version of this figure is available at http://fisica.upc.edu/ca/users/jjose/CRC-Downloads.

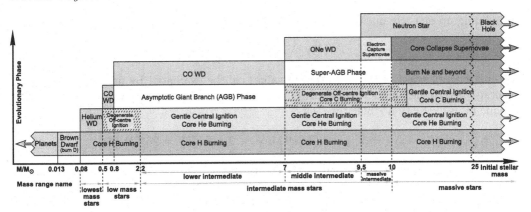

FIGURE 2.10

Summary of the main burning stages and end products of stellar evolution as a function of the stellar mass, based on solar metallicity models. A color version of this figure is available at http://fisica.upc.edu/ca/users/jjose/CRC-Downloads. Figure adapted from Karakas and Lattanzio [949]; reproduced with permission.

already outlined in the discussion of post-main sequence evolution, also continues during helium (and carbon) burning. As a result, the outer layers of the star are gently expelled into the interstellar medium giving rise to a *planetary nebula*, while the stellar core evolves into a white dwarf configuration.

Stars above ~ 10 $M_\odot$ essentially follow all possible burning stages (from H- to Si-burning). Silicon burning synthesizes iron, one of the most tightly bound nuclei (see Figure 2.3). Accordingly, this represents the final nuclear burning stage, since a star cannot obtain additional energy any longer by fusion reactions on Fe. Therefore, while stellar nucleosynthesis up to Fe proceeds through a handful of nuclear burning stages, synthesis of heavier elements requires a totally different mechanism. Indeed, as discussed in Chapter 7, this actually occurs through neutron-capture reactions on seed nuclei, in the so-called s- and r-processes. The final outcome of a $M > 10$ $M_\odot$ star is a *type II supernova explosion*, accompanied by the formation of another compact remnant, either a *neutron star* or a *black hole* (see Chapter 7).

All in all, stellar evolution describes the succession of different burning and contraction episodes that occur along the life of a star, during its search for an energy source that could temporarily maintain its equilibrium with gravity. The timespan of the different stages, from H- to Si-burning, dramatically shortens, as the star progressively relies on more tightly bound fuel, and therefore less energy is released (e.g., the fusion of one gram of hydrogen yields $\sim 4 \times 10^{18}$ erg, while carbon and oxygen burning release about $(4-5) \times 10^{17}$ erg per gram. See Table 2.2 and Sections 2.4.1–2.4.3).

2.4.1 Hydrogen Burning

H-burning in stellar plasmas proceeds through two basic mechanisms, known as the *proton-proton* (or pp) *chains* and the *CNO cycles*[44] [173,177,1871], under hydrostatic ($T \lesssim 50$ MK) or explosive conditions. Through any of these processes, four protons are globally consumed

---

evolution in binary systems, which strongly depend on the adopted recipes for mass transfer [141,871,898, 1437].

[44]See Shaviv [1622] for a detailed account of the discovery of the mechanisms involved in H-burning in stars.

in building up one ^{4}He nucleus,

$$4\,^1\mathrm{H} \rightarrow {}^4\mathrm{He} + 2e^+ + 2\nu. \tag{2.89}$$

The net energy released (i.e., reaction Q-value) can be evaluated by means of the experimentally determined mass excesses through Equation 2.15. Adopting values from the 2012 atomic mass evaluation [1899], one gets

$$Q = 4\,\Delta_{\mathrm{H}} - \Delta_{\mathrm{He}} = 4\,(7.289\,\mathrm{MeV}) - 2.425\,\mathrm{MeV} = 26.731\,\mathrm{MeV}, \tag{2.90}$$

which includes the energy released in the annihilation between the two positrons emitted and two electrons from the environment, $2m_e c^2 = 1.022$ MeV. A fraction of the overall 26.731 MeV released is carried away by the emitted neutrinos, and therefore is not retained locally in the plasma.

2.4.1.1   Proton–Proton Chains

The pp chains, as the name may suggest, begin with the reaction $p + p$. But as mentioned before, the lack of a bound ^{2}He configuration prevents the ocurrence of the strong interaction $\mathrm{p} + \mathrm{p} \rightarrow {}^2\mathrm{He} + \gamma$. Instead, hydrogen burning proceeds through an alternative channel, $\mathrm{p} + \mathrm{p} \rightarrow \mathrm{d} + e^+ + \nu$, in which two protons fuse into a deuterium nucleus. The reaction, as suggested by Atkinson [86] and Bethe and Critchfield [177], relies on the conversion of a proton into a neutron, in a process that mimics a β-decay. The p(p, $e^+\nu$)d reaction is actually mediated by the weak nuclear force, in sharp contrast with most of the nuclear interactions that occur in stellar plasmas, which are ruled by the electromagnetic and strong forces. This dramatically slows down its rate. For illustrative purposes, consider, for instance, the thermodynamic conditions at the center of the Sun, characterized by $T_c = 15$ MK, $\rho_c = 150$ g cm^{-3}, and a hydrogen mass fraction of $X = 0.7$. Theoretical estimates of the p(p, $e^+\nu$)d reaction rate at 15 MK yield a value of $N_{\mathrm{Av}}\langle\sigma v\rangle = 8.1 \times 10^{-20}$ cm^3 s^{-1} mol^{-1} [50]. Accordingly, the characteristic timescale for a proton-proton encounter in the solar interior can be estimated as

$$\tau = \left(\rho\frac{X}{M}N_{\mathrm{Av}}\langle\sigma v\rangle\right)^{-1} = \left(150\frac{0.7}{1.0078}8.1 \times 10^{-20}\right)^{-1} = 1.2 \times 10^{17}s, \tag{2.91}$$

which implies that a proton in the center of the Sun wanders for about 4 Gyr before fusing with another proton. This reaction, in fact, constitutes the slowest link in the pp chains, and, accordingly, dictates the rate at which nuclear energy is actually released through these chains.

An alternative channel to p(p, $e^+\nu$)d is the *pep reaction* $\mathrm{p} + \mathrm{p} + e^- \rightarrow \mathrm{d} + \nu$. Calculations have proved that this process, which involves an electron capture aside from the fusion of two protons, can only compete with p(p, $e^+\nu$)d at high stellar densities, above 10^4 g cm^{-3} [110]. Accordingly, the pep reaction has been frequently ignored in the literature, as it only plays a marginal role in energy production for main sequence stars. However, it may be relevant in other scenarios characterized by much larger densities, such as classical novae and X-ray bursts (see, e.g., Chapter 4).

At the high temperatures that characterize stellar plasmas, the deuterium synthesized by the p(p, $e^+\nu$)d reaction is quickly destroyed, mostly through d(p, γ)^{3}He. Indeed, at the solar central temperature of $T = 15$ MK, this reaction reaches a rate of $N_{\mathrm{Av}}\langle\sigma v\rangle \sim 1.1 \times 10^{-2}$ cm^3 s^{-1} mol^{-1}, which implies that a deuteron can actually survive in this environment for about a second[45] before fusing with a proton. It is therefore clear that the chain p(p,

---

[45]Note that other channels, such as d(d, p)t, have larger cross-sections ($N_{\mathrm{Av}}\langle\sigma v\rangle \sim 240$ cm^3 s^{-1} mol^{-1},

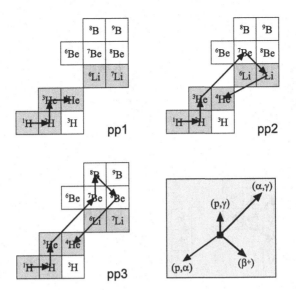

FIGURE 2.11
H-burning through the pp chains. According to the Standard Solar Model [108], the pp chains proceed in the solar interior through chain 1 about 85% of the times, 15% through chain 2, and only 0.02% through chain 3. The horizontal arrow that connects H and d corresponds to p(p, $e^+\nu$)d, while those that connect ^{7}Li and ^{8}Be with ^{4}He correspond to the decays ^{7}Li(p, α)α and ^{8}Be(α)α, respectively. A color version of this figure is available at http://fisica.upc.edu/ca/users/jjose/CRC-Downloads.

$e^+\nu$)d(p, γ)^{3}He is strongly conditioned by the very slow p + p reaction that constitutes the true bottleneck for the process.

The pp1 chain. Different sequences of nuclear processes can occur in a stellar plasma following the synthesis of ^{3}He through the chain p(p, $e^+\nu$)d(p, γ)^{3}He (see Figure 2.11). In principle, the simplest choice relies on nuclear interactions with the very abundant protons, such as ^{3}He + p $\rightarrow$ ^{4}Li + γ. However, the process is hindered by the particle-unbound ^{4}Li nucleus. A reliable alternative is the ^{3}He(^{3}He, 2p)^{4}He reaction, which completes the conversion of four protons to one ^{4}He nucleus through the so-called *pp1 chain*.

The pp2 chain. Other sets of reactions can compete favorably with ^{3}He(^{3}He, 2p)^{4}He in plasmas characterized by large concentrations of ^{4}He and sufficiently large temperatures. In such conditions, ^{3}He can efficiently undergo an α-capture, ^{3}He(α, γ)^{7}Be, followed by an electron capture on ^{7}Be and a (p, α) reaction on the resulting ^{7}Li nucleus, in the form ^{7}Be(e^-, n)^{7}Li(p, α)^{4}He. This completes the sequence of reactions known as the *pp2 chain*. All in all, as for the *pp1 chain*, four protons are transformed into one ^{4}He nucleus (mediated by the presence of an α particle, that acts as a catalyzer for the process).

The pp3 chain. At even higher temperatures, proton captures on ^{7}Be become faster than electron captures. This gives rise to another sequence of reactions known as the *pp3 chain* (Figure 2.11). The ensuing ^{7}Be(p, γ)^{8}B reaction leads to the synthesis of the short-

---

at T = 15 MK) than the d(p, γ)^{3}He reaction. However, as shown by Equation 2.44, the overall reaction rate depends also on the abundances of the interacting species. Therefore, the much lower deuterium abundance compared with hydrogen makes the d + d channel almost irrelevant.

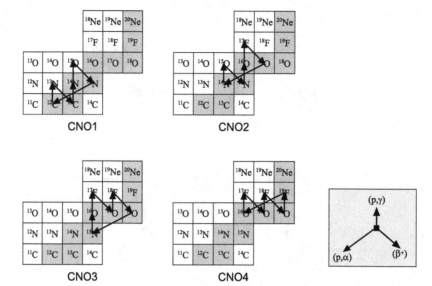

FIGURE 2.12
H-burning through the cold CNO cycles. A color version of this figure is available at
http://fisica.upc.edu/ca/users/jjose/CRC-Downloads.

lived unstable nucleus ^{8}B ($T_{1/2} \sim 770$ ms), which undergoes a β^+-decay into ^{8}Be, followed
by its subsequent breakup, in the form ^{8}Be $\rightarrow$ ^{4}He + ^{4}He.

The specific contribution of the different pp chains to the energy production and nucleosynthesis strongly depends on the composition and thermodynamic conditions of the plasma (see caption in Figure 2.11, for solar conditions). Note that while the overall energy released through any of the chains amounts to 26.731 MeV, the energy carried away by the suite of neutrinos emitted differs substantially. Adopting the neutrino energies from Bahcall [108], one gets

$$Q_{pp1} = 26.73\,\text{MeV} - 2E(pp) \qquad\qquad = 26.19\,\text{MeV} \qquad (2.92)$$

$$Q_{pp2} = 26.73\,\text{MeV} - E(pp) - E(^7\text{Be}) \qquad = 25.65\,\text{MeV} \qquad (2.93)$$

$$Q_{pp3} = 26.73\,\text{MeV} - E(pp) - E(^8\text{B}) \qquad = 19.75\,\text{MeV}. \qquad (2.94)$$

Direct cross-section measurements have been performed for some of the pp chain reactions, frequently at energies above the solar Gamow peak[46] (see reference [878] for details). While useful for evolutionary sequences of stars more massive than the Sun, extrapolation down to the energy range of interest is actually required for solar models[47].

2.4.1.2　CNO Cycles

While energy production in main sequence stars with masses similar to the Sun could be accounted for by the pp chains, more massive stars require a more temperature-dependent mechanism to explain their luminosity. In 1939, Hans Bethe[48] established that proton cap-

---

[46]Exceptions include ^{3}He(^{3}He, 2p)^{4}He [80] and d(p, γ)^{3}He [302], which have been measured at solar Gamow peak energies.

[47]Uncertainties on the S-factor reach up to $\sim 8\%$ near the solar Gamow peak, for reactions such as d(p, γ)^{3}He, ^{7}Be(p, γ)^{8}B, and ^{14}N(p, γ)^{15}O.

[48]See also von Weizsäcker [1871].

ture reactions on C and N nuclei, forming somehow a cycle, could represent the *"most important source of energy in ordinary stars"* [173]. The original *CN cycle* depicted by Bethe was further extended to a set of cycles collectively referred to as *CNO cycles*.

The four *cold* CNO cycles of H-burning that occur under hydrostatic conditions are depicted in Figure 2.12. They consist of a combination of p-capture reactions on stable species and β^+-decays of unstable nuclei (e.g., ^{13}N, ^{15}O, ^{17,18}F). The cycles are closed by energetically allowed (p, α) reactions on ^{15}N, ^{17,18}O, and ^{19}F. Each of these nuclei[49] define a branching point, where the nuclear flow can in principle split according to the competing (p, γ) and (p,α) channels, with a probability given by the ratio of the corresponding reaction rates, or *branching ratio*,

$$B_{\mathrm{p}\alpha/\mathrm{p}\gamma} = \frac{\langle \sigma v \rangle_{(\mathrm{p},\alpha)}}{\langle \sigma v \rangle_{(\mathrm{p},\gamma)}}. \tag{2.95}$$

Each cycle requires the presence of CNO-group nuclei[50] that act as catalyzers. All in all, as in the pp chains, four protons are transformed into one ^{4}He nucleus (mediated by the presence of CNO catalyzers), 4^1H $\rightarrow$ ^{4}He $+ 2e^+ + 2\nu$, with an overall energy release through any of the cycles of 26.731 MeV. The CNO1 (or CN) is, by far, the most relevant among the four CNO cycles, as it involves reactions between species with smaller Coulomb barriers than in any of the other cycles (i.e., ^{12,13}C + p and ^{14}N + p, compared, e.g., with ^{16,17,18}O + p), and therefore, have a larger probability of occurrence.

Detailed calculations show that, for the thermodynamic conditions at the center of the Sun, more than 98.5% of the total energy output is supplied by the pp chains, with only a tiny 1.5% contribution from the CN cycle [111]. This dominance of the pp chains over the CNO cycles extends to temperatures below $T \leq 19$ MK, for a plasma of solar composition [878]. Since the central temperature of a star increases with its mass, this suggests that stars slightly more massive than the Sun (for which $T_{\mathrm{center}} = 15.7$ MK) are predominantly powered by the CNO cycle, with a mass cut around ~ 1.3 M$_\odot$ [1555]. Therefore, in main sequence stars like Vega ($M \sim 2.1$ M$_\odot$) or Altair ($M \sim 1.8$ M$_\odot$), to quote a few of the brightest, visible stars, hydrogen burning proceeds through CNO-cycle reactions. This division between stars powered mostly by pp chains or CNO-cycle reactions has other important consequences. The energy released through pp chains can be efficiently carried by radiation throughout a stellar plasma characterized by moderate temperature gradients. But in sharp contrast, the large temperature gradients established by the highly temperature-dependent CNO-cycle reactions[51], which reach superadiabatic values, power convection as the main energy transport mechanism (see Chapter 1). This in turn, modifies the chemical abundance profile throughout the star, as bubbles of nuclear processed material are efficiently carried by convection toward the outer, cooler layers of the star[52].

For constant temperature, and assuming that all CNO breakout channels are inhibited (i.e., a closed cycle), the net effect of the CNO1 cycle is recycling of CNO-group nuclei into ^{14}N. This is ruled by the rate of ^{14}N(p, γ)^{15}O, the slowest nuclear process in the cycle for $T < 0.1$ GK.

As in the pp chains, a number of neutrinos are released through the operation of the

---

[49]Proton-induced reactions on unstable nuclei are negligible compared to the much faster β^+-decay channels, for hydrostatic H-burning conditions.

[50]Accordingly, primordial stars, almost devoid of CNO nuclei, were exclusively powered by the pp chains during H-burning.

[51]At the Sun's central temperature, the energy generation rate through the pp chains scales as $\propto T^4$, while a dependence $\propto T^{20}$ is inferred for the CN cycle (see a full derivation in Iliadis [878]).

[52]The same mechanism powers convective energy transport in classical nova outbursts. See Chapter 4.

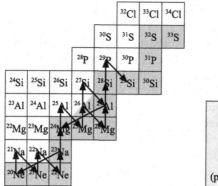

FIGURE 2.13
H-burning in the NeNa–MgAl mass region. Two different states for ^{26}Al—the ground state and a short-lived, isomeric state—are actually distinguished. See text for details. A color version of this figure is available at http://fisica.upc.edu/ca/users/jjose/CRC-Downloads.

different cycles. This translates into a net energy production per cycle given by

$$Q_{CNO1} = 26.73\,\text{MeV} - E(^{13}\text{N}) - E(^{15}\text{O}) \qquad\qquad = 25.03\,\text{MeV} \qquad (2.96)$$

$$Q_{CNO2} = 26.73\,\text{MeV} - E(^{15}\text{O}) - E(^{17}\text{F}) \qquad\qquad = 24.74\,\text{MeV} \qquad (2.97)$$

$$Q_{CNO3} = Q_{CNO4} = 26.73\,\text{MeV} - E(^{17}\text{F}) - E(^{18}\text{F}) \qquad = 25.35\,\text{MeV}, \qquad (2.98)$$

where neutrino energies have been adopted from Bahcall [108], Bahcall and Ulrich [114], and Serenelli and Fukugita [1607].

Most of the reactions of the CNO cycles have been measured at energies above the Gamow peak for hydrostatic conditions, requiring extrapolation of the experimental data down to the energy range of astrophysical interest.

2.4.1.3 H-Burning beyond the CNO Mass Region

H-burning may also involve intermediate-mass elements located beyond the CNO-mass region. At the moderate temperatures that characterize hydrostatic H-burning, CNO-breakout is inhibited by energetically favored (p, α) reactions. Therefore, the nuclear activity above $A \geq 20$ has to be entirely supplied by the presence of preexisting seed nuclei in the stellar plasma (e.g., ^{20,22}Ne, ^{28}Si, ^{24}Mg, ^{32}S).

The main interaction channels correspond to (p, γ) and (p, α) reactions on stable NeNa–MgAl isotopes and to the β^+-decays of unstable species, as depicted in Figure 2.13. As in the CNO cycle, proton-induced reactions on unstable nuclei are much slower than the competing β^+-decay channels. An exception is the radioactive species ^{26}Al, which can be destroyed by both (p, γ) and β^+-decay interactions because of its long half-life ($T_{1/2} = 7.2 \times 10^5$ yr). At temperatures below $T < 0.4$ GK, two different ^{26}Al states must actually be distinguished: the ground state, ^{26g}Al, and an isomeric, short-lived state at $E_x = 228$ keV, ^{26i}Al, characterized by a half-life, $T_{1/2} = 6.4$ s (see Chapters 4 and 7).

The possible existence of closed NeNa and MgAl cycles (and even SiP and SCl cycles) has been repeatedly claimed in the literature, on the basis of energetically allowed (p, α) reactions on ^{23}Na and ^{27}Al, respectively. Unfortunately, the branching ratios $B_{p\alpha/p\gamma}$ for these specific species are not precisely known, and the corresponding uncertainties prevent

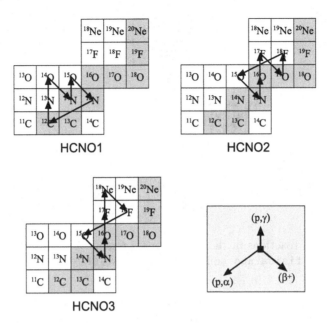

FIGURE 2.14

Explosive hydrogen burning through the hot CNO cycles. A color version of this figure is available at `http://fisica.upc.edu/ca/users/jjose/CRC-Downloads`.

firm statements in this regard. None of the NeNa–MgAl reactions have been measured at the corresponding Gamow peak energies for hydrostatic conditions.

2.4.1.4 Explosive Hydrogen Burning

When the stellar plasma achieves a temperature $T \gtrsim 0.05$ GK, H-burning switches into a regime characterized by fast proton-captures on unstable nuclei. Nucleosynthesis and energy generation mostly takes place through modified versions of the CNO cycles[53] and reactions in the NeNa–MgAl mass regions. This regime is achieved, for instance, in a number of stellar explosions, where H-burning plays a relevant role, like classical novae and a subset of type I X-ray bursts, as described in the following chapters.

The Hot CNO Cycles. The explosive or *hot* mode of the CNO cycles (HCNO) is depicted in Figure 2.14. Overall, the net result is, as in the cold cycles, the transformation of four hydrogen nuclei into one helium nucleus (mediated by the presence of CNO(F) nuclei that act as catalyzers for the process). But differences show up when the specific reactions involved in the hot and cold modes are scrutinized. In cycle 1, for instance, the reaction sequence $^{13}N(\beta^+)^{13}C(p, \gamma)^{14}N$ is bypassed in the hot HCNO1 cycle by $^{13}N(p, \gamma)^{14}O(\beta^+)^{14}N$. In the same way, the sequence $^{17}F(\beta^+)^{17}O(p, \gamma)^{18}F(\beta^+)^{18}O$ of the cold CNO3 cycle is overcome by $^{17}F(p, \gamma)^{18}Ne(\beta^+)^{18}F(p, \alpha)^{15}O$ in HCNO3. This reflects that at the larger temperatures that characterize explosive H-burning, proton-capture reactions on unstable nuclei (i.e., ^{13}N, $^{17,18}F$) become faster than the corresponding decay channels[54].

---

[53]See, e.g., reference [79] for a description of hot pp chains.
[54]Note also that in HCNO2, $^{17}O(p, \gamma)$ becomes faster than $^{17}O(p, \alpha)$.

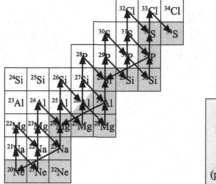

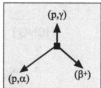

FIGURE 2.15
Explosive H-burning reactions in the $A \geq 20$ mass region. A color version of this figure is available at `http://fisica.upc.edu/ca/users/jjose/CRC-Downloads`.

Another consequence of the higher temperatures involved in the HCNO cycles shows up in the energy production. While the slowest links in the cold CNO cycles are proton-capture reactions (e.g., ^{14}N(p, γ) in CNO1), the hot CNO cycles are limited by β-decays (^{14,15}O in HCNO1), and hence, energy production becomes temperature independent. Accordingly, they are often referred to as β-limited CNO cycles. Moreover, above T $\geq$ 0.4 GK, leakage from the hot CNO cycles proceeds through a number of breakout reactions.

The energy production per cycle is given by

$$Q_{HCNO1} = 26.73\,\text{MeV} - E(^{14}\text{O}) - E(^{15}\text{O}) \qquad\qquad = 24.83\,\text{MeV} \qquad (2.99)$$

$$Q_{HCNO2} = 26.73\,\text{MeV} - E(^{15}\text{O}) - E(^{17}\text{F}) \qquad\qquad = 24.74\,\text{MeV} \qquad (2.100)$$

$$Q_{HCNO3} = 26.73\,\text{MeV} - E(^{15}\text{O}) - E(^{18}\text{Ne}) \qquad\qquad = 24.03\,\text{MeV}, \qquad (2.101)$$

where neutrino energies have been adopted from Bahcall [108] and Bahcall and Ulrich [114].

Experimental data at energies of astrophysical interest is available for most of the reactions that involve stable CNO nuclei. Direct measurements have also been performed for the reaction ^{13}N(p, γ), the first nuclear reaction ever measured with a radioactive ion beam, at energies, however, above the Gamow window [426]. Other direct experiments with unstable CNO(F) nuclei include ^{17}F(p, γ) [327] and ^{18}F(p, α) [123, 147]. The astrophysical interest of some of these rates, together with examples of the operation of the hot CNO cycles, are described, e.g., in Chapter 4.

Explosive H-Burning Beyond the CNO Mass Region. The hot counterpart of the H-burning reactions that involve intermediate-mass elements beyond the CNO-mass region is shown in Figure 2.15. A close inspection reveals far more proton-capture reactions and β^+-decays than for hydrostatic conditions (see Figure 2.13). Unless the plasma exceeds $T > 0.4$ GK, no significant leakage from the CNO-mass region actually occurs, and all the nuclear activity in the $A > 20$ mass region requires once more the presence of preexisting seed nuclei. Closed MgAl, SiP and SCl cycles do not exist, at least below 0.4 GK, while a closed NeNa cycle may develop only at $T \sim 0.1$ GK [1538].

Most of the reaction rates have only been measured by means of indirect techniques, frequently above the energies of astrophysical interest. Exceptions include ^{21}Na(p, γ)^{22}Mg, which plays a key role in the synthesis of ^{22}Na during nova explosions (Chapter 4). This is the first radiative capture reaction ever measured in the nova Gamow window with radioactive

(a)

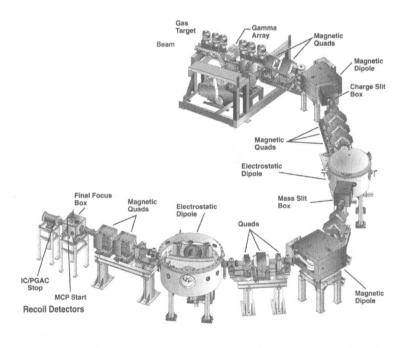

(b)

FIGURE 2.16

(a) DRAGON, a recoil mass separator for the study of nuclear reactions of astrophysical interest, located in the ISAC facility at TRIUMF (Vancouver). Image courtesy of Steven Oates. (b) Schematic layout of the DRAGON facility. Color versions of these figures are available at http://fisica.upc.edu/ca/users/jjose/CRC-Downloads.

ion beams, in a direct experiment performed at the TRIUMF facility in Vancouver (Figure 2.16). Direct measurements of $^{23}\text{Mg}(p, \gamma)^{24}\text{Al}$ [502] and $^{22}\text{Na}(p, \gamma)^{23}\text{Mg}$ [1556] have also been performed at TRIUMF and CENPA (Seattle), respectively.

2.4.2 Helium Burning

2.4.2.1 Hydrostatic He-Burning

When hydrogen is exhausted at the core, the star recontracts in search of another equilibrium configuration. For stars with $M \gtrsim 0.5$ M$_\odot$, the temperature increase following core recontraction leads to helium ignition and nuclear reprocessing into ^{12}C and ^{16}O, the two most abundant elements in the universe after H and He. Helium burning in massive stars ($T = 0.1$–0.4 GK) is, in fact, the main production site of ^{16}O. The isotope ^{12}C is mostly synthesized in massive and AGB stars during this burning stage.

The precise synthesis mechanism of carbon and oxygen during helium burning has, however, puzzled theoreticians for decades. This was mostly motivated by the lack of stable nuclear configurations at $A = 5$ and 8, which in principle prevents reactions like $^4\text{He} + p$ or $^4\text{He} + {}^4\text{He}$ to occur. The possibility of simultaneous fusion of three α-particles was soon discarded because of the low probability of such multiple encounters. Today, helium burning is understood as a sequential, two-step process [1557]. In the first step, two α-particles fuse into ^8Be, an unstable nucleus (by only 92 keV) that quickly decays back into 2 α-particles with a half-life of $T_{1/2} \sim 10^{-16}$ s. However, when the process is repeated multiple times, a small equilibrium population of ^8Be nuclei builds up. This paves the road for the second step, that involves an α-capture on ^8Be leading to ^{12}C. All in all,

$$^4\text{He} + \alpha + \alpha \rightarrow {}^{12}\text{C} + \gamma, \quad Q = 7.275\,\text{MeV}, \qquad (2.102)$$

which is referred to as the *triple-α reaction*[55]. The sequential nature of this reaction represents a challenge for the experimental determination of its rate. This is also hampered by the short half-life of ^8Be, which prevents a direct measurement of the second step, $^8\text{Be} + \alpha \rightarrow {}^{12}\text{C}$. The rate has been inferred theoretically on the basis of experimental information on nuclear masses and partial widths[56].

The synthesis of ^{12}C during hydrostatic He-burning is followed by $^{12}\text{C}(\alpha, \gamma)^{16}\text{O}$,

$$^{12}\text{C} + \alpha \rightarrow {}^{16}\text{O} + \gamma, \quad Q = 7.162\,\text{MeV}. \qquad (2.103)$$

This reaction is considered the "holy grail" of nuclear astrophysics since it determines the final $^{12}\text{C}/^{16}\text{O}$ abundance ratio after He depletion. As such, it has a strong impact on subsequent burning stages in massive stars, on the composition of white dwarfs resulting

---

[55] According to Shaviv [1623], Fred Hoyle visited the Kellogg laboratory in 1953 to find out whether a possible excited state could actually exist in ^{12}C, near the $\alpha + {}^8\text{Be}$ threshold. This was required to account for the production rate of carbon. Preliminary experiments conducted at Kellogg by Ward Whaling and collaborators identified this state, in a beautiful example of the rich interplay between astrophysics and nuclear physics. Results were presented by Hoyle at an American Physical Society (APS) meeting held in Albuquerque (Sept. 2–7, 1953). Shortly afterward, two papers, one by Kellogg's experimentalists, the other by Hoyle, were published with the precise determination of the energy of the excited 7.68 MeV level (today, 7.654 MeV), which is now known as the *Hoyle state* [474, 843]. The abstract of the presentation at the APS meeting appeared in *Physical Review*, even though it is not available online [844].

[56] An example of the rich interplay between nuclear physics and astrophysics was provided by the revision of the triple-α rate, performed by Ogata et al. [1345] on the basis of quantum mechanical calculations of the three-body Schrödinger equation. Around 10 GK, the rate exhibited an increase by 20 orders of magnitude as compared with the NACRE rate [50]. Such results shook the field, since models based on this rate were clearly at odds with observations of red giant branch and He-burning stars in old stellar systems [460]. Ogata's triple-α rate has been actually refuted by new calculations. Differences with respect to NACRE at low temperatures are now reduced to less than an order of magnitude [13, 894].

from the evolution of intermediate-mass stars, as well as on the explosions that involve such compact objects—e.g., classical novae, type Ia supernovae, and double white dwarf mergers (see Chapters 4 and 5).

The ^{12}C(α, γ) reaction rate is dominated by the contribution of the direct capture plus a number of broad, interacting resonance tails (including two subthreshold resonances). It has been measured down to a center-of-mass energy of 1 MeV, above the Gamow window for most situations of astrophysical interest. Extrapolation down to astrophysical energies is actually performed through R-matrix models[57], because of the impossibility of a direct measurement with current techniques (the estimated cross-section at a typical energy of 300 keV is $\sigma \sim 10^{-17}$ b). The experimental uncertainty[58] of the rate amounts to 25%–35% [266, 1834].

The reaction ^{12}C(α, γ)^{16}O is followed by ^{16}O(α, γ)^{20}Ne (Q = 4.730 MeV) whose rate, at the typical temperatures achieved during He-burning, merely relies on theoretical estimates. If the stellar plasma achieves $T \sim 0.4$ GK, this is followed by ^{20}Ne(α, γ)^{24}Mg (Q = 9.317 MeV). See reference [882] for recent evaluations of these rates.

Additional reactions may take place during He-burning in the presence of ^{14}N in the stellar plasma, particularly the sequence ^{14}N(α, γ)^{18}F($\beta^+\nu$)^{18}O(α, γ)^{22}Ne [286]. A fraction of the ^{18}O synthesized in massive stars through this chain can eventually escape destruction as the whole set of reactions do not go to completion. In fact, this is the main production site[59] of the galactic ^{18}O. A branching driven by ^{18}O(p, α)^{15}N(α, γ)^{19}F has been proposed as a feasible path for the synthesis of fluorine in Wolf–Rayet stars[60], where the protons are supplied by the ^{14}N(n, p)^{14}C reaction [1240, 1241, 1378, 1699].

The nuclear activity during hydrostatic He-burning may finally reach Mg through the reaction ^{22}Ne(α, γ)^{26}Mg, and particularly ^{22}Ne(α, n)^{25}Mg, an important neutron source that powers the so-called weak component of the s-process (see Chapter 7). These reactions are important channels in the synthesis of ^{25}Mg and ^{26}Mg in AGB and massive stars.

2.4.2.2 Explosive (Hydrogen and) Helium Burning

The operation of the cold and hot CNO cycles outlined during H-burning is characterized by a very limited breakout at temperatures below T < 0.4 GK. This is caused by the large branching ratios $B_{p\alpha/p\gamma}$ in ^{18,19}F, which are mostly recycled back into lighter species through the prevailing (p, α) channel.

At slightly larger temperatures than those attained during hydrostatic helium burning, the CNO cycles switch into an intertwined set of proton and α-particle induced reactions that produce an important leakage from the CNO mass region toward heavier species. The

---

[57]R-matrix is a technique introduced by Wigner and Eisenbud [1934–1936]. It was originally aimed at describing resonances in nuclear interactions. In the R-matrix method, the configuration space is divided at the *channel radius*, r_{ch}, into internal and external regions. This channel radius is arbitrarily chosen to be large enough, such that $V(r)$ can be approximated by the Coulomb interaction potential for $r > r_{ch}$ [1049]. The internal and external radial wave functions are actually forced to match at the boundary r_{ch} by applying continuity of the wave function and of its first-order derivative (in some traditional presentations of the R-matrix theory, specific boundary conditions are adopted at $r = r_{ch}$ [270]). Reaction rates dominated by the direct capture term plus the contribution of a handful of resonances are frequently extrapolated down to energies of astrophysical interest by application of R-matrix models, which frequently requires a simple parametrization of the data [171]. Other related techniques have been proposed, such as the variational K-matrix method [128]. See reference [433] for a review on the R-matrix model.

[58]See references [67, 1834] for a discussion of the impact of the ^{12}C(α, γ) rate on supernova nucleosynthesis.

[59]He-burning in massive and AGB stars is also the main production site of ^{22}Ne.

[60]The origin of ^{19}F, by far the least abundant of all the stable $12 \leq A \leq 35$ nuclei, is still a matter of debate. A handful of astrophysical scenarios has been suggested, aside from Wolf–Rayet stars. The list includes explosive hydrogen burning sites like classical novae [1933, 1961], thermal-pulsing AGB stars [544, 1162, 1266], and the neutrino process during type II supernovae [1808, 1974, 1989].

main breakout chains at this stage correspond to

$$^{15}\text{O}(\alpha,\gamma)^{19}\text{Ne}(\text{p},\gamma)^{20}\text{Na} \tag{2.104}$$

$$^{14}\text{O}(\alpha,\text{p})^{17}\text{F}(\text{p},\gamma)^{18}\text{Ne}(\alpha,\text{p})^{21}\text{Na} \tag{2.105}$$

$$^{14}\text{O}(\alpha,\text{p})^{17}\text{F}(\gamma,\text{p})^{16}\text{O}(\alpha,\gamma)^{20}\text{Ne}, \tag{2.106}$$

which reveal the existence of a special stage characterized by combined hydrogen and helium burning. The complex network of nuclear interactions combines a series of rapid proton-captures and β^+-decays, known as the *rp-process* [1879], with sequences of (α, p) and (p, γ) reactions (i.e., the αp-process), which extend the main nuclear flow far away from the valley of stability, merging with the proton drip-line beyond $A = 38$ [1578]. This burning regime actually powers some prominent stellar explosions known as X-ray bursts for certain regimes of mass-accretion rates and metallicities, as extensively discussed in Chapter 6.

The rate of the $^{17}\text{F}(\text{p}, \gamma)$ breakout reaction has been evaluated by Iliadis et al. [882]. Below $T = 0.5$ GK, the rate is dominated by the direct capture contribution. For higher temperatures, resonances in the range $E_{\text{c.m.}} = 596\text{--}2226$ keV must be taken into account. Part of the input physics used in the evaluation of the rate relies on experimental information (e.g., excitation energies determined in ^{18}Ne, spectroscopic information of the 600 keV resonance from elastic scattering measurements[61] [124], partial widths). Better constraints on the nonresonant, direct capture contribution are however required [327].

$^{14}\text{O}(\alpha, \text{p})^{17}\text{F}$ has been analyzed through a series of indirect [125, 193, 659, 711, 750, 751, 1390] and direct experiments [1334]. Recent elastic scattering studies of the reaction $^{17}\text{F} + \text{p}$ have improved the spectroscopic information on the resonant levels of interest (i.e., energies, spins, parities, and partial widths). This has been used to infer a new rate for the inverse reaction $^{14}\text{O}(\alpha, \text{p})^{17}\text{F}$ by application of the reciprocity theorem [847].

The status of the additional breakout reactions $^{15}\text{O}(\alpha, \gamma)$ and $^{18}\text{Ne}(\alpha, \text{p})$, including information on the intensive experimental efforts, is discussed in Chapter 6.

2.4.3 Advanced Burning Stages

2.4.3.1 Carbon Burning

Nonexplosive C-burning naturally occurs at the center of intermediate-mass and massive stars ($M \gtrsim 7\text{--}8\ \text{M}_\odot$), following recontraction of the inner layers after H- and He-burning. For hydrostatic conditions, this takes place typically at $T \sim 0.6\text{--}1.0$ GK. Carbon may also burn explosively ($T \sim 1.8\text{--}2.5$ GK), close to the center of a white dwarf in type Ia supernovae (Chapter 5) and in the C-rich ashes piled up on a neutron star after H/He-driven bursts during superbursts (Chapter 6).

The most important C-burning reaction is $^{12}\text{C} + {}^{12}\text{C}$, which involves a relatively heavy projectile and target. In the process, a highly excited ^{24}Mg compound nucleus is formed. The excess energy of the $^{12}\text{C}+{}^{12}\text{C}$ system, that corresponds to a difference between ^{24}Mg and $^{12}\text{C}+{}^{12}\text{C}$ of ~ 14 MeV, is released by emission of light particles (i.e., p, n, α), mostly through the processes [843, 1557]

$$^{12}\text{C} + {}^{12}\text{C} \rightarrow {}^{23}\text{Na} + \text{p}, \quad Q = 2.241\,\text{MeV} \tag{2.107}$$

$$^{12}\text{C} + {}^{12}\text{C} \rightarrow {}^{20}\text{Ne} + \alpha, \quad Q = 4.617\,\text{MeV}, \tag{2.108}$$

as well as the endothermic reaction

$$^{12}\text{C} + {}^{12}\text{C} \rightarrow {}^{23}\text{Mg} + \text{n}, \quad Q = -2.599\,\text{MeV}, \tag{2.109}$$

---

[61]See also reference [847] for additional elastic scattering experiments, and reference [327] for the first direct measurement of the 600 keV resonance.

which can only occur above a threshold energy of $E_{c.m.} \geq 2.6$ MeV. While ^{12}C(^{12}C, p)^{23}Na and ^{12}C(^{12}C, α)^{20}Ne have similar reaction rates, the ^{12}C(^{12}C, n)^{23}Mg rate is much smaller.

At the characteristic temperatures of C-burning, all protons, neutrons, and α-particles released get quickly captured by different species present in the stellar plasma. This gives rise to a suite of secondary reaction chains, such as ^{12}C(p, γ)^{13}N(β^+)^{13}C(α, n)^{16}O(α, γ)^{20}Ne, that significantly contribute to the overall nuclear energy released[62] [72]. All in all, the full set of primary and secondary processes collectively release a specific energy $q_{nuc}[C] \sim 2.5 \times 10^{23} \Delta X[^{12}C]$ MeV g^{-1} ($4.0 \times 10^{17} \Delta X[^{12}C]$ erg g^{-1}) at $T \sim 0.9$ GK, where $\Delta X[^{12}C]$ is the mass fraction of ^{12}C consumed. This corresponds on average to ~ 10 MeV per ^{12}C + ^{12}C reaction [878, 1977].

The ^{12}C + ^{12}C reaction rate has been measured by means of different experimental techniques, including direct measurements of the emitted particles and of the γ-rays released from excited levels in the residual nuclei, down to a center-of-mass energy of $E_{c.m.} \sim 2.5$ MeV [878]. While experimental data around the Gamow peak is available for explosive C-burning conditions, extrapolation down to energies around $E_{Gamow} \sim 1.5$–2.2 MeV is required for core C-burning conditions [1684, 1685]. Plans to extend the experimental data available to lower energies are currently underway, including underground measurements aimed at reducing the environmental background.

2.4.3.2 Neon Burning

Another burning stage encapsulated between two successive core recontractions in massive stars is neon burning. It typically occurs at temperatures in the range $T \sim 1.2$–1.8 GK (for hydrostatic core Ne-burning) and $T \sim 2.5$–3.0 GK (for explosive Ne-burning).

At the end of core C-burning, the stellar plasma consists mostly of ^{16}O, ^{20}Ne, ^{23}Na, and ^{24}Mg. Relying solely on Coulomb repulsion arguments, one would expect that the next nuclear burning stage would involve oxygen. But before the critical temperature to initiate ^{16}O+^{16}O is achieved, reactions between very energetic photons and loosely bound nuclei (i.e., photodisintegration reactions) occur. Neutron, proton and α separation energies for the abundant species ^{16}O, ^{23}Na, and ^{24}Mg range between 7 and 17 MeV. In contrast, ^{20}Ne exhibits a relatively small α-particle separation energy of only 4.73 MeV. Accordingly, for $T > 1$ GK, neon produced during carbon burning can be efficiently photodisintegrated. Therefore, for the first time in the evolution of a star, a photodisintegration process defines a burning stage. As in C-burning, Ne-burning is also dominated by a primary process,

$$^{20}\text{Ne} + \gamma \rightarrow {}^{16}\text{O} + \alpha, \quad Q = -4.730\,\text{MeV}, \tag{2.110}$$

supplemented by a number of secondary reactions[63], mostly involving captures of the α-particles liberated in the photodisintegration of ^{20}Ne, such as

$$^{20}\text{Ne}(\alpha, \gamma)^{24}\text{Mg}(\alpha, \gamma)^{28}\text{Si}, \quad Q = 9.316\,\text{MeV},\ 9.984\,\text{MeV} \tag{2.111}$$

$$^{23}\text{Na}(\alpha, p)^{26}\text{Mg}(\alpha, n)^{29}\text{Si}, \quad Q = 1.821\,\text{MeV},\ 0.034\,\text{MeV}. \tag{2.112}$$

Note that while the primary process is actually endothermic, Ne-burning leads globally to a net energy production. Following Iliadis [878], the two most relevant channels for energy generation, ^{20}Ne + $\gamma \rightarrow {}^{16}$O + α and ^{20}Ne(α, γ)^{24}Mg, can be rewritten in terms of an effective reaction ^{20}Ne + ^{20}Ne $\rightarrow {}^{16}$O + ^{24}Mg, which yields 4.586 MeV. When all secondary

---

[62]The reaction ^{13}C(α, n)^{16}O, and, to some extent, ^{22}Ne(α,n)^{25}Mg, play a key role in the synthesis of neutron-rich elements through the *s-process* during shell C-burning in asymptotic giant branch and massive stars, respectively (see Chapter 7).

[63]It has been proposed that explosive and nonexplosive Ne-burning in massive stars may lead to the synthesis of some proton-rich nuclei beyond Fe, through the *p-process* (see Chapter 7).

processes are taken into account, the specific energy released amounts to $q_{nuc}[Ne] \sim 9.3 \times 10^{22}\Delta X[^{20}Ne]$ MeV g^{-1} ($1.5 \times 10^{17}\Delta X[^{20}Ne]$ erg g^{-1}) at $T \sim 1.5$ GK, which corresponds on average to ~ 6.2 MeV per $^{20}Ne + {}^{20}Ne$ reaction [878, 1977]. Most of the reactions of astrophysical interest that occur during Ne-burning have been directly measured around Gamow peak energies.

2.4.3.3 Oxygen Burning

After Ne-burning, the composition of the cores of massive stars has switched into a plasma made of ^{16}O-, ^{24}Mg-, and ^{28}Si-rich matter. Such composition favors $^{16}O + {}^{16}O$ fusion reactions, which present the lowest Coulomb barriers of all possible interactions between the species mentioned above. This occurs at typical temperatures around $T = 1.5$–2.7 GK in hydrostatic core burning, and in the range $T = 3$–4 GK in explosive environments.

As in $^{12}C + {}^{12}C$, $^{16}O + {}^{16}O$ leads to a highly excited compound nucleus, ^{32}S. The excess energy of the system, which corresponds to a mass difference of ~ 16.5 MeV between ^{32}S and $^{16}O + {}^{16}O$, is again released through the emission of light particles (i.e., p, n, α), in a number of different processes:

$$^{16}O + {}^{16}O \rightarrow {}^{31}P + p, \qquad\qquad Q = 7.678\,\text{MeV} \qquad (2.113)$$

$$^{16}O + {}^{16}O \rightarrow {}^{30}Si + 2p, \qquad\qquad Q = 0.381\,\text{MeV} \qquad (2.114)$$

$$^{16}O + {}^{16}O \rightarrow {}^{28}Si + \alpha, \qquad\qquad Q = 9.594\,\text{MeV} \qquad (2.115)$$

$$^{16}O + {}^{16}O \rightarrow {}^{24}Mg + 2\alpha, \qquad\qquad Q = -0.390\,\text{MeV} \qquad (2.116)$$

$$^{16}O + {}^{16}O \rightarrow {}^{30}P + d, \qquad\qquad Q = -2.409\,\text{MeV} \qquad (2.117)$$

$$^{16}O + {}^{16}O \rightarrow {}^{31}S + n, \qquad\qquad Q = 1.499\,\text{MeV}. \qquad (2.118)$$

Note that both $^{16}O(^{16}O, 2\alpha)^{24}Mg$ and $^{16}O(^{16}O, d)^{30}P$ are endothermic ($Q < 0$), and therefore can only occur above a threshold energy of $E_{c.m.} \geq |Q|$.

As described in previous burning stages with multiple exit channels, the light particles emitted are quickly consumed by secondary reactions involving intermediate-mass elements[64], contributing to the overall nuclear energy released. For O-burning, $q_{nuc}[O] \sim 3.2 \times 10^{23}\Delta X[^{16}O]$ MeV g^{-1} ($5.2 \times 10^{17}\Delta X[^{16}O]$ erg g^{-1}) at $T \sim 2.2$ GK, which corresponds on average to ~ 17.2 MeV per $^{16}O + {}^{16}O$ reaction [878, 1977]. A number of weak interactions that take place during O-burning, such as $^{31}S(\beta^+)^{31}P$, $^{30}P(\beta^+)^{30}Si$, $^{33}S(e^-,\nu)^{33}P$, $^{35}Cl(e^-,\nu)^{35}S$, and $^{37}S(e^-,\nu)^{37}Cl$, deeply influence the neutron excess of the plasma, which rises significantly at the end of O-burning. This paves the road for the onset of the s- and r-processes.

The $^{16}O + {}^{16}O$ reaction rate has been determined through different experimental techniques, including the direct detection of the emitted light particles and of the γ-rays released from excited levels in the residual nuclei, down to energies $E_{c.m.} \sim 6.8$ MeV. This partially covers the Gamow energy window for core O-burning while it entirely reaches astrophysical energies of interest for explosive conditions [878].

2.4.3.4 Silicon Burning

Oxygen burning transforms the inner cores of massive stars into plasmas dominated by the presence of ^{28}Si and ^{32}S. Following oxygen depletion, one may expect a new burning stage characterized by $^{28}Si + {}^{28}Si$ or $^{28}Si + {}^{32}S$ reactions. However, even at the moderately

---

[64]Many nuclei reach an equilibrium between forward and reverse photodisintegration reactions, giving rise to the onset of *quasi-equilibrium clusters*. In particular, nuclei in the range $A = 24$–46, and to a lower extent, Fe-peak nuclei, form two quasi-equilibrium clusters at the end of O-burning.

high temperatures achieved in these stellar plasmas, such nuclear interactions are inhibited by the large Coulomb barriers. Instead, as in Ne-burning, nucleosynthesis relies on the photodisintegration of loosely bound nuclei and on interactions with the suite of emitted light particles (i.e., n, p, α). All in all, the process involves a complex network of intertwined nuclear interactions[65] known as *silicon burning*.

Typical temperatures for core silicon burning in massive stars lie in the range $T = 2.8$–4.1 GK, while explosive Si-burning takes place around $T \sim 4$–5 GK [878]. At such temperatures, the stellar plasma contains a large number of high-energy photons that can disintegrate some of the most abundant species, beginning with ^{32}S. Following Iliadis [878], proton, neutron, and α-particle separation energies (or reaction Q-values) for ^{32}S and ^{28}Si correspond to 8.90 MeV, 15.00 MeV, and 6.95 MeV, for the former, and 11.60 MeV, 17.20 MeV, and 9.98 MeV, for the latter. Accordingly, once the temperature exceeds ~ 2 GK, ^{32}S gets destroyed by the energetically favored ^{32}S$(\gamma, \alpha)^{28}$Si and ^{32}S$(\gamma, p)^{31}$P reactions, which followed by the sequence ^{31}P$(\gamma, p)^{30}$Si$(\gamma, n)^{29}$Si$(\gamma, n)^{28}$Si, efficiently convert ^{32}S to ^{28}Si.

Following the steady temperature rise that accompanies core recontraction, photon-induced reactions begin to act on ^{28}Si, driving a net downward flow that eventually may reach ^{4}He. The main nuclear activity in this flow depends upon the specific thermodynamic conditions of the plasma but mostly involves (γ, p), (γ, α), and (γ, n) interactions. Simultaneously, the recapture of liberated protons, α-particles, and neutrons gives rise to a net upward flow characterized by myriad secondary reactions, extending all the way from Si to the tightly bound Fe-peak nuclei, the truly nucleosynthesis endpoint in Si-burning.

Early numerical models already emphasized that at temperatures $T \geq 3$ GK, nuclear processes such as (p, γ), (p, n), (p, α), (α, γ), (α, n), and (n, γ) are at equilibrium with their reverse reactions, such that two quasi-equilibrium clusters are established [325, 811, 1320, 1792, 1910, 1911, 1984]. One of the clusters forms around ^{28}Si and extends up to $A \sim 40$, while the other forms around the Fe-peak nuclei and involves species with $A \gtrsim 50$. A number of nuclear processes, not in equilibrium, loosely connect both clusters during Si-burning. The boundaries and extent of the clusters change in time, and by the end of the Si-burning stage, they eventually merge into a single quasi-equilibrium cluster [325, 811, 1320, 1910, 1968, 1969].

Estimates of the energy released during silicon burning depend on the specific nuclear interactions involved, which depend, in turn, on the thermodynamic conditions in the stellar plasma. Near 3.5 GK, and assuming that two ^{28}Si nuclei yield one nucleus of ^{56}Fe, the specific energy released amounts to $q_{nuc}[^{28}\text{Si}] \sim 1.9 \times 10^{23} \Delta X[^{28}\text{Si}]$ MeV g^{-1} ($3.0 \times 10^{17} \Delta X[^{28}\text{Si}]$ erg g^{-1}), which corresponds on average to ~ 17.6 MeV per ^{28}Si + ^{28}Si reaction [878, 1977].

The specific rates of most reactions involved in Si-burning are not critical for nucleosynthesis studies because of the existing quasi-equilibrium conditions. Instead, nuclear physics needs mostly rely on nuclear masses (or reaction Q-values) and spins. Exceptions include weak interaction rates (see Section 2.4.3.5) and reactions out of equilibrium, like those mediating between quasi-equilibrium clusters. While a subset of the latter have been measured in the Gamow windows for hydrostatic and explosive conditions (e.g., ^{42}Ca$(\alpha, \gamma)^{46}$Ti, ^{42}Ca$(\alpha, p)^{45}$Sc, ^{42}Ca$(\alpha, n)^{45}$Ti, ^{41}K$(\alpha, p)^{44}$Ca, ^{45}Sc$(p, \gamma)^{46}$Ti), most of the additional mediating reactions are evaluated by means of the Hauser–Feshbach statistical model [878].

2.4.3.5 Road Towards Nuclear Statistical Equilibrium

Recontraction of the stellar core after ^{28}Si depletion translates into a steady temperature increase. At $T \sim 4$ GK, every isotope of the stellar plasma achieves equilibrium through

---

[65]See references [204, 1968, 1969] for pioneering work on Si-burning. See also Hix et al. [809] for a study on the use of reduced nuclear reaction networks during Si-burning.

strong and electromagnetic interactions[66] in a single, huge quasi-equilibrium cluster that extends all the way from H to the Fe-group elements. Such state is referred to as *nuclear statistical equilibrium* (NSE) and can be achieved not only in the cores of massive stars after Si depletion but also during stellar explosions[67].

Under nuclear statistical equilibrium, the abundance of any species, X, is determined by the temperature, density, and neutron excess of the stellar plasma, while being independent of the specific reaction rates (except for weak interactions). It is given by a repeated application of the Saha equation, which in terms of number densities can be written in the form [359, 878]

$$n_X = n_p^Z n_n^N \frac{1}{\theta^{A-1}} \left(\frac{M_X}{M_p^Z M_n^N} \right)^{3/2} \frac{g_X}{2^A} G_X^{\mathrm{norm}} e^{B(X)/kT} , \qquad (2.119)$$

with $\theta = (2\pi ukT/h^2)^{3/2}$. In Equation 2.119, u is the atomic mass unit, $B(X)$ the binding energy of the species $^A_Z X_N$ ($A \equiv Z + N$), M_i is the mass of species i, g_X is the statistical weight (a function of the spin of species X), and G_X^{norm} is the normalized partition function that depends on the energies and spins of the excited levels in species X (see references [50, 1476]).

It turns out that relatively low plasma temperatures lead to a distribution of abundances at nuclear statistical equilibrium that favor ^{56}Ni. At intermediate temperatures, the dominant species is instead ^{4}He, while at higher temperatures, the composition consists mainly of protons and neutrons. For a fixed thermodynamic conditions and neutron excess η, nuclear statistical equilibrium favors species with an individual neutron excess of $(N - Z)/A \approx \eta$ and with the largest binding energy.

2.4.3.6   Nucleosynthesis beyond Iron: The s-, r-, and p-Processes

Elements heavier than Fe cannot be easily synthesized inside stars through charged-particle interactions, since the likelihood of Coulomb barrier penetration decreases with increasing nuclear charge (see, e.g., Equation 2.30 for the dependence of the transmission coefficient with the nuclear charge). Pioneering work by E. M. Burbidge, G. R. Burbidge, W. A. Fowler, and F. Hoyle [268], and independently by A. G. W. Cameron [288], identified neutron captures on intermediate-mass seed nuclei as the nucleosynthesis channel for most of the species beyond the iron peak. Indeed, two mechanisms, known as s- (slow) and r- (rapid) processes[68], were proposed, each accounting for approximately half of the observed solar abundances of species heavier than iron (Sections 7.4.1 and 7.4.2). A handful of extremely underabundant, proton-rich nuclei are actually bypassed against neutron captures (see Section 7.4.3) and therefore require another synthesis channel. The mechanism, known as the p-process[69], was also originally outlined in the seminal papers by Burbidge et al. [268] and Cameron [288], and relies on the combined role of proton-captures and photo-neutron emission reactions. These nucleosynthesis processes will be addressed, together with the rp- (rapid proton), αp-, and νp- (neutrino proton) processes along the following chapters, in connection with different explosive stellar sites.

---

[66] Weak interactions are not part of this equilibrium. Note, for instance, that the neutrinos released through β-decays frequently leave the star without any trace of interaction with the plasma.

[67] An estimate of the time required to achieve nuclear statistical equilibrium was derived by Khokhlov [978] for different thermodynamic conditions, in the form $t_{\mathrm{NSE}} = \rho^{0.2} \exp(179.7/T_9 - 40.5)$, where ρ is the density in g cm^{-3}, $T_9 = T(\mathrm{K})/10^9$, and t_{NSE} is given in seconds.

[68] Cameron originally referred to these processes as "*neutron captures on slow and fast time scales.*"

[69] The p-process refers to the proton-rich nature of the isotopes that this mechanism is responsible for.

Box I. The Nuclear Physics "Hall of Fame"

A selection of facts on the nuclear physics of stars

1. The primordial universe that emerged from the Big Bang consisted of ^{1}H, ^{4}He, and traces of d, ^{3}He, and ^{7}Li.

2. Stars are nuclear furnaces where the chemical abundance pattern of the universe is reshaped. The discovery of technetium in the spectra of several giant stars provided smoking-gun evidence of the active role of stars as nucleosynthesis sites.

3. Nuclear interactions are classified as nonresonant—with smooth, weakly energy-dependent cross-sections—and resonant—with strongly varying cross-sections around specific energies. Resonances result from favorable wave function matching conditions at the nuclear boundary, for specific values of the nuclear potential depth. Isolated resonances are frequently described by the Breit–Wigner formula.

4. Most charged-particle interactions are mediated by the strong nuclear force ($p + p \rightarrow d + e^+ + \nu$ is an important exception). In stellar plasmas, nuclear reactions involve particles whose velocities follow a Maxwell–Boltzmann distribution. Projectiles penetrate the target nuclear potential barrier through quantum tunneling at characteristic energy ranges known as Gamow windows.

5. The presence of free electrons shields the nucleus, reducing the effective Coulomb barrier felt by the incoming projectiles. The net effect, known as electron screening, is an enhancement of the overall reaction rate.

6. Neutron-induced reactions occurring in stellar plasmas are frequently described by cross-sections that are inversely proportional to the neutron thermal velocity, $\sigma \propto 1/v$. This translates into energy-independent (i.e., constant) reaction rates.

7. ^{62}Ni is the most tightly bound species, followed by ^{58,56}Fe. Accordingly, nuclear energy can be released by fusion of nuclei lighter than iron, or by fission of heavier species.

8. Nuclei up to Fe are synthesized through a limited number of fusion stages. Stars up to 0.5 $M_\odot$ only burn hydrogen. Stars up to $\sim 7 - 8$ $M_\odot$ undergo H- and He-burning. Stars up to ~ 10 $M_\odot$ also experience C-burning. Once nuclear burning comes to completion, low- and intermediate-mass stars evolve into compact remnants known as white dwarfs. Stars above ~ 10 $M_\odot$ essentially follow all possible fusion stages, from H- to Si-burning. Their final outcome is a type II supernova explosion, accompanied by the formation of a neutron star or a black hole.

9. Elements heavier than Fe are synthesized in stars mostly through the neutron-capture s- and r-processes (see Chapter 7).

Box II. Mysteries, Unsolved Problems, and Challenges

- Experimental cross-sections are frequently not available for many nuclear processes of astrophysical interest. In many cases, cross-sections are measured at relatively high energies and subsequently extrapolated down to stellar energies (i.e., the Gamow window). Other reactions purely rely on theoretical estimates (e.g., Hauser–Feshbach) or on models that fit partially available experimental information (e.g., R-matrix models).

- Direct measurements have only been performed on a limited number of reactions, since they are frequently hampered by short-lived radioactive targets, the presence of low-energy or subthreshold resonances, or very small cross-sections. Indirect techniques that rely on alternative reactions are frequently used to overcome such limitations.

- Even though $^{12}C(\alpha, \gamma)^{16}O$ has been identified as a key reaction in nuclear astrophysics, as it determines the final $^{12}C/^{16}O$ abundance ratio in stars at the end of He-burning, its rate suffers from an experimental uncertainty that amounts to 25%–35%. Efforts aimed at improving the precision of this rate, particularly at astrophysical energies of interest, are encouraged, probably requiring the use of new techniques to overcome current experimental limitations.

- Weak interactions play an essential role in stellar nucleosynthesis, as they can modify the neutron excess in a stellar plasma. Precise weak interaction rates are needed to understand whether the systematic overproduction of neutron-rich species (e.g., ^{54}Cr and ^{50}Ti) in models of type Ia supernovae are due to uncertainties in those rates. Weak interactions also play a critical role at different stages of the evolution of massive stars (e.g., O- and Si-burning, NSE).

- Need to determine with more precision the mass cuts between the different stellar burning stages and outcomes.

2.5 Exercises

P1. Find the Gamow window of the reactions p + p, ^{12}C(p, γ), ^{12}C(α, γ), ^{12}C + ^{12}C, and ^{64}Ge(p, γ), for different thermodynamic conditions (i.e., Big Bang, Sun's central temperature, classical novae, X-ray bursts, type Ia supernovae).

P2. Derive the transmision coefficient in a 1D model of α-emission from a nucleus, assuming a square-well-plus-Coulomb potential.

P3. Estimate the amount of hydrogen fused to helium per second at the center of the Sun.

P4. In a laboratory experiment, a beam of protons hits a hydrogen target. If the laboratory bombardment energy is 1 MeV, calculate the center-of-mass energy for the reaction p + p.

P5. Prove that the temperature dependence of a nuclear reaction near some energy, $T = T_0$, can be written as [878]

$$N_{\mathrm{Av}} \propto (T/T_0)^{(\tau-2)/3} \,, \tag{2.120}$$

with

$$\tau = \frac{3E_0}{kT} = 4.2487 \left(Z_a^2 Z_b^2 \frac{M_a M_b}{T_9(M_a + M_b)} \right)^{1/3} \,, \tag{2.121}$$

where E_0 is the maximum of the Gamow peak, $T_9 = $ T(K)/10^9, and Z_a, Z_b, M_a, and M_b are the charges and the masses (in amu) of projectile and target. Use Equation 2.120 to estimate the temperature dependence, T^n, of the reactions p + p, ^{12}C(p, γ), ^{12}C(α, γ), ^{12}C + ^{12}C, and ^{64}Ge(p, γ), at T= 10 MK, 100 MK, and 1 GK.

P6. The rate of the reaction ^{22}Na(p, γ) at a temperature of $T = 100$ MK is $N_{\mathrm{Av}}\langle\sigma v\rangle = 6.93 \times 10^{-7}$ cm^3 mol^{-1} s^{-1} [882]. Determine the decay constant and the characteristic timescale for a proton capture on ^{22}Na in a stellar plasma, with $\rho = 500$ g cm^{-3}, X(^{22}Na) $= 1 \times 10^{-4}$, and X(^{1}H) = 0.4.

P7. Prove that the neutron excess, η, comprises values between –1 and 1. Determine the neutron excess in a plasma composed of:
 a) Big Bang nucleosynthesis abundances.
 b) X(^{4}He) = 0.8, X(^{12}C) = X(^{16}O) = 0.1.
 c) Solar system abundances (take, e.g., Lodders [1135]).
 d) X(^{56}Ni) = 0.3, X(^{62}Ni) = 0.3, X(^{56}Fe) = 0.2, X(^{58}Fe) = 0.2.

3

Cosmochemistry and Presolar Grains

"If you can look into the seeds of time,
And say which grain will grow and which will not,
Speak then to me, who neither beg nor fear
Your favours nor your hate."

William Shakespeare, *Macbeth* (1606)

Astrophysics has traditionally relied on electromagnetic radiation, collected by ground-based telescopes and space-borne observatories, as the basic tool to infer stellar properties. But since the 1980s, new methods that rely on matter, rather than on radiation, have also become available.

3.1 Meteorites and Stellar Astrophysics

The chemical content of stars, and their metallicity in particular, provides a wealth of information on their structure, evolution, and, to some extent, age[1]. Unfortunately, unveiling the chemical composition of a distant object is far from trivial. Even the accurate determination of the Sun and Solar System's composition is a complex task. Different sources and techniques, independent but complementary, are used to this end. This includes radiation (mostly photospheric absorption lines from the Sun) and matter samples from meteorites, planets, and moons. Recently, several sample-return missions have for the first time carried material from the interplanetary medium to Earth: While Genesis [271] collected solar wind particles, Stardust [256] gathered particles from comet's Wild 2 coma, along with interstellar dust collected during the journey. The Japanese Hayabusa spacecraft [2014] has also returned samples from asteroid 25143 Itokawa.

Inferring the elemental composition of the Solar System from such a variety of sources is a truly challenging enterprise [1136]. First, the Earth itself does not show a homogeneous chemical abundance pattern. On the other hand, meteorites do not constitute a chemically unique class of objects, since different types exhibit large abundance differences. Meteorites have been traditionally classified into three broad categories: *Stony meteorites*, mostly composed of silicate minerals; *iron meteorites*, mainly made out of metallic iron and nickel; and the somewhat intermediate class of *stony-iron meteorites* that contain a blend of rocky and metallic material. About 84% of the $\sim$ 40,000 known meteorites are stony, and are traditionally subclassified as chondrites and achondrites on the basis of bulk composition and texture. Meteorites are further classified into groups based on their chemistry, O isotopes, mineralogy, and petrography [1028, 1918]. Chondritic meteorites (see, e.g., Figure 3.1) are today considered to be made of more or less undifferentiated, primordial matter that has remained basically unaltered for the last 4.5 Gyr. On the other hand, achondrites have

---

[1]See, e.g., reference [1093] for a recent study that suggests a weak stellar metallicity-age relationship.

FIGURE 3.1
Fragment of the Murchison meteorite, a carbonaceous chondrite that fell in Australia on september 28, 1969. A color version of this figure is available at http://fisica.upc.edu/ca/users/jjose/CRC-Downloads. © Meteorites Australia; reproduced with permission.

suffered some degree of differentiation and reprocessing likely due to melting and recrystallization on or within their parent bodies. Originally, chondrites and achondrites were distinguished by the presence/absence of round chondrules made of silicates[2]. However, the CI group[3] of carbonaceous (or C) chondrites does not contain any chondrules, while a certain type of achondrites shows the presence of such spherules. It is believed that the lack of chondrules in CI chondrites is the result of strong aqueous alteration. Besides, the presence of water suggests that CI chondrites have never been heated above 50 °C, which provides clues on their origin and formation. Only a handful of CI carbonaceous chondrites have been identified to date. Aside from Ivuna (from which 0.7 kg were recovered), the list includes Orgueil (14 kg), Alais (6 kg), Tonk (10 g), Revelstoke (1 g), and possibly a few more recovered from Antarctica [136]. Because of the extreme similarities between the abundances determined in these CI chondrites (through mass spectroscopy, wet chemistry, X-ray fluorescence, and neutron activation) and those inferred from the solar photosphere for non-volatile elements[4], it is believed that this class of meteorites represents the pristine material out of which the Solar System formed about 4.5 Gyr ago, and are used to infer the bulk of Solar System abundances. Certain types of chondrites also contain small amounts of organic matter and *presolar grains* (see Sections 3.2 and 3.3).

---

[2]Such chondrules are small, spherical droplets condensed from the primordial solar nebula. They are composed of magnesium-rich olivine and aluminum-rich pyroxene, surrounded by other silicates in glassy or crystalline form. Other minerals, such as iron sulfide, metallic Fe–Ni, oxides, or phosphates, are often present in small amounts.

[3]Meteorites recovered following observed falls are actually called *falls*, while those that cannot be linked to any observed falls are known as *finds* [1028]. Meteorite groups are designated after a characteristic type specimen and are always named by the places they were found. For instance, chondrites of the CI group refer to the Ivuna meteorite, which fell in Ivuna (Tanzania) in 1938.

[4]In planetary science, materials are classified as volatile (characterized by a low boiling temperature) and non-volatile or refractory (with a high boiling temperature).

Meteorites alone cannot be used to infer the whole set of Solar System abundances, since noble gases and other very volatile species do not easily condense to form solids. This particularly affects the six most abundant elements: H, He, C, N, O, and Ne [84]. Instead, values inferred from solar photospheric spectra are used for C, N, and O, while other techniques are adopted for a suite of special elements, such as Ar, Kr, Xe, or Hg.

Abundances, either meteoritic or photospheric, are normalized to a certain element. Photospheric values are often expressed relative to H. Conversely, meteoritic abundances have to be expressed relative to another element, since H does not easily condense into grains. Silicon is the traditionally chosen species[5], with a defined number abundance of $N(Si)=10^6$. A comparison between the H-normalized photospheric abundances and the Si-normalized meteoritic values is often performed for a number of nonvolatile species (see reference [1136] for details). The difference between photospheric and meteoritic values is more or less constant, and provides the required conversion constant between both scales. Several conversion constants have been published in the literature. For instance, Lodders proposed [1136]

$$A(X) = \log N(X) + 1.533 \qquad (3.1)$$

on the basis of 39 nonvolatile species, where $N(X)$ is the meteoritic abundance of element X normalized to silicon, $N(Si) = 10^6$, and $A(X) \equiv \log[n(X)/n(H)] + 12$ corresponds to the photospheric abundance normalized to hydrogen, $A(H) = 12$.

Photospheric abundances face a number of important drawbacks. First, Li, and probably Be, are depleted in the solar atmosphere; therefore, their current abundances do not reflect those of the primitive Solar System. In addition, the precision currently achieved in photospheric abundance determinations is much lower than that obtained from meteoritic samples. Moreover, it is worth noting that solar photospheric abundances are not directly measured but inferred from observations through state-of-the-art modeling of the solar atmosphere and a detailed account of spectrum-formation processes. This is clearly illustrated by the severe reduction of the abundances of C, N, and O compared to previous estimates [83, 84, 1135], when using improvements in atomic and molecular transition probabilities, coupled with 3D, hydrodynamic models of the solar atmosphere that relax the assumption of local thermodynamic equilibrium in spectral line formation. Indeed, multi-dimensional models proved successful in reproducing a suite of characteristic solar features that 1D models could not account for, including the observed solar granulation topology, the typical length- and timescales, the characteristic values of convection velocities, or the specific profiles of metallic lines [84]. One of the major sources of uncertainty still remaining is the lack of accurate cross-sections for excitation and ionization processes, based on quantum mechanical calculations. Furthermore, abundances cannot be directly inferred through spectroscopy of the quiet Sun for a handful of elements. A suite of alternative measurements are used to overcome such limitation, including infrared spectroscopy of sunspots (used for F and Cl through molecular lines of HF and HCl), helioseismology[6] (He), X-ray and ultraviolet spectroscopy of the solar corona and solar flares (Ne), or solar wind measurements (Ne, Ar), among others.

It is also worth mentioning that solar models computed with revised solar abundances show severe discrepancies with respect to various physical properties of the solar interior determined via helioseismology (e.g., the depth of the convective envelope, the average relative differences of the sound speed and density profiles, or the surface He mass fraction

---

[5]A detailed account on normalization of meteoritic abundances can be found, for instance, in Suess and Urey [1739], one of the first, reliable compilations of Solar System abundances. Many other compilations have been published since then [47, 682, 1135, 1139]. The most recent (re)evaluation of solar abundances has been reported by Pat Scott, Nicolas Grevesse, Martin Asplund, Jacques Sauval, and collaborators [683, 1598, 1599].

[6]Helioseismology studies the propagation of wave oscillations, in particular acoustic pressure waves, in the Sun.

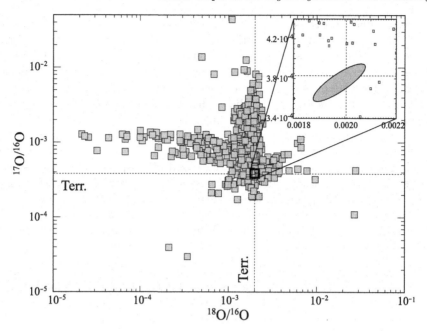

FIGURE 3.2

Oxygen isotopic abundance ratios (by number), displayed on a logarithmic scale, determined in presolar oxide and silicate grains (Section 3.3). The $^{17,18}O/^{16}O$ ratios measured in almost all Solar System samples lie within the ellipse displayed in the inset, on a linear scale. The thin vertical and horizontal lines correspond to terrestrial ratios. Oxygen isotopic ratios measured in presolar grains are shown as squares. They deviate by orders of magnitude from what is found in bulk Solar System matter. Grain data are based on Nittler et al. [1306], Gyngard et al. [703], Nguyen et al. [1292], and references therein. Terrestrial values are based on Clayton [356] and Lodders [1135]. Figure courtesy of Larry Nittler.

[109, 113, 140, 1606]). Therefore, the exact solar composition seems far from being firmly established.

Samples from different Solar System bodies reveal huge variations in *elemental* composition, attributed to a suite of different physical, chemical, and geological processes that took place in their parent bodies. In sharp contrast, the bulk *isotopic* composition throughout the Solar System is homogeneous within a few percent (e.g., the elipse in Figure 3.2). This is fortunate, since isotopic abundances in the Sun cannot be inferred easily. Indeed, direct isotopic measurements can only be performed in the Sun through the analysis of infrared vibrational-rotational lines of the CO molecule. This homogeneity allows the determination of isotopic abundances of the Solar System directly from terrestrial samples, with sufficient accuracy, for most elements. For some volatile species (e.g., $^{3}He/^{4}He$) determinations rely on the solar wind[7], while isotopic information on deuterium[8] ($d/^{1}H$) or nitrogen[9] ($^{14}N/^{15}N$) is based on measurements of the Jovian atmosphere.

Independent estimates of the chemical composition of the Sun can also be obtained through comparison with solar-like stars, massive main sequence O and B stars, H II regions (i.e., large clouds of gas and plasma, like the Orion nebula, characterized by recent episodes

---

[7]See, for instance, reference [271] for the analysis of captured solar wind in the Genesis mission.

[8]See, e.g., reference [633] for estimates of the abundances of deuterium and ^{3}He in the protosolar nebula.

[9]See reference [1201] for new $^{14}N/^{15}N$ estimates from Genesis solar wind samples.

of star formation, with presence of ionized H), or with the interstellar medium in the solar neighborhood. Once the effects of chemical diffusion in the Sun and metallicity enrichment of the Galaxy during the past 4.6 Gyr are taken into account, very good agreement is achieved.

3.2   Grain Formation and Growth

Cosmochemistry studies starting in the 1960s have revealed that a minor fraction of the material embedded in some primitive meteorites, interplanetary dust particles, and comets is characterized by large isotopic anomalies when compared with other Solar System samples[10]. The presence of such isotopic anomalous components in the otherwise homogeneous material that forms the Solar System suggests that they predate the birth of the Sun itself, and therefore were already present in the Solar nebula before the first solids began to condense. Accordingly, they have been coined as *presolar* or *stardust grains*, since they presumably condensed in the shells ejected by stars older than the Sun. Unfortunately, the tiny amounts of long-lived radioactive material trapped in those grains prevents the direct confirmation of this hypothesis through radioactive dating at present[11]. These grains hold an impressive survival record, considering that they escaped destruction from a suite of processes in the neighborhood of their stellar cradles, in the surrounding interstellar medium, along the collapsing cloud that gave birth to the Solar System, the formation and breakup of their meteoritic parent bodies, the atmospheric entry, and their isolation in the lab [417]. Encapsulated in those grains, there is pristine information on the different processes operating in their parent stars, which is partially revealed by the texture, shape, and composition of the grains.

In 1973, A. G. W. Cameron already speculated in a seminal paper that primitive carbonaceous chondrites may host presolar grains[12]: "*During this conference there have been many discussions of methods for determining the properties of interstellar grains by means of electromagnetic observations and by investigation of the properties of possible laboratory prototypes. It has not been realized that it may be possible to study interstellar grains directly within the laboratory*" [287]. Grains (dust) can in principle condense in the cold, low-density envelopes ejected by stars through more or less dynamic events. Such dust-forming episodes have indeed been inferred from a number of different sources, including Wolf–Rayet stars [19,391,708], classical and recurrent novae [632,1644], AGB and post-AGB stars [709,1957], or even from the closest supernova[13] detected since the invention of the telescope, SN 1987A [12,630,631,1157] (see Chapter 7).

The fast decline in temperature and density that characterizes stellar outflows causes the plasma to recombine, allowing the formation of simple and ultimately polyatomic complex molecules. Grain formation is indeed a complex process determined by the local thermodynamical and chemical conditions, requiring[14] relatively low temperatures, below 1500–2000

---

[10]See Black and Pepin [192] for pioneering studies of isotopic anomalies in noble gases. A much more modern account of isotopic anomalies in presolar grains is depicted in Figure 3.2.

[11]Moreover, for many of the long-lived cosmochronometers, like ^{87}Rb–^{87}Sr, the parent species are not expected to nucleate at the high temperatures at which presolar grains condense. The U–Th–Pb offers, however, better perspectives [94].

[12]See also Clayton [346] for a historical perspective on stardust prediction.

[13]See also references [1003, 1550, 1691, 1740] for examples of dust forming episodes—including detection of CO/SiO infrared lines—in recent supernovae.

[14]Moreover, the characteristic timescale for nucleation has to be shorter than the time period for which favorable conditions for grain formation are maintained at the stellar outflow. This essentially restricts the

K, and particle densities above $\log n \geq 10^8$ cm^{-3} [48, 1135, 1601]. Roughly speaking, it can be described as a two-stage process: grain nucleation and growth to macroscopic size. In the first stage, a handful of molecules combine to form small groups or *clusters*, which by effect of additional chemical reactions grow to a critical size. Nucleation requires a certain degree of gas *supersaturation* to overcome the relatively unstable intermediate stages and ultimately reach a solid or liquid configuration[15]. The time evolution of such clusters is ruled by a series of microscopic gain and loss processes. First attempts to comprehensively describe these stages resulted in phenomenological approaches, like the classical nucleation theory[16] [145, 559, 1867], which assumes chemical and thermal equilibrium during the gas phase[17]. This theory already stressed the existence of a bottleneck for grain formation. Molecules of a condensing species randomly combine into small, relatively unstable clusters which after subsequent molecule additions reach a first thermodynamically stable configuration, known as *critical cluster* [517]. The size of such clusters can be determined through maximization of the Gibbs free energy of formation of the system. At the characteristic nucleation temperatures of most astrophysical sites, critical clusters require about $N_c \sim 5 - 20$ monomers and, hence, are quite small [1601]. Critical clusters act as thresholds for stability: Clusters smaller than critical are unstable and likely evaporate, while those with $N \geq N_c$ monomers are thermally stable and likely evolve into homogeneous or heterogeneous macroscopic grains.

Another relevant aspect of the nucleation stage relies on the different condensation conditions that characterize different species. For instance, gases of refractory materials seem to condense at different levels of supersaturation than in volatile materials. Moreover, molecules characterized by a bond energy larger than the corresponding reaction energy required for cluster formation, such as CO or N$_2$, are inert and therefore do not condense. The case of the CO molecule has received particular attention. While most heteronuclear diatomic molecules are characterized by dissociation energies around 3–5 eV, carbon monoxide has a bond energy of 11.2 eV. Therefore, it can only be dissociated by high-energy photons. In the absence of intense, high-energy radiation fields, condensation in an O-rich plasma (with O > C) results in nearly all carbon atoms locked in the form of very stable CO molecules. As a result, C-rich grains like SiC, graphites, or nanodiamonds (see Section 3.3) are not expected to form. Conversely, in a C-rich environment (C > O), all oxygen is consumed in CO formation and, hence, oxidized condensates cannot form, in principle. In these conditions, and on the basis of equilibrium condensation sequences, the list of species that can act as primary condensates is very short: SiO, MgO, MgS, and Fe, for O-rich environments, and C$_2$H$_2$, C$_2$H, MgS, SiC, and Fe, for the case of C-rich gases [581, 1601].

Since theoretical models of nova outbursts (see Chapter 4) yield, on average, O > C, one would expect only oxidized condensates on the basis of the role played by the CO molecule, as discussed above. But preliminary calculations of equilibrium condensation sequences in the nova ejecta suggest that the presence of large amounts of intermediate-mass elements, such as Al, Ca, Mg, or Si may affect the condensation process, allowing the formation of C-rich dust, such as graphite or SiC, even in a O > C environment [922]. Moreover, dust condensation in a nova environment likely proceeds kinetically rather than at equilibrium,

---

list of grain-forming sites to the winds of asymptotic giant branch stars (AGB) and giant stars, as well as to the ejecta from novae and supernovae [1601].

[15] Early estimates of the level of supersaturation in graphite condensates, that is, the excess partial pressure of carbon compared to the overall gas pressure, yielded values of $\geq 3 \times 10^{-7}$ dyne cm^{-2} [458].

[16] See also references [457, 465, 580, 582] for detailed descriptions of nucleation in astrophysical environments, in the framework of the classical nucleation theory.

[17] Kinetic nucleation formulations under nonequilibrium chemical and thermal conditions were subsequently derived. See, e.g., reference [1395].

because of the strong radiation emitted by the underlying white dwarf[18]. This results in a richer mineralogy, with both O-rich and C-rich compounds simultaneously condensing[19].

Clusters satisfying the condition $N \geq N_c$ in supersaturated plasmas likely evolve into macroscopic grains[20] through coalescence, surface growth, and coagulation (sticking) of large clusters [320, 463]. Depending on the local chemistry and thermodynamic conditions, a large variety of configurations, such as monocrystals or polycrystals, amorphous grains, or layered structures, can result[21].

Following Evans and Rawlings[22] [505], the temperature of a dust grain, T_d, located at a distance $r = v_{\text{ejec}} t$ (where v_{ejec} is the velocity of the ejecta) from a star of bolometric luminosity L, can be inferred from a simple balance between grain absorption and emission of radiation, in the form

$$T_d = \left[\frac{L}{16\pi r^2 \sigma} \frac{\langle Q_a \rangle}{\langle Q_e \rangle} \right]^{1/4} = \left[\frac{L}{16\pi v_{\text{ejec}}^2 t^2 \sigma} \frac{\langle Q_a \rangle}{\langle Q_e \rangle} \right]^{1/4}, \tag{3.2}$$

where $\langle Q_a \rangle$ and $\langle Q_e \rangle$ are the Planck mean absorptivity and emissivity of the grain material, calculated as

$$\langle Q(T,a) \rangle = \frac{\int_0^\infty Q(\lambda, a) B(\lambda, T) \, d\lambda}{\int_0^\infty B(\lambda, T) \, d\lambda}, \tag{3.3}$$

with $Q(\lambda, a)$ being the absorptivity (emissivity) of a spherical grain of radius a for light of wavelength λ, and $B(\lambda, T)$ the Planck's function for blackbody radiation (see Chapter 1).

Grain nucleation will begin as soon as the ejecta cools down to a characteristic condensation temperature, T_{cond}, that strongly depends on the chemical composition of the expanding gas. From Equation 3.2, it follows that the time required for a grain to condense, t_{cond}, is given by

$$t_{\text{cond}} = \left[\frac{L}{16\pi v_{\text{ejec}}^2 \sigma T_{\text{cond}}^4} \frac{\langle Q_a \rangle}{\langle Q_e \rangle} \right]^{1/2}. \tag{3.4}$$

Graphite grains, for instance, are characterized by $\langle Q_a \rangle \sim 1$ and $\langle Q_e \rangle \sim 0.01\, a(\text{cm}) T^2(\text{K})$ [195, 354, 464, 505, 645], so that Equation 3.4 can be rewritten as:

$$t_{\text{cond}} \sim 0.078 \left(\frac{L}{L_\odot} \right)^{1/2} \left(\frac{v_{\text{ejec}}}{1000\,\text{km s}^{-1}} \right)^{-1} \text{days}. \tag{3.5}$$

Assuming a typical condensation temperature for graphite, based on equilibrium condensation calculations of the ejecta of neon novae, $T_{\text{cond}} \sim 1900$ K [922], and suitable values for the velocity of the ejecta and luminosity of the underlying white dwarf, $v_{\text{ejec}} \sim 1000$ km s^{-1}, and $L \sim 5 \times 10^4 L_\odot$, Equation 3.5 yields a value of $t_{\text{cond}} \sim 17$ days, for the characteristic time at which grains start to form appreciably in the nova ejecta. Similar estimates, ranging between 10 and 100 days, have been obtained for grain formation in novae by Clayton and Wickramasinghe [354].

---

[18]See Shore and Gehrz [1639] for a study of nonequilibrium condensation through induced dipole reactions in the nova ejecta. See also Cherchneff [319] for a kinetic approach to grain nucleation in supernovae.

[19]Note that infrared observations have revealed the simultaneous formation of both C- and O-rich dust in a number of novae [632], as described in Chapter 4.

[20]See references [116, 117, 846] for early estimates of grain growth rates.

[21]Grains can also be destroyed by a number of physical and chemical processes, including evaporation by radiative heating, sputtering, shattering through grain-grain collisions, or photodesorption [463, 1029]. They can also be destroyed by shocks associated to supernova remnants (see Chapter 7). Grains are not expected to survive in the interstellar medium for more than ~ 0.1–1 Gyr [913, 914].

[22]See also references [354, 590, 1941].

3.3   The Stardust Market

The presolar grain content in primitive meteorites correlates with the fraction of the rock that is composed by the *matrix*, that is, the amalgam of fine-grained amorphous materials and tiny crystals, often soft and porous, that sticks the different meteoritic components together. Presolar grains are actually embedded in the matrix and represent only a fraction of those microcrystals, making up at most a few hundred parts per million (ppm) of the mass of the meteorite.

(Nano)diamonds were the first presolar grains isolated from meteorites [1098], followed by silicon carbides (SiC) [165, 1778, 2036], and graphites [26]. The three carbonaceous phases were identified on the basis of their isotopically anomalous noble gas (Ne, Xe) components[23]. The inventory of presolar grains isolated or identified in meteorites also includes silicon nitrides (Si_3N_4) [1304], silicates [1264, 1279, 1294], and oxides, such as corundum (Al_2O_3) [859, 865, 1308], spinel ($MgAl_2O_4$) [328, 1307], hibonite ($CaAl_{12}O_{19}$) [329, 1306], rutile (TiO_2) [1305, 1310], or alumina (SiO_2) [710] (see Table 3.1).

Those grains, identified and sometimes extracted from meteorites, are systematically analyzed in the laboratory with unprecedented precision (Section 3.4). Such laboratory analyses have revealed a variety of isotopic signatures, suggesting that all SiC grains, approximately half of the graphite grains, $\sim 2\%$ of the spinel grains, and scarcely 0.001–0.02% of the silicates in primitive meteorites, are of presolar origin. Several stellar progenitors, including red giants, asymptotic giant branch stars, novae, and type II supernovae, have been tentatively identified [353, 1137, 1158, 1239, 2029–2031].

3.3.1   Shinning Bright Like a (Nano)Diamond

Typical presolar diamonds have sizes in the range 2–4 nm [417] and, accordingly, are also known as nanodiamonds. They are too small for individual isotopic analyses, and their isotopic characterization is often performed on samples containing a large number of grains. A presolar diamond with a diameter of 2.8 nm, the median of the size distribution, contains

---

[23]See Anders [46] for a review on the discovery of noble-gas anomalies in primitive meteorites, written before the discovery of presolar grains.

TABLE 3.1
Inventory of Known Presolar Grains

Grain type	Size	Stellar sources[*]
Nanodiamonds	2 nm	SN (?)
SiC	0.1–20 μm	AGB, SN, CN, C-stars
Graphites	1–20 μm	SN, AGB, CN
Oxides (corundum, spinel, hibonite, rutile, alumina)	0.2–3 μm	RGB, AGB, SN, CN
Si_3N_4	0.3–1 μm	SN
Silicates (olivine, pyroxene, amorphous silicates)	0.1–0.3 μm	RGB, AGB, SN

Table adapted from Zinner [2031], and Lodders and Amari [1137]. Nanometer-sized carbide (TiC) and Fe–Ni metal subgrains have also been found inside graphite and SiC grains. [*]Acronyms: Asymptotic giant branch stars (AGB); supernovae (SN); classical novae (CN); red giant branch stars (RGB). C-rich stars are characterized by C/O > 1 in their atmospheres and are generally subclassified as N, R, and J stars, according to different nucleosynthesis features.

$\sim$ 1800 atoms of carbon and 18 atoms of nitrogen. Statistically, there is only 1 xenon atom per million nanodiamonds [417]. Surprisingly, presolar diamonds were first identified by their unusual xenon isotopic pattern (Xe-HL), with enrichments in the two lightest xenon isotopes, ^{124,126}Xe, both thought to be produced in the p-process, and also in the two heaviest isotopes, ^{134,136}Xe, which are produced during the r-process [1098] (see Chapter 7). Moreover, nanodiamonds also contain anomalies in a number of r-process nuclides, like ^{110}Pd or ^{128,130}Te [1169,1498], and in the r- and s-process nuclide ^{137}Ba [1097]. The presence of such p- and r-process anomalies strongly points toward a (core-collapse) supernova origin [345, 351]. In sharp contrast, nanodiamonds are also characterized by close-to-solar C and N isotopic ratios, suggesting that a fraction of all nanodiamonds recovered from primitive meteorites may have originated within the Solar System and therefore are not presolar[24] [402]. All in all, data reveal the possible existence of two different populations of meteoritic nanodiamonds, of presolar or solar origin and composition. But how diamonds, a reduced carbon phase, could actually condense in the oxygen-rich protosolar nebula is also a matter of debate [417]. Moreover, nanodiamonds show at least three distinct noble gas components [862] released at different temperatures[25]: The most important components are the one released at around 500°C, characterized by close-to-solar isotopic ratios except for Ne, and a second component released at 1300°C, which is enriched in light and heavy Xe-isotopes. The existence of these different noble gas components reinforces the possibility of different populations.

---

[24]Since SiC X-grains and the fraction of graphite grains likely formed in core-collapse supernovae exhibit a wide range of C isotopic ratios, it may be expected that nanodiamonds, if condensed in the ejecta from supernovae, would show a similar pattern.
[25]In the process, known as *stepwise heating*, a sample is heated in incremental temperature steps, and isotopic analyses of the noble gases released at each step are performed.

3.3.2   Silicon Carbide Grains

SiC grains range in size from a few nm to a few tens of μm. They constitute the most extensively studied class of presolar grains because of a number of reasons [1137]: They are found in different classes of meteorites; they are easier to isolate from the matrix than other types of grains; many SiC grains are $> 1\mu$m in size, and therefore allow individual isotopic studies; and their trace element concentrations are large enough for elemental and isotopic measurements[26]. More than 20,000 SiC grains[27] have been analyzed and classified on the basis of individual carbon, nitrogen, or silicon isotopic ratios. They are characterized by ^{12}C/^{13}C and ^{14}N/^{15}N variations that amount to four orders of magnitude (Figure 3.3), and ^{29}Si/^{28}Si and ^{30}Si/^{28}Si ratios varying by a factor of ~ 2 (Figure 3.4), dwarfing the small variations found among terrestrial samples [417]. SiC grains are studied individually or in the form of aggregated, bulk samples. The morphology of all SiC grains is essentially identical, except for partial damage caused by chemical extraction from their parent meteoritic bodies. While synthetic SiC grains show a suite of different crystal structures, called polytypes (i.e., several hundred configurations, each with its own distinct electrical properties), presolar SiC grains apparently occur only in cubic 3C and hexagonal 2H configurations[28] [409, 410], suggesting condensation at relatively low pressures.

While collective measurements of SiC grains in bulk have proved useful to infer well-defined average compositions, individual grain analyses have allowed researchers to establish different SiC subclasses that share distinctive isotopic features (see Figure 3.3). About $\sim$93% of all SiC grains, the so-called *mainstream population*, characterized by lower ^{12}C/^{13}C and higher ^{14}N/^{15}N ratios than the reference terrestrial (solar) ratios[29], and slightly ^{29}Si- and ^{30}Si-rich (with ^{29}Si/^{28}Si and ^{30}Si/^{28}Si up to 1.2 times solar), are thought to be condensed in the winds accompanying close to solar-metallicity AGB stars [593–595, 1160, 1356].

About $\sim$1% of the SiC grains are known as *X grains*, and are characterized by ^{14}N/^{15}N ratios below solar (that is, with moderate excesses of ^{15}N), large ^{26}Al/^{27}Al ratios (typically ~ 0.1–0.6 [826, 1304]), and excesses of ^{28}Si (up to 5 times solar). Such isotopic patterns suggest a type II supernova origin [28, 835] (see Chapter 7). X grains have been recently subdivided into types X0, X1, and X2 based on distinct silicon and other isotopic features [1115]. Other SiC grains that presumably condensed in the ejecta from (type II) supernova are the unusual ^{29,30}Si-rich grains recently coined as type C grains [389, 701, 830–832, 1421, 2035].

Other populations include *Y* ($\sim$1%) and *Z grains* ($\sim$1%), whose origin is attributed to low-metallicity AGB stars [34, 828]. While Y grains are characterized by ^{12}C/^{13}C > 100, ^{14}N/^{15}N ratios above solar, and ^{30}Si-excesses slightly larger than mainstream grains, Z grains exhibit similar ^{12}C/^{13}C and ^{14}N/^{15}N ratios but somewhat larger ^{30}Si-excesses (and smaller ^{29}Si/^{28}Si ratios) than mainstream grains.

A rare variety of SiC grains ($<$1%), together with a couple of graphite grains, that exhibit a suite of isotopic signatures characteristic of classical nova outbursts, have been reported in recent years by Sachiko Amari and collaborators [24, 25, 27, 33] (see Chapter 4).

Finally, SiC grains with ^{12}C/^{13}C < 10, ^{14}N/^{15}N in the range 40–10,000, and silicon

---

[26]Isotopic measurements of heavy elements, such as Sr, Zr, Mo, Ru, and Ba, have also been performed in some individual SiC grains (see [137] and references therein).

[27]A database on presolar grains is maintained at the Laboratory of Space Sciences, Washington University (St. Louis). See http://presolar.wustl.edu/PGD/Presolar_Grain_Database.html [867].

[28]Evidence of a new polytype, 15R, has been recently reported for the SiC supernova grain Bonanza by Rhonda Stroud and collaborators [1738]. Interested readers can find a comprehensive primer on SiC crystallography and polytypism in Ashcroft and Merrim's *Solid State Physics* [81], a reference book in the field.

[29]$(^{12}$C/^{13}C$)_\odot = 89$. The terrestrial ^{14}N/^{15}N ratio by number, according to Lodders [1135], is 272. The solar ratio, as determined by the Genesis mission, is $(^{14}$N/^{15}N$)_\odot = 459$ [1201].

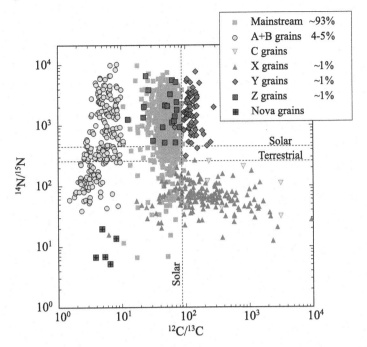

FIGURE 3.3
Carbon and nitrogen isotopic ratios for the different SiC grain populations. A color version of this plot is available at `http://fisica.upc.edu/ca/users/jjose/CRC-Downloads`. Figure courtesy of Ernst Zinner.

abundances similar to mainstream grains, are known as A+B grains (3%–4%). Despite being relatively abundant, their origin is still controversial, and different sites, such as J stars (a variety of carbon-rich stars), born-again AGB stars, or other C-rich stellar types, such as R or CH stars, have been proposed [35, 575, 755]. It is finally worth noting that a few other SiC grains do not fit into the major groups mentioned above.

Nanometer-sized TiC subgrains encapsulated in SiC grains[30] have also been identified through transmission electron microscopy (TEM; see Section 3.4) of thin sections [161].

3.3.3 Silicon Nitrides

A few silicon nitride (Si_3N_4) grains, sharing similar isotopic properties with type X SiC grains, have also been identified [1115, 1304]. They likely condensed in the ejecta of type II supernovae.

3.3.4 Graphites

More than a thousand graphite grains, with sizes ranging between ~400 nm and 20 μm, have been analyzed. Presolar graphite is only present in the least modified and thermally unaltered primitive meteorites [863]. Moreover, they are chemically reactive and therefore their isolation from the matrix is far more complex than for other carbonaceous types like SiC grains.

Imaging of presolar graphites reveal two morphologies (Figure 3.5): *Onion-type* grains,

---

[30]See, e.g., reference [162] for the identification of a TiC subgrain encapsulated in a graphite grain.

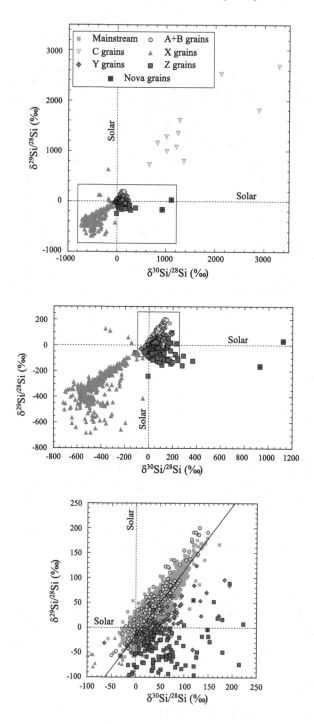

FIGURE 3.4

Silicon isotopic ratios, for the different SiC grain populations, expressed as delta values, or deviations from the solar isotopic ratios in permil, $\delta^{29,30}\text{Si}/^{28}\text{Si} = [(^{29,30}\text{Si}/^{28}\text{Si})/(^{29,30}\text{Si}/^{28}\text{Si})_\odot - 1] \times 1000$. Color versions of these plots are available at http://fisica.upc.edu/ca/users/jjose/CRC-Downloads. Figure courtesy of Ernst Zinner.

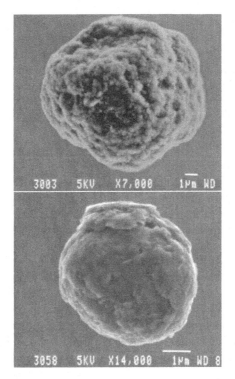

FIGURE 3.5

Scanning electron microscopy (SEM) images of two presolar graphites exhibiting cauliflower (above) and onion configurations (below). Figure from Zinner [2031]; reproduced with permission.

with a layered structure, and *cauliflower-type* grains, likely aggregates of smaller grains [163, 164, 827]. While the former are more abundant in high-density meteoritic fractions, cauliflower-type graphites are largely found in low-density fractions.

Presolar graphite grains exhibit quite variable carbon isotopic ratios, ranging from $^{12}C/^{13}C \sim 2$ to ~ 7200 [38, 827]. In sharp contrast, most graphite grains exhibit nitrogen ratios essentially identical to the atmospheric (terrestrial) value of $^{14}N/^{15}N = 272$, interpreted as evidence that indigenous N in presolar graphites has been isotopically equilibrated (i.e., contaminated) with terrestrial nitrogen, likely during the harsh chemical processing needed for extraction. Some fraction of graphite grains[31] show almost pure ^{22}Ne. Other isotopic ratios measured in graphite grains include high $^{18}O/^{16}O$ (ranging from solar to ~ 16 times solar), $^{29,30}Si/^{28}Si$ (~ 0.5 to 2 solar), and $^{26}Al/^{27}Al$ (up to ~ 0.15) [31, 38, 39, 903]. Isotopic measurements of O, Mg, K, Ca, Ti, Zr, and Mo have also been performed in some individual graphite grains [40, 827, 1299, 1817]. No designation for isotopically distinct graphite subtypes has been established to date, except a subclassification in low-density (likely originated from core-collapse supernovae) and high-density grains.

TiC, Zr–Mo carbides, Fe–Ni metal subgrains, Os, Ru, and Mo-rich refractory metal nuggets, and cohenite ([Fe,Ni,Co]$_3$C) have been identified inside presolar graphites [164, 385, 386, 388, 1694]. These are truly presolar grains within presolar grains. Some interpre-

---

[31]The ^{22}Ne-rich component found in presolar graphite grains is known as Ne-E(L). Such component is released at low (L) temperature. The ^{22}Ne-rich component identified in SiC grains is released at high (H) temperature, and it is therefore known as Ne-E(H).

tations suggest that some of these subgrains formed prior to the graphite and were later incorporated into the grain or served as active nucleation sites.

Different sources have been claimed to explain the origin of presolar graphite grains, including type II supernovae (about 30%), low-mass AGB stars (50%), and different C-rich stars [38,417,825,903], while a handful of presolar graphites have been claimed to be formed in nova ejecta [24,27].

3.3.5 Presolar Oxides

Oxides joined the inventory of discovered presolar grains a few years after carbonaceous grains. Most of the oxide grains found in meteorites actually formed in the early Solar System and therefore are not presolar[32]. Presolar oxide grains are often identified by their anomalous oxygen isotopic ratios (i.e., $^{17}O/^{16}O$ and $^{18}O/^{16}O$), which deviate by orders of magnitude from what is found in bulk Solar System matter (Figure 3.2).

The first clues on the existence of presolar oxide grains were obtained by Zinner and Tang [2038], who reported ^{17}O enrichments in bulk measurements of thousands of oxide grains extracted from the Murray meteorite. Shortly afterward, a single oxide grain isolated from the Orgueil meteorite, highly enriched in ^{26}Mg [861], and with an inferred $^{26}Al/^{27}Al$ ratio much above the upper limit derived for the Solar System, proved to be presolar through oxygen isotopic analysis [865]. Most presolar oxides are, however, routinely discovered through ion imaging of large number of grains, followed by high-resolution mass spectroscopy (see Section 3.4). Indeed, such ion imaging methods resulted in the first unambiguous discovery of a presolar Al_2O_3 grain in a sample of the Murchison meteorite [1301,1312].

The suite of oxide grains discovered to date includes corundum (Al_2O_3) and spinel ($MgAl_2O_4$) [328, 861, 865, 1308], hibonite ($CaAl_{12}O_{19}$) [329, 1306], chromite ($FeCr_2O_4$) [1311], and rutile (TiO_2) [1305,1306]. About 90% of these presolar oxides have a putative origin in AGB stars, while 10% likely come from core-collapse supernovae [703,1306].

The spread in oxygen isotopic composition has led researchers to subclassify oxide grains into four groups. Group I grains are characterized by ^{17}O excesses or modest ^{18}O deficiencies relative to solar[33]. Grains of group II exhibit large ^{18}O deficiencies[34] ($^{18}O/^{16}O < 0.5$ solar) and ^{17}O excesses (up to 3.5 times solar). Grains with moderate depletions in both $^{17,18}O$ are classified as group III, while those with $^{17,18}O$ excesses constitute group IV [1307]. The mean $^{26}Al/^{27}Al$ ratios inferred for the suite of presolar oxide grains range between 0.0004 (group III) and 0.0060 (group II). Corundum, hibonite, and spinel grains are present in all four groups [1306].

3.3.6 Silicates

Presolar silicates have proved much more elusive than other types of grains, and their discovery in interplanetary dust particles [1231] and ancient meteorites [1264, 1279, 1294] had to wait for the advent of very sensitive imaging techniques (see Section 3.4). The reason is threefold [1137]: first, it is not easy to disentangle presolar silicates from the very abundant solar silicates that are major components of meteorites; second, silicates are more sensitive to metamorphic and chemical processing than other types of grains (for instance, aqueous

---

[32] Only 33 out of 1854 oxide grains were identified as presolar in several samples of the Murray and Murchison carbonaceous chondritic meteorites by Ernst Zinner and collaborators [2033].

[33] For oxygen, standard mean ocean water values are frequently adopted: $^{16}O/^{18}O = 499$ and $^{16}O/^{17}O = 2632$.

[34] The large ^{18}O deficits exhibited by group II grains have been suggested to result from extra mixing (i.e., cool bottom processing) [1317, 1908].

alteration efficiently transforms silicates into hydrous silicates, thus erasing any presolar silicate signature in the meteoritic sample); and third, they are efficiently destroyed by the acids used to extract refractory carbon and oxide phases, and therefore require ion imaging of untreated samples.

The limited inventory of presolar silicates discovered to date includes pyroxenes ([Ca, Mg, Fe, Mn, Na, Li][Al, Mg, Fe, Mn, Cr, Sc, Ti][Si, Al]$_2O_6$), olivine ([Mg, Fe]$_2SiO_4$), Al-rich silicates, and amorphous silicates, with typical sizes in the range 0.3–0.9 μm, and similar oxygen isotopic ratios than presolar oxides, suggesting origins either in red giant and asymptotic giant branch stars [1293] or in supernova explosions [2031].

3.4 Experimental Techniques and Instruments

Different techniques are used in the characterization of presolar grains, depending mainly on their size and abundance [2037]. For instance, nanodiamonds, graphites, or SiC grains are extracted from meteorites using chemical and physical separation procedures[35]. Oxide and silicate grains, instead, are mostly located and analyzed in situ using automated ion imaging techniques. Mass spectroscopy is exclusively used for isotopic determinations.

3.4.1 Imaging Techniques

Optical microscopes can magnify an object typically about 100 to 1000 times, with a resolution of about 0.2 μm. This is not enough for proper imaging of stardust, since presolar grains are just a few microns in size (or even smaller).

Better resolution, at the level of a few nanometers, could be achieved by replacing the light beam by an electron beam, like in electron microscopes. Here, electrons from a source are accelerated by means of electric fields and focused onto a sample. For instance, an electron accelerated to a typical energy of 1 keV ($\Delta V = 1$ kV) has an associated wavelength of about 1.2 nm. In the resulting interactions with the sample, a suite of electrons and photons containing key information are released and conveniently analyzed.

3.4.1.1 Scanning Electron Microscopy (SEM)

As in standard electron microscopy, an electron gun releases a small electron beam[36] that is focused on different sections of a sample[37], which is subsequently scanned along different rectangular areas (*raster scanning*). When the electrons, accelerated at typical energies 10–30 keV, hit the target (e.g., a grain), they release energy and different by-products [1158, 1695]: (i) secondary electrons released by atoms of the sample; (ii) back-scattered electrons plus X-rays emitted by the sample, when high-energy electrons fill low-energy states freed by secondary electrons; (iii) Auger[38] electrons. The physical properties of all

---

[35]See Amari et al. [31] for an account of the different stages followed in the isolation of presolar SiC, graphites, and nanodiamonds from primitive meteorites, applied to samples from Murchison.

[36]Most SEMs generate electron beams with spot sizes on sample < 10 nm in diameter, while still carrying sufficient current to form acceptable images.

[37]See Goldstein et al. [657] for a reference book on SEM.

[38]The Auger effect, independently discovered by Lise Meitner (1922) and Pierre V. Auger (1923), occurs when an inner electron in an atom is removed, leaving a vacancy that may be occupied by another electron from a higher energy level. This results in an energy release that can be carried away in the form of a photon or transferred to a second electron—known as Auger electron—which is ultimately ejected from the atom. This Auger electron is therefore emitted with a kinetic energy that corresponds to the difference between the

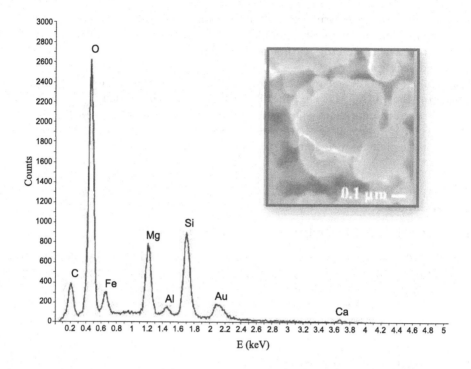

FIGURE 3.6

X-ray spectrum and SEM image of a silicate grain isolated from the Acfer 094 meteorite, identified by the large O and Si peaks. The presence of Mg and Ca reveal that it is a pyroxene. Au features result from the substrate on which the sample was placed, while C possibly indicates sample contamination. Figure from Nguyen and Zinner [1294]; reproduced with permission.

these electrons strongly depend on the composition of the grain (secondary electrons are also influenced by the surface topography). In turn, the energy of the X-rays emitted depend on the specific structure of the atoms of the sample, and, hence, from the X-ray energy spectrum one can infer the major (and some minor) elements present in the grain (Figure 3.6). In SEM, the X-ray energy spectrum can be analyzed by *energy-dispersive X-ray spectroscopy* (EDXS) [2037].

SEM is a nondestructive technique, though for conventional imaging, requires samples to be electrically conductive, at least at a surface level [2039]. This is needed to prevent samples from charging, which will further complicate the analysis (e.g., changing electron trajectories). Therefore, samples are often coated with an ultrathin electrically conducting material (e.g., carbon, gold, platinum, osmium), deposited on the sample either by low-vacuum sputter coating or by high-vacuum evaporation. SiC and graphite grains are conductive and therefore do not require coating. For SEM analysis of nonconducting presolar grains carbon coating is frequently used. Nonconducting samples may also be imaged uncoated under high pressure conditions or in the low-voltage mode of SEM operation [2039].

In the most common detection mode, secondary electron imaging, high-resolution images

---

energy released in the initial electronic transition and the ionization energy of the Auger electron from its shell. *Auger electron spectroscopy*, in which Auger electrons are emitted from a sample after bombardment with X-rays or energetic electrons, is also used in the study of presolar grains, mostly in the determination of the elemental composition of submicrometer grains, such as silicates and oxides [221, 539, 540, 1292, 1866].

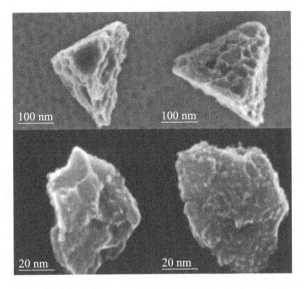

FIGURE 3.7
Images of SiC grains from the Murchison meteorite, after SEM analysis. Figure from Hoppe et al. [832]; reproduced with permission.

of the sample surface can be obtained, revealing, in principle, details ranging from 1.5 nm in size, for electrons accelerated at ≥ 10 keV, and 2.5 nm, for 1-keV electrons. SEM relies on electron interactions with the surface of the sample, and therefore allows imaging of bulk samples with large depth of view. Black and white photographs of presolar grains (see Figure 3.7) are usually obtained in this way. SEM X-ray spectroscopy has also been used to identify presolar grains in situ in meteorite slices and on sample prior to SIMS analysis (see Section 3.4.2.1). Indeed, SEM with X-ray detection was the technique used in the first in situ identification of presolar SiC grains [1340].

3.4.1.2 Transmission Electron Microscopy (TEM)

TEM is another electron microscopy technique[39] used in the analysis of stardust grains (Figure 3.8(a)). In sharp contrast to SEM, TEM relies on a primary electron beam transmitted through a (ultrathin) sample. This requires higher electron energies[40] (200–400 keV), which translate into better resolution (~ 0.1 nm) than SEM. Moreover, since electrons are transmitted through the sample, TEM is better suited than SEM to unveil information on the internal structure of the specimen.

An image, containing physical information of the interaction between the primary incident electrons and the sample (through diffraction, high-angle scattering, or inelastic scattering), is ultimately magnified and focused onto an imaging device, such as a screen, a photographic plate, or an electronic sensor (e.g., a charge-coupled device or CCD). The brightness of the resulting image depends on sample thickness and density, as well as on instrumental conditions, which globally determine the number of electrons transmitted.

The most frequent use of TEM imaging in cosmochemistry includes microstructural studies, high-resolution imaging of lattice structure and atomic planes, or elemental mapping [2037]. The crystal structure of the sample is analyzed by means of electron diffraction. TEM is also used in concert with a number of techniques generally known as *analytical electron*

---

[39]See Williams and Carter [1939] for a reference textbook on TEM.
[40]The electron energies used in SEM are actually limited to prevent beam penetration into the sample.

(a)   (b)

FIGURE 3.8

(a) The Transmission Electron Microscope of the Center for Biotechnology of the University of Nebraska–Lincoln, in the US. The device has a resolution up to 0.5 nm, under ideal conditions. A color version of this picture is available at http://fisica.upc.edu/ca/users/jjose/CRC-Downloads. Image courtesy of the University of Nebraska–Lincoln, reproduced with permission. (b) High-resolution TEM image of presolar nanodiamonds from the primitive carbonaceous meteorite Allende. Dots correspond to columns of pairs of closely spaced carbon atoms. Figure from Daulton et al. [411]; reproduced with permission.

microscopy (AEM), that rely on the detection of X-ray signals generated in the sample by primary incident electrons (EDXS) and on determinations of the energy lost by the incident electrons (*electron energy loss spectroscopy*, EELS[41]). Both spectroscopic techniques provide information on elemental composition (EELS also provides hints on bond configurations and oxidation states). Globally, TEM and AEM allow imaging of grain structure, composition, and mineralogy determinations, with high spatial resolution (see Figure 3.8(b)). A major drawback of this technique is sample preparation. Indeed, TEM requires ultrathin samples for efficient electron transmission. Accordingly, presolar grains have to be cut in very thin slices[42], using special tools, called *ultramicrotomes*, that contain diamond blades.

---

[41]A related method, *electron loss near-edge spectroscopy* (ELNES), is also used with TEM [2039]. Other techniques frequently combined with TEM include cathodoluminescence, Auger spectroscopy, or electron beam-induced current imaging. See reference [689] for an example of correlated NanoSIMS, TEM, and XANES (*X-ray absorption near-edge structure*, another type of absorption spectroscopy) studies of the supernova graphite grain G6 isolated from the Orgueil chondrite.

[42]For an incident electron beam of 200–300 keV, the thickness of the sample must be < 100 nm [1158].

3.4.1.3 Focused Ion-Beam Microscopy (FIB)

Even though FIB has exclusively been used to cut slices for TEM analysis in presolar grain work [675], it can also be used for imaging, with a resolution of about $\sim$ 5–10 nm. FIB microscopy is a similar technique to SEM, but with the focused electron beam replaced by an ion beam [1735]. Sometimes, FIB microscopy is used in combination with SEM (FIB–SEM), allowing the study of a sample with either of the beams. Most FIB microscopes use liquid-metal ion beams, Ga^+ in particular, accelerated to energies $\sim$1–50 keV, and subsequently focused onto the sample. When the ions strike the target, a number of atoms—either neutral or ionized—and electrons are ejected from the surface of the sample. The sputtered ions (or the secondary electrons) are subsequently collected to form an image or used for secondary ion mass spectrometry (SIMS). The sputtered neutral atoms can be resonantly ionized by lasers and analyzed through resonance ionization mass spectrometry (RIMS). Therefore, and in contrast with standard electron microscopy, FIB is a destructive technique[43].

3.4.2 Mass Spectrometry

Mass spectrometers are other distinguished components of the large arsenal of *stardust telescopes*. While analysis of the X-rays produced by electron bombardment in imaging techniques, like SEM, is used to infer elemental abundances, mass spectrometers are particularly well suited for determining the isotopic composition of the sample.

The first stage in mass spectrometry is the production of ions of the sample at the so-called ion source. There is a rich variety of ionization techniques, depending on the phase of the sample (solid, liquid, or gas). These ions are subsequently redirected into a mass analyzer, where they are separated according to their mass-to-charge ratio before being ultimately registered at the detector system, in proportion to their abundance. The mass resolution or resolving power of a mass spectrometer is given by

$$R = \frac{m}{\Delta m}, \tag{3.6}$$

with m being the mass of a given isotope and Δm the mass difference with the next observable species. A high resolution is needed to distinguish isotopes and molecules with similar masses, often requiring previous stages of chemical separation as well as operation in vacuum (to prevent interactions between ions and air molecules). Two main techniques are used to separate ions: spatial separation is achieved by a combination of electric or magnetic fields, which induce slightly different trajectories to isotopes of different masses (e.g., ion beams deflected by a perpendicular magnetic field B follow circular orbits of radii $r = mv/qB$). Separation in time is performed using *time-of-flight* (TOF) techniques, which rely on the different transit times for isotopes of different masses accelerated by the same electric field: Exposed to a voltage V, an ion of charge q will gain a kinetic energy $1/2\, mv^2 = qV$; therefore, lighter species will achieve larger velocities and will travel faster through the device. Magnetic mass spectrometers can be subclassified in single-ion or multicollector systems, depending on the number of isotopes simultaneously measured

---

[43]Other imaging techniques used in cosmochemistry include [2039]: *Scanning transmission X-ray microscopy* (STXM), which yields element maps based on soft X-rays generated from thin samples, with a spatial resolution up to 100 nm; *atomic force microscopy* (AFM), that relies on the electromagnetic interaction between a sharp tip and a closeby surface of a sample, yielding topographic information of the sample with atomic resolution; or *holographic low-energy electron diffraction* (HLEED), a technique aimed at the direct recovery of the 3D structure of the atomic environment around atoms on a surface by analysis of the diffuse diffraction patterns produced when low energy electrons hit the sample. A number of analytical techniques that rely on synchrotron radiation, such as X-ray fluorescence (XRF), X-ray computed microtomography (XRCMT), X-ray diffraction (XRD), *X-ray absorption fine structure* (XAFS), and *X-ray absorption near edge structure* (XANES) have also been used in the study of extraterrestrial samples [2037].

(time-of-flight instruments are inherently single-collector). In the former, a magnetic field whose intensity is varied through a tunable current (*peak-jumping*) is repeatedly changed in order to select different isotopes. In the latter, a magnet is used to spread the ion beam and separate different isotopes.

Different detector systems are used in mass spectroscopy [941], the most standard ones being electron multipliers, microchannel plates, and Faraday cups. Electron multipliers are the most sensitive detectors. They consist of a series of electrodes or dynodes disposed at increasing potentials (up to 1500–3500 V). When a particle hits the first dynode, a few secondary electrons (i.e., 1 to 3) are released. These secondary electrons are accelerated toward the second dynode, where they generate in turn more electrons, ultimately triggering an electron cascade throughout the device. This restricts the use of electron multipliers to moderate ion beams ($< 5 \times 10^6$ counts per second) to prevent fast aging. The number of secondary electrons produced actually depends on the type of primary particle and its energy, as well as on the properties of the dynode surface. Extremely useful for low-counting rates, they face however a number of drawbacks, namely an efficiency < 1 that depends on the element and varies with time. Indeed, they are characterized by short lifetimes, 1–2 yr, even at regular operation. Microchannel plates are arrays of small electron multipliers or channels. Under ideal operating conditions, they can detect count rates as small as 1 count per second (even though with detection efficiencies smaller than 100%). A Faraday cup detector consists of a hollow conducting electrode connected to ground. When charged particles strike the collector, they cause a flow of electrons from/to the ground. The current produced can be measured, and the number of ions striking the cup is inferred. They are not as sensitive as electron multipliers but offer high accuracy because of the proportionality between the number of particles that hit the device and the measured current. A Faraday cup detector is best suited for high-count rates ($> 50,000$ counts s^{-1}). Unlike electron multipliers, they do not discriminate between the type of incident particles or its energy.

3.4.2.1 Secondary Ion Mass Spectrometry (SIMS)

SIMS is the prevalent technique for measuring the isotopic composition of presolar grains. This is achieved by bombardment and subsequent ejection (sputtering), with a focused primary ion beam, and ultimate collection and analysis of the secondary ions released [152]. In a SIMS spectrometer or ion microprobe, an ion gun is used as the source of the primary beam. The choice of the primary beam species (and of the gun itself) depends on the required current and size of the beam, as well as on the nature of the sample. Cs is frequently used for the extraction of negative ions. Its low melting/boiling points and ionization energy favor the production of intense Cs$^+$ beams; it is also a heavy species, which helps increasing the sputtering yield and favors the ejection of negatively charged secondary ions. For the extraction of positively charged ions, a primary beam of O$^-$ is the usual choice, since oxygen tends to take up electrons, increasing the production of positively charged secondary ions[44]. Moreover, O$^-$ ions are cheap and easy to generate. Other materials, such as Ga$^+$ or even fullerenes (C_{60}), are now being used in TOF SIMS.

The primary ion beam is accelerated (typically, at 15–20 keV) and subsequently focused onto the target (Figure 3.9(a)). The interaction between the primary ions and the sample has three basic effects: the outermost layers of the sample get mixed and turned into an amorphous material; a fraction of the primary ions get also implanted into the sample; and finally, a suite of secondary particles, electrically charged or neutral, are sputtered from the sample. Only a fraction of the atoms ejected are actually ionized. These can be easily extracted from the sputtering area by means of an electric field located between the

---

[44]On the other hand, Cs gives off electrons, increasing the yield of negatively charged secondary ions.

sample and an extraction lens. Both the sample and the extraction lens, used to redirect the secondary ions extracted from the sample into a mass spectrometer, are located in a vacuum chamber, with pressures below 10^{-4} Pa. Such vacuum is required to minimize interactions (and possible contamination) between secondary ions and background gases on their way to the detector. Indeed, if the pressure is too high, atoms or molecules from the gas (e.g., N) can be deposited onto the sample and affect the analysis. Finally, a mass analyzer separates and classifies the ions according to their mass-to-charge ratio.

The size of the sputtered area is of the order of the diameter of the incident primary beam (~ 1 μm, in conventional SIMS, and 50 nm–100 nm in NanoSIMS). Therefore, SIMS is limited to the analysis of large grains (> 1 μm), such as SiC and graphite grains. Indeed, the anomalous size of Murchison meteorite's grains [1137, 2031], larger than those isolated from other meteorites (with some SiC grains sizing up to 20 μm, for reasons not fully understood), as well as the large availability of samples, made Murchison one of the favorite targets for laboratory analyses of presolar grains.

With SIMS, most elements of the periodic table can be identified. Noble gases are an exception, because of their difficulties to ionize. Trace elements can be measured at very low concentrations, at the level of a few parts per billion (ppb), with detection limits about $10^{12} - 10^{14}$ atoms per cubic centimeter. It is worth noting, however, that SIMS is a destructive technique, limited by the number of atoms (particularly for small samples). This, for instance, prevents radioactive dating. Following, Kanbach and Nittler [941], a 1 μm–sized SiC grain, for instance, typically contains about 5×10^{10} atoms, 1% of which consist of Al. Assuming a SIMS efficiency[45] for Al of $\sim 10^{-3}$, this yields $5 \times 10^{10} \times 0.01 \times 0.001 = 5 \times 10^5$ Al atoms that can be detected. Presolar SiC grains often display ^{26}Al/^{27}Al ratios of about 10^{-3} [860], implying that only ~ 500 atoms of ^{26}Al (already decayed to ^{26}Mg) are typically present in a SiC grain! The associated uncertainty in the measurement could be estimated from Poisson statistics as the square root of the number of counts, that is $\sqrt{500} \sim 22$, representing a 4% error. Another limitation of SIMS involves its inability to separate isobaric interferences in heavy elements (e.g., ^{92}Zr from ^{92}Mo). Standard techniques are applied to correct or minimize the effect of some of these interferences in light and intermediate-mass elements (e.g., ^{50}Cr and ^{50}Ti [38, 903]).

A breakthrough in secondary ion mass spectrometry has been achieved with the latest generation of ion microprobes: the NanoSIMS (see Figure 3.9(b)) [833, 1696]. Its most relevant feature is the small diameter of the primary beam, which yields a spatial resolution of ~ 50 nm–100 nm. It also offers extremely high sensitivity at high mass resolution and the possibility of simultaneous measurements of up to seven isotopes [2037]. The improved performance of NanoSIMS has been essential in the discovery of submicron-sized presolar silicates (typically, 250–300 nm; see Figure 3.10) [1231, 1294]. It is worth noting that only $\sim$ 0.001%–0.02% of the silicates in primitive meteorites are actually presolar, as revealed by isotopic composition. Therefore, in order to identify presolar silicates, thousands of grains have to be analyzed simultaneously by isotopic raster imaging, in search for anomalous compositions. Isotopic imaging in the ion microprobe operating in automated mode[46] has also been used in the detection of rare grain types (e.g., SiC X-grains; see Section 3.3.2) with the NanoSIMS.

---

[45]See http://www.eag.com/mc/sims-rsf-tables.html and reference [1949]. The secondary ion yield (i.e., fraction of ions detected over atoms sputtered) depends on many factors, such as the bombarding ion species, the composition of the matrix, the instrument, and the instrumental conditions. For instance, while a SIMS efficiency for Al of $\sim 10^{-4}$ may be expected for a Ga$^+$ primary ion beam, bombardment with O$^-$ ions may yield an efficiency of $\sim 10^{-2}$.

[46]A somewhat similar technique has been previously applied to the identification of presolar oxide grains [1301, 1312].

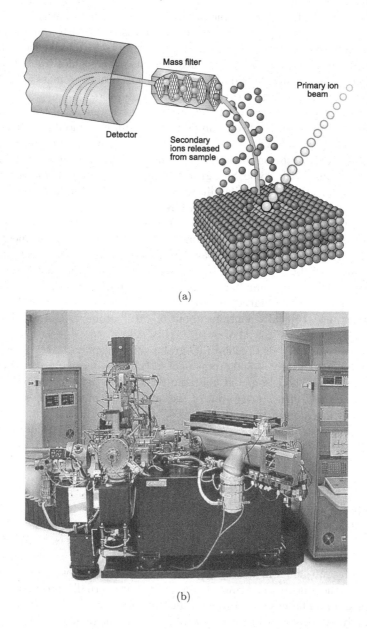

(a)

(b)

FIGURE 3.9
(a) Schematic view of the operation of a SIMS spectrometer. (b) Photograph of NanoSIMS 50, the first of its kind, designed specifically for the study of presolar grains. This ion microprobe, installed in the year 2000 at the Laboratory for Space Sciences of the Washington University at St. Louis (USA), has an advanced multicollection system and achieves a spatial resolution of about 100 nm. Figure courtesy of the Laboratory for Space Sciences (WUST); reproduced with permission. Color versions of these figures are available at http://fisica.upc.edu/ca/users/jjose/CRC-Downloads.

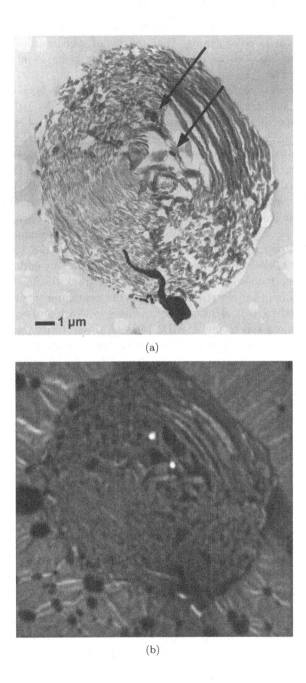

(a)

(b)

FIGURE 3.10

(a) TEM image of two TiC grains (highlighted by the arrows) encapsulated in a graphite grain. (b) NanoSIMS $^{16}O^-$ image of the same graphite slice shown in the TEM image in a). The two TiC grains are clearly visible due to their high O signal. Figures from Stadermann et al. [1694]; reproduced with permission.

3.4.2.2   Resonant Ionization Mass Spectrometry (RIMS)

Some astrophysically important elements, particularly those located beyond the iron peak, are extremely scarce in presolar grains. The analysis of such *trace* elements requires high-sensitivity instruments as well as methods to sort out isobaric interferences that often complicate the analysis.

RIMS technology was created in the 1970s and is now used for a wide variety of applications in physics, chemistry, and biology. In RIMS analysis [858, 1079, 2037], one or more lasers are tuned to specific frequencies that allow the excitation of electrons to higher energy levels and the ultimate ionization of selected types of atoms or molecules, splitting the neutral species into positive ions and free electrons. In short, those atoms are forced to *resonate*. Since the excitation and ionization frequencies are specific for each species, RIMS naturally gets rid of the problem of isobaric interferences. CHARISMA (*Chicago–Argonne Resonant Ionization Spectrometer for Mass Analysis*), a device based on RIMS technology developed at Argonne National Laboratory by Mike Pellin and colleagues [1403, 1569], has been successfully applied to the determination of isotopic abundances of trace elements (e.g., Zr, Mo, Sr, Ba, or Ru) in individual presolar grains [137, 138, 1296, 1298].

For RIMS to work, the sample has to be vaporized. To this end, atoms are thermally desorbed[47] by a laser focused onto the sample. Ions from the misty desorbed material are removed by means of an electric field, while the remaining neutral atoms of interest are resonantly ionized by two or more lasers. These ions are subsequently extracted and analyzed in a TOF mass spectrometer. RIMS has high selectivity and sensitivity (i.e., the fraction of atoms detected over atoms removed from the sample can be 1%–3% [1404], while hardly affecting other elements present in the grain [2037]). Focusing is about 1 μm [1168], which puts constrains on the size of the sample. The major drawback of RIMS is that this is mainly a single-element method[48], requiring laser retuning for measurements of different species.

Other cosmochemistry probes based on RIMS technology[49] include SARISA (*Surface Analysis by Resonance Ionization of Sputtered Atoms* [1856, 1857]), aimed at analyzing elemental and isotopic abundances in solar wind samples returned to Earth by the Genesis mission, and CHILI (*Chicago Instrument for Laser Ionization* [1720, 1721]), a new generation RIMS coming to light, with a planned 10 nm resolution. This resolution is achieved because atoms are not thermally desorbed with a laser but sputtered with a Ga beam.

3.4.2.3   Thermal Ionization Mass Spectrometry (TIMS)

TIMS is another sensitive technique used for bulk isotopic analyses on relatively large samples. In this technique, after chemical separation from the sample, atoms of the element of interest are loaded onto a metal filament. The filament is heated with an electric current causing thermal ionization of the atoms of the sample. Subsequently, the ionized atoms pass through a mass spectrometer where they get separated on the basis of their mass-to-charge ratio. Variations of this technique include ID-TIMS (isotope dilution) and CA-TIMS (chemical abrasion). The main limitation faced by TIMS is that it can only be applied to large samples, and hence only an average composition for many grains is determined.

---

[47] In thermal desorption, heat is used to increase the volatility of certain species allowing to remove them from a solid matrix.

[48] More recently, however, six tunable lasers have been assembled for simultaneous resonance ionization of three elements [1720].

[49] A variation of RIMS used in the analysis of organic molecules is the *microprobe two-step laser mass spectroscopy* (*μL2MS*) [2037].

3.4.2.4 Multicollector Inductively Coupled Plasma Mass Spectrometry (MC-ICPMS)

MC-ICPMS is, together with TIMS, one of the most common techniques used in bulk isotopic analysis of samples. In MC-ICPMS, drops of a purified solution containing sample atoms are sprayed into an Ar plasma at very high temperatures, such that the droplets evaporate and the sample atoms become ionized. Those ions are subsequently redirected towards a multicollector mass spectrometer for isotopic analysis [1358, 2002]. The technique offers a fast data collection and a high ionization efficiency for all elements [941, 2037].

MC-ICPMS instruments can also be equipped with a laser for in situ analysis (*laser ablation multicollector inductively coupled plasma mass spectrometry*, LA-MC-ICPMS). Here, a UV laser is used to ablate a 50–100 μm spot of the sample. The ablated material is then introduced in the Ar plasma, as in standard MC-ICPMS. LA-MC-ICPMS is being increasingly used to overcome some of the limitations of TIMS, in particular for isotopic measurements involving metals like W, Hf, or Zr, which are poorly ionized by TIMS. It is particularly suited to detect very low concentrations of species in large samples, at the level of a few parts per trillion. However, since all ablated elements are transferred to the plasma, unresolved isobaric interferences may result.

3.4.2.5 Accelerator Mass Spectrometry (AMS)

AMS is a sensitive technique used to measure very low concentrations of trace elements in a sample [1119]. In contrast to other types of mass spectrometry, it relies on the acceleration of ions to extremely high velocities (i.e., energies $\sim$ MeV. See Figure 3.11). In AMS, negative ions are typically produced by sample sputtering. This allows suppression of isobaric interferences in the case of isobars that do not form negative ions (like ^{14}N in ^{14}C determinations). Another remarkable feature of AMS is its proven ability to separate a rare species from an abundant neighboring mass (e.g., ^{14}C from ^{12}C).

Accelerator mass spectrometry is often employed in radiocarbon dating (^{14}C) by archaeologists, as well as in biomedical research. In cosmochemistry, one of the recent discoveries performed with AMS has been the detection of the radioisotope ^{60}Fe in the deep-sea, FeMn-rich crust, deposited about 2.8 Myr ago by a nearby supernova explosion [994, 995] (see Chapter 7). Other long-lived radioisotopes detected[50] with AMS include ^{10}Be or ^{26}Al [518, 519]. See references [1357, 1883] for AMS analyses of presolar grains.

A summary of the experimental techniques most frequently used in the analysis of presolar grains is given in Table 3.2.

---

[50]The possible detection of supernova-produced ^{244}Pu in terrestrial samples is still a matter of debate [815, 1043, 1882, 1884].

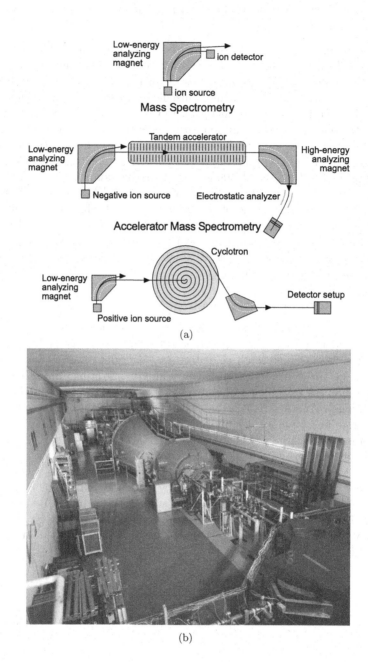

(a)

(b)

FIGURE 3.11

(a) AMS compared with standard mass spectroscopy. In AMS, secondary ions sputtered from the sample are accelerated by means of a Tandem accelerator or a cyclotron. (b) A section of the AMS facility at the Maier–Leibnitz–Laboratory (TU Munich), in Garching, Germany, showing the Tandem accelerator. Image courtesy of Shawn Bishop. Color versions of these figures are available at http://fisica.upc.edu/ca/users/jjose/CRC-Downloads.

TABLE 3.2

Experimental Techniques Used in the Analysis of Stardust Grains

Grain type	Technique	References
Elemental abundances*		
SiC (I/B): Mg, Al, Ca, Ti, V, Fe, Sr, Y, Zr, Ba, Ce	SIMS	[30]
Graphites (I): H, N, O, Al, Si	SIMS	[827]
Diamonds (B): Sc, Cr, Fe, Co, Ni, Ru, Os, Ir	INAA	[1096]
Isotopic abundances		
SiC (I/B): C, N, O, Al-Mg, Si, S, Ca, Ti, Fe, Ni	SIMS	[41, 826, 831, 832, 860, 1115] [1195, 2032]
(B): Sr, Ba, Nd, Sm	SIMS TIMS	[1356, 1435, 1456, 1497, 2034]
(I): Sr, Zr, Mo, Ba, Ru	RIMS	[1295, 1297, 1298, 1567, 1568]
(I/B): He, Ne, Ar, Kr, Xe	GMS	[754, 1094, 1095]
Graphites (I): C, N, O, Al-Mg, Si, K, Ca, Ti	SIMS	[29, 38, 827, 903, 1817]
(I): Zr, Mo	RIMS	[1137, 1299]
(I/B): He, Ne, Ar, Kr, Xe	GMS	[32, 754, 1223]
Diamonds (B): C, N	GMS	[1548]
(B): Pd, Te	TIMS	[1169, 1498]
(B): Pt	AMS	[1357]
Si₃N₄ (I): C, N, Si	SIMS	[1304, 2032]
Oxides (I): O, Al-Mg, Ca, Ti	SIMS	[328, 329, 703, 1306, 1307]
Silicates (I): O, Al-Mg, Fe	SIMS	[540, 541, 1264, 1291, 1997]
Crystal structure		
SiC	TEM, Raman	[161, 409, 410, 866, 1861]
Graphite	TEM, Raman	[162, 164, 386, 387, 1991]
Diamond	TEM, EELS	[166, 411, 1736]
Oxides	TEM	[1737, 2016, 2017]
Silicates	TEM	[1865]
Morphology, geometry...	SEM	[167, 826, 827]

Examples of experimental techniques used in the study of stardust grains, for both individual grains (I) and aggregated, bulk grain samples (B). A few representative references are listed for each technique. *Acronyms: SIMS = secondary ion mass spectrometry; TIMS = thermal-ionization mass spectrometry; RIMS = resonant ionization mass spectrometry; GMS = gas-source mass spectrometry; INAA = instrumental neutron activation analysis; TEM = transmission electron miscroscopy; AMS = accelerator mass spectroscopy; Raman = Raman spectroscopy; EELS = electron energy loss spectroscopy; SEM = scanning electron microscopy. Adapted from Lodders and Amari [1137], and Zinner et al. [2037].

Box I. The Cosmochemistry and Presolar Grain "Hall of Fame"

A selection of facts on solar abundances and stardust grains

1. Solar System abundances rely on (chondritic) meteoritic samples and high-resolution spectroscopy of the solar photosphere. They reveal huge variations in elemental abundances but a remarkable homogeneity in terms of isotopic composition.

2. Dust forming episodes have been inferred for Wolf–Rayet stars, classical and recurrent novae, AGB and post-AGB stars, and supernovae.

3. Simultaneous formation of O- and C-rich dust reported from several classical novae proves that the CO molecule does not totally control the condensation process, particularly in the presence of other abundant intermediate-mass species or strong radiation fields.

4. Grain formation requires gas temperatures ~ 1500–2000 K, and particle densities $> 10^8$ g cm^{-3}. It can be qualitatively described in terms of a two-stage process: nucleation and growth to macroscopic size. Nucleation requires the assembly of a minimum number of seed molecules known as a *critical cluster*.

5. Presolar stardust grains, identified by huge isotopic anomalies with respect to bulk Solar System abundances, have been isolated from primitive meteorites. The inventory of presolar grains discovered includes silicon and titanium carbides, graphites, nanodiamonds, different oxides (corundum, spinel, hibonite, rutile), silicon nitride, and silicates (olivine, pyroxene).

6. Different experimental techniques are used in the analysis of presolar grains, either for imaging and elemental abundance determinations (e.g., SEM, TEM), or for isotopic analysis through mass spectroscopy (e.g., SIMS, RIMS, TIMS). The use of multiple techniques in the analysis of the same grains, in order of increasing destructiveness, has led to significant advances in the study of presolar grains.

Box II. Mysteries, Unsolved Problems, and Challenges

- Need for high-precision Solar System abundances that yield physical properties of the Sun compatible with helioseismology estimates (i.e., surface helium mass fraction, depth of the convective envelope, average relative differences of the sound speed and density profiles [112, 113, 140, 1606]).

- Need for excitation and ionization cross-sections based on quantum mechanical models for solar conditions.

- The physical mechanisms involved in nucleation and grain growth in stellar outflows are not fully understood. Efforts aimed at improving kinetic grain formation models under intense radiation fields are strongly needed. The identification of subgrains embedded in some presolar graphites and silicon carbides may help to unveil the complex processes that characterize heterogeneous nucleation.

- How did presolar grains from many different stellar sources become incorporated into the solar nebula and how did they survive the formation of the Solar System? How long can presolar grains survive in different stellar environments?

- Need to clarify whether nanodiamonds belong to two distinct populations, with a significant fraction not being presolar. Need to understand the stellar progenitors of A+B grains [755].

- Need for multi-isotope data in the analysis of presolar grains for an unambiguous identification of their stellar progenitors.

- Age determination of presolar grains based on long-lived radionuclides has proved difficult since for the proposed cosmochronometers, the parent species are not expected to nucleate at the typical condensation temperatures. Need to better quantify dating possibilities through U–Th–Pb.

- Better resolution in experimental techniques is needed to allow studies of typical stardust grains, with sizes below $\sim 100\,\mathrm{nm}$, rather than large, micron-sized stardust. Efforts should be focused, for instance, on an effective reduction of primary beam diameters (as achived in CHILI).

TABLE 3.3
Rubidium-Strontium Isotopic Composition of the
Juvinas Meteorite (Adapted from Allegre et al. [18])

Sample	$(^{87}Rb/^{86}Sr)$	$(^{87}Sr/^{86}Sr)$
Glass	0.0876	0.70473
Tridymite and quartz	0.0231	0.70063
Plagioclase	0.00301	0.69914
Pyroxene	0.00714	0.69950
Total rock	0.00407	0.69927

3.5 Exercises

P1. Take one of the existing compilations of Solar System abundances by number (e.g., Lodders [1139]), and estimate the corresponding ^{12}C, ^{14}N, ^{16}O, and ^{20}Ne mass fractions.

P2. A C-grain extracted from the Murchison meteorite is characterized by $\delta^{29}Si/^{28}Si = 2600$ and $\delta^{30}Si/^{28}Si = 3250$, in permil. Find the corresponding abundance ratios, $^{29,30}Si/^{28}Si$, in terms of the solar values, $(^{29,30}Si/^{28}Si)_\odot$.

P3. Estimate the number of Si atoms in a 1 μm-sized SiC grain. Assuming a SIMS efficiency for Si of $\sim 3 \times 10^{-3}$, determine the number of Si atoms that can be detected. From the solar value of $^{30}Si/^{28}Si = 0.03347$, estimate the number of ^{30}Si atoms typically present in a SiC grain condensed in the Solar System. Compare the result with the number of ^{30}Si atoms in the presolar SiC grain AF15bB–429–3 [27,605], for which $\delta^{30}Si/^{28}Si = 1118$. Determine the associated uncertainties in these measurements from Poisson statistics.

P4. Two beams of ^{12}C and ^{14}C, accelerated at 1 MeV, are focused onto a mass spectrometer. A magnetic field $B = 1$ T, is applied perpendicularly to the incident beams. Compare the radii of the circular trajectories described by both species.

P5. Consider a species a that decays into species b (a stable isotope), in one or multiple steps. Assume that the total number of atoms of species a and b, $N_a + N_b$, is constant.

a) Show that the variation in the number of atoms of species b as a function of time obeys

$$N_{b,f} - N_{b,i} = \left(e^{\lambda t} - 1\right) N_{a,f}, \tag{3.7}$$

where subscripts i and f refer to times t_i and t_f, and λ is the decay constant (see Chapter 2) of species a. Equation 3.8 is usually expressed in terms of ratios of isotopes, relative to a third (constant) species, c:

$$\frac{N_{b,f}}{N_c} = \frac{N_{b,i}}{N_c} + \left(e^{\lambda t} - 1\right)\frac{N_{a,f}}{N_c} \tag{3.8}$$

b) Table 3.3 illustrates an example of Rb–Sr radioactive dating based on the analysis of five separate phases from the Juvinas meteorite [18]. Plot $^{87}Sr/^{86}Sr$ vs. $^{87}Rb/^{86}Sr$ and infer the age of the meteorite from the slope of the best fit. Compare the result with a value of 4.56 ± 0.08 Gyr obtained independently with the Sm–Nd dating technique [1163]. Assume $\lambda(^{87}Rb) = 1.41 \times 10^{-11}$ yr^{-1}. Estimate the $^{87}Sr/^{86}Sr$ ratios of the five samples after 100 Myr, 1000 Myr, and 10,000 Myr.

4

Classical and Recurrent Novae

"Have you ever realized that every sun, as we travel between them, is a furnace where the very worlds of empire are smelted? Every element among the hundreds is fused from their central nuclear matter. [...] Gold is fusing there right now, and radium, nitrogen, antimony. [...] And there's illyrion there too. [...] Suppose we could stand at the edge of some star gone nova and wait for what we wanted to be flung out and catch it as it flamed by—but novas are implosions, not explosions."

Samuel R. Delany, *Nova* (1968)

Some years before detailed hydrodynamic simulations and multiwavelength observations of novae crystallized in the *thermonuclear runaway model*, Samuel R. Delany, an American novelist, portrayed in the futuristic drama *Nova* (1968) the quest of a team of brave explorers in search of *illyrion*, the most precious substance in the universe, presumably synthesized during such cataclysmic events. Despite in the real world, element production in novae does not involve such precious substances, their interest from an astrophysical viewpoint is actually quite remarkable.

4.1 Stellae Novae: Beacons in the Ocean of Night

Much progress has been achieved in the understanding of the nova phenomenon since the first naked-eye observations carried out by astronomers from China and other cultures more than two millennia ago. The etymological origin of the term *nova*, from the Latin *stella nova* (i.e., new star), reflects somewhat the reasons of the interest that such phenomena awoke in ancient astronomers: The sudden appearance of a luminous object in the sky, at a spot where nothing was clearly visible before, that fades away back to darkness in a matter of days to months (see Figure 4.1).

Until the galactic distance scale was soundly established, both novae and supernovae were misclassified as identical stellar phenomena under the generic term of stellae novae. Light was shed into this matter by George W. Ritchey, Heber D. Curtis, Harlow Shapley, and Knut Lundmark, among others, who reported results from serendipitous discoveries of novae in the so-called *spiral nebulae* (galaxies, in fact) in the early 20th century. These scattered and heroic efforts were soon followed by systematic searches, but Shapley and Curtis were among the first to question the real distance to the spiral nebulae by suggesting their extragalactic nature in 1917. A major step forward was achieved with Nova S Andromedae[1] (S And), discovered and analyzed by the German astronomer Ernst Hartwig on August 31, 1885, and by the new scale distance of Hubble that placed Andromeda outside the Milky Way. This pushed Nova S Andromedae far away, at an incredible distance, and hence its intrinsic luminosity was outstanding. Actually, Lundmark and Curtis, in the early 1920s, were the first to talk about *giant novae* in what was known as *the great debate*: *"It seems*

---

[1] Also noticeable was Z Cen, the *nova* of 1895 in NGC 5253.

FIGURE 4.1
Two optical images of Nova V1500 Cygni (1975) showing decline from peak visual magnitude 2 to magnitude 15. © Lick Observatory/SPL/AGE Fotostock; reproduced with permission.

certain, for instance, that the dispersion of the novae in the spirals, and probably also in our Galaxy, may reach at least ten absolute magnitudes[2], as is evidenced by a comparison of S Andromedae with faint novae found recently in this spiral. A division into two magnitude classes is not impossible." Soon, it was clear that two different classes[3] of stellae novae likely existed. The Swiss astronomer Fritz Zwicky and his German collegue Walter Baade proposed the finally accepted term *supernovae* in 1934, for the most luminous ones.

The understanding of the physics that lay behind a nova outburst motivated large controversies and vivid discussions. Early attempts can be found in the *Almagestum Novum* (1651), by G. B. Riccioli [1496], who compiled thirteen different "theories" of novae. In

---

[2]The apparent magnitude of a star, m, indicates its brightness as seen from Earth. It is based on a logarithmic measure of luminosity, such that a decrease in magnitude, Δm, translates into an increase in luminosity (from L to L') as $L'/L = 10^{\Delta m/2.5}$. Magnitudes are typically measured at certain wavelengths: A B (for blue) magnitude, for instance, is measured at 4000 Å, while a V (for visual) magnitude is measured at 5000 Å in green colors, the center of the visual spectrum. The absolute magnitude, M, is defined as the apparent magnitude that a star would have if it were located at an arbitrary distance of 10 parsecs (32.62 light-years) from Earth. The connection between apparent and absolute magnitudes of a star can be expressed as $m - M = 5 \log (d/10)$, where d is the distance to the star in parsecs. The absolute magnitude and the luminosity of a star are related through $M = M_{sun} - 2.5 \log(L/L_\odot)$, where M_{sun} and $L_\odot$ are the absolute magnitude and the luminosity of the Sun.

[3]A more detailed historical account on stellae novae can be found in Duerbeck [472, 473].

retrospect, an unrelated passage in Newton's *Principia Mathematica* (book 3, prop. 42) contains a statement that, unintentionally, gives clues on the mechanism powering nova explosions: *"So fixed stars, that have been gradually wasted by the light and vapors emitted from them for a long time, may be recruited by comets that fall upon them; and from this fresh supply of new fuel those old stars, acquiring new splendor, may pass for new stars."* Indeed, the concept of revitalization of old stars (i.e., white dwarfs) by fresh supply of new fuel[4] (although not by comets!) is at the base of the thermonuclear runaway model for the nova outburst developed around 1970s, in which mass transfer plays a central role.

Observationally, the nova phenomenon has benefited from multiple efforts[5], including a number of breakthroughs:

- The application of (optical) spectroscopy to the study of nova explosions, first attempted on the occasion of Nova T CrB 1866 (also T Aur 1891) [852].

- The discovery of anomalous features (later identified as neon [NeIII] lines[6] at 3869 and 3968 Å) in the spectra of GK Per [1647, 1648]. Much later, two distinct classes of nova explosions were revealed by the discovery of an intense [Ne II] 12.8 μm emission in Nova Vul 1984 #2 [629]: neon novae, with an underlying ONe white dwarf[7] hosting the explosion, and non-neon novae, likely occurring on CO white dwarfs (see Section 4.3).

- The early suggestion that some of the observed features of the nova spectra could be due to ejection of a shell from a star [1416].

- The identification of Doppler blueshifted lines in the ejecta [291, 358].

- The interpretation of the minimum in the DQ Her light curve as due to dust formation (as previously suggested for the R Coronae Borealis stars and other systems) [1727].

- The discovery of the binary nature of DQ Her [1877], followed by systematic studies of novae revealing that binarity is a common property of most cataclysmic variables, novae in particular [936, 1014, 1015, 1562].

Although a firm observational picture clearly emerged around the idea of ejection from the surface of a star[8], the explanation of the physics behind the outburst had to wait a few decades. Indeed, its thermonuclear origin[9] was first theorized by Evry Schatzman [1580, 1581], although wrongly attributed to nuclear fusion reactions involving ^{3}He. Other significant contributions were made late in the 1950s [285, 694], while attemps to mimic the explosion through the coupling of radiative transfer in an optically thick expanding shell with hydrodynamics were published a decade later [641, 1527, 1681][10]. The idea that

---

[4]The concept of accretion emerged in astrophysics around the 19th century [358].

[5]The Soviet school also significantly contributed to the understanding of nova explosions. See, e.g., V. A. Ambartsumyan [43], E. R. Mustel [1273, 1274, 1276], and V. V. Sobolev [1674].

[6]See also Campbell [292].

[7]First simulations of novae involving ONe white dwarfs were performed in the mid-1980s by Sumner Starrfield and collaborators [1710].

[8]This was clearly stated in the enthusiastic and concise telegram sent by J. Hartmann on occasion of the Nova RR Pic: *"Nova problem solved: star expands, and bursts"* [732]. Probably the shortest scientific communication ever written.

[9]Note that there are other types of novae, such as dwarf novae or nova-like variables, whose outbursts have no thermonuclear origin.

[10]The thermonuclear nature of nova outbursts gave rise to an interesting controversy. Despite his initial support, Kraft concluded that a large fraction of the energy released by thermonuclear reactions would be conducted inward to the core, likely powering a supernova—rather than a nova explosion [1015]. Instead, he supported nonradial, resonant oscillations induced by the asynchronism between the rotational white dwarf motion and the revolution of the secondary, as suggested, among others, by Schatzman [1582]. Shortly after, a series of simulations [641, 1527, 1701, 1702] confirmed instead that electron conductivity plays no

CNO enhancement was critical for the explosion was first proposed by Sumner Starrfield [1701, 1702].

4.2   Classical and Recurrent Novae: The Big Picture

The coupling of multiple and complementary approaches undertaken in the study of the nova phenomenon—spectroscopic determinations of chemical abundances, photometric studies of the nova light curve, as well as pioneering hydrodynamic simulations—paved the road for our current understanding of these stellar beacons. The scenario envisaged assumes a white dwarf star hosting the explosion in a close binary system. The low-mass stellar companion overfills its Roche lobe[11], and matter flows outward through the inner Lagrangian point. The matter transferred (typically at a rate in the range $\dot{M} \sim 10^{-9} - 10^{-10}$ $M_\odot$ yr^{-1}) possesses angular momentum and, hence, does not fall directly onto the compact star. Instead, it forms an accretion disk[12] that surrounds the white dwarf. A fraction of this hydrogen-rich matter lost by the secondary spirals in and ultimately ends up on top of the white dwarf, where it is gradually compressed to large densities by a continuous infall of material, such that it becomes degenerate[13]. Compression heats the envelope and nuclear reactions set in. But because of the degenerate nature of the envelope, it cannot readjust through expansion, and, hence, both the temperature and the nuclear energy generation rate increase continuously, driving a thermonuclear runaway[14].

Observationally, classical novae are known to occur in short period (i.e., 1.4–16 hr)[15], stellar binary systems consisting of a white dwarf star and a low-mass main sequence companion (i.e., a K-M dwarf, although observations increasingly support the presence of more evolved companions in some systems). About a dozen recurrent novae (by definition, novae seen in outburst more than once, but in essence nearly identical to classical novae, with much shorter recurrence times) have been identified so far. They are classified in two different types: The very homogeneous class of long period binaries ($P_{\rm orb} \sim 100$–400 days), like RS Oph, that contain a red giant companion, and the heterogeneous class of short period binaries ($P_{\rm orb} \leq 1$ day), with different subclasses defined by the recurrent novae U Sco, CI Aql, and T Pyx, which consist of a white dwarf[16] and a (slightly evolved?) main sequence star [55].

Novae are very common phenomena (the second most frequent type of stellar ther-

---

role during the thermonuclear runaway, and moreover, the runaway undergoes dynamic effects, similar to those expected in a nova outburst.

[11] Roche lobes are tear-shaped regions around stars in a binary system, such that material within each lobe is gravitationally bound. See Problem 1 at the end of this chapter for an example of a 3D representation of gravitational equipotentials in a binary system, and reference [551] for further details on Roche lobes.

[12] For systems with highly magnetized white dwarfs, the disk is either truncated or nonexistent (in which case matter is directly accreted onto the white dwarf magnetic poles).

[13] When electrons—fermions—are densily packed together, as in the interior of a white dwarf, Pauli's exclusion principle forces them to occupy different energy states in an ordered form (first, ground state, while low-energy excited states becomes successively occupied). At $T = 0$, electrons occupy all low-energy states, while all high-energy levels remain empty. The corresponding configuration is referred to as *fully degenerate*, while the energy that distinguishes occupied and unoccupied states is known as the Fermi energy. In such conditions, the overall pressure is actually insensitive to temperature (therefore, a degenerate stellar plasma cannot react to a temperature increase with its expansion). As temperature increases, thermal effects force some electrons to occupy levels above the Fermi energy, hence reducing the degeneracy of the plasma.

[14] An animation, depicting the most relevant stages of this scenario (`binaries.wmv`), is available at `http://fisica.upc.edu/ca/users/jjose/CRC-Downloads`.

[15] A large fraction of novae, however, have orbital periods in the 2.8–4.1 hr range [439, 1907].

[16] See reference [1360] for an overview on mass determinations in nova systems.

monuclear explosions in the Galaxy after type I X-ray bursts). Although only a handful, 5 to 10, are discovered every year (mainly by amateur astronomers), a much higher nova rate, around 30 ± 10 yr^{-1} [1207, 1609], has been predicted from extrapolation of galactic and extragalactic data (M31, in particular). The reason for the scarcity of detections in our Galaxy[17] is extinction by interstellar dust.

In contrast to type Ia supernovae, in which the white dwarf is fully disrupted by the strength of the explosion, classical nova outbursts are, roughly speaking, restricted to the outer, accreted envelopes. Hence, they are expected to recur, since neither the star nor the binary system are destroyed by the event. Predicted recurrence times for classical nova outbursts are of the order of $10^4 - 10^5$ yr. Typical (observed) recurrence times for recurrent novae range between 10 and 100 yr, likely implying masses for the white dwarf hosting the explosion close to the Chandrasekhar limit and high mass-accretion rates[18] around $10^{-7} - 10^{-8}$ M$_\odot$ yr^{-1}. It is unclear whether recurrence times follow a nearly continuous sequence, ranging from the short values characteristic of recurrent novae to the long values predicted for classical novae.

Both novae and supernovae are characterized by a remarkable energy output, with peak luminosities reaching 10^5 and 10^{10} L$_\odot$, respectively[19]. A basic difference between both explosive phenomena is the mean ejection velocity ($> 10^4$ km s^{-1} in a supernova, while several 10^3 km s^{-1} in a classical nova), as well as the ejected mass (the whole star, ~ 1.4 M$_\odot$, in a thermonuclear supernova versus $10^{-3} - 10^{-7}$ M$_\odot$ for a nova). This suggests that supernovae (rather than novae) are major players in Galactic nucleosynthesis. Nevertheless, the interest in novae extends much beyond their contribution to the Galactic chemical pattern: Whereas the explosion that drives a supernova partially propagates supersonically (likely through a deflagration-detonation transition; see Chapter 5), the ignition regime characteristic of a nova is subsonic (deflagration). This makes the physics of nova explosions complex and numerically challenging. Indeed, hydrodynamic effects (turbulence, instabilities, mixing, etc.) combine with a suite of different nuclear processes in an envelope dominated by convective energy transport, all operating at similar timescales.

4.3 Designing a Nova Outburst

Nova explosions can naturally occur in two types of white dwarfs: carbon–oxygen-rich (hereafter, CO) and oxygen–neon-rich (ONe). The most frequent case is a CO white dwarf, the remnant of a progenitor star with a mass smaller than $\approx$ 7–8 M$_\odot$ [23, 949], after subsequent H- and He-burning. For more massive progenitors, nondegenerate C-ignition leads to the formation of a degenerate core mainly made of O and Ne, with traces of Mg and Na. The mass interval of the progenitor star leading to a particular white dwarf type depends on details of stellar evolution (e.g., the single or binary nature of the progenitor; see Chapter 2). Calculations show that CO white dwarfs are less massive than ONe white dwarfs. The

---

[17] About $\sim$ 400 novae have been detected in the Milky Way. W. Pietsch and F. Haberl maintain a catalogue of all historical optical novae detected in the Andromeda galaxy. As of January 2012, the catalog contains about 900 sources. See http://www.mpe.mpg.de/~m31novae/opt/m31/index.php.

[18] Hydrodynamic simulations of recurrent novae suggest that not all the accreted material is ultimately ejected in the outburst, and hence the white dwarf star is expected to increase in mass. This makes recurrent novae possible candidate sources for type Ia supernovae.

[19] Cecilia Payne-Gaposchkin's classic monograph [1400] has been the standard in the observational classification of novae.

mass cut distinguishing CO and ONe white dwarfs is, however, not well constrained [449]. Approximate values yield ≈ 1.1 M$_\odot$, when binarity is taken into account [643].

The key parameter in determining the strength of a nova outburst is the pressure achieved at the core-envelope interface, $P_\star$, which is a measure of the pressure exerted by the layers overlying the burning shell [572, 1615]:

$$P_\star = \frac{GM_{\rm wd}}{4\pi R_{\rm wd}^4} \Delta M_{\rm env}. \qquad (4.1)$$

To account for mass ejection, $P_\star \geq 10^{20}$ dyn cm^{-2} is required for a solar composition envelope [572]. Similar values were obtained by Jim MacDonald for solar abundances, while smaller pressures were found for a CNO-enhanced composition (e.g., $P_\star \sim 2 \times 10^{19}$ dyn cm^{-2} for $Z_{\rm CNO} = 0.51$) [1170]. For a given value of the pressure, Equation 4.1 reveals that the mass of the accreted envelope, $\Delta M_{\rm env}$, depends only on the white dwarf mass, $M_{\rm wd}$, because of the direct relationship between stellar mass and radius. For instance, a value of $\Delta M_{\rm env} \sim 5 \times 10^{-4}$ M$_\odot$ is obtained for a 1 M$_\odot$white dwarf ($R_{\rm wd} = 5575$ km), assuming $P_\star \sim 10^{20}$ dyn cm^{-2}. Since the white dwarf size is inversely proportional to its mass, $\Delta M_{\rm env}$ decreases as the white dwarf mass increases, and, hence, it becomes easier to produce a nova outburst on a massive white dwarf. Typical accreted envelope masses range between 10^{-3} and 10^{-5} M$_\odot$ (see also reference [1814] for a detailed study of ignition masses and their relation to ejected masses).

At a first glance, Equation 4.1 suggests that, for a given white dwarf mass, the envelope mass required to power a nova explosion is independent of the mass-accretion rate. But detailed hydrodynamic calculations [651, 1450, 1621, 1717, 2001] reveal some influence of the mass-accretion rate on the properties of the outburst. Since high mass-accretion rates result in larger energy released from gravitational compression, the times required to reach ignition conditions get reduced. As a result, as the mass-accretion rate increases, the accreted envelope mass decreases (see Chapter 5). Unfortunately, the mass-accretion rate is not a well-constrained quantity from an observational viewpoint. Mass-transfer rates between components—not mass-accretion rates!—in the range $\dot{M} \sim 10^{-7} - 10^{-11}$ M$_\odot$ yr^{-1}, have been inferred[20] in cataclysmic variables[21], mostly from correlations with observed orbital periods [1394]:

$$\dot{M} = 5.1^{+3}_{-2} \times 10^{-10} \left(\frac{P_{\rm orb}}{4\,{\rm hr}}\right)^{3.2\pm0.2}. \qquad (4.2)$$

According to Patterson [1394], systems in the range $0.7 < P_{\rm orb}({\rm hr}) < 3.3$, are characterized by low mass-transfer rates, $\dot{\rm M} \sim 10^{-10} - 10^{-11}$ M$_\odot$ yr^{-1}, while those with larger orbital periods, $3.3 < P_{\rm orb}({\rm hr}) < 24$, exhibit higher mass-transfer rates, $\dot{\rm M} \sim 10^{-8} - 10^{-9}$ M$_\odot$ yr^{-1}. Additional estimates based on accretion disk models can be found elsewhere [1664, 1665, 1905]. Other attempts include photometric analyses of faint, old novae for which very low mass-transfer rates have also been inferred (e.g., $\leq 10^{-12\pm0.5}$ M$_\odot$ yr^{-1}, for Nova CK Vul 1678 [1619]).

How these mass-transfer rates translate into mass-accretion rates is, however, a matter of debate. According to the semianalytical models of MacDonald, there is a maximum value for the mass-accretion rate that leads to a nova outburst for a given white dwarf mass [1170]. But several hydrodynamic simulations have shown that mass ejection occurs even for higher

---

[20]See, e.g., Shore et al. [1643] for an attempt to infer the mass-accretion rate from the outburst interval and ejected mass in LMC 1990 # 2.

[21]Ofer Yaron, Dina Prialnik, Michael Shara, and Attay Kovetz have computed nova models with very low mass-accretion rates, $\dot{\rm M} = 5 \times 10^{-13}$ M$_\odot$ yr^{-1}. In turn, Ami Glasner and Jim Truran have explored the possibility of CNO-breakout in novae, in the context of low luminosity white dwarfs accreting matter at low rates, $\dot{\rm M} = 10^{-11}$ M$_\odot$ yr^{-1} [651, 2001].

mass-accretion rates and more luminous white dwarfs than the critical values derived by MacDonald. For instance, Yaron et al. report mass ejection from models of 1 $M_\odot$ white dwarfs accreting solar-like material at a rate as high as 5×10^{-7} $M_\odot$ yr^{-1} [804]. For models of very luminous white dwarfs, see also reference [1621].

It is worth noting that a constant mass-accretion rate is assumed in most of the hydrodynamic nova simulations performed to date[22]. Nevertheless, there are reasons to believe that it may change throughout a nova cycle. Indeed, it has been suggested that novae may *hibernate* between outbursts [1127, 1452, 1618]. In this scenario, mass transfer is switched off a few centuries after the explosion, remains off for millennia, and then resumes as soon as angular momentum losses bring the secondary back into contact with its Roche lobe.

The effect of the white dwarf luminosity (or temperature) on the strength of the outburst has also been discussed in a number of papers. Schwartzman, Kovetz, and Prialnik, in particular, pointed out a twofold effect [1594]: in cold, low luminous white dwarfs, heat conduction into the core can delay the ignition. As a result of the longer accretion phase, larger masses, and, hence, larger pressures are achieved, which translate into more violent outbursts[23]. But in hot, luminous white dwarfs, the outermost core layers become convective, and larger levels of mixing through the core-envelope interface are found. Note that if the white dwarf is initially too luminous, the envelope is not highly degenerate when the thermonuclear runaway develops and a mild thermonuclear runaway with no mass ejection may occur. This scenario has been proposed to account for symbiotic novae, explosions that occur in wide stellar binary systems composed of a white dwarf and a red giant companion (see, for instance, Mikolajewska [1243] and references therein).

Degeneracy is a key ingredient for a successful nova outburst. White dwarfs are basically supported by the pressure exerted by electrons, a fermion gas ruled by Pauli's exclusion principle that forces particles to occupy quantum states in a regulated manner (i.e., first, the ground state, followed by ordered low-energy excited states successively occupied). Conditions are such that during the accretion stage, the envelope is degenerate, that is, the thermal energy of the electrons, $3/2\,kT$, is smaller than the Fermi energy, E_F. So the condition for degeneracy can be written, in a crude way, for a fully ionized electron gas, as

$$\frac{3}{2}kT < \frac{\hbar^2}{2m_e}\left(\frac{3\pi^2 (Z/A)\rho}{m_H}\right)^{2/3}, \qquad (4.3)$$

where T, ρ, Z, and A are the temperature, density, (mean) atomic number, and (mean) mass number of the stellar plasma, k is Boltzmann constant, $\hbar$ is the reduced Planck constant (or Dirac constant), m_e the mass of the electron, and m_H the mass of a hydrogen atom. Alternatively, Equation 4.3 can be written as

$$\frac{T}{\rho^{2/3}} < 1.3 \times 10^5 \left(\frac{Z}{A}\right)^{2/3} \text{K cm}^2\,\text{g}^{-2/3}, \qquad (4.4)$$

which points out that the smaller the value of $T/\rho^{2/3}$, the larger the degeneracy. Note that values at the center of the Sun (i.e., 15.7×10^6 K and 162 g cm^{-3}) yield $T/\rho^{2/3} = 5.3 \times 10^5$, meaning that degeneracy is relatively weak in its interior. Conversely, for a 1.3 $M_\odot$ white dwarf with $\rho_c \sim 5 \times 10^8$ g cm^{-3}, and considering a wide range of possible central temperatures (i.e., $10^6 - 10^8$ K), $T/\rho^{2/3}$ is comprised between 2 and 200, so degeneracy is a good approximation for white dwarf interiors.

Figure 4.2 shows snapshots of the evolution of the temperature and density profiles along

---

[22]See, however, reference [2001] for a model with a time-dependent mass-accretion rate based on changes of the stellar masses and binary separation.

[23]Similar effects have been described elsewhere [921, 1450, 1621, 1717, 2001].

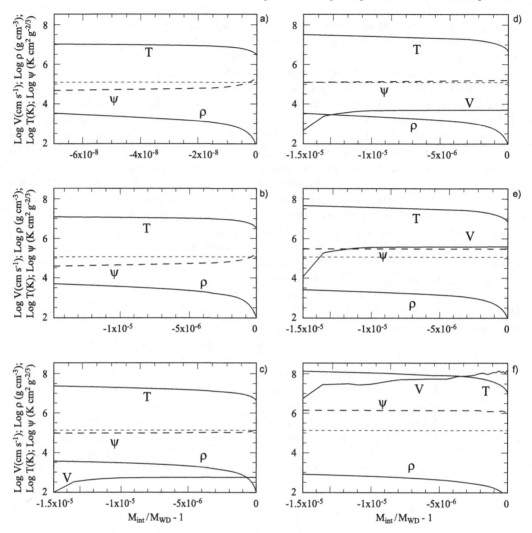

FIGURE 4.2

Snapshots of the progress of a thermonuclear runaway in a nova outburst on a 1.15 $M_\odot$ CO white dwarf, calculated with the hydrodynamic code SHIVA. The plot illustrates the time evolution of the temperature, T, density, ρ, velocity, V, and degeneracy parameter, $\Psi = T\rho^{-2/3}$. Panels correspond to different temperatures at the base of the accreted envelope, T_{base}: (a) 10^7 K; (b) 1.2×10^7 K, at the time when maximum density is achieved at base of the envelope, $\rho_{base} \sim 5170$ g cm^{-3}; (c) 2.4×10^7 K; (d) 3×10^7 K, when all envelope shells no longer fulfill the degeneracy condition given by Equation 4.4; (e) 5×10^7 K; and (f) 1.2×10^8 K. A color version of this plot is available at http://fisica.upc.edu/ca/users/jjose/CRC-Downloads.

the accreted envelope during the course of a nova outburst. For illustrative purposes, a horizontal line given by Equation 4.4, that approximately depicts the transition to degeneracy, is also shown. The early stages of the evolution (i.e., panels a and b) are characterized by large densities and moderate temperatures, such that $T/\rho^{2/3}$ is small, and hence most of the envelope is degenerate. As accretion goes on, compressional heating rises the temperature in the envelope and nuclear reactions ensue (panel c). Because the envelope is degenerate, it does not react to the temperature increase with an expansion, since its pressure is nearly independent of temperature. These conditions pave the road for a thermonuclear runaway. The large energy released by nuclear reactions cannot be transported only by radiation, and convection sets in as soon as superadiabatic gradients are established within the envelope. Convection spreads a fraction of the short-lived β^+-unstable nuclei ^{13}N, ^{14,15}O, and ^{17}F that are synthesized deep in the envelope to the outer cooler regions. The energy released in the β^+-decay of such short-lived species increases the temperature and the entropy of the material. As a result, degeneracy is lifted (panel *d*) and envelope expansion sets in (panels *d* to *f*). Therefore, the explosion is halted by the envelope expansion rather than by fuel consumption.

The effect of the envelope metal content (or the CNO abundance) on the outburst turns out to be similar to the previously described for the mass-accretion rate or the initial luminosity. Indeed, a decrease in the CNO abundance delays ignition, since less nuclear reactions occur (and, hence, less energy is released). This translates into a rise in the characteristic timescale for accretion, subsequently leading to larger accreted masses, larger pressures reached, and more violent outbursts[24]. Even though numerical models have shown that envelopes with solar metallicity can give rise to an explosion resembling *slow novae* [1453,1683], only envelopes with CNO-enhanced abundances (in the range $Z_{\mathrm{CNO}} \sim 0.2 - 0.5$) can reproduce the main properties of a *fast nova* [1714,1716] (see Section 4.5). The origin of such CNO enhancements required by models and inferred spectroscopically has been regarded as controversial. In principle, one may think of two possible sources: nuclear processing during the explosion or mixing at the core-envelope interface. Peak temperatures reached during a nova explosion are constrained by the chemical abundance pattern inferred from the ejecta and do not seem to exceed 4×10^8 K, so it is unlikely that the observed metallicity enhancements can be due to thermonuclear processes driven by CNO breakout. Instead, mixing at the core-envelope interface is a more likely explanation. Several mechanisms have been proposed to date, including diffusion-induced mixing [570,873,874,1006,1449], shear mixing at the disk-envelope interface [477,986,1038,1128,1171,1682], convective overshoot-induced flame propagation [1961], and mixing by gravity wave breaking on the white dwarf surface [15,16,1529][25].

4.3.1   The Roadmap toward Multidimensional Models

The different approaches adopted in the modeling of stellar explosions were briefly introduced in Chapter 1. They include parametrized one-zone models, semianalytical models, spherically symmetric (1D) hydrodynamic simulations, as well as multidimensional (2D and 3D) models. Many nova nucleosynthesis studies performed in the '80s and '90s relied on one-zone models [803,1887,1920,1933] (see also reference [880] for an analysis of the impact of nuclear uncertainties on the final nova yields through a one-zone approach). A semianalytic model, based on MacDonald's [1170], directly coupled to a large nuclear reaction network, was used by Coc et al. [361] in another study of nova nucleosynthesis. The model assumed a *fully convective* envelope in *hydrostatic* equilibrium. Therefore, key aspects of the evolution, such as the way in which convection settles, extends throughout the envelope, and recedes from its surface, were completely ignored.

---

[24]The properties of nova outbursts in metal-poor, primordial-like binaries, have been investigated in reference [918].

[25]These multidimensional studies have focused on the role of shear instabilities in the stratified plasmas that form nova envelopes. They showed that mixing can result from the resonant interaction between large-scale shear flows in the accreted envelope and gravity waves at the interface between the envelope and the underlying white dwarf. However, to account for significant mixing, a very high shear (with a specific velocity profile) had to be assumed.

To date, the state-of-the-art in nova nucleosynthesis relies on 1D, spherically symmetric, hydrodynamic models [432,919,1450,1621,1706,1717,2001], which combine hydrodynamics and nuclear burning through detailed simulations. However, it is becoming increasingly clear that the assumption of spherical symmetry, while capable of reproducing some features of the observed bursts, excludes an entire sequence of events associated with the way a thermonuclear runaway initiates (presumably as a point-like or multiple-point ignition) and propagates[26].

The first study of localized runaways on white dwarfs was carried out by Michael Shara [1616] on the basis of semianalytical models. He suggested that heat transport was too inefficient to spread a localized flame (i.e., the diffusively propagated burning wave may require tens of years to extend along the entire white dwarf surface), therefore concluding that localized, *volcanic-like* eruptions were likely to occur. The analysis relied only on radiative and conductive energy transport, hence ignoring the role played by convection on the lateral thermalization of a thermonuclear runaway. Indeed, as soon as superadiabatic gradients are established (when $T_{base} \sim 2 \times 10^7$ K, for a solar-composition nova envelope), macroscopic mass elements are exchanged between hotter and cooler regions of the envelope through convective transport. These blobs ultimately dissolve in the environment, hence releasing their excess heat. The treatment of heat transfer in convective zones is tackled by means of phenomenological approaches because of its complexity and associated uncertainties. The most commonly used prescription in stellar evolution codes, the *mixing-length theory*, was already described in Chapter 1. Unfortunately, mixing-length was built upon the assumption of hydrostatic equilibrium, which is at odds with the conditions achieved in explosive environments, such as novae and supernovae. Moreover, convection is a truly multidimensional process that hardly can be modeled, for explosive conditions, in spherical symmetry.

The importance of multidimensional effects in explosions occurring in thin stellar shells was revisited by Bruce Fryxell and Stan Woosley [569], who concluded that the most likely scenario for nova outbursts involves runaways propagated by small-scale turbulence. A relation for the velocity of the (subsonic) deflagration front spreading along the stellar surface was derived on the basis of dimensional analysis and flame theory,

$$v_{def} \sim \left(\frac{H_P \, v_{conv}}{\tau_{burn}} \right)^{1/2} , \tag{4.5}$$

where H_P is the pressure scale height, v_{conv} the characteristic convective velocity, and τ_{burn} a characteristic timescale for fuel burning. Typical values for nova outbursts yield $v_{def} \sim 10^4$ cm s^{-1} (i.e., the flame propagates halfway along the stellar surface in about ~ 1.3 days).

Next attempts to address the importance of multidimensional effects on nova explosions in a purely hydrodynamic framework were performed in the 1990s. Anurag Shankar, Dave Arnett, and Bruce Fryxell [1612,1613] evolved a 1.25 $M_\odot$ accreting white dwarf in spherical symmetry, and subsequently mapped the structure into a 2D domain (i.e., a spherical-polar grid of only 25 km×60 km, because of computational limitations). The explosive stage was then followed in 2D with the explicit, Eulerian code PROMETHEUS. Unfortunately, the subsonic nature of the problem, coupled with the use of an explicit code (with a timestep limited by the Courant–Friedrichs–Levy condition; see Chapter 1), posed severe constraints on the simulation. Indeed, the authors had to adopt very extreme conditions, such as huge initial temperature perturbations of about $\sim 100\%$–600% close to the envelope base, driving an unrealistic detonation front in the accreted envelope. Moreover, the total computed

---

[26]Observations of nova shells, in turn, frequently reveal nonspherical ejecta (i.e., multiple shells, emission knots, and chemical inhomogeneities), inferred from line profiles during the early stages of the explosion and from imaging of the resolved ejecta. See also reference [1638] for a study of the effect of asphericity in nova spectral and photometric evolution.

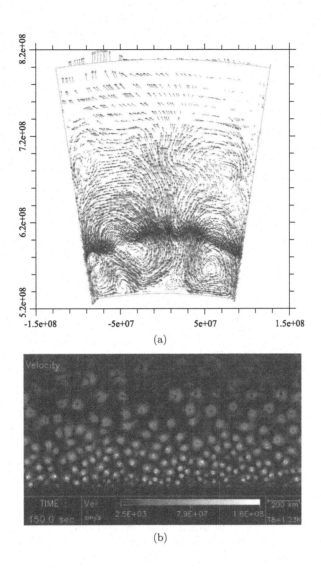

(a)

(b)

FIGURE 4.3

(a) Velocity field of a 2D simulation (5 km×5 km) of mixing at the core-envelope inter-face during a nova outburst on a 1 $M_\odot$ CO white dwarf. Note how the convective eddies have merged into a few, huge cells that extend almost all the way throughout the envelope. Calculations were performed with the VULCAN code. Figure from Glasner, Livne, and Tru-ran [649], reproduced with permission. (b) Same as Panel a, for a higher-resolution (1 km×1 km), 2D simulation computed with PROMETHEUS [971]. Here, a different convective pat-tern, characterized by much smaller, more numerous, and stable cells, was found. A color ver-sion of this plot is available at http://fisica.upc.edu/ca/users/jjose/CRC-Downloads. Figure from Kercek, Hillebrandt and Truran [971]; reproduced with permission.

time was only about 1 second. The simulations revealed that instantaneous, local temperature fluctuations can induce hydrodynamic instabilities whose rapid rise and subsequent expansion (in a dynamical timescale) cools the hot material and halts the lateral spread of the burning front. The study, therefore, favored the local volcanic-like eruptions suggested by Shara, but it was clear that a full hydrodynamic simulation, performed under realistic conditions for nova outbursts, was lacking. This was achieved, shortly afterward, by Ami Glasner, Eli Livne, and Jim Truran [648, 649]. To this end, 2D simulations were computed with the code VULCAN, an arbitrary Lagrangian Eulerian (ALE) hydrocode capable of handling both explicit and implicit steps. As in Shankar et al., only a slice of the star (0.1 π^{rad}), in spherical-polar coordinates with reflecting boundary conditions, was modeled. The adopted resolution near the envelope base was 5 km$\times$5 km. As in previous work, to overcome the early, computationally challenging stages of the runaway, a 1 $M_\odot$ CO accreting white dwarf was first evolved using a 1D hydro code, and then mapped into a 2D domain (when the temperature at the envelope base reached $T_{\mathrm{base}} \sim 10^8$ K).

These 2D simulations showed that the thermonuclear runaway initiates as myriad irregular, localized eruptions, each surviving only a few seconds, that appear close to the envelope base. These localized flames are caused by buoyancy-driven temperature fluctuations (likely driven by an initial model that was not in fully hydrostatic conditions). Turbulent diffusion efficiently dissipates any local burning around the core, and, hence, the flame must spread along the entire envelope shortly afterward. Larger convective eddies (see Figure 4.3(a)) than those reported from 1D simulations were found, extending up to 2/3 of the envelope height with typical velocities $v_{\mathrm{conv}} \sim 10^7$ cm s^{-1}. The core-envelope interface appears to be convectively unstable, and CO-rich material is efficiently dredged up from the outermost layers of the white dwarf core, providing a source for the metallicity enhancement of the envelope through Kelvin–Helmholtz instabilities (see Section 1.3.2) at levels that agree with observations ($\sim$ 20%–30%, by mass [632]).

It is, however, worth noting that despite the differences found in the convective flow patterns in 1D and 2D models, the expansion and progress of the 2D burning front toward the outer envelope quickly becomes almost spherically symmetric, although the initial burning process was not. Moreover, the 2D simulations also show good agreement with the big picture outlined by 1D models. This includes the critical role played by the very abundant, β^+-unstable nuclei ^{13}N, ^{14,15}O, and ^{17}F in the expansion stage, and, consequently, predict the presence of large amounts of ^{13}C, ^{15}N, and ^{17}O in the ejecta.

Another set of multidimensional simulations, aimed at verifying the general trends reported by Glasner et al., were published by A. Kercek, Wolfgang Hillebrandt, and Jim Truran in the 1990s. The simulations were performed with a version of the Eulerian code PROMETHEUS, assuming the same initial model as in Glasner et al., but adopting a Cartesian, plane-parallel geometry to allow the use of periodic boundary conditions. Two different 2D simulations, with a similar computational domain (a box of about 1800 km$\times$1100 km), were performed with a coarser 5 km$\times$5 km grid (as in Glasner et al.), and with a finer 1 km$\times$1 km grid [971]. The simulations produced somewhat less violent outbursts, characterized by longer runaways with lower peak temperatures and ejection velocities, likely resulting from large differences in the convective flow patterns: whereas Glasner et al. found a few, large convective eddies dominating the flow, Kercek et al. found that the early runaway was governed by small, very stable eddies (Figure 4.3(b)), which in turn led to more limited dredge up and mixing episodes.

Such discrepancies were even more striking in 3D [972]: with a computational domain of 1800 km$\times$1800 km$\times$1000 km, and a resolution of 8 km$\times$8 km$\times$8 km, the simulation revealed mixing by turbulent motions taking place on very small scales, and peak temperatures that were slightly lower than in the 2D case (a consequence of the slower and more limited dredge up of core material). Moreover, the envelope attained a maximum velocity that was a factor

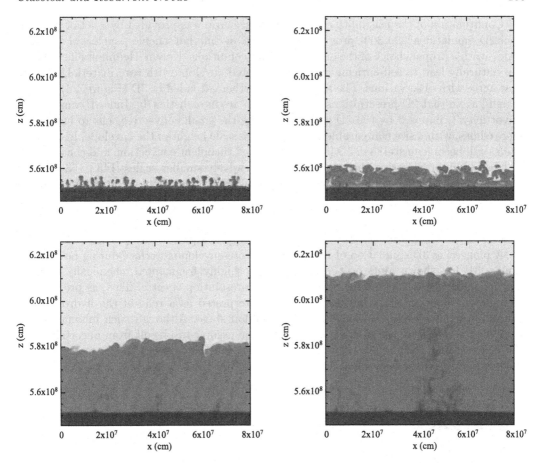

FIGURE 4.4

Two-dimensional snapshots of the development of hydrodynamic instabilities, in a 3D simulation of mixing at the core-envelope interface during a nova explosion, shown in terms of the ^{12}C mass fraction in logarithmic scale. Dredge up of core material by Kelvin–Helmholtz instabilities translates into a mass averaged abundance of CNO-nuclei in the envelope of 0.12 (panel a; t = 152 s), 0.13 (b; 192 s), 0.16 (c; 297 s), and 0.18 (d; 381 s). The mean CNO-mass fraction at the end of the simulation reached 0.20. Simulations were run at the MareNostrum supercomputer, requiring about 150,000 CPU hours, with a maximum resolution of 1.56 km×1.56 km×1.56 km. A color version of this plot, together with a movie, are available at `http://fisica.upc.edu/ca/users/jjose/CRC-Downloads`. Graphic material based on Casanova et al. [301].

∼ 100 smaller than the escape velocity, and, presumably, no mass ejection resulted (except for a possible late wind mass loss phase). In view of these results, Kercek et al. concluded that mixing must take place prior to the runaway, in sharp contrast with the main results reported by Glasner et al. The controversy was carefully analyzed, and partially solved, by Glasner et al. [650], who concluded that the early stages of the explosion, when the evolution is quasi-static, are extremely sensitive to the adopted outer boundary conditions. Indeed, they showed that Lagrangian simulations, in which the envelope was allowed to expand and mass was conserved, led to consistent explosions. In contrast, in Eulerian schemes with a free outflow outer boundary condition (the choice adopted in Kercek et al.), the outburst was artificially quenched.

Confirmation of the feasibility of this mixing scenario was provided by a set of independent 2D simulations [299,300], proving that even in an Eulerian scheme—such as the FLASH code—with a proper choice of the outer boundary conditions, Kelvin–Helmholtz instabilities can naturally lead to self-enrichment of the accreted envelope with core material, at levels that agree with observations. The analysis was further extended to 3D (Figure 4.4), since it is well known that 2D prescriptions for convection are unrealistic [70]. Indeed, conservation of vorticity[27], imposed by the 2D geometry, forces the small convective cells to merge into large eddies, with a size comparable to the pressure scale height of the envelope. In contrast, eddies will become unstable in 3D in fully developed turbulent convection, and consequently will break up, transferring their energy to progressively smaller scales [1439,1634]. These structures, vortices, and filaments will undergo a similar fate down to approximately the Kolmogorov scale,

$$\eta \sim (\nu^3/\epsilon)^{1/4}\,, \tag{4.6}$$

where ν is the kinematic viscosity and ϵ is the energy dissipation rate (see also Chapter 5).

A pioneering 3D simulation of mixing at the core-envelope interface during nova explosions [301] has shown hints on the nature of the highly fragmented, chemically enriched and inhomogeneous nova shells, observed in high-resolution spectra: This, as predicted in Kolmogorov's theory of turbulence, has been interpreted as a relic of the hydrodynamic instabilities that develop during the initial ejection stage. Although such inhomogeneous patterns inferred from the ejecta have usually been assumed to result from uncertainties in the observational techniques, they may represent a real signature of the turbulence generated during the thermonuclear runaway.

4.4   Nova Nuclear Symphony

The most important nuclear processes in nova outbursts involve proton captures, since these have the smallest Coulomb barriers and, thus, the largest reaction cross-sections (see Chapter 2). Consequently, and in sharp contrast with other (related) astrophysical explosive sites, the main nuclear path during classical nova outbursts runs close to the valley of stability and is driven by (p, γ), (p, α), and β^+ interactions. Current nova models predict peak temperatures $< 4 \times 10^8$ K, which prevents both the occurrence of α-capture reactions and the possible extension of the nuclear path through CNO breakout. It is also worth noting that neutron-capture reactions do not play any role in nova outbursts.

From the nuclear physics viewpoint, novae are unique stellar explosions. The somewhat limited nuclear activity, which involves about one hundred relevant species (with $A < 40$) linked through a few hundred nuclear processes, together with the temperatures achieved in the course of the outburst ($T \sim 10^7 - 4 \times 10^8$ K), allow us to rely primarily on experimental information [924].

Measurements of nuclear cross-sections at energies relevant for nova outbursts, characterized by a Gamow peak around ~ 100 keV, although sometimes challenging, are feasible in the laboratory. Indeed, a significant number of nuclear reactions of interest for nova nucleosynthesis have been measured in the laboratory. Unfortunately, direct measurements of reactions involving unstable target nuclei have only begun recently. Indeed, only ^{18}F(p,α), ^{21}Na(p,γ), and ^{26g}Al(p,γ) have so far been measured directly around the nova Gamow window. This is in sharp contrast with hydrostatic burning scenarios (i.e., the Sun), for which

---

[27]Vorticity is a physical magnitude, defined as the curl of the flow velocity vector, and thus measures the spinning of a fluid. In a 2D flow, vorticity is perpendicular to the plane of the flow.

TABLE 4.1
Nova Models

	Model 115CO	Model 115ONe	Model 135ONe
M_{wd} ($M_\odot$)	1.15	1.15	1.35
Pre-enrichment	50% CO	50% ONe	50% ONe
ΔM_{env} (10^{-5} $M_\odot$)	1.73	3.04	0.506
t_{acc} (10^5 yr)	1.08	1.90	0.316
t_{rise} (10^6 s)	0.696	12.3	3.46
P_{max} (10^{19} dyn cm^{-2})	1.15	2.02	5.32
ρ_{max} (10^3 g cm^{-3})	5.17	7.45	14.2
T_{max} (10^8 K)	2.05	2.28	3.13
t_{max} (s)	54	492	175
ΔM_{ejec} (10^{-5} $M_\odot$)	1.40	2.46	0.455
v_{ejec} (km s^{-1})	2460	2290	4220

Gamow peaks are located at lower energies (~ 10 keV). Because of Coulomb barrier considerations, it is almost impossible to measure reactions at such low energies, and often one has to extrapolate from the laboratory values obtained at higher energies.

4.4.1 Nova Nucleosynthesis

Novae involving massive (ONe) white dwarfs, with $M_{wd} \gtrsim 1.1$ $M_\odot$, are thought to host *neon* novae[28], characterized by the presence of intermediate-mass elements (Ne, Al, S...) as inferred from the spectra. On the other hand, novae that take place on white dwarf stars with masses below $M_{wd} \leq 1.1$ $M_\odot$ likely undergo mixing with a CO substrate (see Section 4.3). It is expected that neon novae and non-neon novae will exhibit different characteristics, such as the amount of mass accreted, the maximum temperature achieved in the envelope, the mass and velocity of the ejecta, and the accompanying nucleosynthesis.

For illustrative purposes, the outcome of a nova explosion taking place onto a CO white dwarf (hereafter, model 115CO) will be compared with that involving an ONe white dwarf (model 115ONe), hence analyzing the influence of the substrate on the outburst. Both models assume white dwarf stars of $M_{wd} = 1.15$ $M_\odot$, with the same initial luminosity, 10^{-2} $L_\odot$, accreting mass at a constant rate of 1.6×10^{-10} $M_\odot$ yr^{-1}, and have been evolved with the 1D hydrodynamic code SHIVA. A mixture containing 50% solar material and 50% core (either CO- or ONe-rich) material is assumed to mimic the mixing episodes taking place at the core-envelope interface.

Table 4.1 summarizes the most relevant aspects of models 115CO and 115ONe: ΔM_{env} refers to the envelope mass accreted before the outburst, with t_{acc} being the overall duration of the accretion stage; t_{rise} is the time required for a temperature rise from an arbitrary value of $T_{base} = 3 \times 10^7$ K to 10^8 K at the base of the envelope; P_{max}, ρ_{max}, and T_{max} correspond to the maximum values of pressure, density, and temperature attained at the envelope; t_{max} is the time required to reach peak temperature from $T_{base} = 10^8$ K; and ΔM_{ejec} and v_{ejec} represent the total ejected mass and the mean velocity of the ejecta.

The most relevant nuclear reactions at the innermost, hottest, envelope shells are displayed, at the onset of accretion, in Figure 4.5, in terms of *reaction fluxes* (i.e., number of reactions per unit time)[29].

---

[28] An alternative mechanism to produce moderate-mass CO white dwarfs with Ne- and Mg-rich coatings has been proposed by Mike Shara and Dina Prialnik [1620].

[29] Movies showing the time evolution of mass fractions for the most abundant species (i.e., $X > 10^{-5}$)

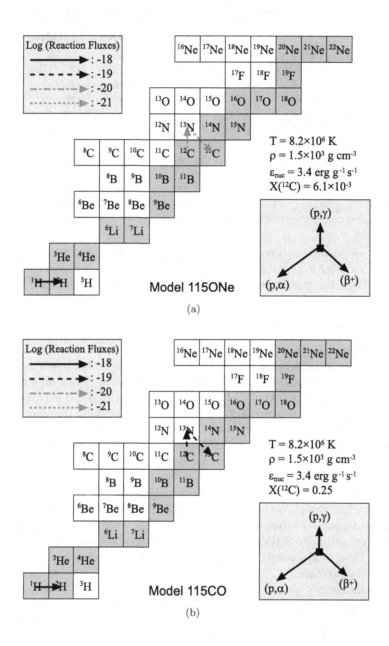

FIGURE 4.5

Main reactions fluxes at the innermost envelope shell for models 115ONe (upper panel) and 115CO (lower panel), at the onset of accretion, when $T_{\text{base}} = 8.2 \times 10^6$ K. Fluxes are dominated by the *pp chain* reactions. Stable isotopes are highlighted as dark squares. A color version of this plot is available at http://fisica.upc.edu/ca/users/jjose/CRC-Downloads.

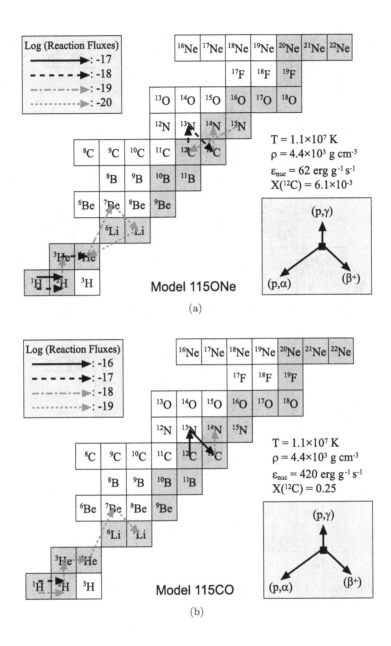

FIGURE 4.6

Same as Figure 4.5, but for the time when $T_{base} = 1.1 \times 10^7$ K. The larger amount of CNO-nuclei present in the envelope of model 115CO translates into a larger nuclear energy release than model 115ONe. Note also that fluxes for model 115CO are dominated by CNO cycle reactions, while pp chains are still the dominant channel in model 115ONe. A color version of this plot is available at http://fisica.upc.edu/ca/users/jjose/CRC-Downloads.

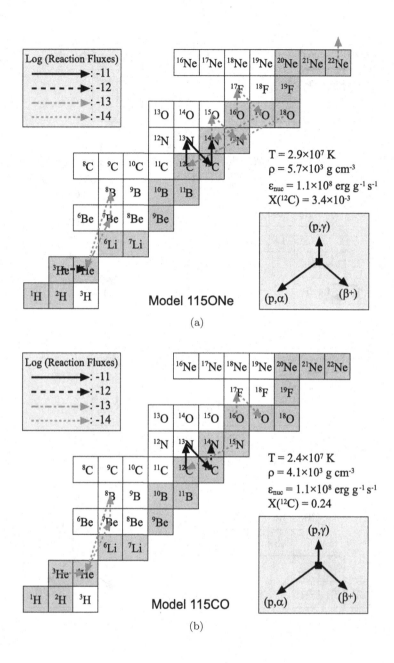

FIGURE 4.7

Same as Figure 4.5, at the end of the accretion stage. Both models are dominated mainly by CNO cycle reactions. Note, however, that model 115ONe requires larger temperatures and densities than model 115CO to release the same nuclear energy, a consequence of the different initial ^{12}C content. A color version of this plot is available at http://fisica.upc.edu/ca/users/jjose/CRC-Downloads.

The early stages of evolution, when temperature at the envelope base is well below 10^7 K, are fully dominated by the pp chain reactions, mostly p(p,e^+ ν)d (see Figure 4.5). A minor contribution from CNO-cycle reactions, such as ^{12}C(p, γ)^{13}N(β^+)^{13}C, is also noticeable, mostly in model 115CO because of its higher ^{12}C as compared with model 115ONe.

As the temperature rises, following compressional and nuclear heating, models progressively exhibit significant differences. Already at $T_{\mathrm{base}} = 1.1 \times 10^7$ K (Figure 4.6), model 115CO becomes dominated by CNO-cycle reactions. Model 115ONe is still governed by the pp chain reactions (e.g., pp1 and pp2). Another reaction, somewhat competing with p(p,e^+ ν)d in model 115ONe, is the so-called *pep* reaction, p(p e^-, ν)d. Although energy production from the *pep* reaction is usually negligible for main sequence stars, such as the Sun, its strong dependence on density[30] makes it suitable for nova envelopes that can naturally reach densities above 10^4 g cm^{-3}. Here, for the models considered, its flux is about 10 times smaller than that from p(p,e^+ ν)d, even at maximum density, but its role in novae hosting more massive white dwarfs could be significant, reducing the evolution time to the peak of the explosion and, thereby, the overall accreted mass, as first pointed out by Starrfield et al. [1706].

The presence of large amounts of ^{12}C in model 115CO favors a much higher activity in the CNO cycle, and, hence, more energy is released (see Figure 4.7). This translates into shorter characteristic timescales for the outburst, as shown in Table 4.1. Notably, shorter accretion times reduce the amount of material that piles up on top of the white dwarf before the dynamic stages of the outburst begin. Indeed, model 115ONe accretes an envelope about twice as massive as that from model 115CO. In turn, somewhat larger densities and pressures are achieved in model 115ONe, with important consequences for the course of the explosion, since the strength of the outburst depends mainly on the presure at the core-envelope interface (see Section 4.3).

A close look at the nuclear activity at peak temperature (i.e., Figures 4.8 to 4.10) reveals that, regardless of the model considered, the main nuclear path proceeds close to the valley of stability and is driven by p-capture reactions (i.e., (p, γ) and (p, α)) and β^+-decays. No relevant n- or α-capture reaction is found (an exception being the pp2 chain reaction ^{3}He(α, γ)^{7}Be).

Model 115CO is mainly powered by reactions from the hot and cold CNO1 and CNO2 (the most important reaction being ^{13}N(p, γ)^{14}O), where all relevant nuclear activity is confined. Figure 4.8 also underlines that no noticeable leakage from the CNO cycle occurs, with a marginal activity in the NeNa–MgAl region being powered by the initial presence of ^{22}Ne. In turn, model 115ONe is mainly powered by reactions of the hot and cold CNO1, CNO2, and CNO3 cycles (dominated by the chain ^{16}O(p, γ)^{17}F(β^+)^{17}O(p, α)^{14}N(p, γ)^{15}O, together with ^{27}Al(p, γ)^{28}Si). As for model 115CO, no significant leakage from the CNO region is found, and remarkable activity in the NeNa–MgAl is indeed driven by the large amounts of ^{20}Ne (and to some extent, ^{21,22}Ne, ^{23}Na, ^{24,25,26}Mg, and ^{27}Al) present in the envelope. As shown in Figure 4.9, the nuclear activity in model 115ONe reaches ^{31}S around T_{peak}. It is worth noting that while the somewhat larger peak temperature reached in this model is due to the larger accreted envelope mass (and, hence, larger pressure), the extent of the nuclear activity is mostly due to the presence of seed ^{20}Ne nuclei.

At nova conditions, two different ^{26}Al states must be distinguished: the ground, long-lived state ($T_{1/2} \sim 0.72$ Myr) and an isomeric, short-lived state ($T_{1/2} \sim 6.4$ s; see Chapter

---

at the innermost envelope shell for models 115CO (`Model-CO115.wmv`), 115ONe (`Model-ONe115.wmv`), and 135ONe (`Model-ONe135.wmv`) are available at `http://fisica.upc.edu/ca/users/jjose/CRC-Downloads`. Remarkably, and regardless of the model, all movies reveal large amounts of H and ^{4}He leftover in the innermost envelope layers, confirming that the nova outburst is not halted by fuel consumption but rather results from envelope expansion.

[30]The *pep* rate includes an extra density term, $\rho(1+X_{\mathrm{H}})/2$, where X_{H} is the hydrogen mass-fraction [304].

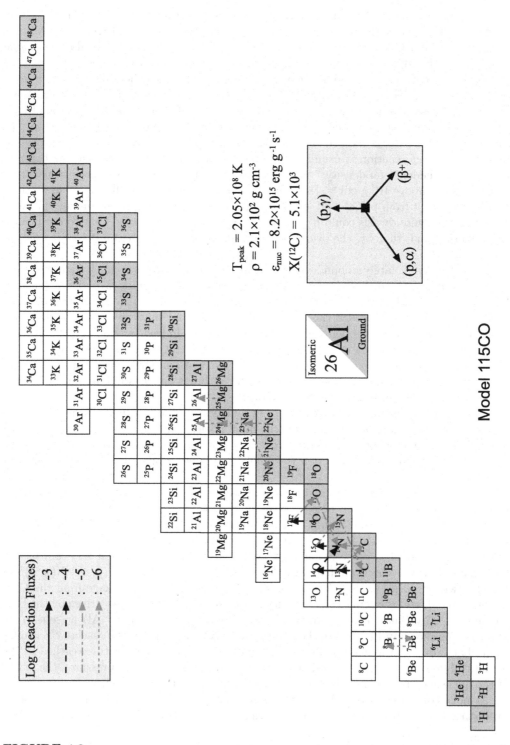

FIGURE 4.8

Main reaction fluxes at the innermost envelope shell for model 115CO, at peak temperature ($T_{\text{base,max}}$ = 2.05 × 10⁸ K). A color version of this plot is available at http://fisica.upc.edu/ca/users/jjose/CRC-Downloads.

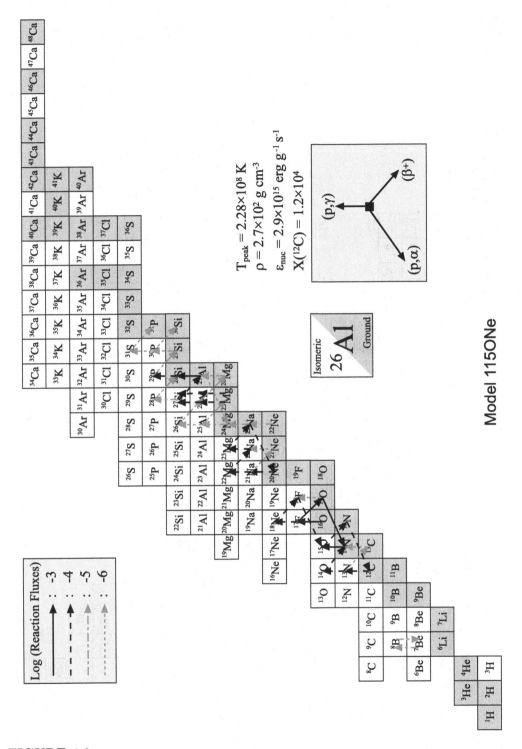

FIGURE 4.9
Same as Figure 4.8 for model 115ONe, at peak temperature ($T_{\text{base,max}}$ = 2.28 × 10^8 K). A color version of this plot is available at http://fisica.upc.edu/ca/users/jjose/CRC-Downloads.

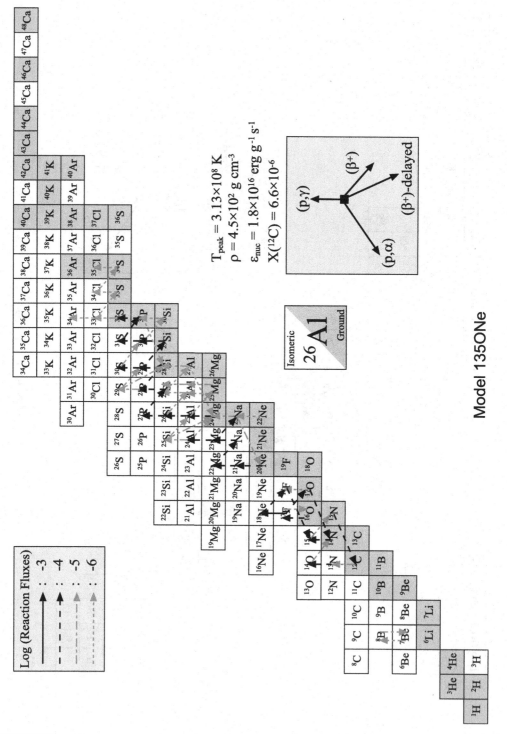

FIGURE 4.10
Same as Figure 4.8 for model 135ONe, at peak temperature
($T_{\text{base,max}}$ = 3.13 × 10⁸ K). A color version of this plot is available at
http://fisica.upc.edu/ca/users/jjose/CRC-Downloads.

2 and Figure 4.11). This, as discussed in Section 4.4.3.3, has strong nucleosynthetic implications, since the fraction of the nuclear flow that goes through the isomeric state bypasses the ground state, deeply affecting the synthesis of ^{26}Al in nova explosions.

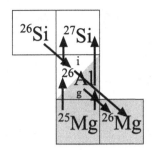

FIGURE 4.11

Main nuclear interactions affecting ^{26}Al ground (g) and isomeric (i) states, for nova conditions. A color version of this figure is available at http://fisica.upc.edu/ca/users/jjose/CRC-Downloads.

Figure 4.10 illustrates the effect of the mass of the white dwarf hosting the explosion. A peak temperature exceeding 3.1×10^8 K is achieved as a consequence of the large pressure reached in model 135ONe (see Table 4.1). The nuclear activity extends all the way up to ^{34}Ar (and reaches, shortly afterward, ^{39}Ca), but as reported for models 115CO and 115ONe little leakage from the CNO region is found. This suggests that the presence of intermediate-mass elements (with $A \geq 10$) systematically inferred from ejected nova shells is driven by mixing at the core-envelope interface rather than by nuclear processing. The main nuclear path is mostly dominated by reactions of the hot CNO1, CNO2, and CNO3 cycles, mainly ^{16}O(p, γ)^{17}F(p, γ)^{18}Ne. This is supplemented by multiple reactions of the NeNa–MgAl region (i.e., ^{20}Ne(p, γ)^{21}Na(p, γ)^{22}Mg) or beyond (a suite of p-capture reactions on several Si and P isotopes). Aside from ^{8}B(γ, p)^{7}Be, which is at equilibrium with the direct ^{7}Be(p, γ)^8 reaction, another photodisintegration reaction, ^{17}F(γ, p)^{16}O, shows up around peak temperature for this model. Also noticeable are the two β-delayed particle decays that connect ^{25}Si $\rightarrow$ ^{24}Mg and ^{29}S $\rightarrow$ ^{28}Si. About 64% of the time, ^{25}Si β^+-decays into an excited state of ^{25}Al, which subsequently γ-decays into its ground state. But 36% of the time, the populated ^{25}Al states decay by emiting a proton, hence leading to ^{24}Mg instead. As for ^{29}S, the probabilities of β-decaying into the ^{29}P ground state or into excited ^{29}P states are very similar. In the latter, the excited levels subsequently decay via proton emission to ^{28}Si.

4.4.2 Novae and the Galactic Alchemy: ^{13}C, ^{15}N, and ^{17}O

The material ejected during a nova outburst has been, according to models, exposed to peak temperatures in the range between 10^8 and 4×10^8 K [919, 1450, 1717, 2001], for several hundred seconds. Hence, the ejecta are expected to show signatures of nuclear processing, with chemical abundance patterns that do not correspond to equilibrium CNO burning [1716]. This has raised the issue of the potential contribution of novae to the Galactic abundances.

A back-of-the-envelope estimate of such contribution can be obtained from the product of the Galactic nova rate (30 ± 10 yr^{-1} [1609]), the average ejected mass per outburst[31] ($\sim 10^{-3} - 10^{-5}$ M$_\odot$), and the Galaxy's lifetime (10^{10} yr):

---

[31]See reference [1621] for a critical study on nova ejected masses.

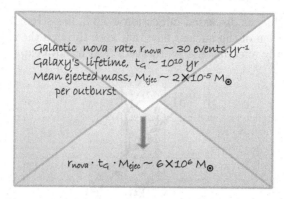

For a mean ejected mass value of 2×10^{-5} $M_\odot$, about 6×10^6 $M_\odot$ injected by nova explosions into the interstellar medium during the overall Galaxy's lifetime are found. This corresponds only to $\sim 1/3000$th of the Galactic disk's gas and dust component, which highlights that novae scarcely contribute to the Galactic chemical abundances (as compared with major sources, such as supernovae or asymptotic giant branch stars). Nevertheless, novae are expected to play a major role in the synthesis of some species, largely overproduced with respect to solar abundances. Here, we define the (over)production factor, f, as the ratio between the mass fraction of a given species in the ejecta over the solar value,

$$f = \frac{X_i}{X_{i,\odot}}. \tag{4.7}$$

Roughly speaking, to overcome the limited amount of mass ejected during outbursts, overproduction factors ~ 1000 are required. Hence, as seen in Figure 4.12, classical novae are expected to be major contributors to the Galactic content of ^{15}N, ^{17}O, and, to some extent, ^{13}C, with a lower contribution in a number of other species with A < 40, such as ^{7}Li, ^{19}F, or ^{26}Al [919, 1008, 1717, 2001]. Figure 4.12 also reveals that the main nuclear activity in CO novae does not extend much beyond the CNO region (see Section 4.4.1). In sharp contrast, ONe models exhibit a somewhat broader activity, extending up to silicon (1.15 $M_\odot$ ONe) or argon (1.35 $M_\odot$ ONe). This, in turn, suggests that the presence of significant amounts of intermediate-mass elements in the ejecta, such as phosphorus, sulfur, chlorine, or argon may indeed reveal the presence of an underlying massive ONe white dwarf hosting the explosion.

Models also suggest that calcium (i.e., A < 40) represents the theoretical endpoint for nova nucleosynthesis. This is in agreement with the abundance patterns inferred from detailed observations of ejected nova shells. Indeed, spectroscopic abundance determinations of nova shells include silicon (Nova Aql 1982 [49, 1673], QU Vul 1984 [49]), sulfur (Nova Aql 1982 [49, 1673]), chlorine (Nova GQ Mus 1983 [1263]), argon and calcium (Nova GQ Mus 1983 [1263], Nova V2214 Oph 1988, Nova V977 Sco 1989, and Nova V443 Sct 1989 [49]), whereas no significant overproduction with respect to solar values has ever been reported for elements above Ca.

The nuclear activity in the Si–Ca mass region is powered by a leakage from the NeNa–MgAl region, mainly through ^{26g}Al(p,γ)^{27}Si and ^{27}Al(p,γ)^{28}Si. The main nuclear reaction that governs the pathway toward heavier species is ^{30}P(p,γ)^{31}S, which is either followed by ^{31}S(p,γ)^{32}Cl(β^+)^{32}S, or by ^{31}S(β^+)^{31}P(p,γ)^{32}S [917]. The current ^{30}P(p,γ) rate is still affected by nuclear uncertainties (Section 4.4.4.3). This is of paramount importance, since this reaction not only determines the abundance pattern of nuclei heavier than P (i.e., S, Cl, Ar, K) but also, in competition with ^{30}P(β^+)^{30}Si, determines the final ^{30}Si/^{28}Si ratio, a valuable tool for identifying presolar nova grains (see Section 4.6.2).

Models of recurrent novae [1708, 1709, 1828], in turn, require large mass-accretion rates

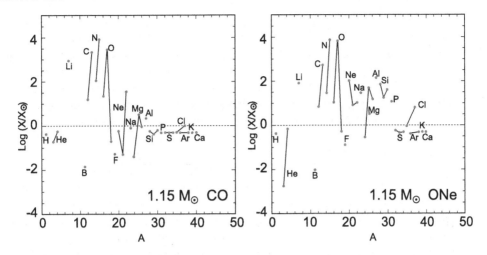

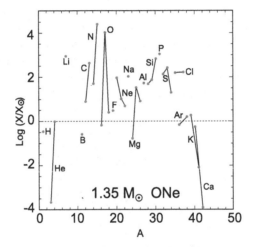

FIGURE 4.12

Mean overproduction factors, relative to solar, in the ejecta of models 115CO (upper left panel), 115ONe (upper right) and 135ONe (lower panel). Color versions of these plots are available at http://fisica.upc.edu/ca/users/jjose/CRC-Downloads.

onto very massive white dwarfs, close to the Chandrasekhar limit, to match the short interoutburst periods observed[32]. Their associated nucleosynthesis is characterized by large H and He abundances in the ejecta, with typical mass fractions around 0.6 and 0.4, respectively. No large overproduction of metals with respect to solar values are found, in agreement with observations. This distinctive chemical pattern reflects limited mixing episodes as well as the fast dynamic stages of the event. In particular, elements like Li, O, or Ne are underproduced, while S and Ar are slightly overproduced [781].

---

[32]See Darnley et al. [407] for observations of the M31 recurrent nova M31N 2008-12a, characterized by a unique recurrence period of ~ 1 yr.

TABLE 4.2

Main Radioactive Species Synthesized During Nova Outbursts [89]

Isotope	Half-life	Decay mode	Nova type	γ-ray emission
^{13}N	598 s	β^+-decay	CO/ONe	511 keV + continuum
^{18}F	110 min	β^+-decay	CO/ONe	511 keV + continuum
^{7}Be	53.2 day	e^--capture	CO	478 keV
^{22}Na	2.60 yr	β^+-decay	ONe	1275 keV + 511 keV
^{26}Al	7.17×10^5 yr	β^+-decay	ONe	1809 keV + 511 keV

4.4.3 γ-Ray Emitters

Among the species synthesized during nova outbursts, several radioactive nuclei have raised particular interest as potential sources of γ radiation [774] (see Table 4.2). A prompt γ-ray emission, consisting of an intense line at 511 keV and a lower-energy continuum, has been predicted from ^{13}N and ^{18}F decay. Models also predict line emission at 478 keV and 1275 keV later in the outburst from ^{7}Be and ^{22}Na decay. Finally, classical novae are also expected to contribute to the overall Galactic 1809 keV ^{26}Al line.

In spite of earlier claims [348], no γ-rays are expected to emerge in connection with the short-lived species ^{17}F ($T_{1/2} = 65$ s), ^{14}O ($T_{1/2} = 71$ s), and ^{15}O ($T_{1/2} = 122$ s), since they are produced too early and cannot yet freely escape the optically thick envelope. Instead, they play a crucial role in the expansion and ejection stages of the outburst.

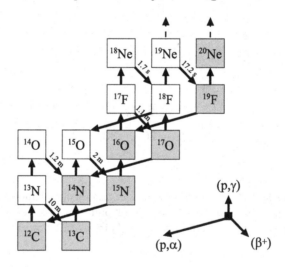

FIGURE 4.13

Nuclear processes relevant for the synthesis of ^{18}F in novae. Half-lifes of the most important β^+-decays are listed. A color version of this figure is available at http://fisica.upc.edu/ca/users/jjose/CRC-Downloads.

4.4.3.1 ^{13}N and ^{18}F

The synthesis of ^{18}F is driven by ^{16}O(p,γ)^{17}F, which is either followed by ^{17}F(p,γ)^{18}Ne(β^+)^{18}F or by ^{17}F(β^+)^{17}O(p,γ)^{18}F (see Figure 4.13). Because of the relatively large half-life of ^{18}F ($T_{1/2} = 110$ min), its dominant destruction channel is mainly ^{18}F(p,α)^{15}O, plus a minor contribution from ^{18}F(p,γ)^{19}Ne. The most uncertain reaction

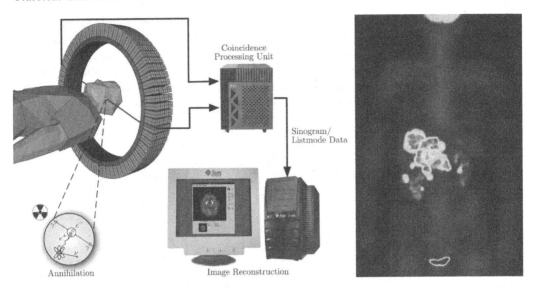

FIGURE 4.14

PET is a noninvasive, nuclear medicine technique used to generate 3D images of functional processes inside the human body. Radionuclides used in PET scanning, such as ^{11}C ($T_{1/2} \sim$ 20 min), ^{13}N ($\sim$ 10 min), ^{15}O ($\sim$ 2 min), or ^{18}F ($\sim$ 110 min), are characterized by short half-lives. These radionuclides are frequently encapsulated into standard compounds regularly used by the body, such as glucose, water, ammonia, or more complex molecules. The most commonly used compound in PET scanning is fludeoxyglucose (FDG), an analogue of glucose that contains ^{18}F. The concentrations of this radiotracer imaged in the body provide an indicator of tissue metabolic activity. This is, for instance, used to unveil cancer metastasis. When the short-lived radiotracer decays, it releases a positron that annihilates with a close-by electron producing a pair of γ-rays. In short, a PET scan (left panel) detects the high-energy radiation indirectly produced by a positron-emitting radiotracer. Shown in the right panel is a maximum intensity projection of a positron emission tomography corresponding to a female after intravenous injection of ^{18}F encapsulated in FDG, 1 hour before measurement. Aside from the expected accumulation of the radiotracer in the heart, bladder, kidneys, and brain, liver metastases of a colorectal tumor are also visible. A color version of this plot is available at http://fisica.upc.edu/ca/users/jjose/CRC-Downloads. Credit: http://en.wikipedia.org/wiki/Positron_emission_tomography, released into the public domain by the author, Jens Maus.

associated with the synthesis and destruction of ^{18}F for nova conditions is ^{18}F(p,α)^{15}O (see Section 4.4.4.1), which translates into a large uncertainty in the expected γ-ray fluxes and in the corresponding detectability distances [9, 419].

In a similar way to positron emission tomography (PET; see Figure 4.14), positrons emitted during the β^+-decay of ^{13}N and ^{18}F are expected to power a prompt γ-ray emission during nova outbursts (Figure 4.15). At the conditions that characterize nova envelopes, these emitted positrons mainly thermalize before annihilating with the surrounding electrons [1070]. While about 10% of the emitted positrons directly annihilate, $\sim$90% form a pseudo-atom called *positronium*, a system consisting of an electron and a positron bound together. One fourth of the positronium atoms are formed in singlet state (para-positronium), characterized by antiparallel electron-positron spins and a mean lifetime of $\tau = 125$ ps. In this configuration, positronium preferentially decays with emission of two 511-keV γ-ray photons. The remaining positronium atoms adopt a triplet state configuration (ortho-positronium), with parallel spins and a mean lifetime of $\tau = 145$ ns, mainly emitting three γ-ray photons (each with an energy below 511 keV). This short γ-ray signal, that lasts only for a few hours after T_{peak}, consists of a 511 keV line and a lower energy continumm (powered both by positronium decay and by Comptonization of 511-keV photons) with a cut-off at ~ 20–30 keV due to photoelectric absorption[33]. It is worth noting that this is the most intense emission in γ-rays predicted for classical novae [348, 660, 782, 1070].

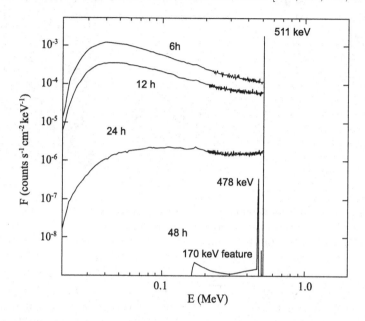

FIGURE 4.15
Theoretical γ-ray spectrum of a 1.15 $M_\odot$ CO novae at $D = 1$ kpc. The prominent 478 keV ^{7}Be line is clearly seen. Note also the feature at 170 keV corresponding to the energy of back-scattered 511 keV photons. Figure adapted from Hernanz and José [779] (see reference [775] for updated fluxes).

Despite the relatively large fluxes predicted for this prompt emission, its detection constitutes a real challenge, since it likely takes place before the nova is discovered optically. This, unfortunately, rules out the possibility of repointing any γ-ray satellite once the nova is found. The only chance relies on a posteriori analysis of data collected from regions where novae were seen in outburst, some days before visual maximum. Indeed, instruments with a wide field of view have a chance to serendipitously discover this prompt emission, provided that they were pointing at the right place at the right time. This type of analysis has already been attempted with the TGRS instrument on board the WIND satellite [729], with the BATSE instrument on board the Compton Gamma-Ray Observatory (CGRO) [787], or with the RHESSI satellite [1208], from which upper limits on the ^{18}F annihilation line were obtained. Estimates of the detectability distance of the 511 keV line with the SPI spectrometer on board the γ-ray observatory INTEGRAL yield a value about 4 kpc [777, 782].

4.4.3.2 ^{7}Be-^{7}Li

Synthesis of ^{7}Li during nova outbursts likely proceeds through the *beryllium-transport mechanism* [284]: It is initiated by ^{3}He$(\alpha, \gamma)^7$Be, which ultimately decays into the first excited

---

[33]The existence of such a sharp cut-off rules out the possibility that the hard X-ray flux observed in novae may result from Compton degradation of γ-rays, as suggested in a number of papers [1124, 1750].

state of ^{7}Li ($T_{1/2} \sim 53$ days) through electron capture. This state de-excites by the emission of a single 478 keV γ-ray photon (see Figure 4.16). No relevant uncertainties in the domain of nova temperatures affect the corresponding reaction rates.

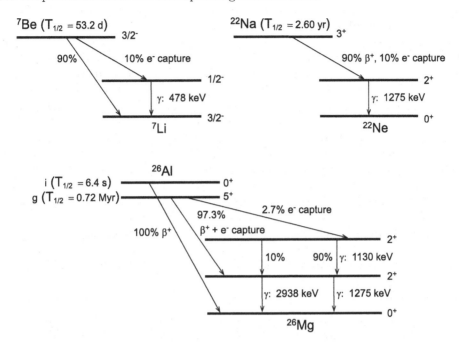

FIGURE 4.16
Decay schemes for ^{7}Be, ^{22}Na, and ^{26}Al (both ground (g) and isomeric (i) states).

Production of ^{7}Li in novae has motivated vivid controversies. Early claims of the feasibility of the beryllium-transport mechanism in novae, based on toy models [78], were later confirmed by hydrodynamic simulations [1715]. Unfortunately, these simulations assumed envelopes in place, thus neglecting the accretion phase as well as the initial stages of the outburst. In sharp contrast, new parametric studies performed in the framework of one- and two-zone models [212], claimed instead that the inclusion of the ^{8}B(p,γ) reaction, not considered in previous work, substantially reduced the final ^{7}Li yields. The scenario was revisited shortly after in a hydrodynamic framework [783,919], taking into account both the accretion and explosion stages, and a large reaction network containing ^{8}B(p,γ). The study concluded that classical novae are likely stellar factories of ^{7}Li (showing in turn the risky conclusions that can be reached on the basis of very simple models) and stressed the key role played by the amount of ^{3}He left over after the initial stages of the explosion. Indeed, a comparison between models of ONe and CO novae (see Section 4.4.1) reveals different timescales to achieve peak temperature, driven by large differences in the ^{12}C content in the accreted envelopes. This translates into larger amounts of ^{7}Be in the ejecta of CO novae, which escape destruction through (p,γ) reactions because of the very efficient ^{8}B(γ,p) photodisintegration reaction.

A rough estimate of the total production of ^{7}Li in novae can be computed as

$$M(^7\mathrm{Li}) \sim f(\mathrm{CO})\, X(^7\mathrm{Li})\, M_{\mathrm{ejec}}(M_\odot)\, R_{\mathrm{nova}}(\mathrm{yr}^{-1})\, t_{\mathrm{Gal}}(\mathrm{yr})\,, \tag{4.8}$$

where $f(\mathrm{CO})$ represents the fraction of CO novae (since ^{7}Li is mostly produced in models of CO novae), $X(^7\mathrm{Li})$ the mean mass fraction of ^{7}Li in the ejecta, M_{ejec} the mean ejected mass per outburst, and R_{nova} the nova rate. Adopting $f(\mathrm{CO}) \sim 2/3$ [1129,1827], $R_{\mathrm{nova}} \sim 30$

yr^{-1} [1609], and a favorable CO nova case ($M_{ejec}[^{7}Li] = 1.1 \times 10^{-10}$ $M_\odot$ [919]), we estimate the contribution of novae to the Galactic ^{7}Li as $\lesssim 20$ $M_\odot$. This turns out to be rather small, since the overall ^{7}Li content in the Galaxy is ~ 150 $M_\odot$. Nevertheless, some models of Galactic chemical evolution seem to require a nova contribution to match the ^{7}Li content [17,1514,1515], although other models do not [429].

Until recently, all attempts to identify ^{7}Li in the optical spectra of ejected nova envelopes proved unsuccessful, and only upper limits have been inferred[34] [560]. In 1999, an observed feature compatible with the Li I doublet at 6708 Å was reported from the spectra of V382 Vel (Nova Velorum 1999) [429]. However, it cannot be clearly disentangled from other sources of low-ionization emission centered at around 6705 Å, like the doublet associated with N I [1641], or some other weak neutral or singly ionized species. Observational evidence of the explosive lithium production in novae was announced in 2015 by Akito Tajitsu and collaborators, who reported the discovery of highly blue-shifted lines in the near-UV spectra of V339 Del (Nova Delphini 2013), attributed to singly ionized ^{7}Be [1770]. If confirmed, these observations will certainly revitalize the interest in novae as important ^{7}Li factories.

Despite its longer duration (~ 2 months) and the possibility of satellite repointing, detection of the 478 keV γ-ray line associated to ^{7}Be decay has been elusive to date. Examples include observations with the GRS instrument on board the SMM satellite [728] or with the TGRS spectrometer on board WIND [730]. The predicted detectability distance for the 478 keV line with INTEGRAL SPI is about 0.2 kpc for a 4 Ms observation [778].

4.4.3.3 ^{22}Na and ^{26}Al

The synthesis of ^{22}Na in novae can proceed through different reaction paths (see Figure 4.17) [916]. In the ^{20}Ne-enriched envelopes, characteristic of ONe novae, it begins with ^{20}Ne(p,γ)^{21}Na. This can either be followed by another proton capture, and then a β^{+}-decay, ^{21}Na(p,γ)^{22}Mg(β^{+})^{22}Na, or can decay first into ^{21}Ne before another proton capture ensues, ^{21}Na(β^{+})^{21}Ne(p,γ)^{22}Na. The main destruction channel at nova temperatures is ^{22}Na(p,γ)^{23}Mg.

The role of ^{22}Na for diagnosis of nova outbursts was first outlined in the seminal work of Clayton and Hoyle [348], and has been extensively discussed elsewhere [919,1316,1438, 1707,1713,1717,1920]. This isotope decays ($T_{1/2} = 2.60$ yr) into a short-lived excited state of ^{22}Ne, which de-excites ($T_{1/2} = 3.7$ ps) to its ground state by emitting a γ-ray photon of 1.275 MeV (Figure 4.16). This powers line emission at 1.275 MeV but also part of the lower energy continuum through Comptonization (see Figure 4.18).

Several attempts to detect these ^{22}Na γ-rays from nearby nova explosions have been undertaken along past decades, including balloon-borne experiments [1082] and detectors on board HEAO–3 [1184], SMM [1072], or CGRO [901,1069], from which upper limits on the total amount of ^{22}Na present in the ejecta have been derived. Indeed, a value of 3.7×10^{-8} $M_\odot$ for the maximum ^{22}Na mass ejected by any nova in the Galactic disk was derived from COMPTEL observations of five Ne novae and other *standard* CO novae [901].

Current detectability distance estimates for the 1.275 MeV line from ONe novae, based on inflight sensitivities of INTEGRAL/SPI, yield a value of 0.7 kpc for a 4 Ms observation [778]. Hence, only (unusually) close-by events would have a chance to be detected by INTEGRAL during its overall mission.

Moreover, it is worth noting that a large number of nova explosions (a fraction of which being Ne novae) will explode in the Galaxy during the relatively long half-life of ^{22}Na, $T_{1/2} \sim 2.60$ yr. Hence, it is, in principle, possible to detect the cumulative ^{22}Na emission

---

[34]Michael Friedjung (1979) reported upper limits to the Li/Na abundance ratio in the range 3.8–4.5 (Li/Na)$_\odot$ from observations of the slow Nova HR Del and the fast Novae IV Cep and NQ Vul.

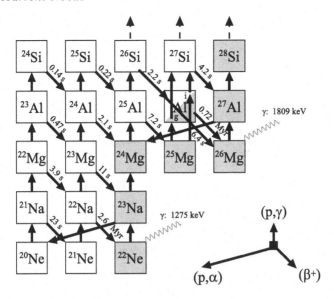

FIGURE 4.17

Nuclear processes in the NeNa- and MgAl-mass regions relevant for nova nucleosynthesis. Half-lifes of the most important β^+-decays are listed. A color version of this figure is available at http://fisica.upc.edu/ca/users/jjose/CRC-Downloads.

during the deep surveys that satellites like INTEGRAL are performing on the Galactic central radian [907].

The synthesis of ^{26}Al in novae requires the presence of some "seed" nuclei, such as ^{24,25}Mg, or to some extent, ^{23}Na and ^{20,22}Ne [916]. The main nuclear reaction path leading to ^{26}Al is ^{24}Mg(p, γ)^{25}Al(β^+)^{25}Mg(p, γ)^{26g}Al, whereas destruction is dominated by ^{26g}Al(p, γ)^{26}Si (see Figure 4.17). A significant nuclear uncertainty affects the ^{25}Al(p, γ)^{26}Si rate [361,916]. This translates into an uncertainty in the predicted contribution of novae to the Galactic ^{26}Al content, since this reaction determines the fraction of the nuclear path that proceeds through the isomeric ^{26i}Al state, thus bypassing ^{26g}Al synthesis.

The synthesis of ^{26}Al requires moderate peak temperatures, $T_{\mathrm{peak}} \leq 2 \times 10^8$ K, and a fast decline from maximum temperatures [1904], conditions that are presumably achieved in many nova outbursts. In the early 1980s, one-zone nucleosynthesis calculations of explosive hydrogen burning [803,1933] suggested already that classical novae might produce significant amounts of ^{26}Al to account for some isotopic anomalies reported from meteorites, although they might not constitute major Galactic sources. Unfortunately, these estimates relied on solar or CNO-enhanced envelopes.

Pioneering simulations in the context of ONe(Mg)-rich envelopes were performed shortly afterward [1316,1920], suggesting in turn that ONe novae might represent important (though not dominant) sources of the Galactic ^{26}Al. These studies were supplemented by hydrodynamic calculations [1438,1713] that stressed the crucial role played by convection in carrying a fraction of the ^{26}Al synthesized at the envelope base to the outer, cooler layers where destruction by (p,γ) reactions could be prevented.

^{26}Al was discovered in the interstellar medium by the HEAO-3 satellite, through the detection of the 1.809 MeV γ-ray line [1184,1185]. This characteristic γ-ray signature is

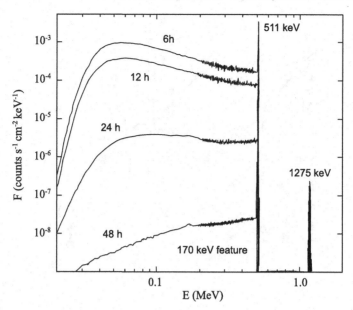

FIGURE 4.18
Same as Figure 4.15, but for the γ-ray spectrum of a 1.15 $M_\odot$ ONe novae. Note the intense 1275 keV ^{22}Na line Figure adapted from Hernanz and José [779] (see reference [775] for updated fluxes).

produced during the β^+ decay of ^{26g}Al ($T_{1/2} = 0.72$ Myr) to the first excited state of ^{26}Mg, which de-excites ($T_{1/2} = 0.49$ ps) to its ground state by emitting a 1.809 MeV photon[35].

The contribution of novae to the overall Galactic ^{26}Al content has also motivated some controversy. Whereas calculations performed by Sumner Starrfield and collaborators [1710, 1711, 1713, 1717] have traditionally assumed a white dwarf composition based on hydrostatic models of C-burning nucleosynthesis [73], highly enriched in ^{24}Mg (with a ratio ^{16}O:^{20}Ne:^{24}Mg of 1.5:2.5:1), calculations by the author and collaborators [916, 919, 923] have adopted estimates based on stellar evolution calculations of intermediate-mass stars [1504], resulting in a smaller ^{24}Mg content (with ^{16}O:^{20}Ne:^{24}Mg being 10:6:1).

A crude estimate of the contribution of novae to the amount of ^{26}Al currently present in the Galaxy can be computed as [1920]

$$M(^{26}\text{Al}) \sim T_{1/2}(^{26}\text{Al})\, f(\text{ONe})\, M_{\text{ejec}}(M_\odot)\, X(^{26}\text{Al})\, R_{\text{nova}}(\text{yr}^{-1})\,, \qquad (4.9)$$

where $T_{1/2}(^{26}\text{Al})$ represents the half-life of ^{26g}Al (0.72 Myr), M_{ejec} the mean ejected mass in a nova outburst, $X(^{26}\text{Al})$ the mean mass fraction of ^{26}Al in the ejecta, $f(\text{ONe})$ the fraction of Ne novae (typically, 1/3 [1129, 1827]), and R_{nova} the nova rate (~ 30 yr^{-1} [1609]). Adopting these estimates and a favorable ONe nova case ($M_{\text{ejec}}[^{26}\text{Al}] \sim 2 \times 10^{-8}$ $M_\odot$ [919]), the contribution of novae to the current Galactic ^{26}Al content ($\sim 2.8 \pm 0.8$ $M_\odot$ [442]) can be estimated as ≤ 0.2 $M_\odot$. This is in agreement with the analysis of COMPTEL/CGRO 1.809 MeV ^{26}Al emission map of the Galaxy [441, 442, 1075, 1433, 1444], which points toward young progenitors (i.e., type II supernovae and Wolf–Rayet stars; see Chapter 7). It is, however, worth noting that this kind of estimates are dramatically affected by uncertainties, since both the mean ejected mass per nova outburst, and the variation of the nova rate along the overall history of our Galaxy are not well constrained quantities.

---

[35]Note that the isomeric state ^{26i}Al does not decay into ^{26g}Al, and therefore does not contribute to γ-ray emission at 1.809 MeV.

4.4.4 Nuclear Uncertainties

Element production in many astrophysical scenarios usually involves a large number of relevant nuclear processes. Their associated reaction rates are often based on theoretical grounds (i.e., statistical models, such as the Hauser–Feshbach theory; see Chapter 2). Classical novae constitute a noticeable exception because of their somewhat limited nuclear activity and range of energies of interest. This allows us to rely primarily on experimental information. Unfortunately, the experimental efforts aimed at providing an accurate determination of these rates are also hampered by nuclear uncertainties. This is due to different reasons, including partial knowledge of spins, parities and energies of the nuclear levels of interest, interferences, or uncertainties in the assignment of analog states between mirror nuclei. Such uncertainties may obviously impact nucleosynthesis predictions. Accordingly, key uncertain reactions have to be conveniently flagged and compiled, providing a guide for future measurements at stable and radioactive ion-beam facilities.

Qualitative evaluations of the impact of nuclear uncertainties on nova yields have been performed in recent years through several, somewhat complementary approaches. Ideally, a series of hydrodynamic tests could be performed, varying each reaction rate individually within uncertainty limits (if upper and lower limit rates are available). This approach, which involves a hydrodynamic code coupled to a detailed nuclear reaction network, clearly becomes computationally intensive and therefore has only been applied to a reduced number of cases [360,548,916,917,1538]. A more feasible approach relies on *postprocessing calculations* with temperature and density versus time profiles extracted from hydrodynamic simulations or semianalytical models. The latter allows for a huge number of tests provided that the uncertainties affecting a particular nuclear reaction will not have an impact on the energetics of the explosion (in which case, the adopted temperature and density profiles would not be valid anymore). A detailed example of this approach can be found in reference [880]. Other studies, performed in the framework of postprocessing calculations, adopted somewhat different strategies, including Monte Carlo simulations in which all nuclear reactions are simultaneously varied with randomly sampled factors [810,1669]. The goal is to properly address the higher order correlations between input rates and nova model predictions because several reactions, not only one, are simultaneously involved in the synthesis and destruction of each element.

It is worth noting that most (but not all) postprocessing calculations ignore the critical impact that convection has on the final nova yields [1384], and, hence, they are not suitable for defining *absolute* final abundances. Moreover, such studies often rely on temperature and density profiles corresponding to the innermost envelope shells, which translates into biased predictions of element synthesis in novae. Indeed, hydrodynamic calculations tend to smooth the results obtained through one-zone approaches because of the role of convective mixing between adjacent shells in an envelope that is numerically treated with a much larger resolution.

One of the important results obtained from the systematic analysis of the impact of nuclear uncertainties is that nova yields are influenced by variations of a very limited number of key reaction rates. Moreover, these studies have also revealed that reaction rate uncertainties of about 30% are small enough for quantitative nova model predictions. Current uncertainties affecting nova yields mainly involve a handful of nuclear reaction rates, particularly $^{18}F(p, \alpha)$, $^{25}Al(p, \gamma)$, and $^{30}P(p, \gamma)$, for which several experiments are being conducted at different facilities around the world. β-decay half-lives of interest for nova nucleosynthesis are experimentally well known and do not suffer significant uncertainties.

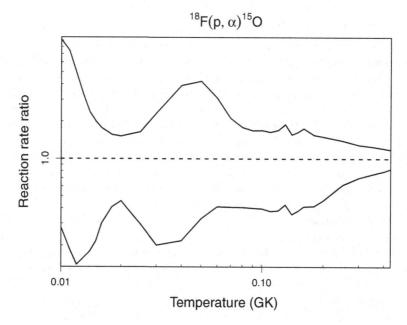

FIGURE 4.19
Nuclear uncertainty in the ^{18}F(p, α) rate, shown in terms of reaction rate ratios (i.e., high/recommended and low/recommended). The main area of uncertainty is located around $T \sim 5 \times 10^7$ K. Figure from Iliadis et al. (2010) [882]; reproduced with permission.

4.4.4.1 ^{18}F(p, α)

Although its associated uncertainty has been reduced through recent neutron and proton-transfer reactions [9], spins, and parities of the relevant levels at E_r(c.m.) $= -121$ keV, 8 keV, and 38 keV are yet not unambiguously determined and need to be confirmed experimentally. Depending on the actual choice of spins and parities for the newly found subthreshold resonance at -121 keV and for the 38 keV one, interference between the tails of these broad resonances and the one at 665 keV may play an important role. This translates into a large uncertainty (by a factor ~ 50) in the overall reaction rate (Figure 4.19).

4.4.4.2 ^{25}Al(p, γ)

The uncertainty associated with the ^{25}Al(p, γ) rate is due to the lack of spectroscopic information (i.e., spins and parities) for some of the levels of interest in the range E_r(c.m.) $= 163 - 965$ keV. The estimated uncertainty amounts to a factor of ~ 2 [882] (Figure 4.20).

4.4.4.3 ^{30}P(p, γ)

As for ^{25}Al(p, γ), the uncertainty associated with this rate is due to the lack of detailed spectroscopic information. Indeed, only constraints on spins and parities are available for relevant ^{31}S states (cf., within ~ 600 keV of the ^{30}P+p threshold in ^{31}S). This results in a factor of 20 uncertainty in the ^{30}P(p, γ) rate [1389] that, coupled to the lack of information on spectroscopic factors, translates into a huge uncertainty (maybe by a factor of 100 or more) on the overall rate (see also reference [1993]).

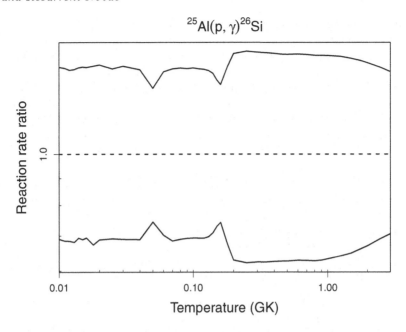

FIGURE 4.20
Same as Figure 4.19, but for ^{25}Al(p, γ). Note the huge uncertainty above $T > 2 \times 10^8$ K.
Figure from Iliadis et al. (2010) [882]; reproduced with permission.

4.5 Nova Light Curve

A key observational tool in the study of variable stars, such as novae, is the light curve,
which displays the amount of energy emitted (in a particular band of the electromagnetic
spectrum) per unit time (i.e., luminosity). Hereafter, we will focus on the main features
that characterize nova light curves, from the rise to bolometric maximum to the late turn-
off stage.

4.5.1 Fast Rise to Bolometric Maximum

Regardless of the nova type (i.e., neon or non-neon), the rise to peak bolometric luminosity
often takes place very rapidly ($\sim$ 1 day). This phase is driven by convective transport
throughout the accreted envelope [36]. During most of the accretion stage, the envelope is
characterized by moderate temperatures ($T \leq 10^7$ K), such that the pp-chains are the
dominant source of energy, which is carried away by radiative transport. Up to this point,
the stellar surface remains basically unaffected, with no observational sign of the ongoing
explosion. The progressive piling up of matter increases temperature. When the envelope
base reaches about $T \sim 2 \times 10^7$ K, superadiabatic gradients are established and convection
erratically sets in, slightly above the core-envelope interface. At this stage, models show
that the surface bolometric luminosity has increased by a factor of 2 with respect to its

---

[36] A number of features in the early nova light curves, such as the presence of prenova near-ultraviolet
and X-ray flashes, or the existence of pre-maximum halts, have been recently attributed to changes in the
convective energy transport [805].

initial value. As the thermonuclear runaway develops, convection extends throughout the envelope. Models suggest that the envelope becomes fully convective at $T \sim 8 \times 10^7$ K. At this stage, the surface bolometric luminosity has increased to about the solar value.

One of the effects of convection is that it transports a fraction of the short-lived species ^{13}N, ^{14,15}O, and ^{17}F, synthesized deep inside the envelope, to its outermost layers (see Section 4.3). Typical convective velocities reach $\sim 10^6$ cm s^{-1}, with a characteristic convective turnover time, at this stage, of about $10 - 100$ s. The time spent by the model to increase surface luminosity from 1 L$_\odot$ to 10^5 L$_\odot$ is also ~ 100 s. Hence, the efficiency of convection in carrying the energy released by nuclear reactions is at the heart of the rise to peak luminosity.

Simultaneously to the increase in luminosity, the nova energy distribution hardens (i.e., displays an increase in the average energy of the photons because of preferential loss at lower energies by absorption, leakage, scattering, or other processes). This results from the fact that little expansion is taking place at this stage, and, hence, $L = 4\pi R^2 \sigma T_{\text{eff}}^4$ translates into an increase of the effective temperature. Typical values, $L_{\text{peak}} = 10^4 - 10^5$ L$_\odot$ and $R \sim 10^9$ cm, yield $T_{\text{eff}} \sim (5 - 9) \times 10^5$ K, which corresponds to the extreme ultraviolet/soft X-rays band of the electromagnetic spectrum.

4.5.2 Rise to Visual Maximum

Optical light curves of classical novae display a spectacular and fast rise in visual luminosity, which is often missed by observers: In a matter of days, the luminosity of a previously faint and undetectable white dwarf increases by several orders of magnitude.

Following the increase in temperature described during the earlier rise in bolometric luminosity, degeneracy is lifted (see Section 4.3), hence the envelope begins to expand. At this stage, the envelope is still optically thick. Based on the ultraviolet (and infrared) temporal evolution, and following a sequence analogous to terrestrial nuclear bombs [1495], this optically thick phase is often subdivided into the *fireball*, *iron curtain*, and *prenebular* stages, with a duration that depends, among other factors, on the mass and velocity of the ejecta [625, 1632, 1635]. These are followed by an optically thin, *nebular* phase when spectra of the expanding ejecta become dominated by collisionally excited nebular lines [625, 1632, 1635, 1637].

The *fireball* stage derives from similarities to the earliest moments after a nuclear explosion, when the heated air creates a spherical, hot, luminous mass of gas that expands progressively with time. The expanding nova envelope is initially small and dense. As the envelope increases in size, it cools adiabatically, resulting in a large increase in the opacity. Since the process takes place while the central source remains at constant bolometric luminosity, $L_{\text{peak}} = 10^4 - 10^5$ L$_\odot$, radiation shifts toward longer wavelengths [1642]. Hence, the optical peak is due to flux redistribution following the initial fireball stage, when the ultraviolet-line opacity increases, and illumination from the central source is reprocessed in the still opaque expanding shell [1642].

The effective temperature reaches its minimum ($\simeq 10^4$ K) when the atmosphere achieves its maximum extent (e.g., $\sim 10^{12}$ cm, for a bolometric luminosity of 10^4 L$_\odot$). At these temperatures, ions of the iron-peak elements recombine, particularly Fe II, hence increasing the overall opacity. These ions have millions of overlapping absorption lines in the ultraviolet and dominate the spectral energy distribution, often referred to as a pseudo-continuum because of severe line blanketing. The true continuum is visible only through gaps in this *iron curtain*, after which this phase is named. The ultraviolet energy absorbed is reemitted in the optical and infrared[37]; thus, visual maximum occurs when ultraviolet absorption is greatest.

The rise time to peak visual luminosity reflects the strength of the explosion and depends on the nova class (i.e., fast or slow novae), ranging from a few days (e.g., V1500 Cyg 1975, a fast nova) to about one hundred days (e.g., HR Del 1867, a slow nova). A key ingredient that dictates the timescale to peak luminosity is indeed the ratio of nuclear energy released at the late stages of the runaway, which depends on the amount of CNO, and the binding energy of the star, $GM_{wd}\Delta M_{env}/R_{wd}$ [1824]. Simulations report shorter rise times to peak visual luminosity than those observed, an issue attributed to the likely localized (point-like) nature of ignition and further extension of the runaway along the stellar surface in the real event, which is not accounted for by current models [1703].

After a nova has declined 2 to 3 magnitudes in the visible, the ultraviolet spectra begin to show strong line emission, which characterizes the start of the *prenebular* phase. The decline from peak visual luminosity can happen at a variety of rates, thus defining the nova *speed class* from the time needed to drop by 2 or 3 magnitudes, denoted as t_2 and t_3, respectively. Very fast novae have $t_2 < 10$ days, for fast novae $t_2 \sim 12$–15 days, and very slow novae have $t_2 \sim 150$–250 days. A number of nova distance indicators, based on light curves, have been proposed in the last decades. A popular one is the *absolute magnitude at maximum rate of decline* relationship (MMRD), which states that brighter novae decay faster, i.e. have shorter t_2 or t_3 [428]. The MMRD has been extensively used as a distance indicator to individual Galactic novae (see also reference [427]). The relation, however, suffers from a large scatter, and an increasing number of studies are refuting its applicability (see, e.g., reference [955]). Shara [1617], and Paresce et al. [1383] have derived similar relations, introducing the effects induced by additional factors, such as the mass-accretion rate[38].

4.5.3   The Constant Bolometric Luminosity Stage

Multiwavelength observations of classical novae have unveiled a characteristic constant bolometric luminosity phase, right after maximum, during the early optical decline [928, 1635]. This phase, first guessed in Nova FH Serpentis 1970, which exhibited flux redistribution from visible light to ultraviolet [591], takes place at values close to the *Eddington luminosity*[39], L_{Edd}, that is the luminosity obtained when radiation pressure pushing outward equals the inward gravitational pull,

$$L_{Edd} \sim \frac{4\pi G M_{wd} m_p c}{\sigma_T} \sim 3.2 \times 10^4 M_{wd}(M_\odot) L_\odot\,, \qquad (4.10)$$

where M_{wd} is the mass of the underlying white dwarf, m_p the proton mass, and σ_T the Thompson scattering cross-section for an electron.

While most of the envelope is actually ejected, the remaining material piles back on top of the white dwarf, where it undergoes steady nuclear burning[40]. This raises the effective

---

[37]Infrared [354, 625, 626, 628] and radio observations [627, 813, 1600] have also significantly contributed to clarify and quantify the role of classical novae in the chemical evolution of the Galaxy [632]. Indeed, the combination of optical, infrared, and radio observations yield estimates of the primary physical parameters that characterize the outburst (i.e., distance, ejected mass, kinetic energy), the elemental abundances present in the ejecta, and the properties of the dust condensed in the outflow (Figure 4.21).

[38]According to Mike Shara, the magnitude of a nova 15 days after maximum light turns out to be a good distance indicator, as first pointed out by Buscombe and de Vaucouleurs [276].

[39]Fast novae often exhibit peak luminosities than exceed the Eddington limit. At L_{max}, nova envelopes are fully convective, and a substantial fraction of the short-lived species ^{13}N, $^{14,15}O$, and ^{17}F are convectively transported to the outermost layers. Their decay releases large amounts of nuclear energy close to the surface, typically at rates $\sim 10^{13} - 10^{14}$ erg g^{-1} s^{-1}, thus powering *super-Eddington* luminosities.

[40]How radiation affects reaccretion of material that does not achieve escape velocity has not been properly addressed to date, however.

FIGURE 4.21
Composite optical image of the shell ejected by Nova GK Persei (1901), located 1500 light-years away, as seen between 2003 and 2011. The shell contains about 10^{-4} M$_\odot$ and is ~ 1 light-year in diameter. Source: http://apod.nasa.gov/apod/ap111105.html. A color version of this plot is available at http://fisica.upc.edu/ca/users/jjose/CRC-Downloads. Credit: Adam Block, Mt. Lemmon SkyCenter, University of Arizona; reproduced with permission.

temperature to about $10^5 - 10^6$ K. The nova energy distribution consequently hardens, and a large fraction of the energy is hereafter emitted beyond the optical region of the spectrum, making novae very bright in the ultraviolet and soft X-rays at this stage [1703]. The emission of novae at high energies can actually be used to probe the late, dynamic stages of the outburst.

4.5.4   Shutting Down a Nova: X-Ray Emission and the Turn-Off Phase

Emission in soft X-rays[41] (E $\leq$ 1 keV) was first dectected in Nova GQ Mus 1983 by EX-OSAT, 460 days after optical maximum [1346]. Other X-ray observations with ROSAT, BeppoSAX, Chandra, XMM-Newton, and Swift reported soft X-ray emission from only a handful of Galactic and M31 novae [767, 768]. The duration of this soft X-ray emitting phase, that persists as long as residual H-burning is on, is short in almost all detected novae, typically < 1 yr except for a few sources with 2–3.5 yr, and, most notably, three sources in the range of 8–12 yr (GQ Mus 1983, N LMC 2005, and V723 Cas 1995). Moreover, soft X-rays were also observed during the last eruptions of the recurrent nova RS Oph in 1985 and 2006 [1636, 1640]: Its short duration ($\sim$ 60 days [205]) suggested the presence of a very

---

[41] Also called *supersoft* X-ray phase.

massive white dwarf [705], in agreement with theoretical estimates of the mass leftover after the explosion [781].

In general, soft X-ray emission from novae is a challenge for the current thermonuclear runaway model of the explosion [785]. All novae are believed to retain a fraction of the H-rich accreted envelope, so why do only a small subset exhibit such soft X-ray emission? Possible explanations involve a short, steady H-burning phase, ending before the ejecta becomes optically thin to soft X-rays, or hypothetical cases where no H-rich envelope is left at all. Furthermore, theoretical estimates [1703, 1825] provide much larger turn-off times than the observed ones,

$$\tau_{\text{nuc}}(\text{yr}) = 400 \frac{M_{\text{H}}/10^{-4} M_\odot}{L/2 \times 10^4 L_\odot}, \tag{4.11}$$

where M_{H} is the hydrogen mass left over in the envelope. The observed short duration of the supersoft X-ray phase in novae suggests that, in general, $M_{\text{env}}^{\text{obs}} < M_{\text{accreted}} - M_{\text{ejected}}$ from hydrodynamical models[42], hence a mechanism responsible for additional mass loss should be invoked. Whereas some mass loss could be induced by radiation-driven winds [957], it is not clear whether this is enough to account for the amounts required. Studies of steady H-burning on white dwarf envelopes [1554] have revealed that the maximum effective temperatures attained depend only on the mass of the underlying white dwarf as well as on the envelope chemical composition (mainly its H content). Typical values range between $T_{\text{eff,max}} \sim 60$–$100$ eV. Experimental determinations of $T_{\text{eff,max}}$, while challenging, provide a method to infer the white dwarf mass and the chemical composition of the envelope independently of the white dwarf luminosity, and thus of the distance to the source. Useful mass-luminosity relations have been obtained by Sala and Hernanz [1554] for ONe models,

$$L(L_\odot) \approx 5.95 \times 10^4 \left(\frac{M_{\text{wd}}}{M_\odot} - 0.536 X_{\text{H}} - 0.14 \right), \tag{4.12}$$

and for CO models,

$$L(L_\odot) \approx 5.95 \times 10^4 \left(\frac{M_{\text{wd}}}{M_\odot} - 0.3 \right). \tag{4.13}$$

Equations 4.12 and 4.13 combined with $T_{\text{eff,max}}$–M_{wd} and M_{wd}–M_{env} plots yield very small envelope masses, in the range $\sim (2 - 30) \times 10^{-7} M_\odot$, in agreement with results previously reported by Tuchman and Truran [1830]. A new analytical expression for the turn-off time was also derived,

$$\tau_{\text{nuc}}(\text{yr}) = 2 X_{\text{H}} \frac{M_{\text{env}}}{10^{-6} M_\odot} \frac{4 \times 10^4 L_\odot}{L}, \tag{4.14}$$

Note that typical values obtained by Sala and Hernanz (i.e., $M_{\text{env}} \sim 10^{-6} M_\odot$ and $L \sim 4 \times 10^4 L_\odot$) yield $\tau_{\text{nuc}} = 0.7$ yr (for $X_{\text{H}} = 0.35$), a value in agreement with estimates inferred from the analysis of the soft X-ray emission as well as from ultraviolet data [664, 1645].

Internal shocks in the expanding envelope, as well as the collision between the ejecta and circumstellar material, can induce heating of the plasma, from which additional X-ray emission can ensue (mainly by thermal bremsstrahlung). Here, the emitted photons are characterized by larger energy than those from the supersoft X-ray phase. Indeed, such (hard) X-ray emission has been detected early after the explosion in some novae. A key example is V838 Her 1991, from which X-rays were observed 5 days after optical outburst, the earliest detection of a classical nova in X-rays [1134]. Hard X-rays up to ~ 50 keV were also detected very early after the 2006 outburst of the recurrent nova RS Oph[43] [205, 1679],

---

[42]Except for models of recurrent novae or for extreme combinations of mass-accretion rates and initial white dwarf masses and luminosities [2001].

[43]See also references [1636, 1640] for X-ray emission during the 1985 outburst.

revealing the interaction between the ejecta and the wind of the red giant secondary. It is also worth noting that post-outburst novae also emit X-rays once accretion resumes (see, for instance, the case of V2487 Oph 1998 [784]).

While the proposed γ-ray emission in the MeV range, associated with radioactive nuclei (i.e., ^{18}F, ^{22}Na), has proved elusive in novae (see Section 4.4.3), high-energy γ-rays at $>$ 100 MeV were unexpectedly detected[44] from the symbiotic binary V407 Cygni [4]. This very high-energy emission, attributed to shock acceleration[45] in the ejected shells after interaction with the wind of the red giant[46], has been subsequently detected in a number of novae (e.g., V407 Cyg, V1324 Sco, V959 Mon, V339 Del, V1369 Cen), by the Large Area Telescope on board the Fermi gamma-ray space observatory (Fermi–LAT). This confirms novae as a distinct class of γ-ray sources [7].

To summarize, the most striking lesson learned here is that the bolometric luminosity is not correlated with any light curve in a specific energy range (i.e., optical, infrared, ultraviolet, or X-rays). Therefore, the observational analysis of the nova outburst should not rely on one band only. Instead, a panchromatic view based on multiwavelength observations is the way to go.

4.6   Observational Constraints

4.6.1   Spectroscopic Abundances

Elemental abundances in the nova ejecta can be spectroscopically inferred by means of complementary methods that provide estimates at different epochs. Detailed model atmosphere codes, such as PHOENIX [737–739] have been used in the analysis of the early, optically thick spectra. Fe-group elements are the main contributors to the early spectra, through ultraviolet[47] absorption and optical emission lines. Spectral fits can yield the metallicity of the system, since the heavy-element abundances are not expected to be modified along the course of the explosion. However, for CNO-group nuclei only rough abundance estimates can be inferred because of fewer line transitions. In contrast, photoionization models like CLOUDY [520, 521] are better suited only for the late, nebular stages of a nova outburst, when the ejecta are already optically thin. Late spectra are characterized by a suite of strong emission lines mostly produced by CNO elements, for which good estimates can be obtained. Most Ne- and Fe-lines are very weak at this stage (except for Ne III–V) and only crude estimates of the overall metallicity of the system can be obtained. It is worth mentioning that model atmosphere and photoionization codes not only provide complementary information on elemental abundances at different stages of a nova outburst[48] but also a crosscheck mechanism to detect possible inconsistencies in the determination of chemical abundances.

---

[44]Diffusive shock acceleration of electrons and protons, with a maximum energy of a few TeV, was actually predicted in the framework of the 2006 outburst of the recurrent nova RS Ophiuchi [1779].

[45]See Shore et al. [1638] for an explanation of the origin of X-ray emission in Nova Mon 2012 as due to internal shocks caused by the collision of filaments that freeze out in the expansion.

[46]See reference [1353] for multidimensional hydrodynamic simulations of the evolution of the 2010 blast wave of the Nova V407 Cyg and its interaction with the stellar secondary. See also reference [1875] for a generic discussion set in the context of the recurrent nova RS Oph.

[47]Since 1996, when the IUE satellite was switched off, no dedicated ultraviolet satellite has been available. Partial coverage, however, has been provided through a number of satellites, such as Hubble/STIS, FUSE, or Spitzer.

[48]See, e.g., reference [1596] for an application of both PHOENIX and CLOUDY codes to the analysis of optical and ultraviolet observations of Nova LMC 1991.

TABLE 4.3

Chemical Abundances Inferred in Neon Novae (Part I)*

	Solar	LMC 1990#1	V4160 Sgr	V838 Her	V832 Vel
X(He)/X(H)	3.85E-1	(4.8 ± 0.8)E-1	(7.1 ± 0.4)E-1	(5.6 ± 0.4)E-1	(4.0 ± 0.4)E-1
X(C)/X(H)	3.31E-3	(3.7 ± 1.5)E-2	(1.43 ± 0.07)E-2	(2.28 ± 0.23)E-2	(2.6 ± 1.3)E-3
X(N)/X(H)	1.14E-3	(1.48 ± 0.42)E-1	(1.27 ± 0.08)E-1	(3.29 ± 0.47)E-2	(2.28 ± 0.54)E-2
X(O)/X(H)	9.65E-3	(2.4 ± 1.0)E-1	(1.35 ± 0.09)E-1	(1.42 ± 0.38)E-2	(4.13 ± 0.38)E-2
X(Ne)/X(H)	2.54E-3	(1.6 ± 1.0)E-1	(1.38 ± 0.05)E-1	(1.22 ± 0.05)E-1	(4.0 ± 0.7)E-2
X(Mg)/X(H)	9.55E-4	(1.37 ± 0.71)E-2	$\approx$8.4E-3	(1.2 ± 0.7)E-3	(2.45 ± 0.14)E-3
X(Al)/X(H)	8.74E-5	(2.3 ± 1.1)E-2	–	(1.8 ± 1.3)E-3	(1.63 ± 0.16)E-3
X(Si)/X(H)	1.08E-3	(4.8 ± 3.9)E-2	(1.09 ± 0.06)E-2	(7 ± 2)E-3	(5 ± 3)E-4
X(S)/X(H)	5.17E-4	–	–	(1.48 ± 0.15)E-2	–
X(Ar)/X(H)	1.29E-4	–	–	–	–
X(Fe)/X(H)	1.81E-3	–	(2.4 ± 0.8)E-3	(2.35 ± 0.63)E-3	–
X(H)	7.11E-1	(4.7 ± 0.9)E-1	(4.65 ± 0.37)E-1	(5.63 ± 0.36)E-1	(6.6 ± 0.4)E-1

*References: LMC 1990#1 [1854]; V4160 Sgr [1597]; V838 Her [1597]; V382 Vel [1641]. Solar abundances are taken from Lodders et al. [1139]. Table adapted from Downen et al. [461].

TABLE 4.4

Chemical Abundances Inferred in Neon Novae (Part II)**

	Solar	QU Vul	V693 CrA	V1974 Cyg	V1065 Cen
X(He)/X(H)	3.85E-1	(4.6 ± 0.3)E-1	(5.4 ± 2.2)E-1	(4.8 ± 0.8)E-1	(5.4 ± 1.0)E-1
X(C)/X(H)	3.31E-3	(9.5 ± 5.9)E-4	(1.06 ± 0.44)E-2	(3.1 ± 0.9)E-3	–
X(N)/X(H)	1.14E-3	(1.61 ± 0.10)E-2	(1.84 ± 0.67)E-1	(6.0 ± 1.5)E-2	(1.40 ± 0.33)E-1
X(O)/X(H)	9.65E-3	(3.2 ± 1.4)E-2	(1.63 ± 0.66)E-1	(1.55 ± 0.85)E-1	(4.7 ± 1.5)E-1
X(Ne)/X(H)	2.54E-3	(5.1 ± 0.4)E-2	(6.7 ± 3.4)E-1	(9.7 ± 4.0)E-2	(5.34 ± 0.98)E-1
X(Mg)/X(H)	9.55E-4	(1.02 ± 0.49)E-2	(9 ± 7)E-3	(4.3 ± 2.8)E-3	(4.4 ± 1.3)E-2
X(Al)/X(H)	8.74E-5	(4.1 ± 1.1)E-3	(5.0 ± 4.6)E-3	>7.8E-5	–
X(Si)/X(H)	1.08E-3	(2.4 ± 1.8)E-3	(2.4 ± 1.8)E-2	–	–
X(S)/X(H)	5.17E-4	–	–	–	(2.3 ± 1.3)E-2
X(Ar)/X(H)	1.29E-4	(4.0 ± 0.3)E-5	–	–	(4.6 ± 1.7)E-3
X(Fe)/X(H)	1.81E-3	(9.53 ± 0.54)E-4	–	(8.8 ± 7.2)E-3	(1.16 ± 0.40)E-2
X(H)	7.11E-1	(6.3 ± 0.3)E-1	(3.8 ± 1.4)E-1	(5.5 ± 0.8)E-1	(3.6 ± 1.0)E-1

**References: QU Vul [1595]; V693 CrA [1853]; V1974 Cyg [1852]; V1065 Cen [762]. Solar abundances are taken from Lodders et al. [1139]. Table adapted from Downen et al. [461].

Although several compilations of classical nova abundances have been published in the last decades [632, 1717, 1887, 1906], a straight comparison between predicted and observed nova abundances is far from simple. In fact, some of the existing compilations are actually incomplete, with elements in the range Na–Fe added together in some sort of equivalent mass fraction. In some unfortunate cases, the sum of all the individual mass fractions yield a value that differs substantially from 1, suggesting possible errors in some of the reported values. But particularly remarkable are the different abundance patterns reported by different authors in their spectroscopic analyses of the same event. For illustrative purposes, let's, for instance mention, the case of nova QU Vul 1984: While Saizar et al. [1553] inferred a mean metallicity of 0.10 in the ejecta, Austin et al. [92] reported a much larger value, 0.44. Such differences, which amount up to an order of magnitude for some elements, make the life of a theorist hard when attempting any fair comparison between models and observations. This reflects an aspect that needs to be stressed: Abundances are not directly observed but inferred from spectroscopic measurements through modeling. A certain choice for the suite of model parameters (e.g., inner radius, thickness, density, and composition of the gas shell, spectrum and luminosity of the underlying ionizing source), together with a number of assumptions, have to be adopted. In this regard, the set of *observed* abundances have to be taken with a grain of salt. Drawbacks and limitations faced in the modeling of chemical abundances from spectroscopic observations include the assumption of spherical and homogeneous shells, incomplete information on ionization stages and filling factors (that is, the fraction of the shell volumes actually occupied by gas), and a number of problems in the detailed characterization of radiative transfer (see reference [928]).

A fair compilation of nova abundances should rely on similar analysis tools. A recent attempt on this regard has been tackled by Lori Downen and collaborators [461] on the framework of neon novae. The study only included novae analyzed with the photoionization code CLOUDY. Nova environments are actually characterized by a diluted ejecta heated and ionized by the radiation field of the underlying white dwarf. The structure of the observed spectra from such regions can be inferred by simultaneously solving the equations of statistical and thermal equilibrium, that account for the different ionization-neutralization and heating-cooling processes [20, 1355]. Photoionization codes like CLOUDY yield fluxes that can be directly compared with measured line fluxes of the observed nebular spectra. By means of numerical algorithms like MINUIT, the full set of CLOUDY model parameters are iteratively selected to optimize the fit. Once the expansion is beyond the limit of photoionization equilibrium, however, the use of photoionization codes can lead to unphysical results.

The mass fraction ratios[49] obtained for elements between He and Fe over H are shown in Tables 4.3 and 4.4, for eight neon novae. Large overabundances relative to solar values are found, in some cases, for N, O, Ne, Mg, Al, Si, S, or Ar (by factors up to ~ 250, for some species). This brings evidence of substantial nuclear processing and confirms that the nuclear activity in nova outbursts does not extend beyond calcium, as predicted by theory. The study also identified a number of nuclear thermometers, N/O, N/Al, O/S, S/Al, O/Na, Na/Al, O/P, and P/Al, that exhibit strong correlations between peak temperature and mass of the underlying white dwarf, leading to the following estimates for the masses of some of the analyzed novae: 1.34–1.35 $M_\odot$ (for V838 Her), 1.18–1.21 $M_\odot$ (V382 Vel), 1.3 $M_\odot$ (V693 CrA), 1.2 $M_\odot$ (LMC 1990#1), and 1.2 $M_\odot$ (QU Vul). The abundance ratios X(CNO)/X(H), X(Ne)/X(H), X(Mg)/X(H), X(Al)/X(H), and X(Si)/X(H) have also been flagged as useful mixing meters for neon novae in another recent study [969]. The comparison between observed and predicted abundances through these mixing meters actually suggests

---

[49]Mass fraction ratios are expected to be less susceptible to systematic errors than individual mass fractions [461].

that mixing at the core-envelope interface in neon novae amounts to about 25%, a smaller fraction than the 50% value usually quoted in the literature.

Most of the predicted nuclear thermometers and mixing meters are unaffected by relevant nuclear uncertainties in the reaction rates involved. An exception is $^{30}P(p,\gamma)^{31}S$ (see Section 4.4.4), whose current uncertainty has a strong impact on the O/S, S/Al, O/P, and P/Al ratios, as well as on the X(Si)/X(H) mixing meter.

It is worth noting that despite the approximations undertaken in the modeling of nova outbursts, the abundances theoretically predicted are qualitatively in good agreement with those spectroscopically inferred, which suggests that the thermal history of the explosion (including maximum temperatures and exposure times) is reasonably well reproduced by current models. However, it must be stressed that spectroscopy yields only atomic abundances (e.g., only the overall oxygen mass fraction, rather than the individual amounts of ^{16}O, ^{17}O, and ^{18}O, is inferred). Other ways to better constrain the predicted abundances, ideally on isotopic grounds, are clearly needed.

4.6.2 Presolar Nova Grains

Infrared [504, 505, 627, 628, 632] and ultraviolet observations [1644] have revealed dust forming episodes in the shells ejected during classical nova outbursts. As first suggested by Stratton and Manning [1727] on the occasion of Nova DQ Her, dust formation is linked to a decline in the optical, several months after maximum luminosity. As observed in many novae, this is accompanied by a simultaneous rise in infrared emission. While CO novae are known to be prolific dust producers, ONe novae scarcely exhibit dust forming episodes. The low-mass, low-density, and high-velocity ejecta that characterize neon novae (Section 4.4.1) may prevent the condensation of appreciable amounts of dust.

The relatively high frequency of nova explosions in our Galaxy has raised the issue of their potential contribution to the different presolar grain populations. Indeed, infrared measurements of a number of novae have shown evidence of C-rich dust (Aql 1995, V838 Her 1991, PW Vul 1984), SiC (Aql 1982, V842 Cen 1986), hydrocarbons (V842 Cen 1986, V705 Cas 1993), or SiO_2 (V1370 Aql 1982, V705 Cas 1993). Remarkable examples, such as Nova QV Vul 1987, exhibited simultaneous formation of all those types of dust [632]. Dust condensation has also been observed in the ejecta from recurrent novae. For instance, silicate dust has been reported from the environment of RS Ophiuchi after its 2006 outburst [506].

Since the pioneering studies of dust formation in novae by Donald D. Clayton and Fred Hoyle [349], most efforts devoted to the identification of potential nova grains have relied mainly on the search for low $^{20}Ne/^{22}Ne$ ratios [803]. Indeed, since noble gases, such as Ne, do not condense into grains, the presence of ^{22}Ne in a grain has been attributed to in situ ^{22}Na decay, a clear imprint of a classical nova explosion. Dust condensation in a nova environment likely proceeds kinetically, through induced dipole reactions [1639], rather than an equilibrium. Nevertheless, preliminary estimates based on purely equilibrium condensation sequences [922] suggest that classical novae may contribute to the known presolar corundum (Al_2O_3), spinel ($MgAl_2O_4$), enstatite ($MgSiO_3$), silicon carbide (SiC), and silicon nitride (Si_3N_4) grain populations.

Clayton and Hoyle [349] pointed out several isotopic signatures (i.e., large overproduction of $^{13,14}C$, ^{18}O, ^{22}Na, ^{26}Al, or ^{30}Si), that may help in the identification of nova candidate grains. Most of the predicted signatures still hold, in view of our current understanding of nova explosions, except ^{14}C, which is bypassed by the main nuclear path in novae, and ^{18}O, which is slightly overproduced by novae, although grains nucleated in this environment are expected to be much more anomalous in ^{17}O [922, 1705].

A major breakthrough in the identification of presolar nova grains was achieved by Sachiko Amari and collaborators [24, 27], who reported several SiC and graphite grains

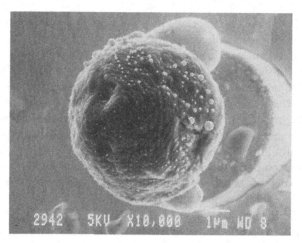

FIGURE 4.22
The nova candidate graphite grain KFC1a-511 after Secondary Ion Mass Spectrometry (SIMS; see Chapter 3). Image courtesy of S. Amari.

(Figure 4.22), isolated from the Murchison and Acfer 094 meteorites, with abundance patterns qualitatively similar to nova model predictions: low $^{12}C/^{13}C$ and $^{14}N/^{15}N$ ratios, high $^{30}Si/^{28}Si$ and close-to-solar $^{29}Si/^{28}Si$, and high $^{26}Al/^{27}Al$ and $^{22}Ne/^{20}Ne$ ratios for some of the grains. However, a major drawback for their unambiguous classification as nova grains is that mixing between material synthesized in the outburst and more than ten times as much unprocessed, isotopically close-to-solar material before grain formation is required to quantitatively match the grain data. This dilution might occur during the impact between the ejecta and the accretion disk, or even with the stellar companion. Additional concerns about the paternity of these grains were raised after isotopic analyses of three additional micron-sized SiC spherules, isolated also from the Murchison meteorite [1303]: While many isotopic ratios showed similar trends (in particular, low $^{12}C/^{13}C$ and $^{14}N/^{15}N$ ratios), some additional imprints (such as nonsolar Ti features) suggested a possible supernova origin. It is, however, not clear if both samples correspond to the same stellar progenitor, since their isotopic signatures are not that similar [921]. Moreover, because of the limited sensitivity of the SIMS microprobe used in the analysis of the original sample, all grains were shattered during measurements. Hence, there was no chance to determine its Ti content afterward. It should however be stressed that the presence of Ti does not necessarily preclude a possible nova paternity: First, because it could be present in the material transferred from the companion star, and second because titanium is close to the nucleosynthetic endpoint for novae (i.e., Ca), and, hence, it might be synthesized in more violent outbursts, driven by explosions on cooler white dwarfs, or under lower mass-accretion rate regimes [651,921]. Furthermore, nova explosions in metal-poor environments, like in primordial binary systems, could lead to similar imprints [918,920].

Other nova candidate grains have also been reported, including a number of SiC and oxide grains [700,702,1301,1309]. While theoretical predictions qualitatively agree with the grain data, it is often challenging to match some of the specific isotopic ratios determined in those grains. One should bear in mind that statistics is still scarce since, all in all, we rely on less than a dozen nova candidate grains. Nevertheless, this is clearly the way to go to constrain nova nucleosynthesis predictions. Moreover, to unambiguously identify the paternity of a grain, cosmochemists will need to rely on a much wider range of isotopic determinations, probably developing new techniques for laboratory analysis that may help to identify key signatures in the grains. For instance, grains condensed in the ejecta of novae involving very massive white dwarfs are expected to exhibit a suite of different sulfur and chlorine anomalies, as well as a severe overproduction of ^{31}P. Unfortunately, H_2SO_4 and HCl are traditionally used during the separation process, when grains are chemically isolated from the matrix. Hence, even though preliminary equilibrium condensation calculations suggest that various trace elements might be incorporated into SiC grains as sulfides[50] [1138], sulfur determinations are highly uncertain because of possible contamination.

---

[50]In contrast, ^{35}Cl is not expected to condense into SiC grains.

Box I. The Nova "Hall of Fame"

A selection of facts on classical and recurrent novae

1. Classical and recurrent novae are thermonuclear explosions that take place on the white dwarf component of a close stellar binary system.

2. Novae represent the second, most frequent type of thermonuclear explosions in the Galaxy (after X-ray bursts), with an estimated frequency of $\sim$ 30 events yr^{-1}.

3. Nova light curves are characterized by a constant bolometric luminosity phase.

4. Theoretical (1D hydrodynamic) models reproduce reasonably well the main observational features of nova outbursts (atomic abundances, light curves).

5. Infrared and ultraviolet observations often reveal dust-forming episodes in the ejected nova shells; presolar nova candidate grains have been identified by low $^{12}C/^{13}C$ and $^{14}N/^{15}N$ ratios, and excesses of ^{26}Mg (from ^{26}Al decay), ^{22}Ne (from ^{22}Na decay), and ^{30}Si.

6. The explosion propagates subsonically (deflagration). The outburst is likely quenched by envelope expansion (rather than by fuel consumption), and is driven by the energy released from the short-lived species ^{13}N, $^{14,15}O$, and ^{17}F, which are convectively transported to the outer envelope layers. Their decay heats the envelope and lifts degeneracy.

7. Nova nucleosynthesis is driven by proton-capture reactions and β^+-decays operating close to the valley of stability.

8. Calcium is the likely endpoint for nova nucleosynthesis.

9. Novae are major contributors to the Galactic abundances of ^{15}N, ^{17}O, ^{13}C, and, to a lesser extent, ^{7}Li and ^{26}Al.

Box II. Mysteries, Unsolved Problems, and Challenges

- Efforts carried to date in the modeling of nova explosions in 2D and 3D must be extended and improved, taking advantage of state-of-the-art parallel computers. In particular, models should explore the ultimate consequences of the mixing episodes that take place at the core-envelope interface in the final stages of the explosion.

- A closely connected problem to mixing is the current discrepancy between theory and observations regarding the overall amount of mass ejected. This, in turn, is relevant for assessing the expected contribution of novae to the Galactic abundances, and to clarify whether the white dwarf mass grows or decreases after a nova outburst.

- Stellar evolution predicts that white dwarf cores are surrounded by buffer layers of unburnt material from previous evolutionary stages. Does mixing penetrate deep enough to inject the required amounts of core material into the accreted envelope, or does the star eventually get rid of these buffers through mass-loss episodes?

- Are recurrent novae the progenitors of (some) type Ia supernovae?

- An observational campaign aimed at analyzing nova explosions in low-metallicity environments (i.e., Galactic halo, LMC, SMC, dwarf galaxies) should be conducted. This, together with theoretical studies of the evolution of the frequency of novae along the Galaxy's lifetime, will help to better constrain the overall contribution of novae to the Galactic abundances.

- Better spectra aimed at providing reliable abundance determinations in the ejecta are needed to constrain current theoretical models of nova outbursts.

- From a nuclear physics viewpoint, efforts should be made to better constrain the rates of key reactions affected by nuclear uncertainties, in particular, $^{18}F(p, \alpha)^{15}O$, $^{25}Al(p, \gamma)^{26}Si$, and $^{30}P(p, \gamma)^{31}S$.

- From a cosmochemistry viewpoint, more presolar nova candidate grains are needed to better constrain model predictions isotopically. This should include extensive measurements, for a large variety of isotopic ratios, aimed at identifying nova grains in the SiC, oxide, and A+B grain populations. New techniques will probably be needed for the interesting and challenging determinations of sulfur abundance ratios.

- The unambiguous detection of γ-ray signatures from classical novae, either lines (478, 511, 1275 keV) or continuum, should be pursued, taking advantage of higher sensitivity detectors planned for future space-borne missions.

4.7   Exercises

P1. Roche lobe surfaces in close stellar binaries (Figure 4.23) can deviate significantly from spherical symmetry. A simple way to characterize Roche lobes, based on numerical integration of the Roche potential for the system, relies on approximate analytical expressions for the radius of a sphere containing the same volume than the Roche lobe.

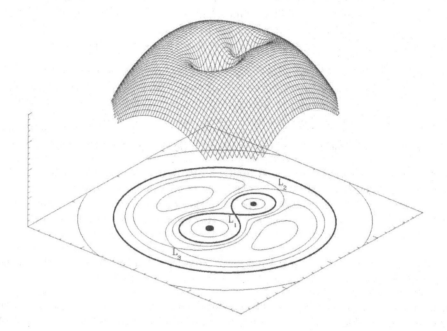

FIGURE 4.23
A 3D representation of the equipotentials in a binary stellar system with a mass ratio $q = 2$. Close to each star, the corresponding equipotential surfaces are approximately spherical and concentric with the nearest star. Far from the system, the equipotentials are approximately ellipsoidal and elongated along the axis joining both stellar centers. The series of equipotential surfaces intersect at the L_1 Lagrangian point of the system, forming a two-lobed figure (the thick, eight-shaped curve highlighted in the figure) that defines the Roche lobes. At the three Lagrangian points shown in the figure, forces cancel out. When a star fills its Roche lobe, mass can flow through the inner Lagrangian point of the system L_1, which corresponds to a saddle point. Credit: Marc van der Sluys, *Formation and evolution of compact binaries* (PhD thesis, 2006); reproduced with permission.

Eggleton [486] derived a formula that relates the Roche lobe radius of the secondary, R_{2L}, the binary separation, a, and the mass ratio of the system, $q = M_2/M_1$,

$$\frac{R_{2L}}{a} = \frac{0.49q^{2/3}}{0.6q^{2/3} + \ln(1 + q^{1/3})}, \tag{4.15}$$

which is valid for all values of q. Other formulae were deduced by Paczyński [1362]. For $0.1 \leq q \leq 0.8$, Paczyński found

$$\frac{R_{2L}}{a} = 0.46224 \left(\frac{M_2}{M_1 + M_2}\right)^{1/3}. \tag{4.16}$$

Expressions for the Roche Lobe of the primary, R_{1L}, can be obtained by simply replacing q by q^{-1}.

Let's assume a binary system containing a 1 $M_\odot$ white dwarf (M_1) and a 0.8 $M_\odot$ secondary (M_2):

a) Using Paczyński's fit and Kepler's third law, demonstrate that $R_{2L} = 6.424 \times 10^7 P_{\text{orb}}^{2/3}\,(s)$.

b) Calculate the maximum orbital period compatible with a low-mass, main sequence secondary filling its Roche Lobe.

c) Calculate the binary separation of the system and compare the result with the Earth–Sun and Earth–Moon distances. Using Plavec & Kratochvil's fit [1430],

$$\frac{b_1}{a} = 0.500 - 0.227 \log q, \qquad (4.17)$$

determine the distance b_1 between the center of the primary M_1 and the inner Lagrangian point of the system L_1.

d) Demonstrate that for $q \leq 0.8$, the mean density of a Roche-lobe filling secondary can be expressed as $\rho_{\text{mean}} \approx 110\, P_{\text{orb}}^{-2}(\text{hr})$ g cm^{-3}.

e) Since low-mass main sequence stars approximately satisfy $M(M_\odot) \approx R(R_\odot)$, demonstrate that

$$\rho_{\text{mean}} \sim 1.4 \left(\frac{M_2}{M_\odot}\right)^{-2} \text{g cm}^{-3} \qquad (4.18)$$

$$\frac{M_2}{M_\odot} \sim 0.11 P_{\text{orb}}(\text{hr}) \qquad (4.19)$$

$$R_2 \sim 0.11 P_{\text{orb}}(\text{hr})\, R_\odot \sim 7.9 \times 10^9 \, P_{\text{orb}}(\text{hr})\,\text{cm.} \qquad (4.20)$$

P2. The pressure at the interior of a degenerate star, such as a white dwarf, can be expressed as

$$P = 6.01 \times 10^{22} f(x), \qquad (4.21)$$

where

$$f(x) = x(2x^2 - 3)\sqrt{1 + x^2} + 3\ln(x + \sqrt{1 + x^2}), \qquad (4.22)$$

with f(0)=0. x is a function of the density, which for a CO white dwarf can be written as $\rho = 1.96 \times 10^6 x^3$.

Integrate numerically the stellar structure equations for a white dwarf in hydrostatic equilibrium:

$$dP = -\rho\, G \frac{m_{\text{r}}}{r^2} dr$$

$$dm_{\text{r}} = 4\pi \rho r^2 dr,$$

with m_{r} being the mass interior to a sphere of radius r ($m_{\text{r}}[r = 0] = 0$). Assume different values for the central density, ρ_{c} (e.g., 10^6, 10^8, and 10^{10} g cm^{-3}), and build a table displaying r(cm), m_{r}(g), ρ(g cm^{-3}) and P(dyn cm^{-2}).

i) Central layer: integration of the innemorst layer can be handled through a Taylor expansion around $r = 0$, in the form

$$r_1 = dr$$

$$P_1 = P_{\text{c}} - \frac{2}{3} G \pi \rho_{\text{c}}^2 dr^2$$

$$m_{\text{r},1} = \frac{4}{3}\pi \rho_{\text{c}} dr^3,$$

which depend on the adopted value of ρ_c and the (spatial) integration step, dr (hint: Try different values in the range $10^4 - 10^6$ cm). Since for a fully degenerate star, P is independent of T, one can infer ρ_1 from P_1 through an iterative procedure.

ii) Intermediate layers: Once the central layer ($i = 1$) has been succesfully integrated, one can proceed with the rest of the star, adopting a suitable integration method (i.e., midpoint, Runge–Kutta). Application of the midpoint method yields, for instance:

$$r_{i+1/2} = r_i + 0.5dr$$

$$P_{i+1/2} = P_i - 0.5(G\rho_i m_{r,i} dr / r_i^2)$$

$$m_{r,i+1/2} = m_{r,i} + 0.5(4\pi\rho_i r_i^2 dr).$$

for the first half step. Once $\rho_{i+1/2}$ is obtained from $P_{i+1/2}$, one can proceed with the second half step:

$$r_{i+1} = r_i + dr$$

$$P_{i+1} = P_i - G\rho_{i+1/2} m_{r,i+1/2} dr / r_{i+1/2}^2$$

$$m_{r,i+1} = m_{r,i} + 4\pi\rho_{i+1/2} r_{i+1/2}^2 dr$$

As before, ρ_{i+1} can be obtained from P_{i+1} through an iterative procedure. The process is iterated for increasing values of i (that is, integrating successive shells outward) up to the point when $P < 0$. For relatively small integration steps, the radius and mass obtained for the last layer represent good approximations to the white dwarf structure.

P3. The kinetic energy per unit mass for a white dwarf, K, is primarily supported by the motion of the electrons, so it can be approximated by $K \sim Np^2/(2m)$, where p is the average electron momentum, m is the electron mass, and N is the number of electrons per unit mass.

a) Since electrons are degenerate, and assuming that their average momentum p is given by the uncertainty principle, demonstrate that

$$K \approx \frac{M^{2/3} N^{5/3} \hbar^2}{2mR^2}, \tag{4.23}$$

where M and R are the white dwarf mass and radius, and $\hbar$ is the reduced Planck constant.

b) Assuming that for a white dwarf in hydrostatic equilibrium, its kinetic and gravitational potential energies should be comparable, demonstrate that white dwarfs obey a mass-radius relationship of the form $R \sim M^{-1/3}$.

c) Compare the numerical $M - R$ values obtained in Problem 2 with the previous analytical mass-radius relationship.

P4. Integration of hydrostatic ONe white dwarf structures for 1.15, 1.25, and 1.35 $M_\odot$ yields radii of 4326 km, 3788 km, and 2255 km, respectively.

a) Assuming a pressure of $P_\star = 10^{20}$ dyn cm^{-2}, determine the envelope masses required to achieve such a pressure for each of the white dwarfs considered. Show that ΔM_{env} decreases as the white dwarf mass increases.

b) Assuming a characteristic mass-accretion rate of 2×10^{-10} $M_\odot$ yr^{-1}, estimate the duration of the accretion stage for each case.

c) Estimate the binding energy, as well as the gravitational potential energy of the envelope, $E_{grav} = GM\Delta M_{env}/R$, for each case.

d) During H-burning, 4 protons are transformed into a single ^{4}He nuclei, with the release of 26.731 MeV per reaction (neglecting the energy carried away by neutrinos, which depends on the specific burning mode—i.e., different pp chains and CNO cycles; see Chapter 2). Calculate the nuclear energy released per gram of consumed H. Assume pure H-envelopes, with the masses and energies calculated in Parts *a* and *c*, and prove that the overall energy released can succesfully lead to envelope ejection (a rough estimate of the energy required for ejection is given by the gravitational potential energy stored in the envelope).

P5. Most 1D hydrodynamic codes used in the modeling of classical novae are written in Lagrangian formulation (that is, with a grid attached to the fluid; see Chapter 1). In this context, the computational domain expands during the ultimate ejection stage. Hence, no single layer *naturally* abandons the computational domain. Describe a criterion for determining which mass is ejected in these simulations.

P6. In a simulation of a nova outburst, the ejecta are predicted to contain radioactive ^{18}F, with a mean mass fraction of 2×10^{-6}, 30 min after T_{peak}. Determine the amount of ^{18}F present in the ejecta, 1 hr, 6 hr, 12 hr, 24 hr, and 48 hr since T_{peak}. What makes ^{22}Na γ-ray lines observable for longer times than the 511 keV ^{18}F-powered line?

P7. The absolute magnitude of the Sun is +4.77. Determine the absolute magnitude of a classical nova characterized by a peak bolometric luminosity of 10^5 $L_\odot$.

P8. Models of recurrent novae suggest that the ejected mass is smaller than that accreted. Estimates for 1.38 $M_\odot$ white dwarfs accreting material of solar metallicity at a rate of 2×10^{-7} $M_\odot$ yr^{-1}, yield a recurrence period of 10.4 yr [781]. Determine the amount of mass accreted. If about 1.3×10^{-6} $M_\odot$ are ejected, determine the net mass gain per outburst, and the number of outburst and time required to reach the Chandrasekhar mass.

5

Type Ia Supernovae

"—Thirteen magnitudes in twenty-four hours! Wow! [...]
—This is too good to be true. Don't start telling everybody about it until we're quite sure. Let's get its spectrum first, and treat it as an ordinary nova until then. [...]
—When was the last supernova in our galaxy?
—That was Tycho's star—no, it wasn't—there was one a bit later, round about 1600. [...]
—And do stars do that sort of thing fairly often?
—Every year about a hundred blow up in our galaxy alone—but those are only ordinary novae. At their peak they may be a hundred thousand times as bright as the sun. A supernova is a very much rarer, and very much more exciting affair. We still don't know what causes it, but when a star goes super it may become several *billion* times brighter than the sun. In fact, it can outshine all the other stars in the galaxy added together. [...] A supernova explosion is the most titanic event known to occur in Nature. We'll be able to study the behaviour of matter under conditions that make the centre of a nuclear explosion look like dead calm. But if you're one of those people who always want a practical use for everything, surely it's of considerable interest to find what makes a star explode? One day, after all, our sun may decide to do likewise."

Arthur C. Clarke, *Earthlight* (1955)

The glitter of Galactic supernovae, bright enough to become visible to the unaided eye, has fascinated human beings since ancient times. Most of the supernovae witnessed before the Renaissance were documented and compiled in the Far East (modern-day China[1], mostly) and later introduced in Western Europe and North America by Edouard C. Biot and Alexander von Humboldt in the mid-nineteenth century. Since then, several astronomers, such as Ernst Zinner, Hans Lundmark, Issei Yamamoto, Ze-Zong Xi, Shu-Ren Bo, Peng Yoke Ho, David H. Clark, Francis R. Stephenson, and David A. Green have followed in their footsteps. The next section reviews the list of supernovae found in historical records by these pioneers.

5.1 Historical Supernovae

5.1.1 SN 185

The *Houhan-shu*, the chronicles of the Han dynasty compiled by the Chinese historian Fan Yeh in the fifth century, reports the appearance of a "guest star" near α-Centauri (between the constellations of Circinus and Centaurus), on Dec. 7, 185. The event, visible for several months (up to 20 months, according to some interpretations of the original documents), is widely regarded as the oldest supernova of any type ever recorded[2]. However, some

---

[1]See, e.g. references [1901, 1995].

[2]In 1976, NASA astronomers suggested that the explosion that resulted in the Vela supernova remnant, 10,000–20,000 years ago, may have been watched by humans of the southern hemisphere. A year later, archaeologist George Michanowsky suggested that some old rock carvings found in Bolivia may represent

reinterpretations of the relevant passages of the *Houhan-shu*[3] have raised concerns on the true nature of the witnessed object. Indeed, Chin and Huang [326] have suggested it was a comet rather than an exploding star (as it is often the case for many recorded "guest stars"). But more recently, Zhao, Strom, and Jiang [2022] have argued that while most references to comets within the *Houhan-shu* indicate their apparent displacement against the background of distant stars, this is not the case of the AD 185 event. Its large duration, together with the identification of several candidate remnants, most notably RCW 86 [340, 1722, 1803], point instead toward a supernova origin. Indeed, estimates based on the expansion velocity and size of RCW 86, inferred in X-rays with the Chandra and XMM-Newton satellites, suggest an age of $\sim$ 2000 yr for the remnant [1860]. Moreover, multiwavelength observations of RCW 86 have also unveiled large concentrations of iron. This and the lack of a neutron star embedded in the remnant suggest that the event was a type Ia supernova (see Section 5.5).

5.1.2   SN 393

While an unconfirmed supernova event may have also been observed in AD 369 (as well as in years 386, 437, 827, and 902), the next widely accepted example of prominent *celestial pyrotechnics* corresponds to SN 393, in the constellation of Scorpius, which was recorded in several Chinese documents[4]. The "guest star" exhibited a maximum apparent magnitude of $\sim -1^{mag}$ and was visible for about eight months [339]. No unambiguous remnant has been linked to this event.

5.1.3   SN 1006

SN 1006 (Figure 5.1) is probably the brightest Galactic supernova ever recorded, with an estimated maximum visual brightness of -7.5^{mag} [1951]. Multiple records from China, Japan, Korea, Arabia, Europe, and perhaps North America[5] provide a good description of its position, in the constellation of Lupus [655, 656]. From this, a posible remnant, the radio source PKS 1459-41, located 7100 light-years away, has been identified [617]. The "guest star" was visible for three months, and, according to some sources, was bright enough to be seen during daylight (and even to cast shadows!). Surveys aimed at finding clues of the SN 1006 progenitor system revealed no companion star [663, 973], suggesting that SN 1006 was the result of the merging of two white dwarfs (i.e., a type Ia supernovae; see Section 5.4.2.1). Indeed, no associated neutron star or black hole has been identified. There is evidence of the presence of several tenths of a solar mass of iron in the remnant of SN 1006 [714, 1994].

---

evidence of the supernova explosion being witnessed by Native Americans. Gerhard Börner, Li Qibin, and Bernd Aschenbach [219] reported possible links between several supernova remnants (hereafter, SNRs), identified in X-rays by the ROSAT satellite, and historical supernovae recorded in ancient Chinese drawings and texts: Most notably, inscriptions on tortoise shells dated from the 14th century BC (some 3500 yr ago) have been interpreted as representing a "guest star" that appeared close to Antares. One of the ROSAT supernova remnants, J1714-3939, is actually located in the vicinity of the star Antares and has an estimated age of 3000–5000 years, similar to that inferred for the tortoise shells.

[3]SN 185 might have also been recorded in Roman literature: in Herodian's *Roman History* (c. AD 250), as well as in *Historia Augusta, Vita Commodi*, attributed to Aelius Lampridius (4th century) [1725].

[4]In particular, *Jin Shu* (The Official History of the Jin Dynasty, AD 635), by X.-L. Fang, *Sung Shu* (The History of the Sung Dynasty, AD 500), by Y. Shen, and *Wen Xian Tong Kao* (Historical Investigation of Public Affairs, AD 1254), by D.-L. Ma [1902].

[5]A petroglyph by the Hohokam peoples in White Tank Mountain Regional Park (Maricopa County, Arizona) has been interpreted by some scholars as the first possible evidence of a North American representation of a supernova.

FIGURE 5.1

Composite image of the SN 1006 remnant, likely a type Ia supernova, located at 7100 light-years from Earth. The object is 65 light-years across. Data has been obtained from NASA's Chandra X-ray Observatory (X-rays), the University of Michigan's 0.9 m Curtis Schmidt telescope at the NSF's Cerro Tololo Inter-American Observatory (optical), the Digitized Sky Survey (optical), NRAO's Very Large Array and Green Bank Telescope (radio). Credit: X-ray: NASA/CXC/Rutgers/G. Cassam-Chenaï, J. Hughes et al.; Radio: NRAO/AUI/NSF/GBT/VLA/Dyer, Maddalena, and Cornwell; Optical: Middlebury College/F. Winkler, NOAO/AURA/NSF/CTIO Schmidt and DSS. Source: http://chandra.harvard.edu/photo/2008/sn1006c/; reproduced with permission. A color version of this image is available at http://fisica.upc.edu/ca/users/jjose/CRC-Downloads.

5.1.4   SN 1054

Shortly afterward, another supernova, SN 1054, was witnessed and multiply recorded in China and Japan. Translations of the original texts were compiled by J. J. L. Duyvendak [478], rediscussed by Ho Peng-Yoke, F. W. Paar and P. W. Parsons [814], and reinterpreted by A. Breen and D. McCarthy [249] (see also reference [367]). The *Sung-shi* (Chronicles of the Sung Dinasty), for instance, reports: *"... in the 1st year of the period Chih-ho, the 5th moon, the day chi-ch'ou* [July 4th, 1054], *a guest star appeared approximately several inches south-east of Tien-kuan* [that is, north-west of the star ζ Tauri on the celestial sphere, probably separated by an angular distance of $0.3 - 0.5$ degrees (see reference [1250])]. *After more than a year, it gradually became invisible.*" An entry in the *Annals of the Sung-shih* also states that: "*On the day hsin-wei of the 3rd moon of the 1st year of the period Chia-yu* [April 17, 1056], *the Chief of the Astronomical Bureau reported that from the 5th moon of the 1st year of the period Chih-ho* [June 9 to July 7, 1054], *a guest star had appeared in the morning in the eastern heavens, remaining in Tien-kuan* [ζ Tauri], *which only now had become invisible.*" The duration of the event, based on the dates compiled in these two documents, July 4, 1054, and April 17, 1056, reveals that the "guest star" was visible to the naked eye for at least 653 days. Other sources, like the *Sung hui-yao*, compiled by Chang Te-hsiang, report as well that the object was visible during the day (like Venus) for 23 days, with a reddish-white color. Its location, close to ζ Tauri, was soon linked to a well-known remnant: The Crab nebula (Figure 5.2), discovered in 1731 by John Bevis. Embedded within the remnant, there is a neutron star, PSR B0531+21 (the Crab pulsar), discovered in 1968 [368, 1481, 1697], which points toward a massive stellar progenitor (i.e., a type II supernova; see Chapter 7).

5.1.5 SN 1181

SN 1181, in the constellation Cassiopeia, is another Galactic supernova witnessed and recorded by Chinese and Japanese astronomers. It remained visible in the night sky for 185 days. The radio and X-ray pulsar J0205+6449 (3C 58), which rotates about 15 times per second, has been identified as a possible remnant of this supernova [1379]. However, VLA radio observations of this pulsar have revealed a low expansion speed, suggesting that 3C 58 may be several thousand years old and therefore not the remnant of SN 1181 [181].

5.1.6 Tycho's Supernova

The last two supernovae that have been unambiguously detected in our Galaxy are SN 1572 (also known as Tycho's supernova) and SN 1604 (Kepler's supernova). On November 11, 1572, a young Tycho Brahe (who, following his uncle's wishes, had studied law, and was only interested in astronomy as a hobby), discovered something unexpected in the night sky (Figure 5.3(a))[6]: "*On the 11th day of November, in the evening after sunset, I was contemplating the stars in a clear sky. I noticed that a new and unusual star, surpassing the other stars in brilliancy, was shining almost directly above my head; and since I had, from boyhood, known all the stars of the heavens perfectly, it was quite evident to me that there had never been any star in that place of the sky, even the smallest, to say nothing of a star so conspicuous and bright as this. I was so astonished of this sight that I was not ashamed to doubt the trustworthyness of my own eyes. But when I observed that others, on having the place pointed out to them, could see that there was really a star there, I had no further doubts. A miracle indeed, one that has never been previously seen before our time, in any age since the beginning of the world.*" The event, also witnessed and recorded in several texts in China, Korea, and Europe, changed Tycho's life (and career!) forever.

In his work *De nova stella*, Tycho Brahe refuted the Aristotelian belief in celestial immutability: While watching the "guest star" of 1572, he realized that the object showed no displacement against the background of distant stars, thus concluding that this was not a near (*sub-lunar*) event but belonged to the distant *sphere* of the "fixed" stars. This shattered

---

[6]Tycho Brahe's *De nova et nullius aevi memoria prius visa stella* (Concerning the Star, new and never before seen in the life or memory of anyone, 1573). Quote extracted from *Burnham's Celestial Handbook: An Observer's Guide to the Universe Beyond the Solar System* (1978), by Robert Burnham.

FIGURE 5.2
Composite image of the Crab nebula, the remnant of a type II supernova explosion, as
seen by NASA's Hubble Space Telescope in 1999–2000. This six light-year wide object,
located 6500 light-years away, is composed of different chemical species, including neutral oxygen, singly ionized sulfur, hydrogen, and doubly ionized oxygen. The rapidly rotating neutron star embedded in the nebula acts as a dynamo, powering the nebula's
interior glow (caused by electrons moving at nearly the speed of light around magnetic field lines of the compact star). Credit: NASA, ESA, J. Hester and A. Loll (Arizona State University). Source: http://hubblesite.org/gallery/album/pr2005037a/,
released into public domain by NASA. A color version of this image is available at
http://fisica.upc.edu/ca/users/jjose/CRC-Downloads.

the traditional interpretation of *stellae novae* as being purely atmospheric, tailless comets.
Tycho also reported that the "guest star" that suddenly appeared in the constellation of
Cassiopeia was about as brilliant as Jupiter and soon rivalled Venus, with a peak visual
luminosity of -4^{mag}. For about two weeks the star could be seen during daylight. At the
end of November 1572, it began to fade, shifting color from bright white to yellow, orange
and red, before fading away by March 1574. Overall, it was visible to the naked eye for
about 16 months.

The remnant associated with SN 1572 (Figure 5.3(b)) was discovered at radio frequencies
(158.5 MHz) in 1952 [724] and confirmed, with a mesurement at 157.8 MHz (1.9 m), in

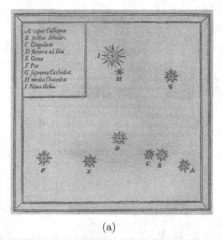

(a)

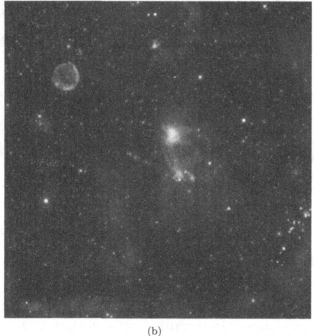

(b)

FIGURE 5.3

(a) Star map of the constellation Cassiopeia showing the supernova of 1572 (the star "I", in the figure). From Tycho Brahe's *De nova stella* (1572). Source: http://commons.wikimedia.org/wiki/File:Tycho_Cas_SN1572.jpg, released into public domain. (b) Mosaic of the constellation Cassiopeia from NASA's Wide-Field Infrared Survey Explorer (WISE), spanning an area of 1.6 × 1.6 degrees on the sky. Tychos's supernova is the circular, nebular object in the upper left corner. Credit: NASA/JPL-Caltech. Source: http://commons.wikimedia.org/wiki/File:SN_1572_Tycho%27s_Supernova.jpg, released into public domain by NASA. A color version of this mosaic is available at http://fisica.upc.edu/ca/users/jjose/CRC-Downloads.

1957 [118]. The remnant has been designed as 3C 10 in the third Cambridge catalogue (2C 34, in the previous catalogue), although it is also known as G 120.1+1.4 in Dave Green's catalogue of Galactic SNRs [681]. The extremely faint optical counterpart was identified on Mt. Palomar photo plates in the 1960s and in X-rays by the satellite Uhuru [1954] (see also reference [715]).

The classification of SN 1572 has been controversial. Reconstruction of its light curve and color evolution, based on reanalysis of historical records, suggested a type Ia (normal or subluminous) supernova explosion [1543,1843], although other possibilities (e.g., a type Ib or II-L core-collapse supernova) cannot be ruled out [448,1573]. X-ray observations of its stratified ejecta [422], and the analysis of light echoes[7] due to scattering and absorption/re-emission of the outgoing supernova flash by nearby interstellar dust [1023], support a type Ia supernova classification (see also references [1154,1491]). In the last decade, efforts aimed at confirming whether the explosion was a type Ia or a type Ib/II have focused on the search for a potential surviving companion star, expected (if any) to be luminous, with unusually high radial velocity, and probably high spin [721,1196]. A G2 stellar companion (labeled as Tycho G) was tentatively identified in 2004, on the basis of a much larger mean radial velocity than other stars in the neighborhood [662,1545]. Moreover, an enhanced Ni abundance was also reported for Tycho G [662], attributed to reaccretion of the supernova ejecta by the companion, secondary star. However, these conclusions are yet regarded as controversial, since other studies [974,975] have revealed that Tycho G is relatively far away from the remnant's geometric center and does not show any significant rotation. Furthermore, no clear overabundance in Ni has been inferred from Keck-I high-resolution spectroscopy. The fact that other stars in the vicinity show similar proper motions suggests instead that Tycho G could be an unrelated background star.

5.1.7　Kepler's Supernova

SN 1604 is the most recent supernova observed in the Milky Way. It was first witnessed on October 9, 1604 in Italy, and a few days later by Chinese, Korean, and other European observers[8] [677,1624,1903]. Johannes Brunowsky, a meteorologist from Prague, comunicated the discovery of a *stella nova* to Johannes Kepler on October 11 [1142]. Unfortunately, it was cloudy in Prague until October 17. Kepler systematically tracked the star during one year (the event was visible during daytime for over three weeks) and published a monograph titled *De Stella nova in pede Serpentarii* (On the new star in Ophiuchus's foot) in 1606. Visible to the naked eye, Kepler's bright supernova overcame any other star in the night sky (and all planets other than Venus), reaching an apparent peak visual magnitude of about -2.5^{mag} [101]. The remnant of Kepler's supernova was identified in the 1940s at Mount Wilson observatory as *"a small patch of emission nebulosity, which is undoubtedly a part of the masses ejected during the outburst"* [101]. Baade's detection of Kepler's SNR in the optical was followed by subsequent observations in radio waves (3C 358), X-rays, microwaves, and infrared. It is also worth noting that its light curve reconstruction (Figure 5.4), on the basis of the 17th century observations, led Baade to classify SN 1604 as a type I supernova (using the adopted designation of the epoch). Even though recent X-ray data of Kepler's SNR taken by Chandra revealed relative elemental abundances typical of a type

---

[7]See Rest et al. [1490] for a review on light-echo techniques.

[8]According to Walter Baade [101], the "guest star" of 1604 was first glimpsed on October 9 by an unknown Italian physician in Cosenza (Calabria), who reported his discovery to the astronomer Chr. Clavius in Rome, and by I. Altobelli in Verona. The day after, it was observed by B. Capra and S. Marius in Padua, and by J. Brunowsky in Prague. The star was also observed by Jan van Heck in Rome (October 11), Giovanni A. Magini in Bologna, and H. Röslin in Hagenau (October 12), and by David Fabricius in Osteel (October 13).

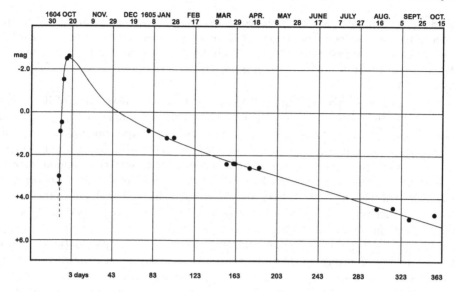

FIGURE 5.4
Visual light curve reconstruction of Kepler's SN 1604 (black dots) based on historical records. A solid line corresponding to the properly adjusted light curve of SN 1947C, a type Ia supernova, is shown for comparison. Figure from Baade [101]; reproduced by permission of the American Astronomical Society.

Ia supernova [1393, 1493], the true nature of this supernova is far for being settled (see reference [194]).

Unfortunately, no Galactic supernova has been detected[9] since the invention of the telescope (around 1608). Astronomical evidence exists, however, for a couple of Milky Way supernovae that could have been detected on Earth since then: Cassiopeia A (Cas A; see Chapter 7), whose light reached our planet around 1680[10], as well as the progenitor of the SNR G1.9+0.3 in Sagittarius (by 1870). This has likely been driven by obscuration of the optical emission by dust along the lines of sight. With the advent of telescopic astronomy, supernovae have been regularly spotted outside the Milky Way, in a large number of galaxies.

5.2   Spectroscopy of Supernovae

In previous pages, differences among historical supernovae have been outlined, and two broad classes, type I(a) and type II, were briefly mentioned. In principle, both classes can be distinguished by the absence (as in type Ia) or presence (type II) of a compact remnant[11],

---

[9]The closest supernova discovered since SN 1604 has been the type II SN 1987A, which occurred in the Large Magellanic Cloud, a *satellite* of the Milky Way galaxy (see Chapter 7).

[10]Although no historical records of any sightings of Cas A exist, it has been claimed that the supernova may have been misclassified as a sixth magnitude star (3 Cassiopeiae) by J. Flamsteed on August 16, 1680 [853].

[11]There is, however, mounting evidence that some subluminous type Ia supernovae, like SN 2002cx or SN 2005hk (see footnote 21, this chapter), may fail to unbind the whole exploding star, and, therefore, would leave a compact remnant [915, 1025].

but this requires certainty in its identification (as for SN 1987A[12], where the existence of a remnant is taken for granted, even though it has not been detected yet!). Nevertheless, there are other photometric and spectroscopic ways to distinguish between the different supernova types.

The first (visual) spectrum of a supernova was obtained from S Andromedae (SN 1885; see Figure 5.5), discovered near the center of M31 in 1885. The results were presented at a meeting of the Royal Astronomical Society a year later by O. T. Sherman [1627]. The marginal spectroscopic observations[13] of S And were reviewed by C. Payne-Gaposchkin [1399], who concluded that its spectral features were *"mainly unidentifiable—except for two possible coincides with emission lines at λ5325 and λ5575 in the spectra of normal novae."* S And was revisited by H. Minkowski [1245], who argued that *"it is not impossible that in the region above λ5000 the spectrum of S Andromedae may have been similar to the spectra of* [type I] *supernovae* [in] *IC 4182 and NGC 1003; the spectra below λ5000 may have been different."* A compilation and reanalysis of 500 magnitude and 100 color estimates, 40 spectral observations, 67 transits, and 136 micrometric measurements of S Andromedae by about 120 observers was published a century after its discovery by G. de Vaucouleurs and H. G. Corwin [420], who concluded instead that the spectrum of S And reveals a surprising agreement with most emission maxima, matching the main lines of typical type I supernovae (see Table 9 in reference [420]). A reconstructed visual spectrum summarizing all the relevant features of S And is shown in Figure 5.5.

Shortly after the discovery of S Andromedae, another supernova, Z Centauri (SN 1895B, probably another Type Ia), was found in NGC 5253. Its spectrum was examined in Dec. 22 and 29, 1895, by W. W. Campbell [293], who concluded that it was continuous, with evidence of "bright lines" somewhat superimposed. In 1936, C. Payne-Gaposchkin [1398] interpreted part of the spectroscopic features of Z Cen as corresponding to greatly widened emission lines, and concluded that *"an interpretation of the spectrum suggesting radial velocities of the order of 10,000 km/sec is not impossible."*

Spectra of two more supernovae, SN 1926A and SN 1936A, were published by M. L. Humason in 1936 [856]. Several redshifted emission lines (e.g., H_β, H_γ, and H_δ) and two absorption lines (Calcium H and K) were identified, from which expansion velocities of several thousand km s^{-1} were inferred. Humason concluded that the existence of emission bands 150–200 Å wide confirms the prediction by W. Baade and F. Zwicky that *"extremely wide emission lines are to be expected in the spectra of super-novae, indicating that gaseous shells are expelled at great speeds"* [103]. The same year, Baade obtained a spectrum of SN 1936A, about 100 days after maximum visual brightness, which revealed an intense, single, wide band at 4640.7 Å, interpreted[14] as due to N III [100]. Furthermore, he considered it proven that the early spectra is due to a superposition of very wide emission lines. Thirty-two additional supernovae were discovered in the period 1885–1937 (six by F. Zwicky, just in 1937)[15].

---

[12]According to the official convention, supernovae are designated by the year of their discovery, followed by one or two letters: an uppercase letter—A to Z—, for the first 26 supernovae discovered in the same year, and pairs of lowercase letters—aa, ab... ba, bb... zz—afterward. Historical supernovae are simply named by the discovery year. The abbreviation SN is an optional prefix.

[13]Some astronomers performed crude spectroscopic measurements of S And visually, at the limit of visibility, since no photographic spectral observations were made in those days. A summary of the contemporary descriptions can be found in reference [420].

[14]The same interpretation was given by C. Payne-Gaposchkin to a broad line reported in Z Cen.

[15]Since then, the number of discoveries escalated almost exponentially: 37 supernovae were found from 1938 to 1950 (with a gap in 1942–1944, when humans were busy with another sort of explosions). A century since S And 1885, the number of discovered supernovae amounted to 623. The year 1954 was so successful that a new nomenclature (with two letters) was introduced to designate the 27th observed supernovae (i.e., SN 1954aa). The overall number of supernovae observed in the period 1885–1990 was 785. Just in the decade 1991–2000, 1085 supernovae were spotted; and in the period 2001–2010, the number raised to 3689. The

Mean Prismatic Spectrum of S And 1885

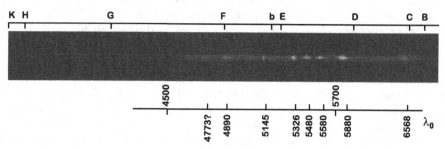

FIGURE 5.5

Reconstructed visual spectra of S And, based on contemporary descriptions and estimated wavelengths. Figure from de Vaucouleurs & Corwin [420]; reproduced by permission of the American Astronomical Society.

Until 1937, spectrograms of supernovae (that is, sequences of spectra) were scarce, and only similarities between supernovae and common novae (and among all supernovae) were reported. The first noticeable differences where identified by D. Popper in microphotometer tracings of SN 1937C. He drew a comparison with available spectroscopic data of Z Cen and SN 1936A, revealing some discrepancies near 423 μm and 475 μm (later identified as H$_\beta$ and H$_\gamma$ absorption lines), two deep minima, which were apparently absent in the spectra of Z Cen and SN 1937C while present in SN 1936A [1440]. But it was not until the observation of SN 1940B, in NGC 4725, when two distinct supernova classes began to be firmly established. As Minkowski reported [1246]: *"the spectrum of this supernova is entirely different from that of any other nova or supernova previously observed."* Similar comments were made by Humason and Minkowski [857] with regard to SN 1941A. A distinction between an extremely homogeneous class of objects (provisionally designated as "type I supernovae"), such as 1937C or 1937D, that lack hydrogen (Balmer) lines, and the more heterogeneous class of "type II supernovae", prominent in those lines, with examples like 1936A, 1940B, or 1941A, was first proposed [1247]. Different interpretations of the spectra were published in the following decades [232]. By 1960s, it was finally established that both type I and type II supernovae initially exhibit low-excitation emission and blueshifted absorption lines in the optical, superimposed to a continuous, thermal spectra. The earliest, broad identification of low-excitation, absorption lines[16] in type I supernovae (e.g., He I, Si II, Fe II, Mg II, Ca II, S II) was made by Yu. P. Pskovskii [1461]. Pioneering work based on synthetic spectra [240] confirmed, shortly afterward, the big picture outlined by Pskovskii.

The current classification of supernovae (see Figure 5.6) mainly relies on optical spectroscopic measurements near maximum light. Reviews on this subject have been published by David Branch [232–235, 237–239, 241], Alexei Filippenko and collaborators [524, 1652–1656], Robert Harkness and J. Craig Wheeler [727, 1927], among others.

The hydrogen-deficient, type I supernova class[17] is further split into several groups: Type

---

total number of supernovae observed since S And 1885 until the end of 2012 was 6094 events, with 2007 the most prolific year, with 573 observations (the last designed as SN 2007va). Several webpages maintain an updated listing of all observed supernovae. An example is the IAU Central Bureau for Astronomical Telegrams, http://www.cbat.eps.harvard.edu/lists/Supernovae.html.

[16]Pskovskii's identification of He I was later assigned to Na I [1275]. The presence of strong He I absorption features were previously suggested in the spectra of SN 1954A [1221], later classified as a type Ib supernova.

[17]The lack of hydrogen in the spectra is an observational constraint that poses limits on the maximum amount of hydrogen that can be present in the expanding atmosphere of the star (i.e., $M_H \leq 0.03 - 0.1 M_\odot$). Some type Ia supernovae are, however, anomalous in this regard. SN 2002ic [717], for instance, while

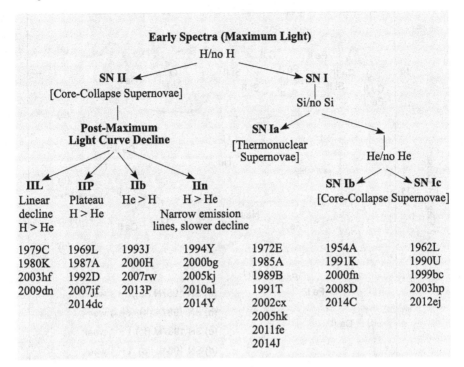

FIGURE 5.6
Supernova taxonomy based on spectroscopic features at early and late times. The type Ia class has been tentatively subdivided in four different subtypes: "normal", 1991T-like, 1991bg-like, and 2002cx-like [1102]. Figure adapted from similar classification schemes [295, 727, 1065, 1836].

Ia supernovae are characterized by a prominent absorption feature near 6150 Å, due to blueshifted SiII $\lambda\lambda 6347, 6371$ (collectively referred to as $\lambda 6355$). This Si II feature is absent in the type Ib and Ic classes, which instead are prominent in oxygen and sodium absorption lines. The presence or absence of strong He I lines (especially He I $\lambda 5876$) distinguishes type Ib from type Ic supernovae. Typical early spectra of the different supernova classes are shown in Figure 5.7.

At later times (i.e., 4–5 months past maximum light, during the nebular phase; see Figure 5.8), type Ia supernovae are characterized by many unresolved Fe emission lines mixed with Co lines, together with Ca II in absorption. In sharp contrast, type Ib and Ic supernovae have relatively distinct emission lines of intermediate-mass elements (i.e., O, Ca). Moreover, type Ib supernovae exhibit narrower and somewhat stronger lines than type Ic supernovae. In turn, type II supernovae are spectroscopically similar to type Ib and Ic, except for the dominance of a strong H_α emission line and the fact that emission lines are narrower and weaker.

Line blanketing[18] of multiple transitions (mainly involving Fe II and Co II) induce a prominent UV deficit relative to a blackbody fit at optical wavelengths in the early spectra

exhibiting the usual features of all type Ia supernovae, unequivocally showed broad-line H_α emission about 90 days post-maximum light. This has been interpreted as proof of a supernova interacting with H-rich circumstellar material. Other SN 2002ic-like events include SN 2005gj, PTF11kx [446], or SN 2008J [1769]. See also reference [1076] for a discussion on upper limits of ejected hydrogen in type Ia supernovae.

[18]Line blanketing refers to the enhancement of the red or infrared regions of a stellar spectrum at the expense of other regions, resulting in an overall diminishing effect on the full spectrum.

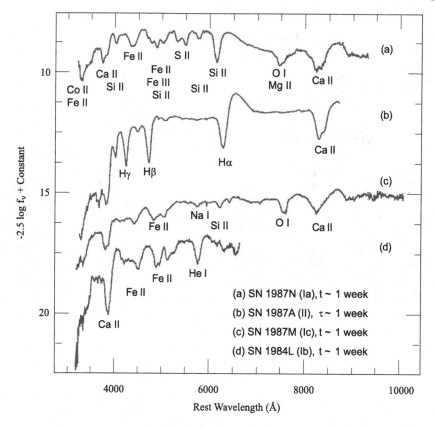

FIGURE 5.7

Early-time spectra of different supernova types [Ia, Ib, Ic, and II]. t and τ indicate time after maximum in the blue band and time after core collapse, respectively. The ordinate units are essentially AB magnitudes, based on flux measurements, f_ν(erg s^{-1} cm^{-2} Hz^{-1}) calibrated in absolute units (see reference [1347] for details). Wavelengths are corrected from redshift and expressed as values at rest. Figure from Filippenko [524]; reproduced with permission.

of all type I supernovae. Moreover, type Ia supernovae (but not type Ib/Ic) also exhibit an IR deficit. Most type II supernovae (SN 1987A being an exception) show neither a UV nor an IR deficit, and can be fitted by a single-temperature Planck function from UV through IR wavelengths.

5.2.1 Spectral Evolution of Type Ia Supernovae

At maximum light, the spectrum of a type Ia supernova is characterized by a thermal continuum with superimposed broad lines (driven by the high velocities of the ejecta), formed by scattering of the photospheric continuum and exhibiting P-Cygni profiles[19], characteristic of expanding atmospheres. Modeling suggests neutral or singly ionized intermediate-mass

---

[19]P-Cygni profiles are spectroscopic features named after P Cygni, a variable star whose spectral line profiles exhibit both emission and absorption features. Such dual characteristics are due to an expanding shell or wind being blown off the star. Where the expanding shell is located between the observer and the star, the emission lobe is redshifted, while the absorption lobe is blueshifted with respect to the spectral lines' wavelength at rest.

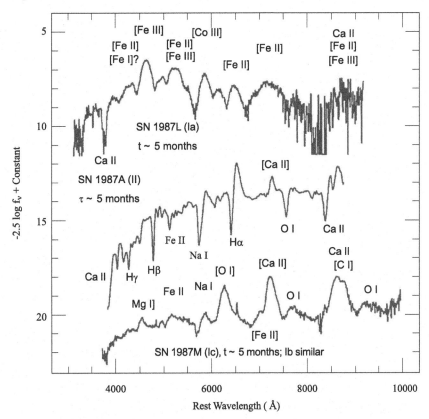

FIGURE 5.8
Late-time spectra of different supernova types [Ia, Ic—Ib—, and II]. Figure from Filippenko
[524]; reproduced with permission.

elements[20] as the species driving most of the spectroscopic features observed at this stage[21],
in particular, the Si II doublet at $\lambda\lambda6347$ and 6371, with a characteristic absorption feature
around 6100 Å (one of the key features defining the type Ia supernova class), the H and K
Ca II lines at $\lambda\lambda3934$ and 3968, Mg II ($\lambda4481$), S II ($\lambda5468$ and $\lambda\lambda$ 5612 and 5654), O I

---

[20]There is no convincing evidence of the presence of He in the spectra of type Ia supernovae, so far
[1216, 1224].

[21]Some type Ia supernovae, such as SN 1991T, 1999ac, 2000cx, 2002cx, 2002ic, or 2005hk, displayed
remarkable spectroscopic surprises. SN 1999T showed strong Fe III lines in the early spectra, rather than
the usual Si II and Ca II [528]. The spectra of SN 1999ac revealed expansion velocities inferred from Fe
lines lower than average, whereas those inferred from Ca H and K lines were higher than average. Moreover,
the expansion velocities inferred from Si II were among the slowest ever reported [606, 1414]. The overall
spectroscopic evolution of SN 2000cx turned out to be quite peculiar. The Si II lines that emerged near
maximum light remained strong until about 3 weeks past maximum. The change in the excitation stages
of Fe-peak elements was slow, and both Fe-peak and intermediate-mass elements were found to be moving
at very high expansion velocities [1101]. The early spectra of SN 2002cx evolved very quickly and were
dominated by lines from Fe-group elements (spectroscopic features from intermediate-mass elements, such
as Ca, S, or Si, were weak or absent). Moreover, its nebular spectrum was also anomalous, consisting of
narrow Fe and Co lines [909, 1100]. The most striking feature in SN 2002ic was the presence of low-velocity,
strong H_α emission [717, 1004]. Finally, late-time spectra of SN 2005hk were not dominated by Fe forbidden
emission lines; instead, several Fe II lines characterized by very low expansion velocities (about 700 km
s^{-1}) have been identified. Low-velocity O I lines have also been reported in the spectra of SN 2005hk,
suggesting the presence of unburned material near the center of the exploding white dwarf [909], and even
the possibility of a deflagration blast [1415].

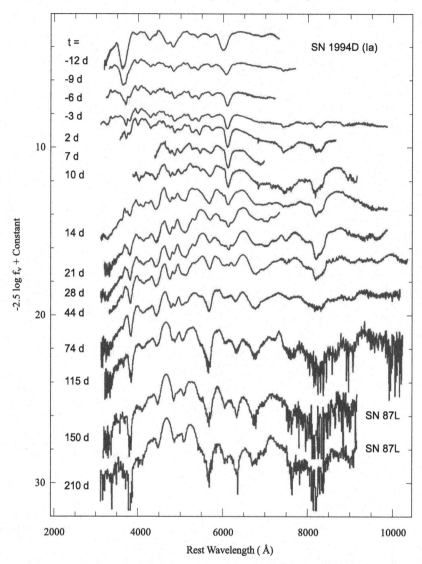

FIGURE 5.9
Time-evolution of the spectra of SN 1994D, a type Ia supernova discovered in NGC 4526. The last two spectra, at days 150 and 210 relative to maximum B brightness, are taken from the similar type Ia SN 1987L, in NGC 2336. Figure from Filippenko [524]; reproduced with permission.

(λ7773), as well as other Si II (λ3858, λ4130, λ5051, and λ5972) and Ca II (λ8579) lines. Intense Doppler-shifted absorption lines (e.g., Si II) are used to infer photospheric expansion velocities, with different lines showing slightly different velocities, up to 25,000 km s^{-1}, rapidly decreasing with time. This suggests a layered distribution of the explosion products, with the outermost ejected shells clearly dominated by the presence of intermediate-mass elements, a proof of incomplete nuclear processing [542,1198,1652,1776] (see Section 5.5.2).

Some contribution from iron-peak elements (Fe, Co), particularly at near-UV wavelengths, increasing progressively as the photosphere recedes and the iron-rich core gets exposed, is also observed. As shown in Figure 5.7, the spectrum below 3500 Å is strongly suppressed. Two weeks after maximum light, the spectrum becomes dominated by Fe II lines, with contributions from intermediate-mass elements (Si II, Ca II) still present. A month past maximum light, when the supernova reaches the so-called *nebular stage*, forbidden and heavily blended Fe and Co emission lines[22], such as [Fe II], [Fe III], and [Co III], become the most prominent spectroscopic features. Absorption Ca II H and K lines, as well as the near-IR triplet at $\lambda\lambda 8498$, 8542, and 8662, remain detectable even at late times.

The near-IR spectrum [1197] brings complementarity to the optical spectrum, since certain elements have no lines at optical wavelengths, or may be heavily blended (superimposed) or saturated in the optical region. Characteristic near-IR lines include Mg II—0.922 and 1.0926 μm—, Ca II—the IR triplet at 0.850, 0.854, and 0.866 μm, as well as a line at 1.268 μm—, Si II—1.691 μm—, and several features attributed to Fe-peak elements (Fe II, Co II, Ni II, and probably Mn II). Few supernovae have been detected outside the optical and near-infrared wavelengths. While SN 2003hv and SN 2005df have been seen in midinfrared (from which Ar lines have been reported [636]), no single type Ia supernova has been detected so far in radio, and only one, SN 2005ke, has been seen in X-rays (likely due to interaction of the supernova shock with circumstellar material [883]).

Another remarkable spectroscopic feature is the progressive decrease in intensity of the Co lines. Indeed, the relative intensities of [Co III] and [Fe III] are interpreted as a signature of ^{56}Co radioactive decay ($T_{1/2} = 77.3$ days), originally in the form of ^{56}Ni ($T_{1/2} = 6.1$ days) [365, 1030, 1826]. It is worth noting that a large fraction of the energy released in the decays from ^{56}Ni and ^{56}Co is in the form of γ-rays, with energies of 158, 812, 750, and 480 keV for ^{56}Ni$(e^-, \nu)^{56}$Co, and 847 and 1238 keV for ^{56}Co$(e^-, \nu)^{56}$Fe. Hence, supernovae have become priority targets for many γ-ray missions, like INTEGRAL (see Section 5.5.2).

Spectroscopic snapshots at different epochs (Figure 5.9) provide valuable clues on the physical mechanism powering these events. The overall picture clearly stresses key nuclear physics issues: The presence of intermediate-mass elements in the early spectra, when only emission from the outermost ejected layers is seen, reveals that the thermonuclear explosion did not incinerate the whole star (incomplete burning). This, in turn, provides clues on the flame propagation regime, a yet unsolved issue in type Ia supernova theory, since the presence of such intermediate-mass elements rules out a pure detonation (see Section 5.5). In sharp contrast, when inner regions of the star become accessible later on, the prominence of the Fe lines clearly points toward full nuclear processing to Fe-peak elements. Finally, the presence of Co lines at late stages strongly supports the hypothesis of a light curve tail powered by ^{56}Co decay.

5.2.2 Spectropolarimetry

A different observational approach has recently opened a new window to probe supernovae at small spatial scales that cannot be resolved through direct imaging. The light emitted by a supernova can become polarized [341], linearly or circularly, due to a suite of different processes, including asymmetries driven by combustion, the spin of the white dwarf, the presence of a stellar companion, or scattering by circumstellar dust, among others [1892, 1896]. Polarized light retains geometrical information of the ejecta, which stresses the interest in *spectropolarimetry*, or the measurement of polarization as a function of wavelength.

Spectropolarimetry offers a chance to infer both the overall shape of the emitting region and that of regions with specific chemical composition, thus providing a new insight into the explosion mechanism together with a wealth of information on stellar winds and

---

[22]The broad emission feature around 6500 Å, which shows up in the spectrum about 2 weeks past maximum light, has sometimes been wrongly identified as H$_\alpha$ (instead of Fe II and later [Fe II]). This has induced misclassification of some type Ia supernovae as being type II.

circumstellar material in the environment. For a typical supernova, characterized by a photospheric radius of $\sim 10^{15}$ cm, located at 10 Mpc, polarimetry can yield an effective spatial resolution of ~ 10 μarcsec, a hundred times better that the resolution attained with optical interferometers [1896]. Polarization from supernovae is expected to be frequency-dependent and to vary across spectral lines. It will also vary in time, in sharp contrast to interstellar polarization. This allows observers to disentagle the intrinsic supernova polarization from the interstellar component.

In 1972, Wolstencroft and Kemp [1952] were the first to suggest that the large, intrinsic stellar magnetic field may induce a mesurable optical polarization on the ejecta, shortly after a supernova explosion. The first, pioneering attempts to detect polarization in the light emitted by a supernova began in the 1970s. They were either heavily contaminated by interstellar polarization or yielded no significant results [773, 838, 1610, 1611, 1614, 1690, 1898, 1952]. Therefore, only upper limits were inferred [1218, 1898]. These studies suggested already that supernovae may exhibit an intrinsic linear polarization resulting from Thomson scattering by free electrons in the asymmetric, expanding envelope (see, e.g., [1610]). The scenario was revisited in 1982 by Shapiro and Sutherland [1614] who conducted a detailed study of linear polarization, expected to result from the nonspherical, scattering-dominated atmosphere of a supernova as a function of its asphericity. The study also analyzed the role of interestellar dust, responsible for switching linear into circular polarization. Shortly afterward, McCall and collaborators [1217], proposed the use of lines with P-Cygni profiles (see Section 5.2.1), induced by scattering in the expanding atmosphere of a supernova, as diagnosis of the degree of polarization, on the basis of their large linear polarization. Detection of the intrinsic supernova polarization may have been marginally achieved in SN 1996X [1897] and SN 1997dt [1077]. Clear evidence of asphericity was first inferred from the type Ia SN 1999by, with an estimated degree of polarization of $\sim 0.3\%$–0.8% [840].

An extensive account of the progress made in spectropolarimetric observations[23] of supernovae can be found in Wang and Wheeler [1896]. Surprisingly, spectropolarimetric studies have revealed that, although core-collapse supernovae are spherically symmetric objects, they exhibit a higher degree of polarization (about 1%; see [840]), that is, a larger deformation, than type Ia supernovae[24], which have naturally aspherical configurations[25].

SN 2001el was the first, normal type Ia supernova for which detailed, high-quality, spectropolarimetric data at different epochs was obtained [1894]. Observations covered a time span of about two months after the explosion (including preoptical maximum measurements). Linear polarization of the continuum at a level $\sim 0.2\%$–0.3% was reported prior to optical maximum. The fact that polarization became undetectable a week after optical maximum suggests that the outermost ejected layers (the first ones to be exposed) are more aspherical than the inner regions (seen later in the spectra). Particularly interesting was the identification of a shell (maybe a ring, with a clumpy distribution of material) of high velocity Ca-rich material (20,000–26,000 km s^{-1}), in the outermost regions of SN 2001el. This was observed as an absorption feature at about 8000 Å, a component of the Ca II IR triplet, from which a 0.7% polarizarion was inferred [527, 1896]. At about 10,000 km s^{-1}, the Ca II and Si II lines show polarizations about 0.3%. The fact that deeper layers are found to be more spherical suggests a structure that favors delayed (supersonic) detonation models while appears in conflict with pure (subsonic) deflagration models [1892, 1895]. Indeed, whereas in pure deflagration models, clumps of different chemical composition are

---

[23]See references [341, 424, 666] for a primer on spectropolarimetric observational techniques.

[24]It is worth noting, however, that peculiar, subluminous supernovae (e.g., 1999by [840]) are characterized by a higher polarization than normal type Ia.

[25]In type Ia supernovae occurring in binary systems containing a low-mass, nondegenerate stellar secondary, polarization may be partially induced by the hole carved by the secondary onto the otherwise spherical ejecta [951].

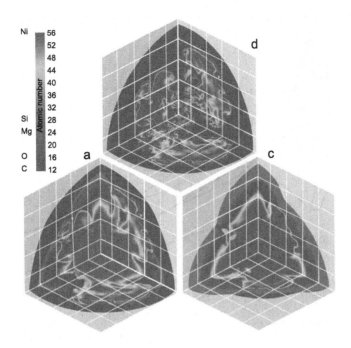

FIGURE 5.10

Distribution of chemical species in the inner regions of an exploding white dwarf, for a pure deflagration model (d) and two delayed-detonation models (a, c). A color version of this plot is available at `http://fisica.upc.edu/ca/users/jjose/CRC-Downloads`. Figure from Gamezo, Khokhlov, and Oran [599]; reproduced with permission.

expected to be found in all layers reached by the flame, the supersonic front resulting in delayed detonation models homogenizes the chemically clumpy structures generated during the deflagration[26], as the detonation wave propagates through the ashes left behind by the former subsonic wave (see Figure 5.10). It is worth noting however that delayed detonation models cannot account for the whole sample of type Ia supernovae spectropolarimetrically observed. In fact, evidence suggests certain spectropolarimetric diversity among type Ia supernovae [1078].

The degree of continuum polarization reported from type Ia supernovae, less than 0.5%, implies flattening (asphericity) of at most 5%. This would contribute by only $\leq 0.05^{mag}$ to the scatter in absolute luminosities in type Ia supernovae [1893]. An observed trend suggests that type Ia supernovae with faster light curve decline rates tend to be more polarized [1895].

---

[26]The structure left behind by a detonation wave is, however, not completely layered because of the formation of cellular structures, especially at densities $\sim 10^7$ g cm^{-3} [601].

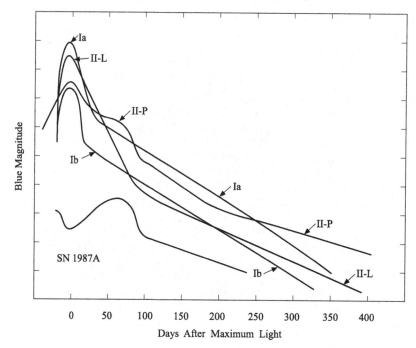

FIGURE 5.11
Light curve comparison between different supernova types (i.e., Ia, Ib, II-P, II-L, and SN 1987A), shown from peak magnitude in the blue up to 400 days past maximum light. Figure from Wheeler and Harkness [1927]; reproduced with permission.

5.3   Light Curves: Supernova Pyrotechnics

Photometrically, type Ia supernovae are characterized by a sudden rise in luminosity[27], up to a maximum visual absolute magnitude $M_v \sim -19.3^{mag}$ ($L_{peak} \sim 10^{10}$ $L_\odot$; see reference [1067]) in about 20 days. This phase is followed by a steep decline in brightness by about 3 magnitudes in ~ 30 days, and later by a second, smoother decline over a period of ~ 70 days [727, 989] (Figure 5.11).

Peak luminosities depend basically on the amount of ^{56}Ni synthesized, which can be inferred from models of the late-time, nebular spectra and through application of *Arnett's rule* [66]: the peak luminosity is proportional to the energy released by radioactive decays within the expanding ejecta. Assuming a rise time to peak luminosity of 19 days, Arnett's rule yields a value of $L_{peak} = 2 \times 10^{43}$ $(M_{Ni}/M_\odot)$ erg s^{-1} [1728]. Estimates[28] of the ^{56}Ni masses from 17 type Ia supernovae, inferred from the application of both methods, range between 0.1–1 $M_\odot$ [1729, 1730]. During the late stages of the event, the light curve is powered by the radioactive decay chain[29] $^{56}Ni \rightarrow {}^{56}Co \rightarrow {}^{56}Fe$ [365, 1826], with two different slopes attributed to the different half-lives of the decaying species, ^{56}Ni and ^{56}Co.

The increasing number of peculiar supernovae discovered (see Section 5.4), with remarkable variations in light curve (i.e., shape, peak luminosity, characteristic timescales),

---

[27] See Piro [1427] for a study of radioactively-powered rising light curves in type Ia supernovae.
[28] See also reference [1570].
[29] One of the first claims of the importance of the radioactive chain $^{56}Ni \rightarrow {}^{56}Co \rightarrow {}^{56}Fe$ in powering type Ia supernova light curve tails appeared in the PhD dissertation of T. Pankey (1962) [1380].

are interpreted as arising from different amounts of ^{56}Ni synthesized in the explosion. Since peak luminosity is proportional to the ^{56}Ni mass, brighter events result when more ^{56}Ni is produced. Conversely, weaker explosions result in less luminous and redder events, showing a faster declining light curve, with a narrower peak, and somewhat smaller expansion velocities (because of the reduced opacity, which strongly depends on the amount of Fe-peak elements synthesized) compared to more energetic supernovae [233]. Whereas SN 1991T is considered one of the most energetic supernovae ever observed, being $\geq 0.6^{\mathrm{mag}}$ more luminous than normal type Ia supernovae [528], SN 1991bg and 1992K are among the reddest, fastest, and most subluminous[30] type Ia supernovae discovered [528, 716, 1066]. Such diversity suggests that type Ia supernovae are not (good) standard candles. This was firmly established by Phillips [1413] on the basis of an empirical relation between maximum brightness and rate of decline during the first 15 days after peak. In fact, this correlation is used to renormalize the observed peak magnitudes, and once all peculiar events are removed from the sample, allow the use of (normal) type Ia supernovae for cosmological applications. The fact that an important fraction of type Ia supernovae exhibit similar light curves after renormalization has prompted their interest as distance indicators, and, broadly speaking, as powerful cosmological tools.

5.3.1 Supernovae and Cosmology

Modern cosmology is deeply rooted in Einstein's nonlinear field equations of general relativity, which describe gravity in terms of a spacetime curvature driven by the presence of matter and energy [297, 731, 1249]. First published by Albert Einstein in 1915, they can be written in tensor form, relating space curvature (left-hand side of Equation 5.1) and energy content (right-hand side):

$$R_{\mu\nu} - \frac{1}{2}g_{\mu\nu}R = \frac{8\pi G}{c^4}T_{\mu\nu} - g_{\mu\nu}\Lambda\,, \qquad (5.1)$$

where $R_{\mu\nu}$ is the Ricci curvature tensor, $g_{\mu\nu}$ the metric tensor, R the scalar curvature, and $T_{\mu\nu}$ the energy-momentum tensor. G, c, and Λ are the gravitational constant, the speed of light in vacuum, and the cosmological constant (introduced by Einstein[31] himself in 1917, in an attempt to reconcile the prevailing belief at the time in a static universe with the predictions inferred from general relativity). The above equation yields the spacetime geometry resulting from the presence of mass-energy and linear momentum. Moreover, it satisfies local energy-momentum conservation and reduces to Newton's law of gravitation for weak gravitational fields and velocities much smaller than the speed of light. Exact solutions for Einstein's field equations can be derived only under certain simplifying assumptions. In this framework, space and time intervals are determined by means of a *metric*, which reveals in turn the geometry (flat or curved) of spacetime.

The Soviet mathematician Alexander Friedmann was the first to derive a metric for a model of the universe that satisfies the *cosmological* principle (i.e., homogeneity and isotropy at very large scales). Unfortunately, his work, published between 1922 and 1924 in the journal *Zeitschrift für Physik* [561, 562], remained mostly unnoticed[32]. In 1927, the Belgian priest and astronomer Georges Lemaître independently obtained Friedmann's results. His work, originally published in the *Annales de la Societé Scientifique de Bruxelles* [1073],

---

[30]In terms of bolometric luminosity, events like SN 1991bg or SN 1992K are likely a factor > 5 dimmer at maximum than the most luminous type Ia supernovae.

[31]Hubble's discovery of the expansion of the universe consigned the cosmological constant to oblivion for many years. Indeed, Einstein is often quoted for considering its inclusion as his *biggest blunder*. The story, however, is much more complicated. For a full account, see reference [1633], p. 775.

[32]English translations of Friedmann's foundational papers were published in 1999 [563, 564].

aroused the interest of Arthur Eddington and other astronomers[33]. Further insight into the geometry of homogeneous and isotropic universes was provided during the 1930s by Howard P. Robertson [1508] and Arthur G. Walker [1876], who proved that Friedmann's metric was the most general metric for such a model universe. The resulting *Friedmann–Lemaître–Robertson–Walker* metric can be written in the form:

$$ds^2 = c^2 dt^2 - a^2(t) dl^2 \,, \tag{5.2}$$

with dl^2 being the metric of a homogeneous and isotropic 3D space, and $a(t)$ a scale factor that describes how the universe unfolds with time. In spherical coordinates, the metric can be written[34] as:

$$ds^2 = c^2 dt^2 - a^2(t) \left(\frac{dr^2}{1 - kr^2} + r^2 d\theta^2 + r^2 \sin^2\theta \, d\phi^2 \right) , \tag{5.3}$$

where k is a curvature constant reflecting whether the universe is open (that is, a universe with a mass-energy content insufficient to halt its expansion; $k = -1$), flat (also a universe that expands forever but at a decelerating rate, with expansion asymptotically approaching zero; $k = 0$), or closed (a universe whose expansion eventually stops, followed by a contraction until all matter collapses to a singularity known as the *Big Crunch*; $k = 1$). A key ingredient in the understanding of the large-scale structure of the universe is the time-evolution of the scale factor, a(t). For a homogeneous, isotropic universe it is described by Friedmann equations, which can be obtained by combining the Friedmann–Lemaître–Robertson–Walker metric and Einstein's field equations. For a universe characterized by a cosmological constant Λ, they can be written as:

$$\left(\frac{1}{a(t)} \frac{da(t)}{dt} \right)^2 = \frac{8\pi G\rho}{3} + \frac{\Lambda c^2}{3} - \frac{kc^2}{a(t)^2} \tag{5.4}$$

$$\left(\frac{1}{a(t)} \frac{d^2 a(t)}{dt^2} \right) = -\frac{4\pi G}{3} \left(\rho + \frac{3P}{c^2} \right) + \frac{\Lambda c^2}{3} \,, \tag{5.5}$$

where ρ and P are the density and pressure of the multicomponent fluid that forms the universe (i.e., baryons, radiation, leptons, dark matter) and are actually linked through an equation of state of the form $P = \omega\rho$. These equations constitute the basis of the standard Big Bang model.

Friedmann's equations are often expressed in terms of two important cosmological quantities: the *Hubble parameter*,

$$H(t) = \frac{1}{a(t)} \frac{da(t)}{dt} \tag{5.6}$$

and the *deceleration parameter*,

$$q(t) = -a(t) \frac{d^2 a(t)/dt^2}{(da(t)/dt)^2} = -\left(1 + \frac{dH(t)/dt}{H(t)^2} \right) , \tag{5.7}$$

which is a measure of the change[35] in the expansion rate of the universe.

---

[33]The paper was later translated into English and published in the *Monthly Notices of the Royal Astronomical Society* [1074].

[34]A full derivation of the metric in terms of tensor algebra can be found in Shore [1633]. For those with a fainter heart, a more phenomenological approach can be found in Shu [1646].

[35]The term *deceleration* reveals the firm belief in a universe that was actually slowing its expansion. The current view, first revealed by observations of distant type Ia supernovae, suggests instead that the expansion rate of the universe accelerates.

In a universe made out of matter, radiation, and dark energy (considered in terms of a cosmological constant[36]), Equation 5.4 can be formally rewritten as

$$H^2 = \frac{8\pi G}{3}(\rho_{\mathrm{m}} + \rho_{\mathrm{rad}} + \rho_\Lambda) - \frac{kc^2}{a(t)^2}, \qquad (5.8)$$

where ρ_{m} is the matter density (baryonic plus dark), ρ_{rad} the radiation density, and $\rho_\Lambda = \Lambda c^2/8\pi G$ the density formally associated to the cosmological constant. For an homogeneous and isotropic flat universe (i.e., $k = 0$) this results in

$$(\rho_{\mathrm{m}} + \rho_{\mathrm{rad}} + \rho_\Lambda) = \frac{3H^2}{8\pi G}. \qquad (5.9)$$

The quantity $3H^2/8\pi G$ corresponds to the *critical density*, ρ_{crit}, or amount of matter (and energy) required to progressively slow down and eventually halt an expanding universe after achieving an infinite size. All in all, we have

$$\frac{\rho_{\mathrm{m}} + \rho_{\mathrm{rad}} + \rho_\Lambda}{\rho_{\mathrm{crit}}} = 1, \qquad (5.10)$$

which, in terms of the dimensionless density parameter, $\Omega = \rho/\rho_{\mathrm{crit}}$, can be written as

$$\Omega_{\mathrm{m}} + \Omega_{\mathrm{rad}} + \Omega_\Lambda = 1. \qquad (5.11)$$

Equation 5.11, derived in the framework of a flat universe, can be generalized for an arbitrary curvature. Recalling Equations 5.4 and 5.8, one has:

$$\Omega_{\mathrm{m}} + \Omega_{\mathrm{rad}} + \Omega_\Lambda - 1 = \frac{kc^2}{H^2 a^2}. \qquad (5.12)$$

And defining $\Omega_{\mathrm{k}} = -kc^2/H^2 a^2$, one finally obtains:

$$\Omega_{\mathrm{m}} + \Omega_{\mathrm{rad}} + \Omega_\Lambda + \Omega_{\mathrm{k}} = 1. \qquad (5.13)$$

These dimensionless density parameters can actually be inferred from a combination of various types of observations (e.g., galaxy clusters, cosmic microwave background radiation, and type Ia supernovae).

5.3.1.1 Type Ia Supernovae and the Accelerated Expansion of the Universe

The main application of type Ia supernovae for cosmology has been as distance indicators. This relies on the definition of *luminosity distance*, d_{L}, or the distance to an object with luminosity L, from which a given flux, F, is measured[37]

$$F = \frac{L}{4\pi d_{\mathrm{L}}^2}. \qquad (5.14)$$

The luminosity distance is usually expressed in terms of the *distance modulus*, $m - M$, or difference between apparent and absolute magnitudes of the object,

$$m - M = 5\log_{10} d_{\mathrm{L}}(\mathrm{Mpc}) + 25. \qquad (5.15)$$

---

[36]It is worth noting that two different forms of dark energy have been proposed to date: One is the cosmological constant, a constant energy density that fills the space-time homogeneously, and scalar fields (such as quintessence) characterized by an energy density that is time- and space-dependent.

[37]Other methods rely on a comparison between angular diameter (determined from the radial velocity of the expaning photosphere) and the observed brightness.

A useful quantity in the study of distant astronomical objects, type Ia supernovae in particular, is the cosmological redshift, z, which can be spectroscopically determined. Physicists are familiar with different types of redshift. The Doppler redshift, for instance, occurs whenever a light source moves away from an observer. The gravitational redshift, instead, is a relativistic effect in which the light emitted from a source in the presence of a gravitational field exhibits a reduced frequency when observed in a region characterized by a weaker gravitational field. The cosmological redshift is due to the expansion of space itself. It can be expressed in terms of the scaling factor $a(t)$ as

$$1 + z = \frac{a(t_o)}{a(t_{em})} = \frac{\lambda_o}{\lambda_{em}}, \tag{5.16}$$

where t_o and t_{em} represent the current and emission times. A related expression,

$$1 + z = \frac{\Delta t_o}{\Delta t_{em}}, \tag{5.17}$$

accounts for the longer duration of any event in the observer's reference frame than in the source's rest frame (that is, the event is "stretched"). First confirmation of the cosmological nature of the redshifts inferred in supernovae was provided by high-precision photometric measurements of the type Ia SN 1995K, at a redshift $z = 0.479$: Its light curve was actually stretched in the observer's frame by a factor $1 + z$, compared with the object's rest frame [1068], as predicted by theory [1947]. This unambiguously proved time dilation of a time-dependent phenomenon at relatively large redshifts. Similar effects have been reported for other supernovae (e.g., references [653, 654]).

Note that, in terms of redshift, Equation 5.15 reduces to the usual linear Hubble relation,

$$m - M = 5 \log cz - 5 \log H_o + 25, \tag{5.18}$$

for nearby supernovae (i.e., supernovae with small recession velocities).

The luminosity distance can be expressed in terms of the curvature density, Ω_k, the Hubble parameter, $H(t)$, and redshift, z, in the form (neglecting the radiation density term, $\Omega_{rad} \sim 10^{-5}$) [298, 665, 1064, 1411]:

$$d_L = \begin{cases} (1+z)\frac{1}{\sqrt{-\Omega_k}} \sin(\sqrt{-\Omega_k}I), & \Omega_k < 0 \\ (1+z)I, & \Omega_k = 0 \\ (1+z)\frac{1}{\sqrt{\Omega_k}} \sinh(\sqrt{\Omega_k}I), & \Omega_k > 0 \end{cases}, \tag{5.19}$$

where the following quantities have been defined:

$$I = \int_0^z \frac{dz'}{H(z')} \tag{5.20}$$

$$H(z) = H_o\sqrt{(1+z)^3\Omega_m + f(z)\Omega_\Lambda + (1+z)^2\Omega_k} \tag{5.21}$$

and

$$f(z) = \exp\left[3 \int_0^z dz' \frac{1 + \omega(z')}{1 + z'}\right]. \tag{5.22}$$

Several cosmological parameters can be inferred from accurate luminosity distances, on the basis of plots of the distance moduli (or apparent magnitudes) versus recession velocities known as *Hubble diagrams*[38]. Early diagrams applied to nearby supernovae [121, 231, 236, 1009, 1772] suggested that peak magnitudes followed the Hubble law, $V =$

---

[38]See reference [850] for the original Hubble's paper that unveiled the expansion of the universe.

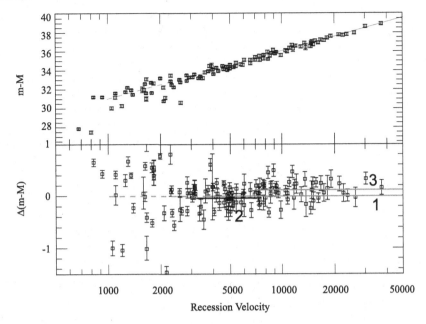

FIGURE 5.12
(Top) Hubble diagram of nearby type Ia supernovae depicting distance moduli (derived from light curve shape corrected luminosities) vs. recession velocities (in km s^{-1}, corrected to the rest frame of the cosmic microwave background, CMB). Fits to different recession velocity ranges are also shown: $v > 3000$ km s^{-1} (1), 3000 km s^{-1} $< v <$ 8000 km $^{-1}$ (2), and $v > 8000$ km s^{-1} (3). (Bottom) Dispersion in distance moduli after substraction of the expansion field from the data shown in the upper panel. Figure from Leibundgut [1065], based on data from Jha et al. [910]; reproduced with permission. Color versions of these plots are available at `http://fisica.upc.edu/ca/users/jjose/CRC-Downloads`.

H_oD, but with considerable scatter. The underlying assumption in those plots was that type Ia supernovae behaved like standard candles, and hence their apparent brightness directly reflected distance. More recent examples of Hubble diagrams for type Ia supernovae can be found, for instance, in references [115, 134, 369, 604, 976, 1010, 1956]. A representative diagram for a sample of nearby supernovae is shown in Figure 5.12.

The slope of the Hubble diagram depicted in Figure 5.12 can be used to infer the expansion rate of the local universe. An isotropic universe in linear expansion would be characterized by a constant slope, since Hubble's law, $V = H_oD$, simply translates into a straight line in a Hubble diagram. Moreover, plugging Hubble's law into the distance modulus expression (Eq. 5.15) yields

$$m - M = 5 \log_{10} V - 5 \log_{10} H_o + 25 , \qquad (5.23)$$

where V is the recession velocity expressed in km s^{-1} and H_o the Hubble constant given in km s^{-1} Mpc^{-1} (which, according to the latest estimates based on Planck mission best fits, together with results from WMAP polarization and baryon acustic oscillations, has a value of 67.80 ± 0.77 km s^{-1} Mpc^{-1} [8]. See also references [485, 557]). The intercept at zero recession velocity yields the value of the local Hubble constant, provided that the absolute (normalized) luminosity of the supernova is known. Another interesting feature of the Hubble diagram is the scatter around the linear fit, which is a measure of the accuracy of the relative distance determination, measurement errors and possible deviations from

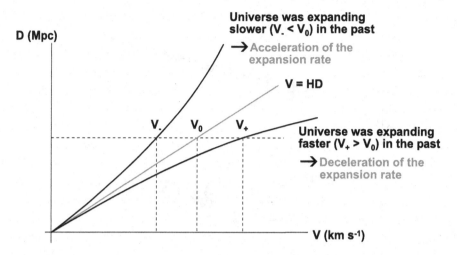

FIGURE 5.13

Expansion rates in three possible models of the universe. Measurements at low redshifts (low recession velocities) are in agreement with Hubble's law, $V = H_o D$. If the expansion rate has always been the same, all objects will follow a linear fit in a Hubble diagram, regardless of their redshift (central line). In a universe now slowing its expansion rate (and therefore characterized by a faster expansion in the past), an object at given distance D would show a velocity larger than $H_o D$. Conversely, in a universe now speeding up, an object at distance D would show a velocity smaller than $H_o D$. A color version of this figure is available at http://fisica.upc.edu/ca/users/jjose/CRC-Downloads.

linear expansion. Figure 5.12 shows very small scatter for recession velocities $v > 3000$ km s^{-1}. Below this value, the corresponding Hubble flow (i.e., the motion of the supernovae as due exclusively to the expansion of the universe) clearly exhibits a different pattern, characterized by a larger scatter induced by the local matter distribution. These supernovae are conveniently flagged and excluded for further cosmological applications. It is important to stress that a number of corrections (i.e., reddening, light curve shape, extinction, filter corrections) have to be applied to each individual supernova before they can be used as cosmological probes (see reference [665]).

Nearby type Ia supernovae follow a rather linear pattern in a Hubble diagram, but what can be expected at much larger distance? The idea to use supernovae to outline the expansion history of the universe dates back from the pioneering proposal of O. C. Wilson in 1939 [1947], further elaborated by R. V. Wagoner [1874] and S. A. Colgate [363], among others. At high redshifts, one would naively expect to find evidence of the pulling effect of gravity on the large-scale structure of the universe. Most likely, the expansion rate of the universe would have slowed down with time. To this end, two independent projects, aimed at detecting the deceleration of the expansion rate of the universe using distant type Ia supernovae, were arranged in the 1990s: the *Supernova Cosmology Project* [1411, 1412] and the *High-z Supernova Search Team* [1502]. If the expansion of the universe was faster in the past, distant (older) supernovae would exhibit larger recession velocities than those inferred from Hubble's law, with the current Hubble's constant value. Therefore, they whould appear somewhat below the $v = H_o D$ line in a Hubble diagram. This is schematically illustrated in Figure 5.13. Surprisingly, distant type Ia supernovae revealed the opposite behavior: The

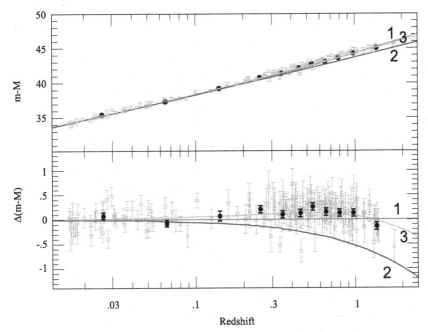

FIGURE 5.14

(Top) Hubble diagram of distant type Ia supernovae, showing distance moduli (derived from light curve shape corrected luminosities) vs. redshift (corrected to the rest frame of the cosmic microwave background). Results assuming different model universes are shown for comparison: (1) an empty universe ($\Omega_\Lambda = \Omega_m = 0$); (2) the Einstein–de Sitter model ($\Omega_\Lambda = 0$, $\Omega_m = 1$); (3) best fit to data, corresponding to $\Omega_\Lambda = 0.7$ and $\Omega_m = 0.3$. (Bottom) Dispersion in distance moduli after substraction of the expansion field from the data shown in the upper panel. Figure from Leibundgut [1065], based on data from Davis et al. [418]; reproduced with permission. Color versions of these plots are available at http://fisica.upc.edu/ca/users/jjose/CRC-Downloads.

expansion rate of expansion of the universe actually accelerates[39] [1410, 1501, 1587] (Figure 5.14).

A simple way to reconcile the data with our current conception of the nature of the universe is to assume the existence of a dominant, dark energy component, responsible for the acceleration of the expansion rate. Best fits to type Ia supernova data suggest that dark energy is the dominant component of the universe, with $\Omega_\Lambda \sim 0.713$, while ordinary and dark matter would amount to $\Omega_m \sim 0.287$ (assuming $\Omega_k = 0$ [1010]), in good agreement with the values inferred from WMAP data (see also reference [1489]).

---

[39]In 2011, Saul Perlmutter, of the *Supernova Cosmology Project*, and Brian Schmidt and Adam Riess, of the *High-z Supernova Search Team*, were awarded with the Nobel Prize in Physics "for the discovery of the accelerating expansion of the universe through observations of distant supernovae."

5.4 Progenitors

Type Ia supernovae are routinely observed in all types of galaxies, in sharp contrast with other supernova classes (i.e., type Ib/c and type II; see Chapter 7), which are only found in spiral and irregular galaxies. This suggests that type Ia supernovae are likely associated with old stellar populations[40], while the progenitors of the other classes are much younger.

The energy required to power a type Ia supernova can be estimated from the kinetic energy of the expanding ejecta, $E_{kin} \sim 10^{51}$ erg (the typical velocities range between 5000–10,000 km s^{-1}), as well as from the energy integrated over the light curve, $E_{rad} \sim 10^{49}$ erg. Note that $E_{kin} \gg E_{rad}$, therefore what is seen in a type Ia supernova is basically the fallout from a thermonuclear explosion [1961]. Fred Hoyle and Willy Fowler [845] were among the first to propose that type Ia supernovae result from the ignition of degenerate nuclear fuel (CO-rich material, in fact[41]) in low-mass stars evolved into white dwarfs[42] (see also reference [1967]). Indeed, the incineration of ~ 1 M$_\odot$ of a C-O mixture, which releases $\sim 10^{18}$ erg g^{-1}, can account for the required 10^{51} erg.

The first observational surveys suggested that type Ia supernovae constituted a rather homogeneous class of objects, characterized by similar properties (i.e., light curves and spectra), from which a unique progenitor was inferred. This, in fact, was the expected outcome of an exploding white dwarf, destabilized by accretion from a companion star when approaching its maximum mass (the so-called *Chandrasekhar limit*, $M_{Ch} \sim 1.4$ M$_\odot$). Until recently, the accepted paradigm was that about 85% of all observed type Ia supernovae shared similar spectral features, absolute magnitudes and light curve shapes. These are known as *Branch normals* or *normal type Ia*, with canonical examples such as SN 1972E, SN 1989B, SN 1994D, or SN 2005cf. The scatter among this class is only $\leq 0.3^{mag}$ around peak luminosity [233, 281]. The intensive observational campaigns undertaken in the last two decades have, however, reported an increasing number of anomalous supernovae, casting doubts on the historically postulated uniqueness of the progenitor system. More recent classifications [1102, 1653] suggest instead that the number of *peculiar* supernovae[43], deviating from the class of *Branch normals*, may be as high as $\sim 25\%$–30%. This heterogeneity likely suggests multiple progenitors or explosion mechanisms[44].

Broadly speaking, two basic scenarios have been proposed to explain the origin of normal[45] type Ia supernovae [799, 800, 839, 891, 1122, 1126, 1455, 1544, 1891]: a single degenerate channel, consisting of a low-mass, nondegenerate star that transfers hydrogen- or helium-rich matter onto a carbon-oxygen white dwarf ([875, 1930]; see also references [1121, 1227, 1392]), and a double degenerate channel, in which two carbon-oxygen white dwarfs merge as a re-

---

[40]Note, however, that the frequency of type Ia supernovae appears to be be much larger in spirals than in elliptical galaxies. This suggests that some type Ia supernovae are actually associated with younger progenitors.

[41]ONe cores, for instance, ignite at such high densities that they always tend to collapse. Therefore, their explosion is unlikely [697, 1323].

[42]Early observations of the recent SN 2011fe discovered in M101, for instance, strongly point toward a (CO) white dwarf as the most likely progenitor [202, 1337]. The presence of a red giant secondary or a Roche-lobe overflowing main sequence companion have both been ruled out [202, 255, 836, 1099, 1337]. Moreover, the lack of early shocks also suggests the presence of a main-sequence stellar companion.

[43]See also references [201, 678, 1176].

[44]This trend was previously outlined in a number of articles (e.g., references [234, 952, 1103]).

[45]Several progenitor systems and explosion models have been proposed to account for the different types of peculiar type Ia supernovae: The characteristic features of the low-velocity, subluminous SN 2002cx/SN 2008ha-like class can be reproduced in the framework of weak deflagrations in near-Chandrasekhar mass (CO) white dwarf stars leaving bound remnants [915]. The spectra, colors, and low expansion velocity of the subluminous SN 1991bg-like class can be instead reproduced by the merger of two, relatively light CO white dwarfs of nearly equal mass [1374].

sult of energy and angular momentum losses driven by gravitational wave radiation, with the total mass of the system exceeding the Chandrasekhar limit ([875, 1912]; see also references [456, 1200, 1429]).

Other possible channels leading to type Ia supernovae have been proposed in the literature in recent years. These include the core-degenerate model, produced in the merger between a white dwarf and the core of an AGB star during the common envelope phase [1677,1678], the eccentric double white dwarf merger, produced in triple stellar systems [713], the spin-up/spin-down model, in which the transfer of material carrying angular momentum spins up the accreting white dwarf that may even exceed the Chandrasekhar mass limit [316, 438, 707], or double-detonations in sub-Chandrasekhar mass white dwarfs.

5.4.1   Single-Degenerate Scenario

The single-degenerate channel was, in the past, widely thought to be the most promising path toward type Ia supernova explosions. Potential candidates for this channel include cataclysmic variables (formed by a white dwarf and a close, low-mass, main sequence companion) and symbiotic variables (a white dwarf and a red giant star in a wide binary system). Cataclysmic variables (e.g., classical novae) are characterized by H-rich accretion. This poses constraints on the maximum mass of hydrogen that can be accumulated on top of the white dwarf without being detected. But its major drawback is related to the difficult pathway faced by a CO white dwarf to increase its mass and reach the Chandrasekhar limit [876, 1321, 2012]. Stellar evolution predicts that the maximum mass of a CO white dwarf[46] is about ~ 1.1 $M_\odot$ [23] In the canonical model for type Ia supernovae, the white dwarf would need to accumulate roughly 0.3 $M_\odot$ before getting disrupted by the explosion. This is difficult to achieve and, in particular, low mass-accretion rates leading to nova explosions (see Chapter 4), for which the white dwarf hosting the outburst actually loses (rather than accumulates) mass, must be avoided [1326, 1540] (but see reference [1704]).

In symbiotic variables, the accreted matter is He-rich[47] instead [437, 1155]. In certain classes of symbotic variables, such as recurrent novae (e.g., RS Oph; see Chapter 4) [705, 706, 1104, 1890, 2011], the white dwarf is expected to grow in mass. Some studies, however, argue against symbiotic systems as likely supernova progenitors [970, 1488], on the basis of the limited mass available for transfer, thus preventing the white dwarf to reach the Chandrasekhar limit. Nevertheless, the main drawback faced by the single-degenerate scenario is that the theoretically predicted type Ia supernova rate is usually lower than the empirically inferred value [644, 1540].

5.4.1.1   Fate of Sub-Chandrasekhar Mass Explosions

Although the canonical scenario for type Ia supernovae relies on the explosion of a Chandrasekhar mass CO white dwarf, He-accretion onto low-mass white dwarfs that never reach the Chandrasekhar limit before exploding has also received a particular interest. In these

---

[46]White dwarfs with M> 1.1 $M_\odot$ are the endproducts of the evolution of intermediate-mass star (i.e., $M \sim 10 - 12$ $M_\odot$), which undergo C-ignition, and, hence, form ONe-rich cores.

[47]A somewhat related subclass, often referred to as AM Canum Venaticorum (AM CVn) binaries, are ultracompact systems consisting of a white dwarf accretor and a He-rich donor exchanging matter via Roche-lobe overflow [430, 1284]. The mass-transfer rates onto the white dwarf are expected to be sufficiently high at early times, so that the accumulating He layer ignites and burns into carbon and oxygen in a bright flash that may be observable as a helium nova. Because the mass-transfer rate drops progressively in those systems, and the accumulated mass increases as the mass-accretion rate decreases, the system will inevitably undergo a strong flash, likely leading to some sort of dim, ".Ia" supernova [188] (see also reference [1408]). Indeed, it has been claimed that AM CVn systems may contribute up to 25% to the type Ia supernova rate [1285], although later estimates have lowered this value to <1% [1680].

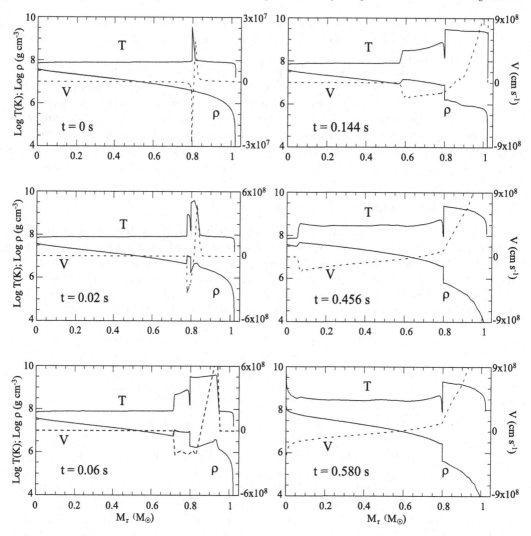

FIGURE 5.15

Temperature (T), density (ρ), and velocity (V) profiles resulting from He-rich accretion onto a 0.8 M$_\odot$ CO white dwarf, at a rate of $\dot{M} = 3.5 \times 10^{-8}$ M$_\odot$ yr^{-1}, following the progress of the detonation front originating at the envelope's base. The horizontal axis corresponds to the (internal) mass coordinate. Notice how the compressional wave that propagates inward (i.e., $v < 0$) provokes carbon ignition near the center (last panel). Simulation performed by the author with the SHIVA code.

models, a He-layer piles up on top of an accreting CO white dwarf, for specific values of the mass-accretion rate. Following the temperature rise, helium ignites at the core-envelope interface, powering a double shock front (Figure 5.15): a detonation wave that moves outward, reaching the stellar surface after incinerating most of the envelope to ^{56}Ni, and an ingoing compressional wave that propagates toward the center, where it induces carbon ignition [1988]. A crucial issue is the precise He-ignition density, which determines the strength of the explosion, and therefore the range of temperatures attained in the envelope. Early models [822, 1988] suggested that the overall amount of ^{56}Ni produced in these sub-Chandrasekhar models was smaller than in standard C-deflagration/detonation models, leading to fainter events and, hence, providing a possible explanation for the origin of sub-luminous supernovae. From a nucleosynthesis viewpoint, these models were characterized by a significant overproduction of a number of nuclear species, including ^{44}Ca (from the β-decay of ^{44}Ti) and ^{47}Ti, systematically underproduced in other type Ia and type II supernova models. Moreover, they have also been proposed as potential sites for the synthesis of p-nuclei[48] [670, 671, 1823].

The scenario was seriously challenged because of the stratified chemical structure of the ejecta: an outer, high-velocity Ni/He-rich layer, followed by a region containing intermediate-mass elements, and, finally, the inner Fe/Ni-rich core [1988]. An object with such structure would exhibit high-velocity Ni and He in the early spectra, a pattern clearly at odds with observations [823]. Furthermore, the presence of significant amounts of Ni in the outer envelope would induce substantial heating in the outer ejecta. As a result, such supernovae would appear too blue at maximum, showing probably very steep light curves, both at rise and decline [820, 823]. Another criticism focused on the feasibility to induce a central C-ignition, which in spherically symmetric models may be imposed by the adopted geometry, favoring the convergence of the inward propagating wave. And finally, the range of mass-accretion rates that result in a central C-detonation is very narrow (i.e., $\sim 10^{-9} \leq \dot{M}_{acc}[M_\odot$ yr$^{-1}] \leq 5 \times 10^{-8}$), and, thus, models require fine tuning[49] (for instance, mass-accretion rates above that range drive He-ignition well above the envelope base; the resulting thermonuclear runaway is quenched by expansion and C-ignition is prevented).

Despite of the abovementioned drawbacks, this scenario has gained momentum in recent years, after more and more peculiar type Ia supernovae were discovered. This is because the sub-Chandrasekhar models have basically only one adjustable parameter (i.e., the mass of the white dwarf at the onset of the explosion) that determines the strength and brightness of the explosion [1657]. In fact, a number of studies have revealed that central C-detonations can be triggered for thinner He-envelopes than previously anticipated [188, 531]. Simulations yielded light curves covering both the range of brightnesses and the rise and decline times of observed type Ia supernovae, although with colors and spectra that again do not match the observations [1027]. The scenario appears to be particularly sensitive to details of the convective transport operating during the last stages prior to the explosion, proving that unless a shallow temperature gradient is maintained and unless the density is sufficiently high, the accreted helium would not detonate [1980]. Moreover, only the models with the hottest, most massive white dwarfs, together with the smallest helium layers, show reasonable agreement with the light curves and spectra of normal type Ia supernovae. In fact, recent 2D and 3D simulations [1255] suggest that a helium detonation cannot easily reach the core unless it sets in well above the hottest layer of the He-rich envelope. Furthermore, it appears that helium detonations are favored if starting at an extended sheet, of the size of a characteristic convective cell, rather than at a point. Truly 3D simulations with an accurate treatment of convection are needed to shed light into the feasibility of this mechanism.

5.4.2 Double-Degenerate Scenario

The double-degenerate channel is partially supported by a number of observational facts. First, it is clear that a (double CO) white dwarf merger would naturally account for the lack of hydrogen in the spectra. Moreover, ejected masses larger than the Chandrasekhar limit have been inferred in a number of bright type Ia supernovae, such as SN 2003fg [841], SN 2006gz [794], SN 2007if [1571], or SN 2009dc [2000]. On the other hand, the single

---

[48]See also references [1820, 1821] for studies on the synthesis of p-nuclei in delayed-detonation models.

[49]See, however, reference [1541] for a study stressing that the adopted prescription for He-accretion in sub-Chandrasekhar mass models is not that critical.

degenerate scenario is somewhat disfavored by the lack of emission[50] expected during the interaction between the supernova ejecta and the secondary star [180, 202, 603, 749, 950], the lack of radio detections [330, 725, 836], as well as the nondetection of companion stars, neither in supernova remnants [484, 975, 1574] nor in preexplosion archival images of nearby explosions [1099].

Theoretical estimates based on population synthesis codes [875, 892, 1287, 1540, 1542, 2011, 2013] yield, in principle, an appropriate frequency of white dwarf mergers to account for the expected type Ia supernova rate (~ 0.3 per century in the Milky Way galaxy [437]). In addition, studies of the distribution of times elapsed between a hypothetical burst of star formation and the subsequent supernova explosions, the so-called *delay time distribution* (DTD), suggest a power-law of the form t^{-1} [120, 584, 679, 1191, 1193, 1561], precisely the type of distribution expected for a supernova population dominated by double degenerates. However, when delay times are taken into account, population synthesis codes often yield rate estimates that fall short[51] by factors ~ 10 [318, 337, 1192, 1193, 1228, 1540, 1811, 2009]. Some related models actually suggest the existence of two contributors to account for the longest and shortests delay times [217, 337, 1190, 1471, 1572], the former likely dominated by double degenerates.

A major problem in the double-degenerate scenario relies on the scarcity of candidates detected, capable of driving a Chandrasekhar-mass explosion after a merging episode in less than the age of the universe (which requires binary periods shorter than ~ 13 hr). On theoretical grounds, double degenerate scenarios have been historically disfavored, since numerical simulations do not necessarily support an explosive fate for such systems: Carbon is, in fact, expected to ignite close to the surface after the merger episode [1252], transforming the CO-rich material into an ONe mixture [960]. The resulting merger products likely evolve toward an accretion-induced collapse (AIC) to a neutron star due to electron captures, rather than powering a normal explosion [294, 1321, 1323, 1552, 1986]. The double-degenerate merger scenario poses, however, a serious challenge for modelers, since accretion has to be followed over many orbital cycles while the time-step is severely constrained by the Courant condition[52]. Recent work suggests, however, that double white dwarf mergers can succesfully lead to type Ia supernova explosions under certain circumstances [2003], including detonations driven by He-rich accretion stream instabilities [405, 693], violent mergers involving massive white dwarfs [1026, 1373, 1375, 1376], head-on collisions of white dwarfs (see Figures 5.16 and 5.17) [615, 1036, 1472, 1534], or detonations driven by magnetorotational instabilities [911]. Detailed multidimensional simulations that explore the large parameter space can also be found in references [97, 404].

5.4.2.1 When White Dwarfs Merge: Gravitational Waves and Nucleosynthesis

So far, white dwarf mergers have been exclusively addressed in the context of possible mechanisms powering type Ia supernovae, but their astrophysical significance also encompasses

---

[50]Daniel Kasen [950] suggested that the interaction between the supernova ejecta and the secondary, in the framework of the single degenerate channel, would power a strong shock, potentially detectable in X-rays less than a day after the explosion, as the companion star carves a hole in the expanding supernova ejecta. The shock is also expected to produce excess emission at ultraviolet (and blue) wavelengths. Because of the restrictions imposed by the solid angle, about 10% of the supernovae produced through the single degenerate channel are expected to show signatures of this interaction (see also references [610, 1196, 1377]). However, Brian Hayden and collaborators [749] have analyzed a large number of type Ia supernova light curves and found no evidence of such interaction. On this basis, they concluded that, if the single degenerate channel dominates type Ia supernova progenitors, this constrains the mass of the companion stars to be $\leq$ 6 $M_\odot$ on the main sequence and strongly disfavors the presence of red giant secondaries.

[51]Only if mergers of double CO white dwarfs are considered.

[52]To date, such simulations are followed by means of explicit SPH codes.

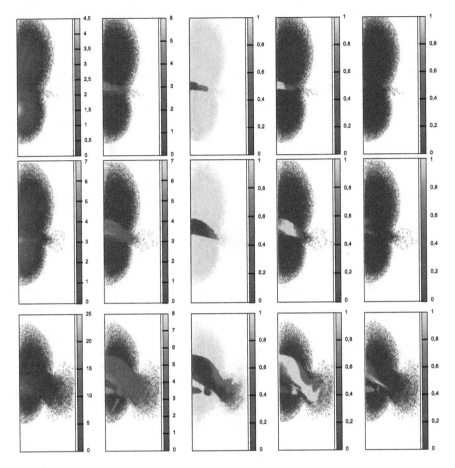

FIGURE 5.16

Snapshots of the evolution of a head-on collision of two CO white dwarfs, of masses 1.0 $M_\odot$ and 0.81 $M_\odot$, resulting in a (super-Chandrasekhar) supernova explosion with an overall ^{56}Ni production of 1.02 $M_\odot$. Panels show, left to right, the density (in units of 10^7 g cm^{-3}), temperature (10^9 K), and the mass fractions of ^{4}He + ^{12}C + ^{16}O, intermediate-mass elements (^{20}Ne to ^{40}Ca), and Fe-group elements (^{44}Ti to ^{60}Zn), at different times since the beginning of the simulation (from top to bottom, t = 16.9, 17.1, and 17.5 s). The simulation was performed in 2D with the axisymmetric SPH code AxisSPH using 88,560 particles. A color version of this plot is available at http://fisica.upc.edu/ca/users/jjose/CRC-Downloads. Figure from García-Senz et al. [615]; reproduced with permission.

other domains. First, they have been invoked to explain certain nucleosynthetic anomalies in hydrogen-deficient stars that cannot be reproduced through standard stellar evolution, such as HdC (a type of H-deficient, supergiant C-rich stars), EHe (low-mass supergiant stars, almost devoid of hydrogen while extremely enriched in helium), or R CrB (R Coronae Borealis, stars that are H-poor, C- and He-rich, and very luminous, resembling simultaneously erupting and pulsating variable stars) [355, 1147, 1551]. Second, double white dwarf mergers can result in the formation of high-field magnetized white dwarfs [607, 911, 1932] or give rise to magnetically powered outbursts [151]. Third, they have been suggested as sources of cosmological fast radio bursts [954]. And fourth, they are considered potential sources of gravitational waves, together with double neutron stars and neutron star-white

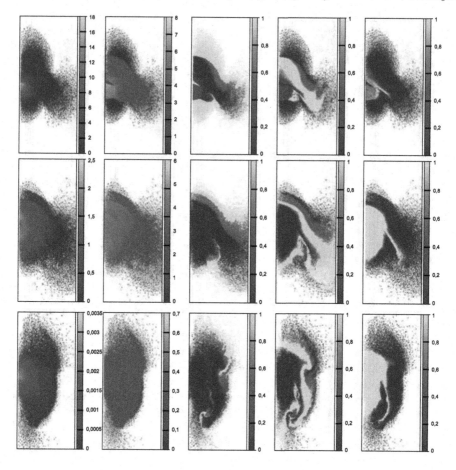

FIGURE 5.17
Same as Figure 5.16, for $t = 17.6$ (top), 17.9 (middle), and 21.7 s (bottom).

dwarf binaries [5, 54, 958]. In fact, the emission from Galactic (close) white dwarf binary systems is expected to be the dominant source of the background noise in the low frequency range [42, 1150].

Simulations of double white dwarf mergers [98, 691, 693, 1150] have received little attention compared to the coalescence of double neutron stars[53] (see references [669, 1532, 1535],

---

[53] Another interesting example of stellar merger involves systems composed of a white dwarf and a neutron star. About 40 binaries of this type have been discovered to date [1848], although the number of systems expected to merge in less than the age of the universe within 3 kpc from the Sun could be as high as ~ 850 [483]. This results from the estimated formation rate of neutron star–white dwarf binaries that is about 10–20 times larger than for binary neutron star systems [1780]. Two classes of neutron star–white dwarf binaries are actually distinguished [290, 1698]: *intermediate-mass binary pulsars*, characterized by very eccentric orbits and masses of the secondary stars ranging between 0.5 $M_\odot$ and 1.1 $M_\odot$, and *low-mass binary pulsars*, with almost circular orbits and masses of the secondaries between 0.15 and 0.45 $M_\odot$. Such systems are expected to merge due to emission of gravitational waves, with an estimated frequency of 1.4×10^{-4} mergers per year [1286]. The list of neutron star–white dwarf binaries expected to merge in less than the age of the universe includes the sources PSR 30751+1807, PSR 31757-5322, and PSR 31141-6545. It has been estimated that the mass ejected during a merger amounts to about 0.01% of the overall system mass, assuming a 1.4 $M_\odot$ neutron star and a 0.6 $M_\odot$ white dwarf [609]. Further studies aimed at determining the nucleosynthesis impact of these neutron star–white dwarf mergers on the Galactic abundances are clearly needed.

and Chapter 7). A merging episode can be considered the final fate of a significant fraction of this type of binaries, with an estimated rate of $\sim 8.3 \times 10^{-3}$ yr^{-1} in the Galaxy. In these systems, mass transfer ensues as soon as the less massive dwarf fills its Roche lobe (Figure 5.18). Actually, mass transfer can be stable for some combinations of masses of the system [1199]. Since the white dwarf radius scales as $M^{-1/3}$, when the secondary loses mass, its radius increases, and, hence, its mass-loss rate would increase, reinforcing the overall mass transfer process. An accretion arm forms, extending all the way from the secondary to the surface of the more massive white dwarf. The orbital motion of the coalescing white dwarfs reshapes the arm into a spiral form. The secondary eventually gets fully disrupted, and a Keplerian disk surrounding the primary star forms after some orbital periods, depending on the initial conditions.

Models involving a He white dwarf undergo thermonuclear fusion as the temperatures achieved during the merging episode exceed 10^8 K. This translates into noticeable amounts of Ca, Mg, S, Si and Fe, particularly in the hot corona that surrounds the primary star. This has been claimed as a feasible explanation for the origin of metal-rich DAZd white dwarfs, a subclass with strong H lines and dusty disks [608]. Recent studies have also focused on the nucleosynthesis produced during the merging of a He white dwarf with a CO white dwarf [1147]. The resulting abundances, particularly for oxygen and fluorine, are in qualitative agreement with the chemical abundance pattern inferred from R Coronae Borealis stars. Moreover, it has also been proved that lithium can also be synthesized in those mergers [1146]. The expected gravitational wave emission[54] [98,1150,1152] would be characterized by an initial, almost sinusoidal, pattern of increasing frequency (the so-called *chirping phase*), followed by a sharp cut-off of the gravitational wave signal, on a timescale of the order of the orbital period as the merger proceeds. All such merging episodes could be detectable by the planned LISA mission with different signal-to-noise ratios.

5.5 Blowing Up Stars in the Laptop: The Modeling of Type Ia Supernovae

About once per century in our Galaxy—maybe once per second in the observable universe—a white dwarf gets fully disrupted by a supernova blast. The confrontation between the chemical abundance pattern inferred from high-resolution spectra of supernovae and the yields resulting from state-of-the-art simulations has helped to rule out some of the proposed explosion models while provided clues to infer the characteristics of the model(s) *used by nature* [74].

The roadmap for a successful supernova model was already outlined by Fred Hoyle and Willy Fowler, who in their seminal paper [845] envisaged the ignition of fuel under high electron degeneracy conditions as the likely mechanism powering the explosion. Whether these explosions take place in single or binary stellar systems was unclear. Indeed, electron degeneracy can even occur during the evolution of single stars. Main sequence stars, however, burn hydrogen steadily. For low-mass stars, this is followed by a degenerate He-flash that leaves the structure of the star unaffected. Thus, the first stage in the evolution at which an explosion may occur is C-burning, which for low- and intermediate-mass stars (up to ~ 8 M$_\odot$) takes place at high densities, such that their stellar cores are degenerate.

Evry Schatzman [1583] already suggested in the 1960s that type Ia supernovae resulted from accreting white dwarfs in binary systems (the same environment in which classical

---

[54]See reference [1503] for a review on gravitational waves.

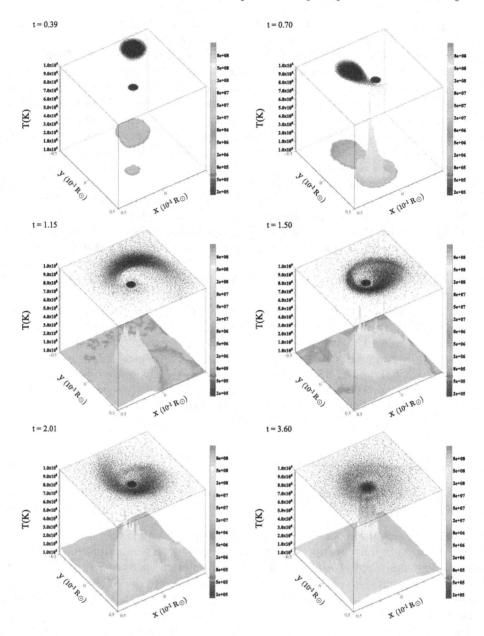

FIGURE 5.18
Snapshots of the merging episode of two white dwarfs of masses 1.2 and 0.4 M$_\odot$, simulated with a *smoothed-particle hydrodynamics* code. Each panel shows the equatorial temperature contours as well as the location of the SPH particles. When the first particles of the secondary star hit the surface of the primary (top right panel), the shocked regions reach $T \sim 5 \times 10^8$ K. The colliding particles are reflected almost instantaneously by the hard boundary of the primary. As mass transfer proceeds (central panels) the accreted matter is shocked and heated on top of the surface of the primary. This matter is actually ejected and a hot toroidal structure forms (right middle panel). Temperatures exceed 10^9 K, and nuclear reactions transform the accreted He into C and O. A color version of this plot is available at http://fisica.upc.edu/ca/users/jjose/CRC-Downloads. Figure from Guerrero, García-Berro and Isern [691]; reproduced with permission.

novae take place—see Chapter 4), but the lack of irrefutable observational evidence allowed a variety of competing alternatives. These included intermediate-mass ($4 \leq M[\text{M}_\odot]$ ≤ 9), single stars, destabilized by the continuous supply of matter from shell burning into their CO-rich cores. Nevertheless, the observed occurrence of type Ia supernovae in elliptical galaxies strongly pointed toward old progenitors and accreting white dwarfs in binary systems soon became the canonical scenario [500, 695, 1211, 1930].

The first published numerical studies of degenerate C-ignition, aimed at explaining the nature of type Ia supernovae, were performed by Dave Arnett [60, 62, 74][55]. Calculations relied on a CO-rich white dwarf core, with a mass close to the Chandrasekhar limit. A central density around 2×10^9 g cm^{-3} was adopted, and the outer He-burning shell was numerically replaced by suitable boundary conditions. A detonation front developed[56], leading to the total disruption of the star[57]. About 40 numerical mass shells were used in this pioneering hydrodynamic simulation. Nucleosynthesis, calculated in 8 mass shells with a network containing 81 nuclear species, revealed full incineration, with total disruption of the star (i.e., no remnant leftover), and production of iron-peak elements (remarkably, ^{56}Ni), ejected at average speeds of $v \sim 6000$–$10{,}000$ km s^{-1}. Three major drawbacks were soon identified in Arnett's C-detonation model: On one hand, the lack of any gravitationally-bound remnant was, at the time, at odds with the prevailing theory of pulsar formation, which strongly supported low- and intermediate-mass stellar progenitors (2–10 M$_\odot$); furthermore, galactic chemical evolution models predicted unrealistically high abundances of iron-peak nuclei if type Ia supernova yields from the C-detonation model were implemented; and finally, the lack of sufficient amounts of intermediate-mass elements (particularly silicon, sulfur, and calcium, which characterize the type Ia class) was clearly in contradiction with the chemical abundances inferred spectroscopically.

The feasibility of C-ignition in powering a (supersonic) detonation, in which the burning front is driven by shock waves, became a matter of debate at the time. A detonation propagating through a medium initially at rest is described by the set of Rankine-Hugoniot continuity equations (see Chapter 1). In a detailed study, Thaddeus Mazurek, David Meier, and Craig Wheeler [1215] concluded that the formation of a planar Chapman–Jouguet detonation, for the typical densities at which C-ignition initiates at the central layers of a degenerate star ($\rho \sim 10^9$ g cm^{-3}), requires the accumulation of as much energy as released by nuclear fusion. The energy released by C-burning ($\sim 3 \times 10^{17}$ erg g^{-1}) is, however, smaller than the internal energy of the degenerate electrons. Therefore, even though a shock wave likely develops, the overpressure resulting from nuclear burning is small, and a detonation is not produced [1215, 1328, 1743, 1925] (see also [264, 265, 897]).

Instead, a (subsonic) deflagration, catalysed by the thermal conduction of the degenerate electron gas, looked like a promising alternative. Ken Nomoto, Daiichiro Sugimoto, and Sayuki Neo [1328] analyzed the fate of a white dwarf star undergoing degenerate C-ignition

---

[55]See also references [746, 747] for previous studies of C-ignition in the framework of stellar evolution models.

[56]The mass-accretion rate determines the central ignition density and the outcome of the explosion. However, many type Ia supernova simulations have not properly modeled the accretion stage. Instead, a static white dwarf is constructed already with a mass close to the Chandrasekhar limit, with a value of the central density around $\sim 10^9$ g cm^{-3}. In this framework, the onset of a detonation can be imposed, for instance, by artificially incinerating a certain volume of the star or through the implementation of a specific recipe for the speed of the burning front.

[57]S. W. Bruenn showed that total disruption occurs unless the central density of the star unrealistically exceeds 1.5×10^{10} g cm^{-3}, above which core implosion is expected [257]. Inclusion of Coulomb corrections to the equation of state have lowered the minimum density for collapse of a white dwarf down to $\sim 6 \times 10^9$ g cm^{-3} [242]. The exact value actually depends on the flame speed (in 3D models, it also depends on the ignition conditions—i.e., a single or multiple ignition spots, distance from the center).

through a deflagration front propagating at a speed

$$v_{\text{def}} = \alpha \left(l_{\text{m}} \frac{GM_{\text{r}}}{4r^2} \Delta \ln \rho \right)^{1/2}, \tag{5.24}$$

where l_{m} is the mixing length, α is an efficiency parameter, and $\Delta \ln \rho$ represents the density contrast at both sides of the front, which in turn produces a buoyancy force. Simulations using this phenomenological recipe for the speed of the deflagration front revealed that a slow deflagration (e.g., $\alpha = 0.05$) results in total disruption of the star, with the ejection of 11% of core mass as iron-peak nuclei, and about 7% as intermediate-mass elements (i.e., O–Si). A faster deflagration (e.g., $\alpha = 1.0$) yielded 73% of iron-peak nuclei and 20% of intermediate-mass elements. In none of the cases did the initial deflagration turn into a detonation. Interestingly, the study revealed that the poisoning overproduction of iron accompanying pure detonation models could be somewhat alleviated (and even avoided) with a proper choice of the deflagration speed[58].

5.5.1   Preexplosive Evolution: The Accretion Phase

A number of studies have focused on the physical conditions for which an accreting white dwarf is expected to grow in mass, eventually reaching the Chandrasekhar limit. Different nuclear burning regimes have been identified, depending on the composition of the accreted material, the mass-accretion rate or the mass and composition of the underlying white dwarf [571, 572, 780, 1319, 1326, 1704, 1742] (Figure 5.19).

5.5.1.1   H-Rich Accretion

Accretion of hydrogen-rich material at very high rates, above the *Eddington limit*,

$$\dot{M} \geq \dot{M}_{\text{Edd}} = 4\pi cR/\kappa_{\text{sc}} \simeq 3 \times 10^{-14} R(\text{cm})/(1+X) \ \text{M}_\odot \text{yr}^{-1}, \tag{5.25}$$

results in an intense radiation pressure that overcomes the strong gravitational pull exerted by the white dwarf star, thus inhibiting further accretion. In Equation 5.25, c is the speed of light, κ_{sc} the opacity due to Thomson scattering, R the stellar radius, and X the hydrogen mass fraction.

Below the Eddington limit, a narrow range of mass-accretion rates drive the expansion of the envelope to a size resembling a red giant. On the basis of Paczyński's core mass-luminosity relation [1361], and for a hydrogen mass fraction of X = 0.7, Ken'ichi Nomoto and collaborators obtained the following relationship[59] (for $0.6 \leq M_{\text{wd}}[\text{M}_\odot] \leq 1.39$) [1319, 1325]:

$$\dot{M}_{\text{RG}} \geq 8.5 \times 10^{-7} \left[\frac{M_{\text{wd}}}{M_\odot} - 0.5 \right] \text{M}_\odot \text{yr}^{-1}. \tag{5.26}$$

For intermediate rates, hydrogen is burned steadily[60] [961, 1125, 1364, 1382, 1650, 1651] or through weak flashes [571, 572, 870, 925, 1007, 1364, 1370, 1450, 1451, 1661–1663, 1741, 2001]. Weak shell flashes mark the boundary between dynamic events with massive ejection episodes (e.g., nova outbursts) and steady burning, where hydrogen is quietly converted into helium.

---

[58]See also Section 5.5.2 for improvements due to a major revision of stellar weak interaction rates.

[59]A similar relationship, $\dot{M}_{\text{RG}} \geq 6.68 \times 10^{-7}$ [$(\text{M}_{\text{wd}}/\text{M}_\odot) - 0.445$], has been recently published with updated steady state models [1326]. Similar results have also been reported by Ken Shen and Lars Bildsten [1625] in the framework of one-zone models.

[60]Supersoft X-ray sources, characterized by a prominent X-ray emission below 1 keV, are examples of accreting white dwarfs with hot photospheres powered by steady H-burning [939, 1326].

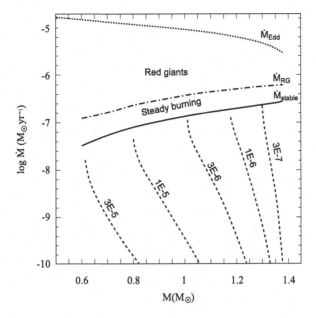

FIGURE 5.19

Nomoto diagram showing different H-burning regimes in accreting white dwarfs, as a function of the mass-accretion rate. Dashed lines trace the loci of envelope masses (in $M_\odot$). Figure from Nomoto et al. [1326]; reproduced with permission.

The upper limit leading to H-shell flashes, inferred from numerical simulations, is given by the fit[61] [870],

$$\dot{M}_{\text{flash}} \leq 1.32 \times 10^{-7} \left[\frac{M_{\text{wd}}}{M_\odot} \right]^{3.57} M_\odot \text{yr}^{-1}. \tag{5.27}$$

As shown in Figures 5.20 and 5.21, flashes that do not result in any significant mass loss are quasi-periodic. Their recurrence time increases as the mass-accretion rate decreases, as expected from the larger amount of mass accumulated before ignition conditions are reached. Moreover, the strength of the flashes increases as the mass-accretion rate decreases, since larger accreted masses imply higher degeneracy and, therefore, stronger explosions result [870, 925, 1364]. This is illustrated in Figure 5.21 through larger differences between maximum and minimum shell temperatures, and larger areas in the $T-\rho$ diagrams. Another indication of the strength of the flash is given by the ratio between the duration of the explosion over the recurrence period: The stronger the flash, the smaller the ratio.

Finally, low enough mass-accretion rates lead to nova outbursts (see Chapter 4). According to the semianalytical models of MacDonald [1170], this regime is achieved for

$$\log \dot{M}_{\text{nova}}(M_\odot \text{yr}^{-1}) \leq -8.775 - 15.008 \left[\frac{M}{M_\odot} - 1.459 \right]^2 \tag{5.28}$$

whenever

$$\log L_{\text{nova}}(L_\odot) \leq -0.629 - 5.923 \left[\frac{M}{M_\odot} - 1.766 \right]^2 \tag{5.29}$$

---

[61]Similar fits have been reported by Livio [1125], $\dot{M}_{\text{flash}} \simeq 0.4 \dot{M}_{\text{RG}} \leq 3.3 \times 10^{-7}$ [$(M_{\text{wd}}/M_\odot) - 0.5$], and by Nomoto and collaborators [1326].

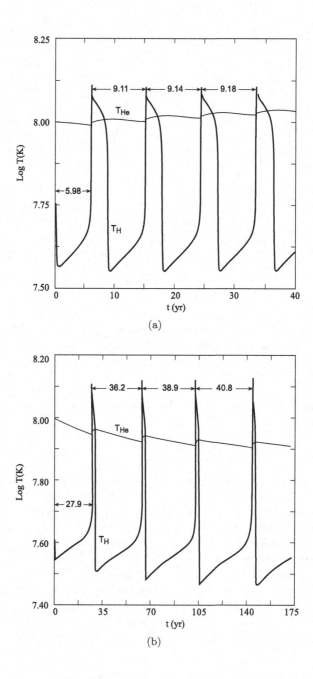

FIGURE 5.20

(a) Time-evolution of the temperature (T) at the H- and He-burning shells in a semiana-lytical model of accretion of solar composition material at a rate of 2×10^{-7} M$_\odot$ yr^{-1}onto a 1.2 M$_\odot$white dwarf, leading to quasi-periodic, weak shell flashes. (b) Same as a), but for $\dot{M} = 5 \times 10^{-8}$ M$_\odot$ yr^{-1}. Figures from José, Hernanz, and Isern [925]; reproduced with permission.

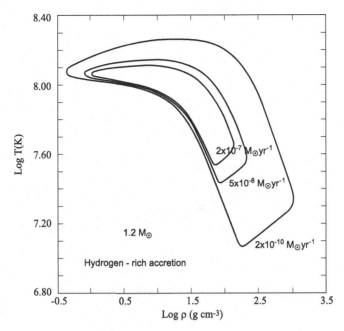

FIGURE 5.21

Temperature (T) vs. density (ρ) diagrams showing the increasing strength of shell flashes in a 1.2 $M_\odot$ white dwarf, as the mass-accretion rate decreases, for three representative cases: $\dot{M} = 2 \times 10^{-7}$ $M_\odot$ yr^{-1}, 5×10^{-8} $M_\odot$ yr^{-1}, and 2×10^{-10} $M_\odot$ yr^{-1}. Figure from José, Hernanz, and Isern [925]; reproduced with permission.

for accretion of solar composition material ($Z = 0.02$), and for

$$\log \dot{M}_{\text{nova}}(M_\odot \text{yr}^{-1}) \leq -8.632 - 4.596 \left[\frac{M}{M_\odot} - 1.334 \right]^2 \tag{5.30}$$

whenever

$$\log L_{\text{nova}}(L_\odot) \leq -1.375 - 7.027 \left[\frac{M}{M_\odot} - 1.308 \right]^2 \tag{5.31}$$

for material with a higher metal content ($Z = 0.51$).

It is worth noting that some hydrodynamic simulations [1450, 1451, 1710] have reported mass ejection for higher mass-accretion rates (and more luminous white dwarfs) than those derived by MacDonald. Moreover, Sumner Starrfield has recently published an extensive series of models of accretion of solar composition material onto a white dwarf, for a wide range of mass-accretion rates, that always lead to thermonuclear runaways [1704]—that is, steady burning never occurs[62]!

Attempts to quantify the long-term evolution of white dwarfs accreting H-rich material at moderate rates have been performed in the framework of spherically symmetric, quasi-static codes [303, 1417], stressing the difficulties encountered by a white dwarf to reach the Chandrasekhar limit by accretion at any realistic rate. Moreover, it has been suggested that in those models that experience recurrent mild flashes or burn hydrogen steadily for

---

[62]Starrfield and collaborators [1712] have also reported stable conversion of hydrogen to helium (i.e., steady burning) for a much wider range of mass-accretion rates than previously predicted (between 1.6×10^{-9} $M_\odot$ yr^{-1} and 8×10^{-7} $M_\odot$ yr^{-1}). However, according to Nomoto [1326], these results were artificially driven by a too-coarse zoning adopted for the envelope.

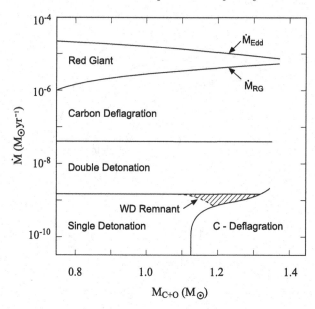

FIGURE 5.22
Nomoto diagram for different He-burning regimes in accreting white dwarfs, as a function
of the mass-accretion rate. Figure from Nomoto [1319]; reproduced with permission.

long enough time, a thick helium layer ($\sim 0.1 - 0.2$ M$_\odot$) piles up underneath. This layer
may undergo a dynamic, violent flash that could eventually drive mass ejection. For specific
combinations of white dwarf masses and mass-accretion rates, a sub-Chandrasekhar-mass
explosion may also result [1417].

5.5.1.2   He-Rich Accretion

The difficulties faced by a white dwarf accreting H-rich material in order to reach the Chandrasekhar limit and eventually explode as a normal type Ia supernova directed the interest
toward helium accretion, which can be directly trasferred by a helium stellar companion
or, indirectly, as a He-rich layer left over from steady H-burning or through mild flashes
accumulates on top of the white dwarf core. Early exploratory studies of the fate of white
dwarfs accreting He-rich material were performed for He white dwarfs [1327] and CO white
dwarfs [574,1758,1759]. In these works, rapid (steady) He-rich accretion onto helium white
dwarfs resulted in a weak, off-center helium flash, whereas slow accretion induced a central
He-detonation and the subsequent disruption of the star, powering a supernova-like event
characterized by large amounts of freshly synthesized ^{56}Ni in the ejecta. Other models
hosting CO white dwarfs resulted in very dynamic He-flashes that triggered supernova-like
explosions.

As discussed for hydrogen, different helium-burning regimes have also been identified.
The upper part of the *Nomoto diagram* for He-burning (Figure 5.22) is very similar to
the one described for hydrogen. Accretion is inhibited at rates above the *Eddington limit*
(Equation 5.25). This is followed by a range of mass-accretion rates driving the expansion
of the envelope toward a red giant configuration, as given by

$$\dot{M}_{\rm RG} \geq 7.2 \times 10^{-6} \left[\frac{M_{\rm wd}}{M_\odot} - 0.6 \right] {\rm M}_\odot {\rm yr}^{-1}, \tag{5.32}$$

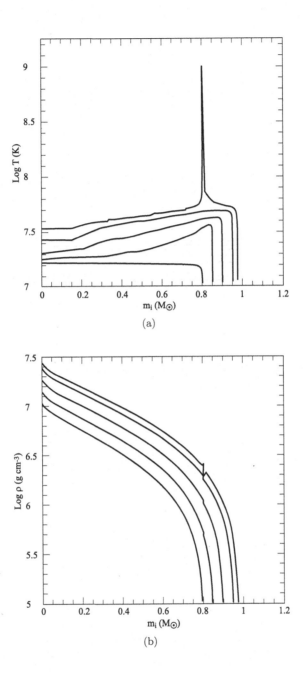

FIGURE 5.23

Temperature (T) and density (ρ) profiles for a 0.8 $M_\odot$ CO white dwarf, accreting He-rich material at a rate of 3.5×10^{-8} $M_\odot$ yr^{-1}. Profiles (from bottom to top) correspond to 0, 1.40, 2.83, 4.25, and 4.93 Myr since the beginning of the accretion phase. The density jump marks the location where shock waves initiate. Calculations performed by the author with the SHIVA code.

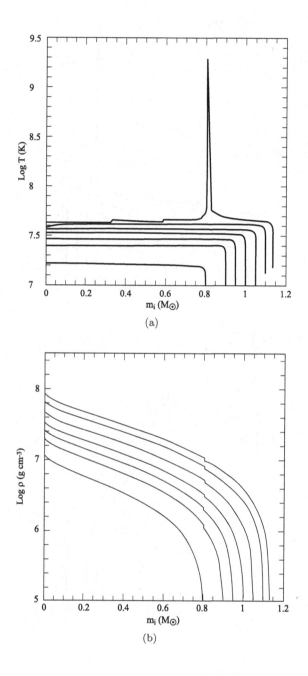

(a)

(b)

FIGURE 5.24

Same as Figure 5.23, but for a mass-accretion rate of 5×10^{-9} $M_\odot$ yr^{-1}. Profiles (from bottom to top) correspond to 0, 26.7, 27.5, 35.3, 44.1, 53.5, and 59.1 Myr since the beginning of the accretion phase.

whenever $0.75 \leq M(M_\odot) \leq 1.38$ [1319]. Steady He-burning occurs for mass-accretion rates roughly 10 times larger than for hydrogen [571, 572, 870, 925].

For lower mass-accretion rates, several types of unstable nuclear burning may occur. Supernova-like explosions are expected to result from intermediate mass-accretion rates ($\dot{M} \leq 4 \times 10^8$ $M_\odot$ yr^{-1}), regardless of the white dwarf mass, as well as from low mass-accretion rates ($\dot{M} \sim 10^{-10} - 10^{-9}$ $M_\odot$ yr^{-1}), for white dwarfs masses below ≤ 1.1 $M_\odot$, as illustrated in the lower-right corner of the corresponding Nomoto diagram (Figure 5.22). In both cases, the event is triggered by an off-center helium detonation. Accretion at higher rates ($\dot{M} \geq 4 \times 10^{-8}$ $M_\odot$ yr^{-1}) induces He ignition off-center, at relatively low densities[63] ($\sim 10^6$ g cm^{-3}; see reference [1319]). The resulting flash is very weak, and no dynamical effects are produced. The phenomenon can recur periodically[64], as more and more He accumulates, leading to the growth of the CO white dwarf mass and eventually powering a C-deflagration supernova. The same explosive outcome is expected for massive CO white dwarfs ($M_{wd} \geq 1.1$ $M_\odot$) undergoing slow He-accretion ($\dot{M} \sim 10^{-10} - 10^{-9}$ $M_\odot$ yr^{-1}; see Figure 5.22). The latter requires the presence of thin He-envelopes ($< 6 \times 10^{-4}$ $M_\odot$) to avoid an off-center detonation, whereas thick He-envelopes (up to ~ 0.25 $M_\odot$) are expected in the former.

Other important aspects for the outcome of He-accreting white dwarfs can be learned from a careful analysis of the overall effects of accretion. To this end, let's focus on the progress of a 0.8 $M_\odot$ CO white dwarf, accreting He-rich matter at two different rates, 3.5×10^{-8} $M_\odot$ yr^{-1} (Figure 5.23) and 5×10^{-9} $M_\odot$ yr^{-1} (Figure 5.24). The accreted matter piles up on top of the white dwarf core and progressively forms a thick He-rich envelope, leading to a significant increase of the overall white dwarf mass. Material is compressed and the accompanying release of gravitational energy tends to increase the temperature of the star, an effect that depends critically on the mass-accretion rate adopted. Following Nomoto [1319], the compressional rate due to accretion, $\lambda \equiv d\ln\rho/dt$, can be expressed as a term related to the density increase at a fixed mass[65] $q \equiv m_i/M_{wd}$, λ^M (which is proportional to the mass-accretion rate, $\dot{M}$), plus the contribution of a term that describes how matter moves inward in q-space, λ^q. λ^M turns out to be rather spatially uniform throughout the star, while λ^q appears to be very large near the surface as a result of the steep density gradient present in the outermost layers. Therefore, it is expected that the evolution at high or moderate mass-accretion rates will result in steep temperature profiles (see Figure 5.23, upper panel) because of the higher compressional heating near the surface (due to the dominance of λ^q) and also because the time scale for compressional heating, $\tau_{comp} = \lambda^{-1}$, is shorter than the characteristic time scale for heat transport to the interior, $\tau_{heat} \sim 3\kappa\rho^2 C_p l^2/64\sigma T^3$, where c_p is the heat capacity at constant pressure and l a characteristic length [765, 776]. In sharp contrast, since $\tau_{comp} = \lambda^{-1} \propto \dot{M}_{acc}^{-1}$, sufficiently low mass-accretion rates, for which $\tau_{comp} \geq \tau_{heat}$, will result in rather flat temperature profiles [776] (see Figure 5.24, lower panel).

The most relevant feature from the accretion phase is the formation of thicker He-rich envelopes for lower mass-accretion rates as a result of the lower compressional heating. Evolution in the $\rho - T$ plane of the He-ignition shell, for a sample of models, is shown in Figure 5.25. It is worth noting that ignition takes place well above the base of the

---

[63]Another nuclear process competes with the triple-alpha reaction at $\rho \sim 10^6$ g cm^{-3}. At such densities, electron-captures can efficiently proceed on ^{14}N, becoming ^{14}C. As the temperature rises, α-captures can proceed onto the newly formed ^{14}C nuclei and the dynamic explosive stage sets in. The effect of the chain ^{14}N(e^-,ν)^{14}C(α,γ)^{18}O on the total mass accreted by the star has been analyzed in a number of papers [735, 786, 1113, 1417, 1988]: It leads to an overall reduction of the mass of the He layer that piles up on top of the accreting white dwarf, particularly at low mass-accretion rates.

[64]Typically, helium flashes exhibit recurrence times about 400 times larger than hydrogen flashes, for the same mass-accretion rate.

[65]Note that, as defined in Chapter 1, m_i represents the mass interior to the ith-interface of the computational domain.

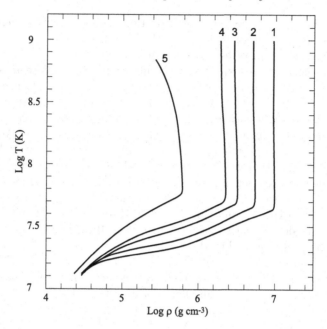

FIGURE 5.25

Temperature (K) vs. density (ρ) diagrams at the location of the He-ignition shell, for a suite of models of He-rich accretion onto a 0.8 $M_\odot$ CO white dwarf at a rate of 5×10^{-9} (1), 1×10^{-8} (2), 2.5×10^{-8} (3), 3.5×10^{-8} (4), and 1×10^{-7} $M_\odot$ yr^{-1}(5). Note that model 5 has a different initial $T - \rho$ value, since He-ignition takes place in a different location, well above the envelope's base, in this particular case. Simulations performed by the author with the 1D hydrodynamic code SHIVA.

formerly accreted envelope when high mass-accretion rates are considered (see, for instance, model 5, evolved with $\dot{M}_{acc} = 1 \times 10^{-7}$ $M_\odot$ yr^{-1}. In this particular model, He ignites at 0.03 $M_\odot$ above the envelope's base, almost 1/3 of the total envelope's mass). This results from temperature inversions in the He-envelope driven by the rapid release of gravitational energy [1319, 1988].

5.5.2   Explosive Evolution and Nucleosynthesis

Ken Nomoto, Friedel Thielemann, and Koichi Yokoi [1331, 1798] reported in the 1980s a series of C-deflagration models in accreting CO white dwarfs, for a range of mass-accretion rates and mixing length parameters, $\alpha = l/H_P$. Among those, the most successful one, Model W7 (named after the adopted value of $\alpha = 0.7$), has become a standard in the field for decades. The model, computed with a spherically symmetric, implicit, hydrodynamic code, relies on a 1.0 $M_\odot$ CO white dwarf accreting at a rate of 4×10^{-8} $M_\odot$ yr^{-1}. For simplicity, the white dwarf is assumed to grow in mass at the same rate as accretion proceeds (that is, in steady H/He-burning regime). As matter piles up on top of the white dwarf, the central density and temperature of the star progressively increase. Simultaneously, heat flows from the outer H/He layers into the core by conduction. The temperature rise drives

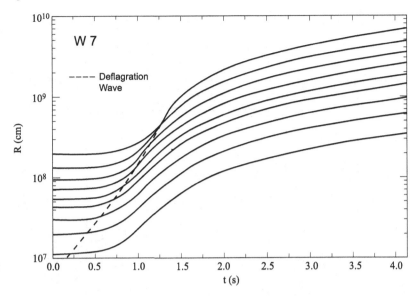

FIGURE 5.26
Propagation of the C-deflagration front (dashed line) and expansion of the white dwarf in Model W7. Solid lines correspond to different Lagrangian shells located at $m_i(M_\odot) = 0.007$, 0.03, 0.10, 0.25, 0.41, 0.70, 1.00, 1.28, and 1.378. Figure from Nomoto, Thielemann, and Yokoi [1331]; reproduced with permission.

central C-ignition[66] at a density $\rho_{ig} = 2.6 \times 10^9$ g cm^{-3}, when the white dwarf mass has reached a value of 1.378 M$_\odot$.

While main sequence stars react to an internal increase of temperature by expansion (i.e., hydrostatic equilibrium), the strong degeneracy of matter inside a white dwarf results in a thermonuclear runaway. Heat transport is initially carried out by convection, up to a temperature of $T_c \sim 8 \times 10^8$ K, when the characteristic timescale for a temperature rise, $\tau_n = c_p T/\varepsilon$ (with c_p denoting the specific heat and ε the nuclear energy generation rate), becomes shorter than the dynamical timescale, $\tau_d \sim 1/\sqrt{G\rho}$, and a shock front results [1318]. Even though spherically symmetric models, such as W7, assume by construction that this takes place simultaneously all along a sphericall shell, this is likely to happen only in one or more spots. Whether ignition actually occurs in a single or multiple spots, at or close to the white dwarf center, is not well known, since both the convective C-burning stage and the physical properties of the flame at the time of ignition are very difficult to model [1517]. These are critical issues for multidimensional models of the explosion [543, 611, 614, 802, 1131–1133, 1521, 1526, 1588].

---

[66] The ocurrence of a supernova may be delayed to ignition densities $\sim (4-6) \times 10^9$ g cm^{-3} with the inclusion of the *convective Urca process* [383, 1363]. In the process, a baryon (e.g., a nucleus) absorbs a free lepton (for instance, an electron) and releases a neutrino. The nucleus is convectively carried to the outer, cooler layers of the white dwarf, where it undergoes a β-disintegration (i.e., it emits an electron together with an antineutrino). Convection then carries the nucleus back to the interior of the star. The neutrino–antineutrino pair leaves the star carrying away energy. If a large number of pairs are emitted, the white dwarf may cool down and recontract to larger densities. The process was first discussed by George Gamow and Mario Schoenberg in 1941 while visiting the *Cassino da Urca* in Rio de Janeiro. Schoenberg is frequently quoted as saying that "the energy disappears in the nucleus of the supernova as quickly as the money disappeared at that roulette table." There is, however, no consensus on the efficiency of the Urca process. Moreover, the suggested increase in density may even result in the collapse (rather than the explosion) of the white dwarf.

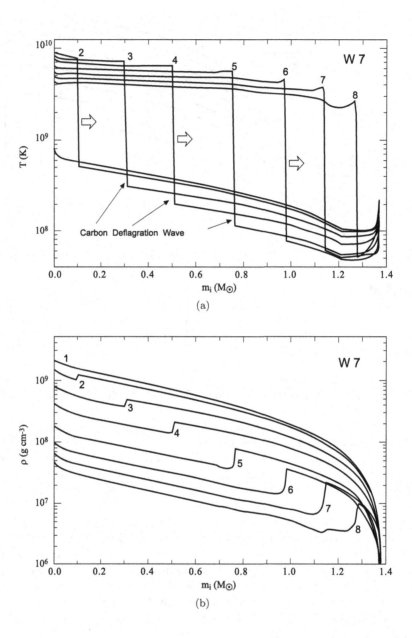

FIGURE 5.27
Time evolution of the temperature (panel a) and density profiles (panel b) throughout the white dwarf as a function of the mass interior to the ith-interface of the computational domain, m_i, during the advance of the deflagration front, in Model W7. The different labeled profiles correspond to times 0.0 (1), 0.60 (2), 0.79 (3), 0.91 (4), 1.03 (5), 1.12 (6), 1.18 (7), 1.24 (8), and 3.22 s (9). Figures from Nomoto, Thielemann, and Yokoi [1331]; reproduced with permission.

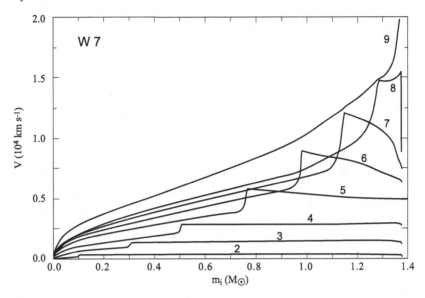

FIGURE 5.28
Same as Figure 5.27, but for the velocity profile.

In the reaction front reported in Model W7, the nuclear energy released, $\sim 3 \times 10^{17}$ erg g^{-1}, amounts only to 20% of the internal energy of the electrons, and, therefore, falls short in powering a Chapman–Jouguet detonation (see Chapter 1). Instead, a deflagration front, that propagates subsonically at the characteristic convective timescale, is born (Figure 5.26). In the mean time, compression waves travel ahead of the flame at the speed of sound, preexpanding the star and lowering the fuel density prior to ignition, with dramatic consequences for the final nucleosynthesis. At the early stages of the runaway (i.e., at t = 0.60 s; see Figures 5.27 and 5.28), the front advances at barely 8% of the local sound speed, $v_{def} \sim 0.08c_s$. But later on, the large density jump experienced by the deflagration front as it moves forward increases its speed (note the curved shape of the deflagration line in Figure 5.26). Indeed, at t = 1.18 s, $v_{def} \sim 0.3c_s$. A detonation never formed, however.

The expasion of the star, driven by the subsonic nature of the front, decreases the density and temperature of the white dwarf, progressively weakening the nuclear energy released at the location of the front, and ultimately quenching C-burning at $M = 1.3$ M$_\odot$ ($\rho = 10^7$ g cm^{-3}) before the front hits the surface of the star. The overall nuclear energy released in Model W7, $E_n = 1.8 \times 10^{51}$ erg, exceeds the binding energy of the star, $E_{bind} \sim 5 \times 10^{50}$ erg, and, therefore, the white dwarf gets fully disrupted. The kinetic energy of the exploding star is about 1.3×10^{51} erg (see also reference [1794]). A short neutrino burst, lasting only $\sim$1.5 s, carries away a small fraction of the overall energy, $\sim 4 \times 10^{49}$ erg.

From a nucleosynthesis viewpoint (see Figure 5.29), as the deflagration front propagates outward, the white dwarf experiences different explosive burning stages (i.e., C-, Ne-, O-, and Si-burning). The inner regions of the star ($M < 0.7$ M$_\odot$), which achieve $T_{max} = 6 \times 10^9$ K and $\rho_{max} = 9 \times 10^7$ g cm^{-3}, undergo almost full incineration to nuclear statistical equilibrium (NSE). At the center of the star, the high densities[67] favor electron captures and, therefore, the most abundant species correspond to the neutron-rich, iron-peak nuclei ^{56}Ni, ^{56}Fe, and ^{54}Fe. The lower density region $M \geq 0.1$ M$_\odot$ is, however, fully dominated by ^{56}Ni. On top

---

[67]The ^{22}Ne abundance is also critical in the overproduction of the neutron-rich species ^{54}Fe and ^{58}Ni [1798]. Indeed, the production of these elements can be reduced if ^{22}Ne sinks away from the outer regions through gravitational settling [187, 247].

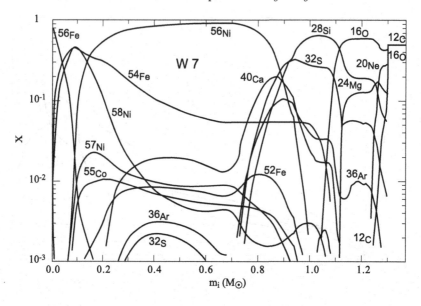

FIGURE 5.29
Final distribution of chemical species (in mass fractions) after the freezing of nuclear reactions, as a function of the white dwarf mass, in Model W7. Figure from Nomoto, Thielemann, and Yokoi [1331]; reproduced with permission.

of it, the region $0.3 \leq M(M_\odot) \leq 0.6$, while still dominated by large amounts of iron-peak nuclei, exhibits a moderate presence of intermediate-mass elements, such as ^{40}Ca, ^{36}Ar, or ^{32}S, suggesting partial breakdown of NSE.

The region $0.7 \leq M(M_\odot) \leq 0.9$ barely achieves $T_{max} \sim (5-6) \times 10^9$ K and only partial Si-burning takes place, with nucleosynthesis products dominated by the presence of intermediate-mass elements, such as ^{28}Si, ^{32}S, ^{36}Ar, ^{40}Ca, ^{54}Fe, or ^{56}Ni. The region $0.9 \leq M(M_\odot) \leq 1.1$, reaches $T_{max} \sim (4-5) \times 10^9$ K, not enough to drive Si-burning. Only C- and O-burning ensue, and the yields are therefore dominated by ^{28}Si, ^{32}S, or ^{36}Ar. The region immediately above, $1.1 \leq M(M_\odot) \leq 1.25$, achieves $T_{max} \sim (3-4) \times 10^9$ K, and only C- and Ne-burning occur. These layers are rich in ^{16}O, ^{24}Mg, and ^{28}Si. Last but not least, carbon-burning products appear in the region $1.25 \leq M(M_\odot) \leq 1.3$, while the outer $M(M_\odot) \geq 1.3$ essentially shows unburned, CO-rich matter[68]. All in all, 0.86 $M_\odot$ of Fe-peak nuclei are produced in Model W7 (0.58 $M_\odot$ in the form of ^{56}Ni).

Several estimates of the frequency of the different supernova types have been published in the last decades. Sidney van den Bergh and Gustav Tammann [1845], for instance, suggest the following supernova rates for the for the expected for the expected Milky Way galaxy: 0.0041 yr^{-1} for type Ia, 0.0065 yr^{-1} for types Ib/c, and 0.033 yr^{-1} for type II supernovae. More recent estimates (e.g., reference [437]) have somewhat lowered the expected frequency of type Ia supernovae to values within $\sim$ 0.2–0.3 events per century. Nevertheless, since the amount of ^{56}Ni synthesized and ejected by the different supernova types is about 0.6 $M_\odot$(type Ia), 0.3 $M_\odot$(type Ib/c), and 0.07 $M_\odot$(type II), it turns out that type Ia supernovae

---

[68]Nomoto and his collaborators already pointed out that the huge temperature jump existing between burned and unburned regions (see Figure 5.27(a)) is accompanied by a density inversion that, most likely, would induce Rayleigh–Taylor hydrodynamic instabilities, resulting in mixing between hot (burned) and cold (unburned) material. This poses limitations to the accuracy of the final abundances in the outmost layers of the star reported in that work [1331].

produce about half of the iron content in the Milky Way[69]. In fact, this role of type Ia supernovae as prominent Galactic ^{56}Fe factories[70] was anticipated by Fred Hoyle[71]. His hypothesis relied on the large binding energy of ^{56}Fe. Unfortunately, Hoyle changed his view a few years later. Together with Fowler [845], Hoyle estimated the yields expected in type I supernova explosions, but the high densities employed in their calculations, which favored electron captures, led them to predict an ejecta dominated by the presence of ^{58}Fe, rather than[72] by ^{56}Ni–^{56}Fe. Since this particular isotope is actually quite rare in nature, Hoyle and Fowler concluded that the rate of type Ia supernova explosions must be low, and, therefore, they could not be the major sources of Galactic ^{56}Fe, invoking instead type II supernovae (a possibility also outlined by Walter Baade in the late 1950s)[73].

A major success of W7, and of C-deflagration models in general, with respect to pure detonation models is its ability to qualitatively match the abundance pattern inferred from observations (that is, reduce the large overproduction of iron and yield a moderate synthesis of intermediate-mass elements like ^{40}Ca, ^{36}Ar, or ^{32}S). Since the main nucleosynthesis product is indeed ^{56}Fe (Figure 5.30), a successful type Ia supernova model must reproduce the isotopic abundance pattern of the Fe-peak elements, as inferred for the Galaxy. This requires Galactic chemical evolution models fed with combined yields from all possible Fe-peak production sites[74]. For decades, C-deflagration models systematically overproduced several neutron-rich species, such as ^{54}Cr or ^{50}Ti, with respect to solar values [1322, 1797, 1963, 1966]. This has greatly improved (but not yet solved) after the major revision of stellar weak interaction rates [1050], a key ingredient in any simulation aimed at determining the yields of type Ia supernovae[75]. Nevertheless, most of the inaccuracies found in C-deflagration models likely arise from the assumption of a pure deflagration, or more in general, from the poor knowledge of the specific regime in which the flame propagates [1961, 1970]. For the typical conditions that characterize white dwarf interiors, such deflagration fronts are highly subsonic [1807], allowing the star to preexpand. This ultimately quenches the burning at a time when the white dwarf is still gravitationally bound. This issue has been explored in the framework of high-resolution, multipoint ignition simulations (e.g., [1521]), including relevant multidimensional effects, such as hydrodynamic

---

[69]See reference [6] for recent estimates of iron production by the different supernova types.

[70]Not directly synthesized as ^{56}Fe, but as a decay product of ^{56}Ni [204, 547, 1826].

[71]See references [842, 843], as well as the famous B^2FH paper [268], one of the foundational works of nuclear astrophysics.

[72]Actually, ^{58}Fe is highly abundant in the deeper layers of the exploding star but not in the outer part, which is ^{56}Ni-dominated.

[73]The wrong nucleosynthesis predictions motivated other mistakes [1967]. Since ^{56}Ni was scarce in the ejecta, Hoyle and Fowler proposed ^{254}Cf as the likely source powering the light curve tail (see also reference [102]). Moreover, they also wrongly concluded that type Ia supernovae were the site of the r-process (see Chapter 7).

[74]See reference [1330] for early fits of the solar abundances with combined yields from type Ia and type II supernovae.

[75]Several studies have focused on the effect of electron captures, and of the different parameters used in 1D C-deflagration models, on type Ia supernova nucleosynthesis [229, 899, 1794, 1965]. The ignition density, ρ_{ig}, determines the extend of electron captures in the central regions of the white dwarf (with lower densities reducing the probability of electron captures), and, hence, the electron fraction, Y_e. Values in the range Y_e = 0.47–0.485 result in chemical abundance patterns dominated by ^{54}Fe and ^{58}Ni; values between 0.46 and 0.47 predominantly yield ^{56}Fe; ^{58}Fe, ^{54}Cr, ^{50}Ti, and ^{64}Ni are mostly synthesized for $Y_e = 0.45 - 0.43$, while ^{48}Ca requires $Y_e \leq 0.43$ [1794]. The speed of the deflagration front, v_{def}, determines the time available for nuclear reactions and weak interactions (e.g., electron captures) to proceed, as well as the dynamic response of the star before the front reaches a specific point. From a comparison with solar Fe-group abundances, together with chemical evolution models, Iwamoto and collaborators [899] concluded that ρ_{ig} must lie in the range $\sim (2-3) \times 10^9$ g cm^{-3}, while $v_{def} \sim (1.5\% - 3\%)c_s$. Larger ignition densities are required to account for ^{48}Ca, ^{50}Ti, ^{54}Cr, ^{58}Fe, ^{64}Ni, or ^{66}Zn. For some of these nuclei, no other production site seems to exist [1965].

instabilities[76] [600, 1144, 1482, 1519, 1520, 1522, 1588, 1822]. Simulations revealed the inability to obtain ^{56}Ni masses in excess of 0.7 M$_\odot$ and kinetic energies $> 0.7 \times 10^{51}$ erg, and thus cannot account for the bulk of normal type Ia supernovae. Indeed, pure deflagrations can only reproduce the least energetic/luminous events. Despite of the overall improvement on the resulting yields, large amounts of unburnt C and O (> 0.57 M$_\odot$) are left behind [1589]. Indeed, the presence of unburnt C and O is expected to yield O I and C I lines at the nebular stage of the event, a feature never observed so far in type Ia supernovae [1013]. Two additional problems associated with C-deflagration models include the lack of chemical stratification in the ejecta as well as the presence of big clumps of radioactive ^{56}Ni in the photosphere around peak luminosity [891].

5.5.3   Deflagration to Detonation Transitions and Other State-of-the-Art Models

New generations of models, aimed at reducing the shortcomings of the C-deflagration models, were developed mostly in the 1990s. The most successful ones, the *delayed detonation models*, rely as well on a deflagration front that propagates and preexpands the star, but subsequently switches into a detonation at some characteristic density[77], via a deflagration-detonation transition (DDT) [980, 1300]. Several versions of these models exist, with [897] or without [977, 978] a pulsating stage. One-dimensional simulations of delayed detonations have proved successful in reproducing many observational features of type Ia supernovae. In particular, the combination of the deflagration and detonation regimes allows for the synthesis of intermediate-mass elements (the large amounts of unburnt C and O leftover in the C-deflagration models are incinerated by the late detonation [612, 979]; see Figure 5.30) and provide the required energy budget for normal type Ia supernovae. Moreover, these models account for the expected light curves and photospheric expansion velocities [820]. It is, however, worth noting that switching from 1D delayed detonation models to truly multidimensional models results in a different stratification pattern for the ejecta [1178, 1524, 1604] (Figure 5.31). While mostly Fe-peak nuclei are synthesized near the white dwarf center, the outer, low-density layers of the star are mostly incinerated into intermediate-mass elements, thus producing a chemically stratified structure. But in a truly multidimensional framework, the late detonation also burns downward in funnels and cavities leftover between the ash plumes of the deflagration stage. While the characteristic temperatures and densities of the fuel encountered by the downward detonation are such that iron-group nuclei are produced, they are not high enough for electron captures to proceed efficiently. As a result, the stratified ejecta resulting from multidimensional delayed detonation models is characterized by innermost regions dominated by ^{56}Ni and outermost layers composed of intermediate-mass elements, encapsulating a central part dominated by the presence of stable, Fe-group nuclei [78].

The main uncertainty affecting the delayed detonation models is the physical mechanism driving the deflagration-detonation transition. Recent studies suggest that this could naturally occur once the flame reaches the so-called *distributed burning regime* [1524], at typical densities $\sim 10^7$ g cm^{-3}, right at the time when the laminar flame width equals the Gibson scale[79], that is, the scale at which turbulent velocity fluctuations equal the lami-

---

[76]Indeed, several hydrodynamic instabilities, acting on various scales, lead to a geometric increase of the flame surface, resulting in larger flame speeds, about a few percent of the sound speed in the central regions of the star, and about a significant fraction of the sound speed far from the center [600, 800, 1525].

[77]See, however, reference [336].

[78]See also references [200, 244, 599, 658] for additional studies of multidimensional delayed detonations.

[79]The flame thickness during a type Ia supernova ranges between $10^{-4} - 10$ cm [82, 1518, 1807, 2026]. At $\rho \sim 10^7$ g cm^{-3}, the flame thickness, $l_{\text{flame}} \sim 0.1$ cm, is 9 orders of magnitude smaller than the radius of

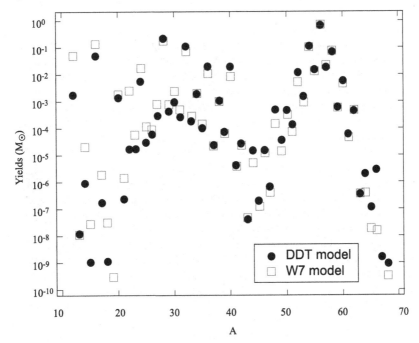

FIGURE 5.30
Final yields (in solar masses) obtained in two models of type Ia supernovae, computed with standard nuclear reaction rates: The W7 C-deflagration model (squares) and a delayed detonation model (circles). Figure from Parikh, José, Seitenzahl, and Röpke [1388]; reproduced with permission.

nar flame speed (see Chapter 1). It is also worth noting that the incineration of the outer, CO-rich layers of the star, solve some of the reported inconsistencies with observations but also generate new puzzles. Indeed, the lack of substantial amounts of unburnt C leftover in these delayed-detonation models is at odds with the reported C II absorption line at 6580 Å, observed in the premaximum spectra of ∼ 30% of type Ia supernovae [542].

The *gravitationally confined detonation* is another, somewhat related model that has been proposed in the framework of multidimensional simulations [1432]. It assumes asymmetric deflagration flame ignitions that push burnt fuel toward the stellar surface. The collision of these ashes on the far side of the white dwarf triggers a detonation front that propagates inward and incinerates the stellar core, potentially leading to a very energetic event [1431, 1815], with an ejecta rich in intermediate-mass elements. However, the feasibility of this scenario has been questioned by other studies. Fritz Röpke, Stan Woosley, and Wolfgang Hillebrandt [1526] reported, for instance, that the maximum temperature achieved in these collisions is sensitive to the level of burning and expansion in previous stages. Moreover, detonations are only triggered in some 2D simulations, but not in 3D.

Alternative models based on pulsational delayed detonations have also been proposed to account for the bulk of type Ia supernovae [243, 245, 246]. As in the canonical delayed

---

the star—about 2000 km for a Chandrasekhar-mass white dwarf. Moreover, the characteristic length scale at which the turbulent kinetic energy cascade is dissipated into heat, the Kolmogorov scale, is less than a millimeter. This vast range of characteristic length scales cannot be resolved in current computational models. Since the maximum resolution attained in 3D type Ia supernova models is about 1 km, that is, 4 orders of magnitude larger than the flame width at ∼ 10^7 g cm^{-3}, special numerical techniques (e.g., effective flame models and small-scale simulations) have to be adopted [336, 1517, 1518].

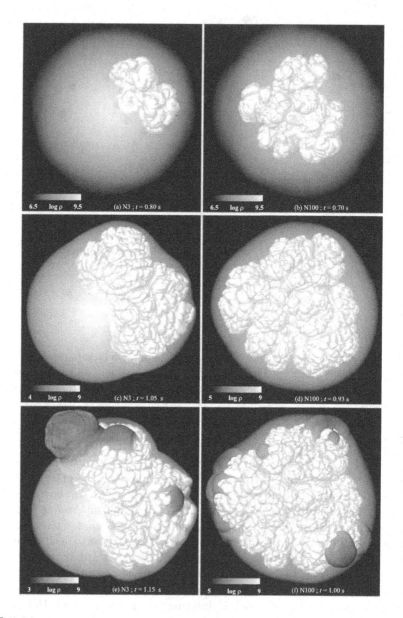

FIGURE 5.31

Snapshots of the evolution of two Chandrasekhar-mass, delayed detonation models, with 3 ignition points (left panels) and 100 ignition points (right panels). The central density is assumed to be 2.9×10^9 g cm^{-3}. The top panels display the rising plumes of the deflagration front (white) during the Rayleigh–Taylor unstable stage of the deflagration phase, superimposed to a density map (in g cm^{-3}). The middle panels correspond to the time when the first detonation sets in. The subsequent spreading of the detonation front (dark grey) is depicted in the bottom panels. A color version of this plot is available at http://fisica.upc.edu/ca/users/jjose/CRC-Downloads. Figure from Seitenzahl et al. [1604]; reproduced with permission.

detonation models, an initial deflagration wave preexpands the star. Because of the subsonic nature of the flame, the burning front is quenched and fails to unbind the star. But during recontraction, compressional heating at the interface between burnt and unburnt material may ultimately trigger a detonation front. Recent multidimensional models [600, 1482] suggest, however, that the star may instead get unbound rather than recontracting. This depends on the amount of mass incinerated (and the nuclear energy released) by the deflagration, with $m_{def} \leq 0.3$ $M_\odot$ leading to bound remnants [245]. Moreover, detailed spectral modeling of these 3D pulsating reverse detonation simulations reveals too much iron-peak elements (nickel, in particular) in the ejecta, resulting in events intrinsically far redder than the bulk of observed type Ia supernovae [133].

5.5.4 γ-Ray Emission

Potentially detectable γ-ray fluxes are expected when the supernova ejecta becomes sufficiently transparent [44, 273, 347, 440, 621, 661, 821, 1032, 1179, 1745, 1789]. The predicted emissivity in the γ-ray range, while model-dependent, is determined by a handful of interaction processes (i.e., pair production, Compton scattering, and photoelectric absorption [1244]) and mainly depends on the density, velocity, and composition of the ejecta, as well as on the mass and distribution of different radioactive species. It provides a valuable tool to discriminate between the set of proposed explosion models[80]. This is particularly relevant, since a direct comparison with observables, on the basis of optical, near-peak luminosity measurements, does not necessarily favor a particular model [1179, 1523].

From the myriad possible radioactive chains, only $^{56}Ni \rightarrow {}^{56}Co \rightarrow {}^{56}Fe$ [347], and, to some extent, $^{57}Ni \rightarrow {}^{57}Co \rightarrow {}^{57}Fe$ [660], play a key role in the γ-ray output from type Ia supernovae[81] (see Figure 5.32, for the corresponding decay schemes). The predicted spectral evolution of the γ-ray emission, for four different type Ia supernova models (three Chandrasekhar-mass models, with flame fronts propagating as a pure deflagration, a pure detonation, or through a combined deflagration/detonation—i.e., a delayed-detonation model—together with a sub-Chandrasekhar-mass model) are shown in Figure 5.33 [661]. The best chances to discriminate between models rely on energies < 1 MeV and are restricted to the first weeks after the explosion. Indeed, some 20 days after the explosion, the γ-ray spectra of all models (except for the pure deflagration one) exhibit strong 158 keV, 750 keV, and 812 keV ^{56}Ni lines, together with the 847 keV ^{56}Co line, which are particularly intense in those models with large amounts of ^{56}Ni freshly synthesized in the outermost layers of the ejecta (like in the pure detonation and sub-Chandrasekhar-mass models). In sharp contrast, the C-deflagration model displays a rather smooth continuum, with only a few, faint lines superimposed. All models are characterized by a sharp cut-off due to photoelectric absorption, at energies $\sim$ 40–80 keV (the exact value depends on the chemical composition of the outer layers, where most of the continuum originates). Already two months after the explosion (see Figure 5.33(b)), most of the low-energy ^{56}Ni lines are gone. Some faint lines at 122 keV and 136 keV due to ^{57}Co decay are, however, visible in the γ-ray spectra, but the best chances to discriminate between models have quickly vanished. After four months,

---

[80]Early line emission in the 6–8 keV X-ray range can also provide hints to discriminate between several proposed models—in particular, Chandrasekhar-mass vs. sub-Chandrasekhar-mass models. This emission originates from Fe, Co, and Ni nuclei excited by γ-rays released from ^{56}Ni and ^{56}Co decay, as well as by thermalized photons with energies > 7 keV [891]. Indeed, while ^{56}Ni is expected to be synthesized deep inside the star in Chandrasekhar-mass models, explosions hosting sub-Chandrasekhar-mass white dwarfs likely produce large amounts of nickel in the outermost layers. Such asymmetry translates into distinct spectra [1422]. See also reference [1179] for ^{56}Ni-mass constraints based on the hard X-ray emission of type Ia supernovae.

[81]Note, for instance, that the amount of ^{44}Ti synthesized during type Ia supernova explosions is 4–5 orders of magnitude smaller than for ^{56}Ni [443].

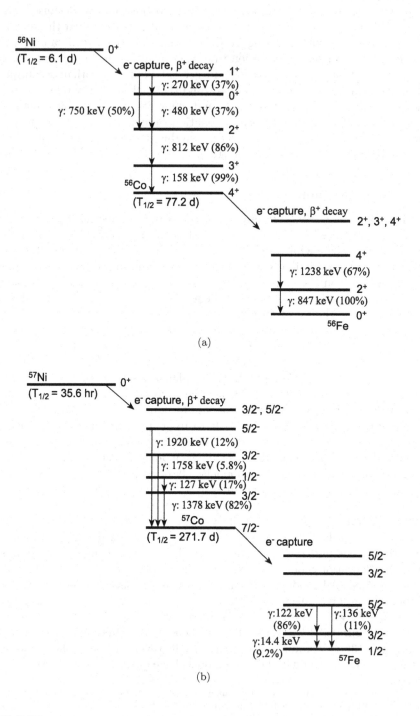

(a)

(b)

FIGURE 5.32

Simplified decay schemes for the radioactive chains ^{56}Ni → ^{56}Co → ^{56}Fe (panel a), and ^{57}Ni → ^{57}Co → ^{57}Fe (panel b), showing the most important energy levels and gamma transitions. See http://www.nndc.bnl.gov/nudat2/reColor.jsp?newColor=dm, for details.

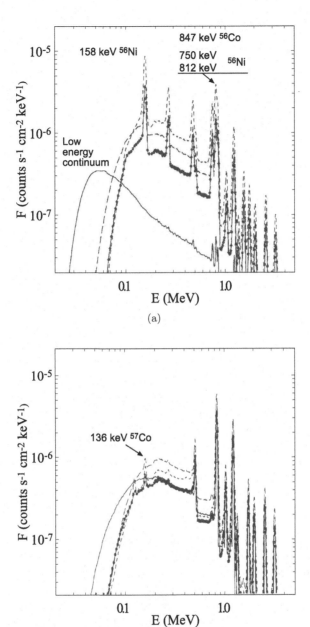

FIGURE 5.33
Spectral evolution of the predicted γ-ray emission in four different type Ia supernova models: a deflagration (solid line), a pure detonation (short-dashed line), a delayed-detonation (long-dashed line), and a sub-Chandrasekhar-mass model (starred line). Flux estimates, for a distance of 5 Mpc and a 10^6 s exposure, are given for 20 days (panel a) and 60 days (panel b) after the explosion. Figures from Gómez-Gomar, Isern, and Jean [661]; reproduced with permission.

TABLE 5.1

Most Uncertain Nuclear Processes in Different Models of type Ia Supernova Nucleosynthesis

Reaction	W7 model	DDT model
$^{12}C(\alpha, \gamma)^{16}O$	^{36}Ar, ^{40}Ca, ^{45}Ti	^{36}Ar, ^{39}K, $^{40,41}Ca$, $^{44,45,46}Ti$, ^{45}Sc, ^{52}Cr, ^{54}Mn, ^{56}Fe
$^{12}C(^{12}C, \alpha)^{20}Ne$	$^{20,21}Ne$, ^{24}Na, $^{25,26}Mg$	^{17}O, ^{20}Ne, $^{23,24}Na$
$^{12}C(^{12}C, p)^{23}Na$	^{18}O, $^{20,21}Ne$, $^{23,24}Na$, ^{26}Mg, ^{26}Al	$^{23,24}Na$, ^{26}Al, ^{28}Mg, ^{31}Si
$^{16}O(n, \gamma)^{17}O$	^{17}O, ^{21}Ne, ^{24}Na, ^{26}Mg	
$^{16}O(\alpha, \gamma)^{20}Ne$	^{21}Ne, ^{26}Al, ^{35}S	
$^{20}Ne(\alpha, p)^{23}Na$	^{18}O, ^{23}Na, ^{26}Al	^{23}Na, ^{28}Mg, ^{31}Si
$^{20}Ne(\alpha, \gamma)^{24}Mg$	^{24}Mg, ^{27}Al, ^{30}Si, ^{32}P, ^{35}S, $^{35,36}Cl$, $^{36,37}Ar$, ^{39}K, ^{41}Ca, ^{46}Ti	$^{24,26}Mg$, $^{26,27}Al$, ^{30}Si, ^{33}P, $^{36,37}Ar$, $^{39,40}K$, $^{40-43}Ca$, ^{45}Sc, $^{45-47}Ti$, ^{56}Fe, ^{60}Ni
$^{22}Ne(p, \gamma)^{23}Na$	^{18}O, $^{23,24}Na$	^{24}Na
$^{22}Ne(\alpha, n)^{25}Mg$	$^{17,18}O$, ^{24}Na	^{24}Na
$^{23}Na(\alpha, p)^{26}Mg$		$^{23,24}Na$, ^{53}Cr
$^{25}Mg(n, \gamma)^{26}Mg$	^{26}Mg	^{21}Ne, ^{24}Na, $^{25,26}Mg$
$^{27}Al(\alpha, p)^{30}Si$	^{24}Mg, ^{27}Al, ^{36}S	^{24}Mg, ^{27}Al
$^{30}Si(p, \gamma)^{31}P$	$^{31-33}P$, $^{35,36}S$	
$^{30}Si(\alpha, \gamma)^{34}S$	^{30}Si, $^{32,33}P$	^{33}P
$^{45}Sc(p, \gamma)^{46}Ti$		^{44}Ca, ^{45}Sc, ^{45}Ti

the spectra below 1 MeV show only a faint continuum due to positronium annihilation and Comptonized photons, with the exception of a 170 keV feature, corresponding to the energy of back-scattered 511 keV photons. Maximum detectability distances for some of the most intense lines have been reported in different studies. For instance, Jordi Gómez-Gomar, Jordi Isern, and Pierre Jean [661] yield detectability distances with the SPI spectrometer on board the INTEGRAL satellite up to 10 Mpc for the 158 keV ^{56}Ni line, and 16 Mpc and 12 Mpc for the 847 keV and 1238 keV ^{56}Co lines, respectively. More recent estimates by Keiichi Maeda and collaborators, again for INTEGRAL/SPI, yield 4 Mpc for the 158 keV and 812 keV lines and up to 8 Mpc for the 847 keV line.

Attempts to detect these early γ-ray lines associated to ^{56}Ni and ^{56}Co decay from nearby type Ia supernovae include two events analyzed with the COMPTEL instrument on board the CGRO satellite: SN 1991T (located at a distance of d $\sim$ 13 Mpc), from which a marginal detection of the ^{56}Co lines at 847 keV and 1238 keV was reported [1108], and SN 1998bu (d $\sim$ 17 Mpc), from which upper limits for the same ^{56}Co lines were obtained [635]. Also, the recent SN 2011fe (d $\sim$ 6 Mpc) was observed by INTEGRAL, but again only upper limits, this time for the 158 keV and 812 keV ^{56}Ni lines, were obtained [893]. Very recently, the first unambiguous detection of the 847 keV and 1238 keV ^{56}Co lines has been finally achieved through INTEGRAL observations of SN 2014J, the closest (d $\sim$ 3.5 Mpc) type Ia supernova detected since the dawn of γ-ray astronomy. Observations were performed between 50 and 100 days since its outburst [335]. Line fluxes suggest an overall 0.62 $\pm$ 0.13 $M_\odot$ of ^{56}Ni likely synthesized during the explosion of a Chandrasekhar-mass white dwarf. But other interpretations, based on an earlier-than-expected detection of ^{56}Ni lines, suggest a more exotic scenario involving the accretion of a He-rich belt onto a white dwarf as the mechanism triggering the subsequent supernova explosion [444]. In any case, the detection of the early γ-ray emission from type Ia supernovae opens new perspectives to constrain models and gives full credit to the basic picture envisaged for the origin of such thermonuclear explosions.

5.5.5 Nuclear Uncertainties

Nucleosynthesis in type Ia supernovae depends critically on the peak temperature achieved and the density at which the explosion occurs. Nevertheless, the composition of the white dwarf hosting the explosion (in particular, the amount and distribution of ^{12}C and ^{22}Ne) plays also a key role as it affects the ignition density, the overall energy released, the flame speed, or the specific density at which the deflagration-to-detonation transition occurs [309, 310, 1816].

First attempts to address the impact of uncertainties in the nuclear processes involved in type Ia supernova nucleosynthesis mostly focused on ^{12}C+^{12}C [248, 315, 868, 1684], since this reaction[82] triggers the explosion when the temperature exceeds $\sim 7 \times 10^8$ K. Recently, Anuj Parikh and collaborators [1388] have investigated the sensitivity of the predicted nucleosynthesis to variations of both thermonuclear reaction and weak interaction rates involved in the explosion, in the framework of two representative models: the C-deflagration W7 model [1331], decades, and a 2D delayed-detonation model (see Figure 5.30). Postprocessing calculations with a network containing 443 species, ranging from n to ^{86}Kr, were performed with temperature and density versus time profiles extracted from a representative number of shells of the C-deflagration model and trace particles of the delayed detonation 2D simulations [1388]. All thermonuclear reaction and weak interaction rates were individually varied by a factor of ten, up and down. Several million, postprocessing calculations were performed to recalculate the yields accordingly, which were subsequently compared with those obtained with standard, recommended rates. The study revealed that out of the 2305 thermonuclear reactions included in the network, only the uncertainties in 53 reactions affect the yield of any species with an abundance $\geq 10^{-8}$ $M_\odot$, by a factor ≥ 2. A subset of these reactions, whose variation by a factor of ten affects the yields of three or more species by a factor ≥ 2 in any of the models, is listed, for illustrative purposes, in Table 5.1. The study revealed that variation of the rates of the ^{12}C(α, γ), ^{12}C+^{12}C, ^{20}Ne(α, p), ^{20}Ne(α, γ), and ^{30}Si(p, γ) reactions have the largest impact on the yields. Furthermore, the individual variation of 658 weak interaction rates by a factor of ten, underlined that only variations of the stellar ^{28}Si(β^+)^{28}Al, ^{32}S(β^+)^{32}P, and ^{36}Ar(β^+)^{36}Cl rates have a significant effect on the yields in any of the models considered.

According to this study, nucleosynthesis in the two adopted models turns out to be relatively robust to variations in individual nuclear reaction and weak interaction rates. Laboratory experiments aimed at reducing the nuclear uncertainties of the subset of reactions identified (particularly at T $\geq 1.5 \times 10^9$ K) are, however, needed for a better account of the nucleosynthesis expected in type Ia supernovae and for further constraining model predictions. In addition, a detailed, consistent treatment of all relevant stellar weak interaction rates is clearly needed, since simultaneous variation of all these rates (as opposed to individual variations) turns out to have a significant effect on the yields.

---

[82]The possible existence of a resonance around E = 1.5 MeV, close to the energies of interest during the C simmering phase in type Ia supernovae and with a strong impact on the overall rate, has been reported in recent years [1684], but follow-up studies have not confirmed its presence.

Box I. The Thermonuclear Supernova "Hall of Fame"

A selection of facts on type Ia supernovae

1. A handful of historical supernovae (i.e., SN 185, SN 393, SN 1006, SN 1054, SN 1181, SN 1572, and SN 1604) have been identified in ancient records, some of which likely corresponding to type Ia events.

2. Type Ia supernovae are spectroscopically identified by the absence of H and the presence of a prominent Si absorption line near 6150 Å in the early optical spectra (around peak luminosity). At later times, 4–5 months past maximum light, they are characterized by unresolved Fe and Co emission lines. SN 1885 (S And) was the first supernovae in which marginal spectroscopic measurements were made.

3. The spectral evolution of type Ia supernovae suggests a layered distribution of explosion products. The outermost layers of the ejecta are dominated by the presence of intermediate-mass elements while the inner layers are prominent in Fe-peak nuclei.

4. Spectropolarimetric measurements have been obtained from a sample of type Ia supernovae. They display modest continuum polarization but strong line polarization prior to peak luminosity. The fact that polarization declines after maximum strongly suggests that the outer ejecta is more aspherical than the inner regions.

5. About 25%–30% of all observed type Ia supernovae exhibit remarkable variations in light curve (i.e., shape, peak luminosity, characteristic timescales) with respect to the group of standard or normal type Ia events. They are flagged as "peculiar", and excluded for most applications.

6. Type Ia supernova light curves are characterized by a sudden rise to peak luminosity, with an absolute visual magnitude reaching $\sim -19^{\mathrm{mag}}$, followed by a steep decline powered by the radioactive chain $^{56}Ni \rightarrow {}^{56}Co \rightarrow {}^{56}Fe$. About ~ 0.6 $M_\odot$ of ^{56}Ni are likely synthesized by normal type Ia supernovae.

7. Observations of type Ia supernovae at high redshift have helped to establish the acceleration of the expansion rate of the universe. High-precision photometric measurements of SN 1995K first proved the stretching (i.e., time dilation) of supernova light curves.

8. Type Ia supernovae result from the explosion of a CO white dwarf in a binary system, with an estimated frequency around 0.3 events per century in the Milky Way. No compact remnant is left after the explosion.

9. C-ignition likely starts propagating as a (subsonic) deflagration front but eventually may switch into a (supersonic) detonation. The resulting nucleosynthesis, rich in Fe-peak nuclei, is qualitatively in agreement with the abundance pattern inferred spectroscopically and with Galactic chemical evolution estimates. Type Ia supernovae are indeed prominent Fe factories in the Galaxy.

10. Double white dwarf mergers, one of the proposed scenarios for type Ia supernovae, are expected to be the dominant sources of the background noise in the low-frequency range of gravitational waves.

11. Unambiguous detection of the early (i.e., the 158 keV and 812 keV ^{56}Ni lines) and late (the 847 keV and 1238 keV ^{56}Co lines) γ-ray emission has been reported for SN 2014J.

Box II. Mysteries, Unsolved Problems, and Challenges

- Need to provide a theoretical basis to the empirical calibration methods on which cosmological applications of type Ia supernovae rely.

- Studies aimed at understanding how dependent (and why) are type Ia supernovae on metallicity, stellar population, age (redshift), or environment must be further explored.

- What causes the diversity among type Ia supernovae?

- Spectropolarimetric studies of the outermost, high-velocity ejecta of type Ia supernovae do not support pure deflagration models but also are not fully described by delayed-detonation models. Which fraction of type Ia supernova explosions are actually aspherical? No direct observational evidence of the binary nature of type Ia supernovae has been obtained so far through spectropolarimetry.

- Two, somewhat competing scenarios have been proposed to explain the origin of type Ia supernova explosions: the single degenerate channel (mostly focused on the explosion of individual, Chandrasekhar-mass white dwarfs) and the double degenerate channel (through double white dwarf mergers). Efforts aimed at clarifying whether one or both scenarios can actually lead to type Ia supernovae and account for their expected frequency are strongly needed. In particular, systematic searches aimed at detecting possible surviving companion stars from such explosions must be conducted. Do all proposed scenarios (e.g., sub-Chandrasekhar-mass models) actually lead to supernova-like events? If not, how to prevent the explosion in those systems whose predicted observables do not match type Ia supernova properties?

- Need to elucidate how does the ignition start in type Ia supernovae (i.e., in a single spot or multiple spots, how far from the center).

- The current paradigm for type Ia supernova explosions relies on delayed-detonation models, but the physical mechanism driving the deflagration-detonation transition and the specific density at which such transition occurs are not well understood.

- Modern parallel computers allow for multidimensional simulations of type Ia supernovae, but even on the largest existing platforms, serious constraints remain. Current efforts must be followed with increased resolution and improved techniques to capture physics on unresolved length-scales.

- Nuclear uncertainties in the rates of the $^{12}C(\alpha, \gamma)$, $^{12}C+^{12}C$, $^{20}Ne(\alpha, p)$, $^{20}Ne(\alpha, \gamma)$, and $^{30}Si(p, \gamma)$ reactions, together with the weak interactions $^{28}Si(\beta^+)^{28}Al$, $^{32}S(\beta^+)^{32}P$, and $^{36}Ar(\beta^+)^{36}Cl$, have the largest impact on type Ia supernova yields. Efforts aimed at better constraining those rates are clearly needed.

- Need for a consistent treatment of weak interaction rates for stellar conditions.

5.6   Exercises

P1. Estimate the number of type Ia supernova that explode per second across the observable universe.

P2. Show that a universe filled with a uniform density fluid satisfies

$$H^2 = \frac{8}{3}\pi G\rho - \frac{kc^2}{R^2}$$

in the framework of Newtonian cosmology. Compare the physical meaning of k in Newtonian and relativistic cosmologies.

P3. Prove that the critical density of the universe corresponds to

$$\rho_{\text{crit}} = \frac{3H^2}{8\pi G}$$

and estimate its current value.

P4. A type Ia supernova is discovered at redshift $z = 4$. Determine the scaling factor that describes the expansion of the universe (i.e., how much bigger is the universe now compared with the time when light left the object).

P5. Assume that a $1.4\,M_\odot$ white dwarf undergoes a type Ia supernova explosion. Determine how much energy is released by nuclear fusion (see Chapter 2 for estimates of the energy released during C, O, and Si burning) during the explosion if the white dwarf is fully processed into Fe-peak nuclei, for the following initial compositions:
    a) 50% ^{12}C and 50% ^{16}O, by mass.
    b) 60% ^{16}O and 40% ^{20}Ne, by mass.

P6. A standard type Ia supernova produces about $0.6\,M_\odot$ of radioactive ^{56}Ni. Plot a qualitative light curve during decline from peak luminosity on the basis of the energy released through the chain ^{56}Ni $\rightarrow$ ^{56}Co $\rightarrow$ ^{56}Fe ($T_{1/2}[^{56}$Ni$] = 6.1$ days; $T_{1/2}[^{56}$Co$] = 77.3$ days).

P7. The peak visual luminosity in a typical type Ia supernova reaches $10^{10}\,L_\odot$.
    a) Determine the absolute visual magnitude at maximum.
    b) Assume an exponential decline from peak, of the form $L = L_{\text{max}}\exp[-t/\tau]$, where $\tau \sim 70$ days is the rate of decline, as determined empirically. Calculate the energy emitted by the supernova during the decline phase (≤ 100 days).
    c) Compare the result obtained in b) with an estimate of the total energy released in Problem 6 in the first ~ 100 days.
    d) Estimate the amount of ^{254}Cf required to power the light curve during decline, as originally suggested by Hoyle and Fowler [845].

6

X-Ray Bursts and Superbursts

"The Enterprise, accompanied by Daystrom Institute research astrophysicist Doctor Paul Stubbs, is observing a red giant/neutron star binary system. Sensor analysis indicates that the neutron star formed some ninety thousand years ago in the aftermath of a supernova explosion. [...] The companion star has been sporadically losing mass to the compact but more massive neutron star ever since. Over the course of some twenty years of research, Doctor Stubbs has developed a highly sophisticated computer model of the matter flow in such systems. His model attempts to predict with unprecedented accuracy the moment at which a mass of stellar material sufficient to trigger a nova-class explosion will accumulate on a neutron star. No probe or starship has ever directly observed such event. We successfully observed the accretion of a critical mass of matter from the red supergiant star onto the neutron star on Stardate 43128.6, resulting in the initiation of a runaway nuclear fussion effect and subsequent nova-scale explosion."

Andre Bormanis, *Star Trek — Science Logs* (1998)

The first, primitive X-ray instruments, launched in the 1940s and 1950s on board balloons and rockets, revealed that the Sun emits a modest amount of X-rays. The surface temperature of the Sun is relatively cold (i.e., about 5780 K), and therefore it emits most of the radiation at larger wavelengths. In this framework, the discovery of extrasolar sources, extremely prominent in X-rays, came as a real surprise. Scorpius X-1, later identified as an accreting neutron star [1630], was the first of these sources. It was discovered in 1962 by a team of researchers led by Riccardo Giacconi [640]. Its X-ray power output, 2.3×10^{38} erg s^{-1}, represented about 60,000 times the overall luminosity of the Sun integrated for all wavelengths [230]. Such astonishing figures motivated the launch of different space probes equipped with X-ray detectors, such as Solrad 1, Kosmos 215, and several Vela, OSO, and OGO satellites. But it was not until the launch of NASA's satellite Uhuru—not Uhura!—in 1970, the first satellite specifically suited for X-ray astronomy, when astronomers opened a new window to observe the violent side of the universe with X-ray eyes.

Since then, a large variety of astrophysical sources have been identified in X-rays, from stars (e.g., accreting white dwarfs—classical novae—, neutron stars—X-ray bursts, pulsars—, and black holes—X-ray novae—; plasma ejection episodes from the Sun and other stars; early-type stars; supernova explosions and supernova remnants; gamma-ray bursts) to galaxies (quasars and other active galactic nuclei), or clusters of galaxies [1608]. Among these, X-ray bursters have deserved particular attention, becoming unique natural laboratories that serve as probes of the neutron star structure and of the properties of matter under extreme conditions. Indeed, X-ray burst analyses provide valuable information on neutron stars (e.g., mass-radius relation, equation of state, thermal state), and on different burning regimes and flame spreading (which are also relevant for other thermonuclear-driven explosions, such as classical novae or type Ia supernovae).

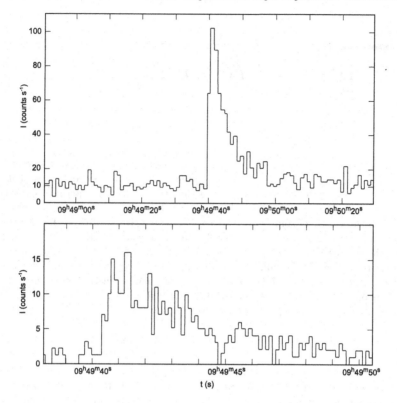

FIGURE 6.1

Intensity histograms of one of the first X-ray burst ever discovered, the one identified in the source 3U 1820-30. It was registered with the hard X-ray detector on board the ANS satellite, for 1 s integration intervals (above), and with the soft X-ray detector for 0.125 s integration intervals (below). Figures adapted from Grindlay et al. [686]; reproduced with permission.

6.1   Discovery of X-Ray Bursts

X-ray bursts were serendipitously discovered in the 1970s, independently by Babushkina et al. [106], Grindlay et al. [686], and Belian, Conner, and Evans [150]. Such new cosmic X-ray phenomena[1] were characterized by brief, bursting episodes, typically lasting from 10 to 100 s (see Section 6.2, for details). In contrast, standard transient events are characterized by much longer periods of activity, ranging from weeks to months.

The events reported by Grindlay et al. [686] (Figure 6.1) were detected by the Astronomical Netherlands Satellite (ANS) from the X-ray source[2] 3U 1820-30, located in a globular cluster. Similar bursting episodes were reported by Belian et al. from observations per-

---

[1] In retrospect, the first observed X-ray burst was probably the one detected from Cen X-4 in July 1969 by the Vela-5B satellite [149]. The event, however, was not recognized as a new type of source until 1976. It was a bright burst, with an energy released of 5×10^{39} erg, an unusually long duration (~ 10 min), and a rise time to peak luminosity of about 60 s [1041].

[2] Most X-ray sources are named after the satellites that discovered them (e.g., 3U stands for the 3rd Uhuru catalogue). The accompanying numbers indicate their celestial coordinates in right ascension (e.g., 1820 stands for 18 hr 20 min) and declination (30 deg). However, they can also be named after the constellation

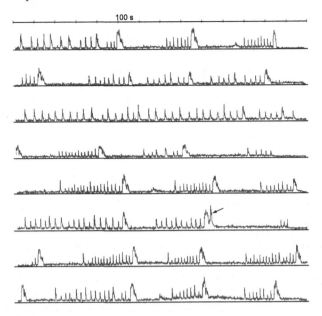

FIGURE 6.2

Series of type II bursts from the Rapid Burster, based on SAS-3 observations performed in March 1976. The burst pinpointed with an arrow is actually a type I burst. Image from Lewin [1084]; reproduced with permission.

formed with two military Vela-5 satellites on sources in the Norma constellation, covering the 15-month period from May 1969 to August 1970. Two additional bursts, detected in June 1971 by Kosmos 428, were also reported by Babushkina et al. [106].

As of August 1976, 15 bursting sources were already identified along the Galactic equator, frequently associated with globular clusters [1084]. One of them, the enigmatic *Rapid Burster* (MXB 1738-335) [1085,1087], characterized by an unprecedented succession of quick bursts, with recurrence times as short as ~10 s, was clearly at odds with the general pattern exhibited by all other bursting sources[3]. A major breakthrough in the understanding of the nature of these events was the discovery of two different types of bursts in the Rapid Burster (Figure 6.2): A classification in type I and type II bursts was immediately established [818], the former associated with thermonuclear flashes, the later linked to accretion instabilities (see Section 6.3). The sequence of type I and type II bursts observed in the Rapid Burster varies along a single "outburst", or period of activity characterized by a fast increase in X-ray luminosity every 100–200 days, followed by a smoother decay over two to four weeks. Two distinct phases, dominated by type I bursts and a strong persistent emission component (Phase I), or by type II bursts (Phase II), have been identified [692]. The type II class, univocally linked to the Rapid Burster during many years, includes also a second member: the bursting pulsar GRO J1744-28 [530,1005]. As of January 2014, 104

---

where they are located and the order of discovery. As a result, a source may have more than one name. For instance, 3U 1820-30 is also known as Sgr X-4.

[3]It seemed that less—rather than more!—observations were needed to unveil the real nature of these sources. Such ironic comment was made in a review on X-ray bursts submitted for publication by the author of this book. Unfortunately, it was not fully appreciated by one of the referees, who considered it "a nonsensical statement. No serious scientist would indicate or prefer such a thing". The comment was finally removed from the text.

Galactic type I X-ray bursting sources have been identified[4], representing the most frequent type of thermonuclear stellar explosion in the Galaxy (the third, in terms of total energy output after supernovae and classical novae[5]).

While this chapter mostly focuses on the properties of type I X-ray bursts, longer duration bursts (e.g., intermediate-duration bursts and superbursts) will be also addressed. The interested reader will find complementary information on the physics of X-ray bursts in a number of review papers (see, e.g., references [182, 1089, 1091, 1092, 1387, 1460, 1579, 1731]).

6.2   Observational Constraints

Type I X-ray bursting sources show a strong concentration around the Galactic center[6] [598], with a spatial distribution matching that of *low-mass X-ray binaries*, that is, stellar binary systems that contain either a neutron star or a black hole. In such systems, the secondary is less massive than the compact star, and frequently corresponds to a main sequence star, a red giant, or even a white dwarf star. A noticeable fraction of such bursting sources is actually found in globular clusters, which strongly suggests the implication of old population stars in the events [1091].

6.2.1   Orbital Periods and Masses

Typically, type I bursting sources have short orbital periods[7], in the range 0.2–15 hr. In those systems, mass-transfer episodes are likely driven by Roche-lobe overflow of the secondary star, resulting in the formation of an accretion disk that surrounds the neutron star. Mass-accretion rates are typically inferred from the persistent X-ray flux between bursts, F_{per}:

$$\dot{M}(M_\odot \mathrm{yr}^{-1}) = 1.33 \times 10^{-11} \left(\frac{F_{per} \, C_{bol}}{10^{-9} \, \mathrm{erg \, cm}^{-2} \, \mathrm{s}^{-1}} \right) \left(\frac{D}{10 \, \mathrm{kpc}} \right)^2 \left(\frac{M_{ns}}{1.4 \, M_\odot} \right)^{-1}$$

$$\left(\frac{1+z}{1.31} \right) \left(\frac{R_{ns}}{10 \, \mathrm{km}} \right), \tag{6.1}$$

where C_{bol} is the bolometric correction that applied to the detected X-ray flux, F_{per}, yields the bolometric flux[8]. The maximum mass-accretion rate[9] is set by the Eddington limit ($\dot{M}_{Edd} \sim 2 \times 10^{-8} \, M_\odot \, \mathrm{yr}^{-1}$, for H-rich accretion onto a 1.4 $M_\odot$ neutron star).

The masses of the neutron stars[10] inferred from X-ray bursting systems are actually quite

---

[4]See http://www.sron.nl/~jeanz/bursterlist.html for an updated list of known Galactic type I X-ray bursting systems.

[5]Gamma-ray bursts, arguably the brightest electromagnetic events occuring in the universe, have only been observed in distant galaxies (see Chapter 7).

[6]The first extragalactic type I X-ray bursts have been reported from two globular cluster source candidates of the Andromeda galaxy (M31) [1418]. Plans to extend the observational search to other members of the Local Group are underway.

[7]Exceptions include GX 13+1 (P_{orb} = 592.8 hr), Cir X-1 (398.4 hr), and Cyg X-2 (236.2 hr).

[8]See, e.g., Galloway et al. [598] for the determination of other observables.

[9]A number of observations have found evidence for mass-accretion rate variations during type I X-ray bursts (see, e.g., reference [1992]).

[10]The upper limit to the mass of a neutron star is given by the Tolman–Oppenheimer–Volkoff limit, a concept analogous to the Chandrasekhar limit for white dwarfs (i.e., maximum mass of a compact star that can be supported by the pressure of degenerate electrons—in the case of a white dwarf—or neutrons—in a neutron star). The Tolman–Oppenheimer–Volkoff limit, first estimated by J. Robert Oppenheimer and George Volkoff in 1939 [1350], depends on the choice of an equation of state for the neutron star interior,

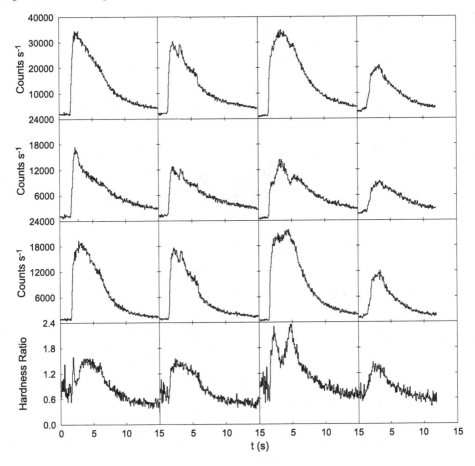

FIGURE 6.3
A suite of X-ray burst light curves from the source 4U 1728-34 as observed with the RXTE satellite. Each sequence (top to bottom), shows the overall count rate in the energy bands 2–60 keV, 2–6 keV, and 6–30 keV, together with the hardness ratio (6–30 keV)/(2–6 keV). Figure adapted from Strohmayer and Bildsten [1731]; reproduced with permission.

uncertain[11]. Therefore, the maximum value of the neutron star mass has only been determined within a range, $M_{\mathrm{TOV}} \sim 1.4$–$3$ $M_\odot$ [1055, 1056, 1693]. Recent estimates of neutron stars masses in two pulsars have reported values of 1.97 ± 0.04 $M_\odot$ (PSR J1614-2230 [431]) and 2.01 ± 0.04 $M_\odot$ (PSR J1903+0432 [52]), thus confirming the existence of massive neutron stars through independent techniques.

which is not well constrained. A suite of different equations of state for the neutron star interior have been proposed [1055, 1056]. Integration of the stellar structure equations (see Chapter 1) with a specific choice of the equation of state yields a value of the neutron star radius for a given mass. Therefore, the different equations of state proposed can be tested by means of observational constraints on masses and radii. Bursts characterized by a significant envelope expansion are actually used to infer neutron star masses from the bolometric luminosity, assuming that the maximum observed luminosity corresponds to the Eddington limit. Moreover, the neutron star radius can be inferred from spectral fits during the decline from peak luminosity, assuming that the emitting area corresponds to the entire neutron star surface [403, 698, 699, 1718]. Note, however, that relativistic effects have to be taken into account for a proper determination of neutron star radii, and, hence, local versus observer's frame values have to be distinguished.

[11] See references [1359, 1719].

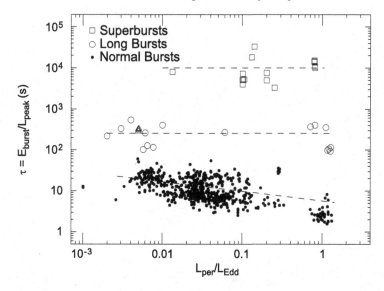

FIGURE 6.4
Bursts effective duration as a function of persistent luminosity, for normal bursts (points), intermediate-duration bursts (open circles, triangles), and superbursts (open squares). Figure adapted from Falanga et al. [512]; reproduced with permission.

6.2.2 Light Curves and Spectra

X-ray burst light curves are characterized by a fast rise to peak luminosity followed by a power law-like decay[12] (Figure 6.3). Some, like those reported from the so-called "clocked burster" GS 1826-24 [597], are remarkably regular; others vary from burst to burst, sometimes with single, double[13], or even triple peaks [2021] whose origin is not well known (see Section 6.3.2). Deviations from the general pattern are due to a suite of different phenomena, including disturbances driven by the accretion flow, nonspherical emission, or the extent of the rp-process.

Light curves provided one of the first pieces of evidence of the thermonuclear origin of type I bursts. Indeed, the ratio between time-integrated persistent and burst fluxes, α, typically in the range ~ 40–100, was soon associated with the ratio between the gravitational potential energy released by matter falling onto a neutron star during the accretion stage ($G\ M_{ns}/R_{ns} \sim 200$ MeV per nucleon), and the nuclear energy generated in the burst (about 5 MeV per nucleon, for a solar mixture burned all the way up to the Fe-group nuclei).

The largest subset of type I X-ray bursts, or *normal* bursts, is characterized by fast rise times between 1 and 10 s, burst durations ranging from ~ 10 to 100 s, a total energy output of about 10^{39} erg, and recurrence periods from hours to days[14] [598,967,1040]. More recently, longer duration bursts have also been identified[15] (see Figure 6.4) [598,967,1040].

---

[12]See, e.g., Galloway et al. [598] for an in-depth analysis of observational properties of 1187 X-ray burst light curves.

[13]See Fisker, Thielemann, and Wiescher [538], for the possible link between nuclear waiting point impedance in the thermonuclear reaction flow (at, e.g., ^{22}Mg, ^{26}Si, ^{30}S, or ^{34}Ar) and the origin of double peaks in the bolometric light curves of type I bursts.

[14]See Linares et al. [1116] for the shortest recurrence times reported from *normal* thermonuclear bursts, down to just a few minutes in Terzan 5 X-2.

[15]Also some thermonuclear bursts have been reported with extremely short recurrence times, in the range ~ 4–10 min, whose ignition may be driven by rotational mixing [964].

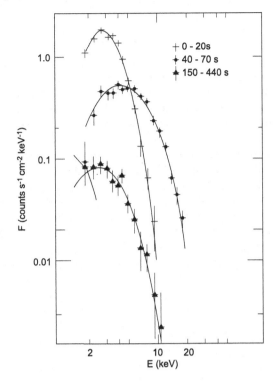

FIGURE 6.5

Spectral evolution of an X-ray burst, showing the initial burst (10–20 s), the hardest X-ray radiation interval (40–70 s), and the tail of the decline (150–440 s). Solid curves correspond to blackbody fits. Figure adapted from Swank et al. [1752]; reproduced with permission.

Intermediate-duration bursts can last for about 15 to 40 min and are characterized by a total energy output of about 10^{40}–10^{41} erg, and recurrence periods of tens of days [512, 513, 596, 886, 1117, 1118]. Superbursts, in turn, have typical durations of about 1 day, a total energy output of $\sim 10^{42}$ erg, and recurrence periods of about 1 yr [380, 889, 1732]. Such differences have been interpreted as due to different fuels and ignition depths (see Sections 6.3.2 and 6.4).

In general, the observed burst spectra can be approximated by a Planck's law (i.e., a blackbody[16]; see Figure 6.5). The blackbody fit provides estimates of the bolometric flux, F_{bol}, and the blackbody temperature, T_{bb}. Moreover, the neutron star radius, R_{ns}, can be inferred by assuming isotropic emission from a spherical surface. In the observer's frame at infinity, this corresponds to

$$F_{bol} = \frac{\sigma T_{bb}^4 R_{ns}^2}{f_c^4 D^2}(1+z)^2 \rightarrow R_{ns} = \left(\frac{F_{bol} f_c^4 D^2}{\sigma T_{bb}^4}(1+z)^{-2}\right)^{1/2}, \qquad (6.2)$$

where f_c is the color correction factor, D the distance to the source, z the gravitational redshift, and σ the Stefan–Boltzmann constant. The gravitational redshift z can be determined

---

[16]Deviations from a blackbody fit are, for instance, discussed in references [223, 887, 1143]. They are partially driven by electron scattering in the hot neutron star atmosphere, which suppresses part of the emission [1849].

from the expression

$$1 + z = \left(1 - \frac{2GM_{\mathrm{ns}}}{R_{\mathrm{ns}}c^2}\right)^{-1/2}, \tag{6.3}$$

which, for a typical neutron star of mass $M_{\mathrm{ns}} = 1.4\,\mathrm{M}_\odot$ and radius $R_{\mathrm{ns}} = 10$ km, yields a value of $1 + z = 1.31$. The color correction factor, $f_{\mathrm{c}} = T_{\mathrm{bb}}/T_{\mathrm{eff}}$, relates the effective temperature, T_{eff}, to the blackbody temperature inferred from the spectral fit and can be obtained from detailed calculations using atmosphere models [1175, 1186, 1744] (see discussion in reference [1387]).

Oscillations with a frequency close to the neutron star spin frequency, ranging between 11 and 600 Hz, have also been identified[17] in the X-ray light curves of $\sim 25\%$ of all bursting sources [1270, 1731, 1734]. Oscillations in the early stages of an X-ray burst, during the rising phase to peak luminosity, have been interpreted as due to the spreading of a hot spot on a rotating neutron star [1733]. Those observed during decline exhibit a drift in frequency, increasing by a few Hz and approaching an asymptotic value, characteristic for each bursting source, as the burst progresses [1271, 1272]. The origin of such late oscillations is not well settled, even though the possibility of a confined radiating region has already been suggested[18].

While X-ray light curves have provided a wealth of information on the physics of X-ray bursts, less has been achieved in other wavelengths. The first simultaneous detection of a burst in the optical and in X-rays was reported from the source 1735-444 in 1978 [688]. The optical fluence, or time-integrated flux, was actually $\sim 2 \times 10^{-5}$ times that in the X-ray band. This is simply too large to be explained by the low-energy tail of the blackbody emission [1091]. Moreover, the optical burst was delayed by ~ 3 s with respect to the X-ray burst [1219]. Similar time delays have also been reported from Ser X-1 [709], and, later, from other sources [1091], suggesting that the optical emission results from reprocessing of X-rays in material within a few light-seconds from the source—either at the accretion disk that surrounds the neutron star or at the hemisphere of the secondary star directly illuminated by the source. In this scenario, the delay in the optical emission would result from travel-time differences between the X-rays directly leading to the observer and the fraction that first hits the disk—or the stellar companion—loses energy down to optical wavelengths, and finally reaches the observer. Infrared and radio emission have also been predicted to accompany type I X-ray bursts, but early claims [743, 1031] have not been unambiguously confirmed [1086]. A panchromatic, multiwavelength study of type I X-ray bursts is definitely needed to shed new light into these questions.

6.3   A Spark to a Flame: Outlining the Explosion Mechanism

6.3.1   Clues on the Nature of X-Ray Bursts

Shortly after the discovery of the neutron by James Chadwick [307], Walter Baade and Fritz Zwicky proposed[19] in 1934 that during a supernova explosion, the nucleons present

---

[17] Also, 3–9 mHz quasi-periodic oscillations (mHz QPOs) have been observed prior to a burst, providing a promising tool to predict the occurrence of a burst. Such mHz oscillations may result from marginally stable nuclear burning on the neutron star surface [22, 1116].

[18] An interesting alternative involves global surface modes that may be excited by the advance of the deflagration front across the neutron star surface [160, 394, 792, 793, 1061, 1428].

[19] According to D. G. Yakovlev and collaborators [1999], Lev Landau somehow outlined the concept of neutron stars before the discovery of the neutron by Chadwick, in a paper finished in February 1931, in which he calculated the maximum mass of white dwarfs.

in the stellar plasma are transformed into neutrons, forcing the star to adopt a closely packed configuration coined as a neutron star [104]. The interest in neutron stars escalated after the discovery of the first pulsars [652, 791] and of a number of X-ray sources [639, 1593, 1777], interpreted as spinning neutron stars that were accreting mass from a nearby stellar companion. Estimates of the amount of nuclear energy released from accretion and fusion of H-rich material deposited on a neutron star were first made by Rosenbluth et al. [1528]. The scenario was revisited by Van Horn and Hansen [726, 1847], who pointed out that nuclear burning on the surface of neutron stars may actually be unstable. The connection between thermonuclear runaways driven by unstable nuclear burning and the nature of X-ray bursts was first suggested by Woosley and Taam [1983] (in the framework of He- or C-burning driven bursts), and independently by Maraschi and Cavaliere [1194] (for H-burning bursts)[20].

All these early scenarios assumed that large amounts of gravitational energy were released as X-rays by the matter infalling into a compact star, but the precise nature of the object hosting the burst was not immediately associated with a neutron star. Early explanations, for instance, pointed instead at the presence of an underlying giant, supermassive black hole (> 100 M$_\odot$ [685])[21]. The "smoking gun" was provided by a series of observations of bursting sources in globular clusters, from which reasonably accurate distance estimates can be obtained. In particular, the spectral evolution of one of the longest bursts observed from the source 4U 1724-30 by the OSO-8 satellite [1752], best fitted with a blackbody spectrum with kT $\sim$ 0.87–2.3 keV, suggested a much smaller object (i.e., a neutron star or a stellar black hole). Indeed, a blackbody radius of $\sim$ 10 km was inferred, assuming a distance of $\sim$ 10 kpc to the source [816, 817]. A number of additional observational features accompanying X-ray bursts, such as a characteristic spectral softening during the decline from peak luminosity, or the masses inferred, strongly supported the presence of underlying neutron stars[22] [1850].

6.3.2   X-Ray Fireworks: Modeling the Bursts

The feasibility of accreting neutron stars as the likely site where type I X-ray bursts occur was originally explored in a series of semianalytical calculations by Joss [929], and Lamb and Lamb [1045], on the basis of the seminal models developed by Hansen and Van Horn [726]. Helium and carbon were identified as the most likely nuclear fuels powering such cosmic explosions[23]. Peak luminosities about $L_{peak} \sim 10^{37}$ erg s^{-1}, light curve rise times of $\sim$ 0.1 s, burst durations ≥ 10 s, an overall energy release of 10^{39} erg per burst, and ratios of persistent over burst luminosities about $\alpha \geq 100$ were estimated in fairly good agreement with the values inferred in a number of bursting sources.

Such pioneering efforts paved the road for the first detailed numerical simulations of X-ray bursts performed by Joss, for a suite of different mass-accretion rates, and neutron star central temperatures [930] (Figure 6.6). The study confirmed that unstable He-burning can account for the main observational features of type I X-ray bursts, including the charac-

---

[20]See also Lewin and Joss [1088].

[21]See also Lamb and Lamb [1044] for a thorough comparison between models involving compact stars and massive black holes.

[22]Unfortunately, the erratic behavior exhibited by the Rapid Burster, characterized by a quick succession of bursts, made it hard to accomodate all observed bursting sources under the umbrella of accreting neutron stars. A number of alternative scenarios were explored in-depth, including instabilities in the interaction between the infalling matter and the magnetosphere of the accreting neutron star [56, 105, 743, 763, 837, 933, 1047, 1751], flare-like eruptions driven by the release of magnetic energy in the accretion disk [1924], convective-driven instabilities in the disk [1106, 1107], or thermal instabilities driven by Compton heating of the accretion flow [687].

[23]See also Taam and Picklum [1764].

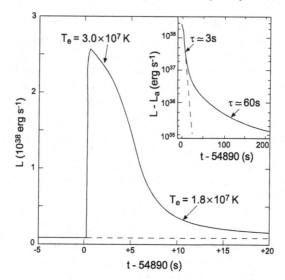

FIGURE 6.6

An early hydrodynamic simulation of a type I X-ray burst, for a model of a 1.41 $M_\odot$ neutron star (with an adopted size of R_{ns} = 6.57 km), accreting He-rich matter at a rate of 3×10^{16} g s^{-1}. Figure adapted from Joss [930]; reproduced with permission.

teristic light curves (i.e., peak luminosities, rise and decay times, the presence of low-energy tails), total energies, spectral features, and recurrence times between bursts. Moreover, Joss's work also revealed that the available nuclear fuel gets virtually consumed during the explosion (likely being transformed into Fe-peak nuclei), while the overall energy released during the explosion is preferentially emitted in the form of X-rays.

The modeling of thermonuclear shell-flashes on accreting neutron stars also experienced a *burst* during the 1980s, with a suite of complementary studies that combined semianalytical approaches[24] [135, 263, 401, 501, 573, 1364] and hydrostatic/hydrodynamic simulations in spherical symmetry [96, 931, 1365, 1760, 1761, 1763, 1765, 1881, 1985].

The effect of the most influential paramaters in the properties of X-ray bursts was analyzed in-depth by Ayasli and Joss [96], through a series of models with different neutron star masses and radii, central temperatures and magnetic fields[25], mass-accretion rates, and metalliciticies of the accreted material. The study pioneered the inclusion of general relativistic corrections to the equations of stellar structure and evolution, and revealed a number of dependencies. For instance, an increase in the mass-accretion rate translates into bursts of shorter duration (with a stable burning regime obtained for high mass-accretion rates). Moreover, a reduction of the overall metallicity of the accreted material delays the ocurrence of the bursts, increasing the amount of mass piled up on top of the star, and, in turn, the strength of the explosion.

In retrospect, a major drawback[26] shared by all the numerical simulations performed in the 1980s was the use of reduced nuclear reaction networks to limit, at the time, the

---

[24] See also Cooper and Narayan [374, 376] for recent two-zone models of type I X-ray bursts, including the first simulation of pure He bursts triggered by accumulation of the ashes from a series of weak hydrogen bursts (see Figure 6.7).

[25] A number of numerical simulations (e.g., reference [931]) suggested that in highly magnetized neutron stars (B $\geq 10^{12}$ G), matter transferred from the stellar companion would be funnelled onto the neutron star magnetic poles, enhancing the local mass-accretion rate in those regions by a factor of 1000.

[26] Ayasli and Joss assumed a radius of 6.57 km in most of the models computed for 1.4 $M_\odot$ neutron stars.

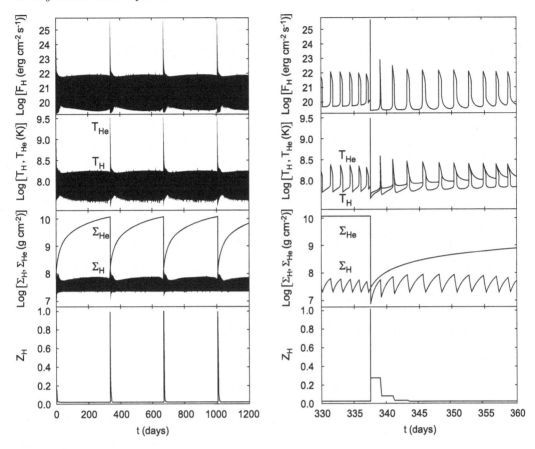

FIGURE 6.7

(Left panels) Time evolution of a sequence of type I X-ray bursts computed by means of a two-zone model. The different panels display relevant physical quantities: F_H is the overall flux emitted by the neutron star; Σ_H (Σ_{He}) corresponds to the depth at which H (He) is depleted through nuclear processes; T_H (T_{He}) and Z_H denote the temperature and CNO mass fraction at Σ_H (Σ_{He}). (Right panels) Time evolution of the system just before and after a pure helium flash, following a large sequence of hydrogen bursts. Figure adapted from Cooper and Narayan [376]; reproduced with permission.

overwhelming computational load. Moreover, most of the results obtained were exclusively based on the analysis of a single burst because of computational constraints. In this regard, a major step forward was achieved with the modeling of full series of consecutive bursts, which provided a better insight into the long-term evolution of these systems (particularly, since the properties of the first burst may be affected by the initial conditions). The simulation of a sequence of repeated X-ray bursts is relatively easy to handle in a Lagrangian framework. Indeed, models suggest that no mass is directly ejected by the explosion (i.e., no numerical shell achieves escape velocity and therefore needs to be removed from the computational domain). Therefore, and in sharp contrast with other astrophysical scenarios (e.g., classical novae), freshly accreted material continuously piles up on top of previously accreted layers.

This value falls short compared with current estimates that favor a radius in the range between 10 and 15 km.

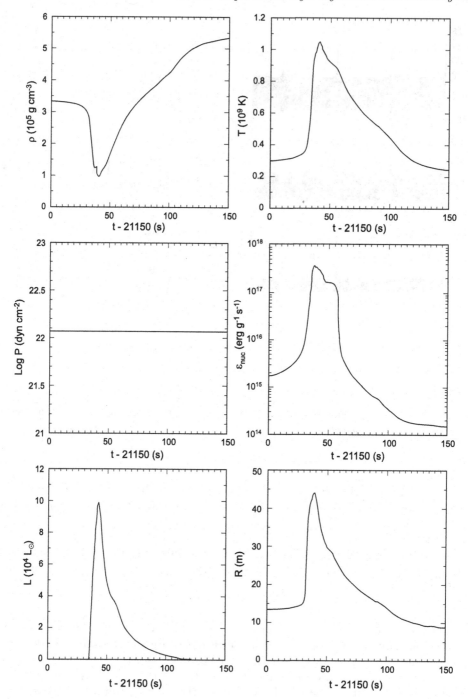

FIGURE 6.8

Time evolution of the density (upper left panel), temperature (upper right), pressure (middle left), and nuclear energy generation rate (middle right), at the innermost envelope shell, in 1D hydrodynamic simulations of X-ray bursts. Lower panels display the overall neutron star luminosity (lower left) and the envelope size as measured from the core-envelope interface (lower right). The model assumed a 1.4 $M_\odot$ neutron star (with $R_{ns} = 13.1$ km), accreting solar composition material ($Z = 0.02$) at a rate of 2×10^{-9} $M_\odot$ yr^{-1}. Figure adapted from José, Moreno, Parikh, and Iliadis [927]; reproduced with permission.

Early studies of repeated bursts also revealed the existence of both thermal [1760] and compositional inertia [1985]. While thermal inertia reflects the role played by the energy released during a burst—and the subsequent heating of the surface layers—on the critical mass required to power the next burst, compositional inertia reveals that burst properties are sensitive to the chemical abundance pattern of the ashes of previous bursts onto which accretion and explosion will occur in the next bursting episode. In particular, compositional inertia reduces the influence of metallicity on burst properties.

At the very early stages of accretion (when the temperature of the envelope is moderately low), the envelope piled up on a neutron star is mildly degenerate. As in classical nova outbursts, a small increase in temperature is actually enough to lift degeneracy during X-ray bursts. A simple estimate can be obtained by means of Equation 4.4 (see Section 4.3 for details): For a chemical mixture characterized by $Z/A \sim 0.5$, and even for a density at the envelope base of about 10^5 g cm^{-3} (close to $\rho_{\max}$), degeneracy is lifted (i.e., the thermal energy of the electrons becomes comparable to the Fermi energy) at a temperature of $T \geq 1.8 \times 10^8$, which corresponds to $\sim 0.1 T_{\mathrm{peak}}$ in a typical type I X-ray burst. Therefore, during the early stages of the burst, the envelope is mildly degenerate at most (or not degenerate at all).

With a neutron star hosting the explosion, the nuclear energy released is not enough to power significant mass ejection, in sharp contrast to a nova outburst. This can be easily understood from the characteristic escape velocities from the surface of these compact objects. While a typical neutron star ($M_{\mathrm{ns}} \sim 1.4$ M$_\odot$, $R_{\mathrm{ns}} \sim 10$ km) requires $v_{\mathrm{esc}} = \sqrt{2GM_{\mathrm{ns}}/R_{\mathrm{ns}}} \sim 190,000$ km s^{-1}, a typical white dwarf ($M_{\mathrm{wd}} \sim 1$ M$_\odot$, $R_{\mathrm{wd}} \sim 5600$ km) is characterized by an escape velocity about 6900 km s^{-1}. The huge escape velocity from a neutron star causes, in general, a limited envelope expansion during bursts. Consequently, X-ray bursts are halted by fuel consumption (due to efficient CNO-breakout reactions; see Section 6.3.3) rather than by expansion. Therefore, X-ray bursts are characterized by a nearly constant pressure at ignition depth (see Figure 6.8).

It is however worth noting that simulations for suitable values of the mass-accretion rate reveal a dramatic photospheric radius expansion in models characterized by large envelope masses. The high pressures and densities achieved at the envelope base lead to strong bursts, characterized by short periods of super-Eddington luminosities, frequently accompanied by the presence of precursors in the X-ray light curve [1365, 1763], together with mass-loss episodes through radiation-driven winds, as inferred from some bursting sources [704, 1090, 1781, 1782]. The origin of these radiation-driven winds is explained by the fact that the radiation flux difusing outward from the burning regions may exceed the local Eddington limit in the outer, cooler layers of the star, and, therefore, hydrostatic equilibrium is broken in those layers. Pioneering models of radiation-driven winds from neutron stars were developed in the 1980s by Kato [956], Ebisuzaki, Hanawa, and Sugimoto [481], and Quinn and Paczyński [1464], assuming Newtonian gravity. General relativistic effects were introduced shortly after by Paczyński and Proszynski [1369], and Turolla, Nobili, and Calvani [1838]. More refined treatments of radiative transfer in quasi-static winds from neutron stars, including the effect of bremsstrahlung and Compton scattering, can be found in references [932, 1314, 1916, 1998], in which mass-loss rates in the range $\dot{M}_{loss} \sim 10^{17} - 10^{20}$ g s^{-1} ($10^{-9} - 10^{-6}$ M$_\odot$ yr^{-1}) were derived.

Different regimes of unstable burning on the surface of accreting neutron stars have also been identified (see Table 6.1), including combined H/He bursts and pure He flashes. The existence of different burning regimes leads to a large spread in possible burst properties (see, e.g., references [573, 579, 1731, 1761]). Particular efforts have been devoted to the study of mass-accretion rates leading to stable nuclear burning (Figure 6.9; see, e.g., refs. [963, 1762, 2015]). To date, simulations predict that the transition between stable and bursting regimes occurs at about 2–10 times higher mass-accretion rates than observed. Attempts

TABLE 6.1

Different Burning Regimes* in Accreting Neutron Stars [967]

$\dot{M}/\dot{M}_{Edd}$	Burning regime
≤ 0.005	Mixed H/He flashes (initiated by H-ignition)
$\sim 0.005\text{--}0.03$	He flashes (with stable H-burning)
$\sim 0.03\text{--}1$	Mixed H/He flashes (initiated by He-ignition)
≥ 1	Stable H/He burning

*Pure hydrogen flashes have been predicted, but not yet observed, for a narrow range of mass-accretion rates around $\sim 0.1\%\text{--}1\%$ of the Eddington value. They would be characterized by modest peak luminosities (about 10^{-3} times those of helium-fueled bursts) and recurrence times of ~ 1 day [376,1405]. See in't Zand et al. [884], for prospects of detectability of these pure H bursts with the planned LOFT mission (Large Observatory for X-Ray Timing). The specific values of the mass-accretion rates at which transition between different burning regimes occurs are expressed in terms of the Eddington value, and depend on the adopted metallicity. The values given in the table have been obtained for models of 1.4 $M_\odot$ neutron stars accreting matter of solar composition (X = 0.7, Z[CNO] = 0.01).

TABLE 6.2

Characteristic Features in Normal and Intermediate-Duration Bursts and Superbursts** [967].

	Normal bursts	Intermediate bursts	Superbursts
Duration	10–100 s	15–40 min	1 day
Energy	10^{39} erg	$10^{40}\text{--}10^{41}$ erg	10^{42} erg
Recurrence period	hr–days	tens of days	1–2 yr
Observed bursts	$\sim 12{,}000$ in 104 sources	20 in 8 sources	22 in 13 sources

**Data on observed number of bursts as of January 2014. Note, however, that classification of a burst as normal or intermediate-duration depends upon the definition of duration, and, hence, the absolute number of observed bursts in each category has to be taken with caution.

to reconcile theoretical and observed values include variations of key nuclear reaction rates (e.g., the triple-alpha reaction, $^{15}O(\alpha, \gamma)^{19}Ne$, and $^{18}Ne(\alpha, p)^{21}Na$ [963]), or the inclusion of a base heating flux in models of accreting neutron stars [2015]. Moreover, marginally stable nuclear burning expected for mass-accretion rates close to transition has been predicted to drive oscillations in the X-ray burst light curve [396, 757]. This has been identified with the mHz quasi-periodic oscillations (mHz QPOs) discovered in neutron stars accreting H-rich matter at rates in the range $0.05\ \dot{M}_{Edd}\text{--}0.5\ \dot{M}_{Edd}$ [22, 1116, 1492]. Transition to stable burning has also been invoked to account for the observed quenching of type I X-ray bursts following a superburst (see Section 6.4) [397, 398, 966, 1042].

Observations have revealed a spread in burst properties, which have been explained in terms of different fuels and ignition depths. This defines a number of burst subtypes[27] (i.e., normal and intermediate-duration bursts, and superbursts), whose main observational properties are summarized in Table 6.2. Models for the most common He-powered bursts

---

[27]Note, however, that Linares et al. [1117] have shown that there is a continuous distribution in energy between normal and intermediate-duration bursts, which suggests that they may not correspond to two distinct classes of events.

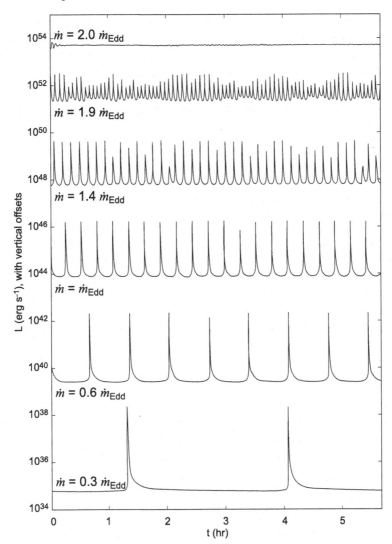

FIGURE 6.9
Light curve profiles in accreting neutron stars, as computed with the hydrodynamic code
MESA, showing the transition between stable nuclear burning and several bursting regimes
for different mass-accretion rates (see Linares et al. [1116] for a comparison with observational light curves). Figure adapted from Zamfir et al. [2015]; reproduced with permission.

suggest that the burst duration is determined by the characteristic cooling timescale of
the burning shell (typically ~ 10 s), which in turn is set by the depth at which burning
occurs. In the presence of hydrogen, ignition occurs at similar depths, but rapid proton
captures (rp-process; see Section 6.3.3) during the decay from peak luminosity can extend
the duration of a burst up to ~ 100 s.

Bursts characterized by longer durations, as in intermediate-duration bursts and superbursts, suggest ignition at larger depths, and, therefore, at higher pressures. While
intermediate-duration bursts have been associated with ignition in thick He layers on cold
neutron stars—resulting either from direct He-accretion in ultra-compact X-ray binaries,

or through accumulation of ashes from a series of hydrogen bursts— $[376, 393, 399, 573, 886, 1405, 1881]$, superbursts are likely driven by carbon burning (see Section 6.4). Accumulation of thicker envelopes, as required to account for the longer duration of these bursts, explains as well the longer recurrence periods observed in such systems.

6.3.3   The X-Ray Philosopher's Stone: Nucleosynthesis in Type I X-Ray Bursts

With a neutron star as the underlying object hosting the explosion, temperatures, and densities in the accreted envelope during a normal type I X-ray burst reach relatively high values (i.e., $T_{peak} \sim 10^9$ K, $\rho_{max} \sim 10^5 - 10^6$ g cm^{-3}). As a result, detailed nucleosynthesis studies require hundreds of isotopes, up to the SnSbTe mass region [1575] or beyond (the nuclear activity in the X-ray burst nucleosynthesis studies of Koike et al. [998] reached ^{126}Xe), and thousands of nuclear interactions. The maximum extent of the rp-process in type I X-ray bursts is still a matter of debate[28].

Because of computational limitations, early studies of type I X-ray burst nucleosynthesis have been performed with limited nuclear reaction networks truncated around Ni (see Woosley and Weaver [1985], and Taam et al. [1766, 1767], all using 19-isotope networks), Kr (Hanawa et al. [723], 274 isotopes; Koike et al. [997], 463 isotopes), Cd (Wallace and Woosley [1880], 16 isotopes), or Y (Wallace and Woosley [1879], 250 isotopes). More recently, Schatz et al. [1575, 1578] have carried out very detailed nucleosynthesis calculations with a network containing more than 600 isotopes (up to Xe, in reference [1575]), but using a one-zone approach (sometimes referred to as a *0-D model*)[29]. Recent attempts to couple hydrodynamic models in 1D and detailed nuclear reaction networks include Fisker et al. [534, 536–538] and Tan et al. [1774] (with networks containing ~ 300 isotopes, up to ^{107}Te), José et al. [927] (1392 nuclear reactions and 325 isotopes, up to ^{107}Te), and Woosley et al. [1976] (up to 1300 isotopes in an adaptive network). The inclusion of such large nuclear reaction networks proved essential to understand and reproduce the evolution of X-ray burst light curves during decay [183, 722, 756], which are powered by a series of proton captures by elements heavier than Fe [536, 927, 997, 1575, 1976].

The main nuclear reaction path in type I X-ray burst has been extensively discussed in the literature[30]. From a nucleosynthesis viewpoint, the most interesting case study corresponds to mixed H/He bursts, because of the complex nuclear reaction interplay that combines the *rp-process* (i.e., a series of rapid proton-captures and β^+-decays [1879]), together with the 3α-reaction[31], and the *αp-process* (a sequence of (α,p) and (p,γ) reactions). As shown in Figures 6.10 and 6.11, the main nuclear flow is expected to proceed far away from the valley of stability, merging with the proton drip-line beyond A = 38 [1578].

The nuclear activity is driven by efficient breakout from the CNO region, first through ^{15}O$(\alpha, \gamma)^{19}$Ne, and once enough ^{14}O is synthesized through the triple-α reaction, followed by ^{12}C$(p, \gamma)^{13}$N$(p, \gamma)^{14}$O, an alternative path through ^{14}O$(\alpha, p)^{17}$F dominates the flow [1976]. Indeed, ^{14}O$(\alpha, p)^{17}$F bypasses the ^{15}O$(\alpha, \gamma)^{19}$Ne path to ^{21}Na (which requires two consecutive proton captures on ^{19}Ne and ^{20}Na, plus a β-decay), through ^{17}F$(p, \gamma)^{18}$Ne$(\alpha,$

---

[28]Recent mass measurements in the Sn–I mass region have revealed the difficulty in reaching the SnSbTe region during X-ray bursts [494].

[29]Koike et al. [998] have also performed detailed one-zone nucleosynthesis calculations with an extended nuclear reaction network that contained 1270 isotopes, from H to ^{198}Bi. The temperature and density profiles used in these simulations were extracted from 1D hydrodynamic models.

[30]See, e.g., references [308, 536, 723, 734, 788, 878, 926, 927, 1576, 1578, 1579, 1851, 1879, 1881, 1916] for studies of nucleosynthesis in pure He and mixed H/He type I bursts.

[31]See references [1205, 1406] for studies of the effect of the proposed new 3α-reaction rate on type I X-ray burst properties.

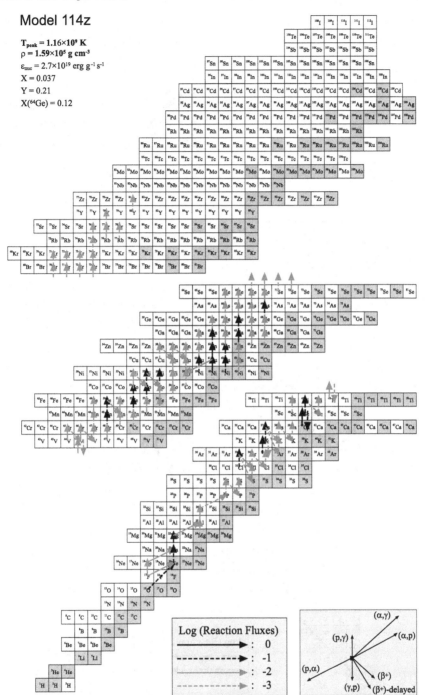

FIGURE 6.10

Main reaction fluxes at the innermost envelope shell for Model 114z, at peak temperature ($T_{base,max} = 1.16 \times 10^9$ K), for the first computed burst. Calculations have been performed with the code SHIVA, and rely on a 1.4 $M_\odot$ neutron star ($L_{ini} = 1.6 \times 10^{34}$ erg s^{-1} = 4.14 $L_\odot$), accreting mass a rate $\dot{M}_{acc} = 1.8 \times 10^{-9}$ $M_\odot$ yr^{-1} (0.08 $\dot{M}_{Edd}$). The composition of the accreted material is assumed to be solar-like (X = 0.7048, Y = 0.2752, Z = 0.02). A color plot and a movie portraying the time evolution of the main reaction fluxes for this model are available at http://fisica.upc.edu/ca/users/jjose/CRC-Downloads.

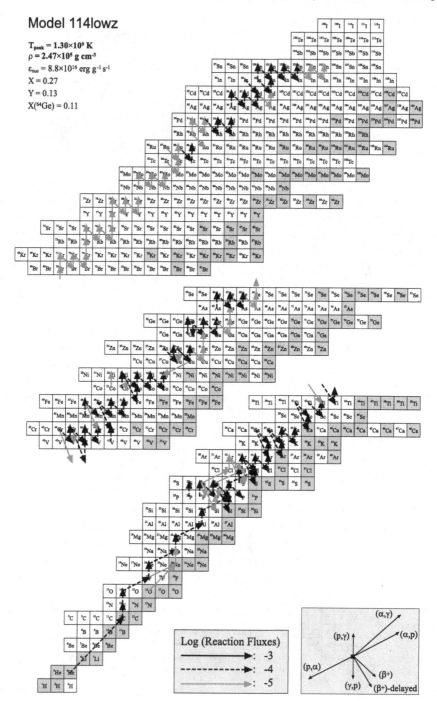

FIGURE 6.11

Same as Figure 6.10, for Model 114lowz, at peak temperature ($T_{base,max} = 1.30 \times 10^9$ K). Conditions are similar to Model 114z, except for the composition of the accreted material, here assumed to be metal-poor (i.e., X = 0.759, Y = 0.240, and Z = 0.001 = $Z_\odot/20$). A color plot and a movie portraying the time evolution of the main reaction fluxes for this model are available at http://fisica.upc.edu/ca/users/jjose/CRC-Downloads.

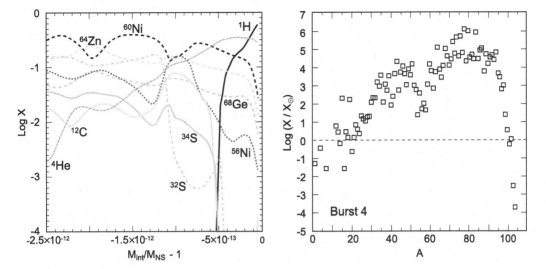

FIGURE 6.12
(Left panel) Mass fractions of the most abundant, stable isotopes (or with half-lifes > 1 hr) after 4 mixed H/He bursts, in Model 114z (Figure 6.10). (Right panel) Same as in *left panel*, for overproduction factors with respect to solar abundances. Color versions of these figures are available at http://fisica.upc.edu/ca/users/jjose/CRC-Downloads. Figures adapted from José et al. [927]; reproduced with permission.

p)^{21}Na, with ^{18}Ne$(\alpha,$p$)^{21}$Na governing the main path toward heavier species. The composition of the envelope at the end of the burst depends on a number of parameters in the simulations (mainly, the mass-accretion rate and the adopted initial metallicity), but tends to favor species in the mass range $A = 60$–70, including stable and radioactive species with half-lifes > 1 hr, such as ^{56,60}Ni, ^{64}Zn, ^{68}Ge, ^{72}Se, as well as ^{32}S and some leftover ^{4}He [927, 1575, 1976]. Models with smaller initial metallicites yield additional species with masses near A$\sim$100 [927, 1575], whose relevance is progressively reduced because of compositional inertia (see Figures 6.12 and 6.13).

Pure helium bursts can also result, either from direct accretion of helium (as suggested for ultra-compact X-ray binaries [967]) or for low mass-accretion regimes (see Table 6.1), where the hot CNO cycle burns all the hydrogen into helium prior to He-ignition[32]. Early semianalytical models with approximate nucleosynthesis were developed by Hashimoto, Hanawa, and Sugimoto[33] [734]. The model included a detailed nuclear reaction network containing 181 species (up to ^{62}Cu) but assumed constant pressure (with log P[dyn cm^{-2}] ranging between 19 and 25). The study revealed that below log $P = 22$, the main nuclear path simply follows an α-chain (up to ^{36}Ar), but for log $P > 22$, secondary channels initiated through $(\alpha,$p$)$ and (p,γ) reactions become progressively important[34]. The most abundant elements produced depend strongly on the specific choice of P (i.e., ^{40}Ca, for

---

[32]Sedimentation of He-rich ashes at large depths, leftover after a large sequence of weak hydrogen bursts, can enhance the effect [1405].

[33]Wallace, Woosley, and Weaver [1881] also explored pure He bursts but in the framework of a simplified 19-isotope network truncated around ^{56}Ni.

[34]Timmes, Hoffman, and Woosley [1805] have performed simulations of hydrostatic and explosive He-burning, using different nuclear reaction networks: 7- and 13-isotope α-chains, plus a 489-isotope network. The study stressed the important role played by $(\alpha,$p$)(p,\gamma)$ and $(\gamma,$p$)(p,\alpha)$ links in determining reasonably accurate energy generation rates and chemical abundances (see also references [734, 1578]).

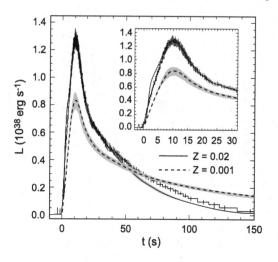
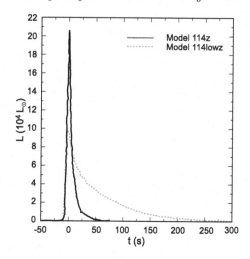

FIGURE 6.13

(Left panel) Light curve comparison between observed and simulated type I X-ray bursts, for the extremely regular bursting source GS 1826-24. Displayed in the figure are the average light curve from the bursts observed during the year 2000, when the recurrence time was 4 hr, as well as average bursts from models computed with two different metallicities (Z = 0.02—solid line—and Z = 0.001—dashed line—). Magnification of the rise and the early part of the decline are shown in the inset. Figure adapted from Heger et al. [756]; reproduced with permission. (Right panel) Light curve comparison between models computed with solar metallicity, Z = 0.02 (solid line—hereafter, model 114z) and Z = 0.001 (dashed line—model 114lowz). Figure adapted from José et al. [927]; reproduced with permission.

log P = 22; ^{56}Ni, for log P = 24). A handful of analytical models of pure He burning, for several mass-accretion rates, have also been compared with models of mixed H/He burning by Weinberg, Bildsten, and Schatz [1916]. In this case, nucleosynthesis was handled by means of a 394-isotope network, ranging from H to Sr. For pure He bursts, the main reaction flow is mainly characterized by a sequence of α-captures into the ^{36}Ar-^{44}Ti region, supplemented by a number of reactions off the α-chain, in the form of (α,p) and (p,γ) sequences. For instance, at T > 10^9 K the ^{12}C(α,γ) reaction is bypassed by the much faster ^{12}C(p,γ)^{13}N(α,p)^{16}O chain. While a pure α-chain network would yield ^{12}C as the main nucleosynthesis product in these pure He burning models, a full nuclear reaction network results in an envelope dominated by the presence of intermediate-mass elements such as ^{40}Ca, ^{44}Ti, and ^{36}Ar, plus some unburned ^{4}He, with negligible amounts of carbon. Woosley and collaborators [1976] have also shown that some models accreting solar composition material at a rate of $\dot{M} \sim 3.5 \times 10^{-10}$ M$_\odot$ yr^{-1} ignite in a hydrogen-free, helium layer, resulting in bursts characterized by briefer and brighter light curves, with shorter tails, very rapid rise times, and ashes dominated by elements lighter than the iron group, such as ^{24}Mg, ^{28}Si and ^{40}Ca. It is, however, worth noting that the resulting nucleosynthesis is deeply affected by vigorous convection that induces mixing between the He layer and an overlying H-rich shell.

The potential impact of X-ray burst nucleosynthesis on the Galactic abundances is still a matter of debate. Despite the large gravitational potential of neutron stars, mass loss through radiation-driven winds has been inferred during photospheric radius expansion. Whether the material ejected contains traces of nuclear-processed material synthesized during the burst is not yet well settled and strongly depends on the efficiency of convective mixing. Even though it has been claimed that X-ray bursts may help to explain, for instance, the Galactic abundances of the elusive light *p-nuclei* [412, 1575, 1576], detailed simulations suggest instead that the concentration of p-nuclei in the outermost layers of a neutron star envelope is orders of magnitude below the required values[35] [927, 1916].

Whether or not X-ray bursts efficiently contribute to the Galactic abundances, reliable estimates of the nucleosynthesis produced during type I X-ray bursts are still important, since the specific abundance pattern critically determines a suite of thermal [1248, 1578], radiative [1364], electrical [254, 1578], and mechanical properties [185, 186] of the neutron star. Moreover, it has also been claimed that the nuclear ashes produced by X-ray bursts (both in the ejecta and in the neutron star surface) may provide observable features, in the form of gravitationally-redshifted atomic absorption lines that could be identified through high-resolution X-ray spectra. This may represent a valuable tool to constrain X-ray burst models [184, 311, 312, 1916]. Preliminary determinations of such gravitationally-redshifted absorption lines were reported by Cottam et al. [381] on the basis of high-resolution spectra of 28 X-ray bursts detected from the source EXO 0748-676 after 335 ks of observations with the XMM-Newton X-ray satellite. The work tentatively identified lines of Fe XXVI (during the early phase of the bursts), Fe XXV, and perhaps O VIII (at later stages). But no evidence for such spectral features was found neither during the analysis of 16 bursts observed from GS 1826-24 [999], nor from another series of bursting episodes detected from the original source after 600 ks of observations[36] [382, 1473].

6.3.4 Nuclear Uncertainties

Most of the reaction rates used for X-ray burst nucleosynthesis studies rely on theoretical estimates obtained from statistical models, and therefore may be affected by significant uncertainties. Efforts to quantify the impact of such nuclear uncertainties have been undertaken by different groups[37]. The most extensive work to date has been performed by Anuj Parikh and collaborators [1385, 1386] by means of a twofold approach. First, the effect of individual reaction-rate variations was quantified in the framework of postprocessing calculations for different temperature and density versus time profiles. An extensive nuclear network containing 606 isotopes (ranging from H to ^{113}Xe), and linked through a suite of 3551 nuclear processes was used to this end, and each reaction of the network was arbitrarily varied, to account for the associated uncertainties. As summarized in Table 6.3, only a handful of reactions, out of the 3551 nuclear processes considered, have an impact [38] on the final yields larger than a factor of 2 (on, at least, 3 isotopes of the network, in any of the models computed), when their nominal rates are varied by a factor of 10, up and down. This includes mostly proton-capture reactions, such as ^{65}As(p, γ)^{66}Se, ^{61}Ga(p, γ)^{62}Ge, ^{96}Ag(p, γ)^{97}Cd, ^{59}Cu(p, γ)^{60}Zn, ^{86}Mo(p, γ)^{87}Tc, ^{92}Ru(p, γ)^{93}Rh, or ^{102,103}In(p, γ)^{103,104}Sn, as

---

[35]Moreover, the new experimental half-life of ^{96}Cd [144] translates into a lower overproduction of the p-nucleus ^{96}Ru. This suggests that X-ray bursts are not the major ^{96}Ru factories in the Galaxy.

[36]Strong absorption edges have been reported from two bursts exhibiting strong photospheric expansion. The spectral features have been attributed to iron-peak elements with abundances about ~ 100 times solar, which may suggest the presence of heavy-element ashes in the ejected wind [890].

[37]A number of sensitivity studies (see, e.g., Figure 6.14) have analyzed the influence of the size of the adopted nuclear reaction network, the use of different reaction-rate and mass compilations, and the impact of varying reaction rates within uncertainty limits. See, e.g., references [45, 252, 375, 400, 414, 534, 537, 538, 881, 997, 998, 1576, 1774, 1793, 1879, 1976].

[38]In fact, the single, most influential nuclear process turned out to be the triple-α reaction. However, when more realistic uncertainty limits were adopted ($\pm 12\%$ [1833]; but see references [1205, 1406]), no effect, neither on the final yields nor on the overall energy output, was found. The same applies to individual variations of the β-decay rates included in the network.

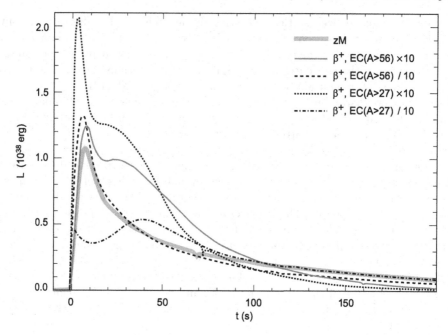

FIGURE 6.14

Example of sensitivity of a type I X-ray burst light curve to different choices of weak interaction rates: Nominal, 10 times nominal for $A > 56$, 10 times below nominal for $A > 56$, and similar modifications above $A = 27$. Figure adapted from Woosley et al. [1976]; reproduced with permission.

well as a few α-capture reactions[39] like $^{12}C(\alpha, \gamma)^{16}O$, $^{30}S(\alpha, p)^{33}Cl$, or $^{56}Ni(\alpha, p)^{59}Cu$. Variations of these reaction rates mostly affect isotopes with masses and atomic numbers similar to those of the projectile + target system.

The study also identified 17 reactions affecting energy production[40] by more than 5% (as well as the yield of at least one isotope; see also Table 6.3), when individually varied within a factor of 10, up and down. This has to be taken as a warning of the limitations of postprocessing techniques. Indeed, a self-consistent analysis would require computationally intensive hydrodynamic simulations capable of self-adjusting both the temperature and the density of the stellar envelope.

A second, somewhat complementary approach to the individual-reaction rate variation study relied on the simultaneous variation of all reaction rates through a Monte Carlo approach. This intends to mimic the complex interplay between multiple nuclear processes in the highly coupled environment of an X-ray burst. In this approach, all reaction rates were arbitrarily multiplied by a random factor that follows a Log-Normal distribution, with an expected value of 1, and a probability of 95.5% to range between 0.1 and 10 (the same uncertainty limits adopted in the previous method). In this procedure, the identification of

---

[39]Note, however, that the $^{12}C(\alpha, \gamma)^{16}O$ rate is constrained to better than a factor of ~ 10 for X-ray burst conditions.

[40]The impact of uncertainties in reaction Q-values on the nucleosynthesis accompanying type I X-ray bursts has also been explored by Parikh and collaborators [1385]. Only the reactions $^{26}P(p,\gamma)^{27}S$, $^{45,46}Cr(p,\gamma)^{46,47}Mn$, $^{55}Ni(p,\gamma)^{56}Cu$, $^{60}Zn(p,\gamma)^{61}Ga$, and $^{64}Ge(p,\gamma)^{65}As$ showed an effect on the final yields when their Q-values were varied between 1σ uncertainty bounds. $^{64}Ge(p,\gamma)^{65}As$, whose influence extends between ^{64}Zn and ^{104}Ag, exhibited the largest effect, and has motivated a number of direct mass measurements on ^{65}As and other species (see, e.g., references [1829, 1878]).

TABLE 6.3

Most Uncertain Nuclear Processes in X-Ray
Burst Nucleosynthesis [1386]

Reactions affecting yields*	Reactions affecting energy production**
$^{12}C(\alpha, \gamma)^{16}O$	$^{15}O(\alpha, \gamma)^{19}Ne$
$^{18}Ne(\alpha, p)^{21}Na$	$^{18}Ne(\alpha, p)^{21}Na$
$^{25}Si(\alpha, p)^{28}P$	$^{22}Mg(\alpha, p)^{25}Al$
$^{26g}Al(\alpha, p)^{29}Si$	$^{23}Al(p, \gamma)^{24}Si$
$^{29}S(\alpha, p)^{32}Cl$	$^{24}Mg(\alpha, p)^{27}Al$
$^{30}P(\alpha, p)^{33}S$	$^{26g}Al(p, \gamma)^{27}Si$
$^{30}S(\alpha, p)^{33}Cl$	$^{28}Si(\alpha, p)^{31}P$
$^{31}Cl(p, \gamma)^{32}Ar$	$^{30}S(\alpha, p)^{33}Cl$
$^{32}S(\alpha, \gamma)^{36}Ar$	$^{31}Cl(p, \gamma)^{32}Ar$
$^{56}Ni(\alpha, p)^{59}Cu$	$^{32}S(\alpha, p)^{35}Cl$
$^{57}Cu(p, \gamma)^{58}Zn$	$^{35}Cl(p, \gamma)^{36}Ar$
$^{59}Cu(p, \gamma)^{60}Zn$	$^{56}Ni(\alpha, p)^{59}Cu$
$^{61}Ga(p, \gamma)^{62}Ge$	$^{59}Cu(p, \gamma)^{60}Zn$
$^{65}As(p, \gamma)^{66}Se$	$^{65}As(p, \gamma)^{66}Se$
$^{69}Br(p, \gamma)^{70}Kr$	$^{69}Br(p, \gamma)^{70}Kr$
$^{75}Rb(p, \gamma)^{76}Sr$	$^{71}Br(p, \gamma)^{72}Kr$
$^{82}Zr(p, \gamma)^{83}Nb$	$^{103}Sn(\alpha, p)^{106}Sb$
$^{84}Zr(p, \gamma)^{85}Nb$	
$^{84}Nb(p, \gamma)^{85}Mo$	
$^{85}Mo(p, \gamma)^{86}Tc$	
$^{86}Mo(p, \gamma)^{87}Tc$	
$^{87}Mo(p, \gamma)^{88}Tc$	
$^{92}Ru(p, \gamma)^{93}Rh$	
$^{93}Rh(p, \gamma)^{94}Pd$	
$^{96}Ag(p, \gamma)^{97}Cd$	
$^{102}In(p, \gamma)^{103}Sn$	
$^{103}In(p, \gamma)^{104}Sn$	
$^{103}Sn(\alpha, p)^{106}Sb$	

*Reactions affecting the yields of, at least, 3 isotopes when their nominal rates are varied by a factor of 10.

** Nuclear processes affecting the total energy output by more than 5 %, as well as the yield of at least one isotope, when their nominal rates are individually varied by a factor of 10.

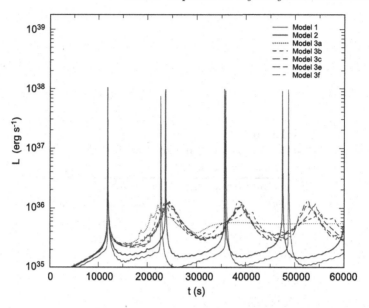

FIGURE 6.15

Example of sensitivity of a type I X-ray burst light curve to different choices of the $^{15}O(\alpha,\gamma)$ reaction rate: (1) upper limit; (2) nominal rate; (3a to 3f) lower limit, with different resolutions [75, 103, 129, 154, and 175 numerical shells, respectively]. A color version of this plot is available at http://fisica.upc.edu/ca/users/jjose/CRC-Downloads. Figure adapted from Fisker et al. [534]; reproduced with permission. See reference [537], for an updated version.

the most influential reactions relies on the determination of correlation coefficients between each isotope and all reaction rates: The larger the correlation coefficient, the more influential the reaction is. Moreover, the impact of a given reaction on a specific isotope is quantified by means of a linear fit between the final yield and the corresponding (random) variation factor, whose slope indicates the strength of this dependence. Like in the individual reaction-rate variation study, only a small subset of the thousands of reactions considered had an impact on the final yields. Indeed, all reactions flagged as important in the Monte Carlo simulations were previously identified in the individual reaction-rate variation study. When not too restrictive conditions were applied to the Monte Carlo studies, a total agreement on the results obtained with both techniques was achieved. While both techniques yielded similar results, the individual reaction-rate variation study turned out to be better suited for flagging reactions that affect the overall energy output or for handling specific reactions whose rates are known with better precision.

6.3.4.1 $^{15}O(\alpha, \gamma)$

$^{15}O(\alpha, \gamma)^{19}Ne$ is one of the important channels triggering the breakout from the hot CNO cycles and the subsequent rp-process in type I X-ray bursts. Early theoretical calculations of the reaction rate suggested that it is dominated by a 500 keV resonance, that corresponds to the $3/2^+$ state at 4.034 MeV in ^{19}Ne [1052], with a negligible contribution of higher energy states. An extensive number of experiments as well as new theoretical estimates have however revealed that contribution from the 4.143 MeV and 4.200 MeV levels could dominate the rate at some temperature ranges [414, 1775].

Currently available ^{15}O beam intensities are not enough to allow direct measurement of the strength of the resonances of interest, which have been analyzed only through indirect approaches (see, e.g., references [416, 1183, 1774, 1864]). Indeed, resonance strengths have been determined from measurements of α-decay branching ratios and mean lifetimes for some levels of interest, but still with substantial uncertainty [414, 1775]. Note that no statistically significant detection has been achieved to date for the most critical 500 keV resonance. Further experimental investigations (in particular, more precise α-decay branching ratios) are needed to reduce the remaining uncertainty in the rate (exceeding a factor of 1000 at some temperatures; see reference [414]). Moreover, further hydrodynamic simulations are also required to disentangle the apparent discrepancy between different studies with regard to the importance of the ^{15}O(α, γ) reaction in models of type I X-ray bursts, in particular whether a low value of the reaction rate leads to stable burning (Figure 6.15) [534, 927].

6.3.4.2 ^{18}Ne$(\alpha,$ p$)$

^{18}Ne$(\alpha,$ p$)^{21}$Na plays also an important role in CNO breakout during type I X-ray bursts. It operates at higher temperatures than ^{15}O$(\alpha, \gamma)^{19}$Ne, because of the higher Coulomb barrier. The reaction rate is dominated by contributions from several resonances in the compound nucleus ^{22}Mg, with energies in the range 8.6 MeV–11.0 MeV. The first theoretical estimate of the rate relied on limited spectroscopic information [672]. Since then, a number of indirect[41] studies (see, e.g., references [314, 1204]) have constrained the energies of the levels of interest in ^{22}Mg. Spins and parities have also been experimentally determined for a number of states [752]. Two new reaction rates have been recently reported [752, 1254] on the basis of different spin-parity assignements. Preliminary estimates of the impact of these different rates, on the basis of postprocessing, one-zone models, yield no differences in the nuclear energy generation rate. This suggests that the reaction rate may be sufficiently constrained for X-ray burst nucleosynthesis calculations [752]. Nevertheless, future measurements aimed at improving the spectroscopic information (i.e., partial widths, spectroscopic factors) are encouraged[42].

6.3.4.3 ^{65}As$($p, $\gamma)$

One of the nuclear processes whose uncertainty has a large impact on X-ray burst nucleosynthesis is ^{65}As$($p, $\gamma)^{66}$Se. The reaction rate has been obtained by means of statistical, Hauser–Feshbach models, on the basis of 3 states in the mirror nucleus ^{66}Ge, with energies of 2.71 MeV, 2.50 MeV, and 2.90 MeV. The rate derived by van Wormer and collaborators [1851] relied on proton widths based on the assumption of spectroscopic factors of C^2S = 0.1 for all levels of interest (γ-widths were taken from the mirror levels). With a half-life of only $T_{1/2}(^{65}$As$) \sim 190$ ms [1950], little spectroscopic information is actually known (see, however, Obertelli et al. [1339], for the first spectroscopic study on ^{65}As and ^{66}Se), and therefore the theoretical rate is poorly constrained.

6.3.5 Multidimensional Simulations of X-Ray Bursts

To date, no self-consistent multidimensional full simulation of an X-ray burst, for realistic conditions, has been performed, neither in 2D nor in 3D. So far, a number of efforts have focused on the analysis of flame propagation on the envelopes accreted onto neutron stars

---

[41]See also reference [1560] for cross-section measurements of the reaction ^{21}Na$($p, $\alpha)^{18}$Ne in inverse kinematics at about Gamow peak energies.

[42]See reference [1253] for a recent evaluation of the ^{18}Ne$(\alpha,$ p$)^{21}$Na rate.

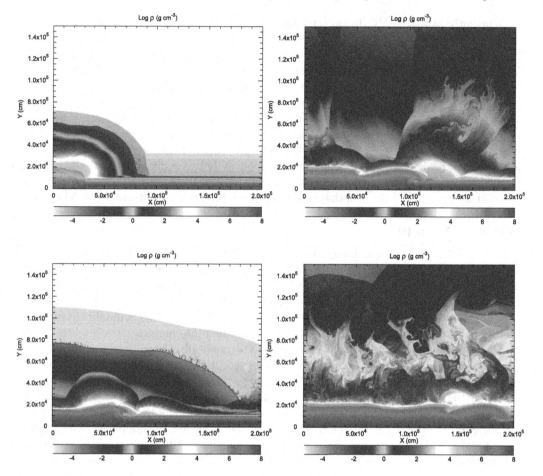

FIGURE 6.16

Snapshots of the propagation of a detonation front in a He-pure envelope on top of a neutron star. Panels depict a density map at times 30 μs (upper left), 60 μs (lower left), 90 μs (upper right), and 150 μs (lower right) since the onset of the detonation. Color versions of these plots are available at http://fisica.upc.edu/ca/users/jjose/CRC-Downloads. Figures adapted from Zingale et al. [2025]; reproduced with permission.

or on convection-in-a-box studies aimed at characterizing convective transport during the stages prior to ignition[43].

Some of the pioneering studies of thermonuclear flame propagation on neutron stars[44], in the framework of X-ray bursts, were performed by Shara [1616]. He found that, while localized runaways on white dwarfs resulted in volcanic-like eruptions rather than in de-

---

[43]Other multidimensional calculations have also focused on turbulent combustion during the transition of a neutron star into a strange quark star. Even though this scenario has been invoked as a possible nucleosynthesis site for r-process elements, recent 3D simulations suggest that mass ejection from the star during such transitions is unlikely [790]. See also references [462, 1372, 1409] for the connection between the transition of neutron stars into strange quark stars and the origin of energetic core-collapse supernovae and γ-ray bursts (see Chapter 7).

[44]See also Nozakura et al. [1335] for one-zone models of the lateral propagation of helium shell flashes in accreting neutron stars.

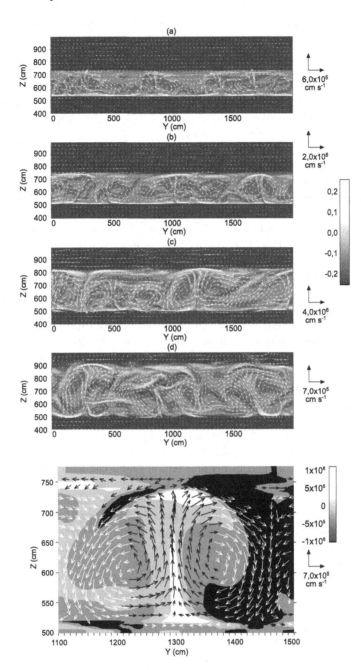

FIGURE 6.17
Flow velocity fields superimposed over contours of adiabatic excess (i.e., difference between the actual temperature gradient and the adiabatic gradient), at four stages of an X-ray burst, corresponding to maximum energy generation rates of log $\varepsilon_{nuc} = 16$ (panel a), 17 (b), 18 (c), and 18.5 (d). (Bottom) Development of a Bénard cell in the flow velocity field, for log $\varepsilon_{nuc} = 17$. Bénard cells are buoyancy-driven structures that naturally occur in plane, horizontal fluid layers heated from below. Figures adapted from Lin et al. [1114]; reproduced with permission.

flagrative spreads (see Section 4.3.1), a localized ignition on a neutron star would likely propagate as a deflagration front, incinerating the whole envelope in a timescale of ~ 100 s.

The scenario was also analyzed by Fryxell and Woosley [569], who concluded that two different propagation regimes were actually possible: If ignition occurs deep inside the envelope, at densities $\rho \sim 10^8$ g cm^{-3}, a detonation front propagating at $v \sim 9000$ km s^{-1} will likely occur. But if the density is $\rho < 10^7$ g cm^{-3} a subsonic front (i.e., a deflagration) will ensue. Estimates of the characteristic speed of the deflagration front yielded a value of $v \sim 5$ km s^{-1}. In such a deflagrative regime, the front would horizontally spread, with a characteristic timescale for a halfway propagation across the envelope of about 8 s.

Simultaneously, Fryxell and Woosley also addressed multidimensional studies of flame propagation on neutron stars [568]. The work included pioneering two-dimensional hydrodynamic simulations of the propagation of a detonation front in a thick envelope on top of a neutron star, during ~ 50 milliseconds. Unfortunately, the conditions leading to the onset of the detonation were imposed in order to qualitatively match the observational properties of gamma-ray bursts, and, therefore, do not represent a realistic X-ray burst simulation.

The next multidimensional study of detonation flames on neutron stars was performed by Zingale et al. two decades later [2025]. A 2D simulation was performed with the FLASH code up to hundreds of milliseconds (Figure 6.16), assuming a Chapman–Jouguet detonation propagating at a speed $\sim 10^9$ cm s^{-1} (see Section 1.3.3.1). Again, the physical conditions adopted in the simulation bear little resemblance with those expected during X-ray bursts. Indeed, the thick layers required to initiate a detonation are unlikely to accumulate between bursts, so that the flame should propagate as a deflagration (see, e.g., references [1114,1188] and Figure 6.17).

The dicotomy between detonations and deflagrations was subsequently explored in two complementary papers[45] by Simonenko and colleagues [1658, 1659]. Two-dimensional simulations were performed with the Eulerian/Lagrangian code TIGR-3T for two different densities, 1.8×10^8 g cm^{-3} and 1.8×10^7 g cm^{-3}. In the high-density simulation, the flame propagated as a supersonic, detonation front, in a similar way as described by Zingale et al. Ignition was initially confined in a small region along the neutron star surface. As the detonation proceeded, ashes rapidly rose and spread through the upper layers due to turbulent convection. The piling up of additional matter drove compression and ignition. The low-density regime, however, did not develop into a steadily propagating flame.

Inclusion of rotational effects in flame propagation[46] have been considered by Cavecchi et al., first through the analysis of the role of a constant Coriolis force in longitudinal flames propagating at speeds $\sim 10^5$ cm s^{-1} [305], and, more recently, through a latitude-dependent Coriolis force—maximum at the neutron star poles, zero at the equator—in meridional flame propagation [306]. These studies proved that while the suppression of the Coriolis force at the equator affects the structure of the flame, it does not prevent its advance from one hemisphere to the other. On the other hand, the simulations revealed that flame propagation strongly depends on the angular velocity and heat conductivity of the fluid.

The early development of the convective stages preceding thermonuclear ignition in X-ray bursts has been recently analyzed in a multidimensional framework, in an attempt to clarify whether a fully-turbulent convection can actually modify the expected nucleosynthesis, and to assess the possibility of dredge up of ashes enriched in heavy elements to the neutron star photosphere [179, 890]. A pioneering effort in this regard was performed in 2D by Lin et al. [1114]. Other studies have been conducted with the Eulerian code MAESTRO,

---

[45] See also Timmes and Niemeyer [1806] for a thorough analysis of the different He-burning regimes (i.e., detonation vs. deflagration) in X-ray bursts.

[46] See Spitkovsky, Levin, and Ushomirsky [1686] for analytical and two-layer shallow water models of the effect of the Coriolis force in the propagation of deflagration fronts in rapidly rotating neutron stars during type I X-ray bursts.

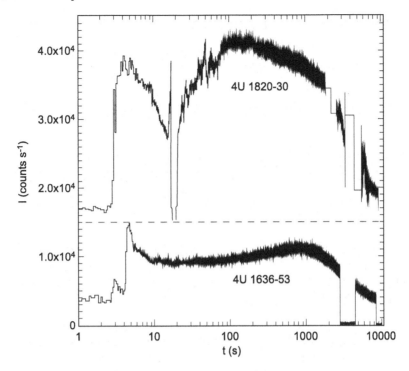

FIGURE 6.18
Two superbursts observed with the RXTE satellite, in the (2–30) keV band. Figure from
Strohmayer and Bildsten [1731]; reproduced with permission.

including 2D simulations of pure He bursts [1188] and mixed H/He bursts [1189], as well as
a pioneering 3D simulation of mixed H/He bursts [2024]. The latter, in particular, assumed
an outer envelope made of (accreted) mixed H/He, slightly overabundant in CNO nuclei
with respect to solar values, on top of an inert ^{56}Ni substrate. A plane-parallel geometry on
a uniform grid was adopted, with a spatial resolution of only 6 cm—the same used in reference [1189]. Comparison between 2D and 3D turbulent convection results in similar peak
temperatures and Mach numbers, but different convective velocity patterns, with evidence
of the energy cascade into smaller scales that characterizes 3D convection (see Chapter 4).

Further studies linking current progress in the physical description of the early convective
stages and state-of-the-art flame propagation are needed to properly address X-ray burst
modeling under realistic conditions.

6.4 Superbursts

Surprises in the arena of accreting neutron stars did not end with the discovery of the type
II bursts exhibited by the Rapid Burster. In the year 2000, Cornelisse and co-workers [380]
reported the first observation of an extremely energetic and long-lasting event from an

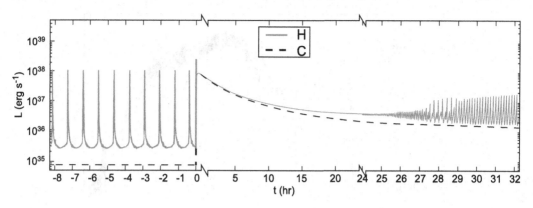

FIGURE 6.19

Light curves of two, 1D hydrodynamic models of superbursts, for different compositions of the accreted material (20% ^{12}C and 80% ^{56}Fe, for model C; 71% ^{1}H, 27% ^{4}He, and 2% ^{14}N, for model H; see text, for details). Figure from Keek, Heger, and in't Zand [966]; reproduced with permission.

otherwise normal, type I bursting source 4U 1735-44. Such *superbursts*[47] represent some sort of extreme X-ray bursts. They are characterized by an overall energy release of $\sim 10^{42} - 10^{43}$ erg (that is, about $\sim$ 1000–10,000 times larger than a normal type I burst), long durations, with a typical (exponential) decay time ranging from 1 to 3 hours[48] (Figure 6.18), and long recurrence periods (with an average value of $2^{+2}_{-0.7}$ yr [889]). Although superbursting sources also exhibit normal type I bursts, their occurrence is quenched for $\sim$ days after each superburst (see, e.g., Figure 6.19), after which a regime of marginal burning, characterized by mHz quasi-periodic oscillations, initiates [398, 966]. To date, about 22 superbursts from 13 different bursting sources have been identified, including GX 17+2 and 4U 1636-536, for which 4 events have been recorded [966].

The duration and energetics of superbursts suggest that fuel ignites at deeper layers ($\sim$ 100 m) than normal type I bursts, at densities typically exceeding 10^9 g cm^{-3} [397], close to the neutron star crust. Therefore, superbursts can be used as probes of the thermal properties of the crust[49]. Since neither hydrogen nor helium likely survives at those depths, Cumming and Bildsten [397] and Strohmayer and Brown [1732] suggested that superbursts are likely powered by ignition of carbon-rich fuel[50]. In this framework, carbon may be already present in the accreted material or synthesized through the rp-process during H/He burning in the preceding type I bursting phase. Controversy remains as to how much carbon is actually consumed during a type I burst, and whether enough carbon is leftover to power

---

[47]The term *superburst* was first used by Wijnands [1937] to characterize these long-lasting bursts. It is, however. worth noting that, historically, the same denomination was given to a relatively strong type I X-ray burst reported from the source 4U 1728-34 by Basinska and collaborators [139], back in 1984.

[48]An extreme event recorded from the source KS 1731-260 lasted for more than 10 hours [1042].

[49]Superbursts can be used to constrain the physics of neutron star interiors and the properties of dense matter through a number of observables—i.e., energetics, recurrence, and cooling times. See, e.g., references [253, 373].

[50]In their search for a model that could account for the origin of γ-ray bursts, Woosley and Taam [1983] analyzed the properties of C-burning in the envelopes of accreting neutron stars. They argued that C-ignition may result in a bursting episode—a flash—but their analysis lacked detailed simulations. The scenario was revisited by Taam and Picklum [1764] who, in the framework of 1D models of H/He accreted envelopes on top of a neutron star, concluded that carbon-driven flashes could occur whenever $\dot{M} \sim 10^{-10} - 10^{-9}$ M$_\odot$ yr^{-1}.

a superburst, a possibility not directly favored by current models [372,927,1575,1578,1976]. Cumming and Bildsten [397] have shown, however, that even small amounts of carbon could power a superburst in layers enriched with heavy ashes synthesized by the rp-process. Cooper, Steiner, and Brown have also analyzed the impact of a hypothetical 1.6 MeV reasonance in the $^{12}C + {}^{12}C$ reaction rate on the minimum amount of carbon required to power a superburst [377]. But even a strong resonance would only reduce the minimum carbon abundance needed by a small amount. Therefore, the existence of such hypothetical resonance would alleviate but not solve the problem[51]. Alternative models rely on combined stable and unstable burning regimes of the accreted H/He mixture [889,1723], or even invoke the presence of strange quark matter stars hosting the explosions [1371].

While most superbursting sources are observed to accrete continuously at a high rate, around 10% of the Eddington limit, a number of recent observations clearly at odds with thermonuclear ignition models have challenged the standard scenario. For instance, a superburst from the X-ray transient source 4U 1608-522 was detected only 55 days after the onset of an accretion outburst, too short to warm up the neutron star to carbon ignition conditions [968]. Moreover, the superbursting event reported from Terzan 5 is even harder to explain, since the neutron star is very cold and the onset of the explosion occurred at a period of only low-level accretion, or no accretion at all [423,1938] (see also reference [21], for another challenging superburst observed in the source EXO 1745-248). Such observations lead us to question whether carbon ignition is actually the truly driving mechanism in superbursts[52].

It is finally worth noting that in the few cases where the initiation of a superburst has been observed (e.g., 4U 1820-30), a short precursor burst has been identified. Models indicate that burning of accreted helium is insufficient to account for the observed properties of such precursors, which turn out to be more energetic than normal type I bursts [962]. Instead, shock heating has been proposed to account for the energy shortage. In this scenario, ignition of a sufficiently large column of carbon may trigger a detonation[53] [1915]. The shock front would subsequently propagate to the surface, resulting in a short-duration peak in the corresponding light curve. The shock would eventually heat the neutron star atmosphere, driving a precursor burst[54], while triggering hydrogen and helium ignition [965,966,1914]. Interested readers can find further insights into the physics of superbursts in a number of reviews [395,1039,1731].

---

[51]Interestingly, the existence of such resonance would have a minor impact for ignition in type Ia supernova models. In contrast, it may affect the predicted nucleosynthesis accompanying the explosion of massive stars, in the form of enhanced ^{26}Al and ^{60}Fe yields [618].

[52]For instance, Schatz, Bildsten, and Cumming [1577] have shown that, at the high temperatures reached during superbursts ($T > 10^9$ K), photodisintegration through (γ, n) reactions of the heavy rp-process ashes yields a larger nuclear energy release than carbon fusion itself.

[53]See also reference [1315] for 1D C-detonation models of superbursts.

[54]Two exceptionally short precursors recently observed in intermediate-duration X-ray bursts from 4U 0614+09 and 2S 0918-549 [888] have been interpreted as driven by detonations. Indeed, the precursor rise times determined, with only $\sim$ 1 ms, are inconsistent with thermonuclear flames spreading as deflagrations [888].

Box I. The X-Ray Burst "Hall of Fame"

A selection of facts on normal type I X-ray bursts, intermediate-duration bursts, and superbursts

1. Type I X-ray bursts take place on the neutron star component of a close stellar binary system and represent the most frequent type of thermonuclear explosions in the Galaxy.

2. Different types of X-ray bursts (normal, intermediate-duration, and superbursts) can be distinguised in terms of duration, energetics, and recurrence period, reflecting ignition of different fuels (H/He, He, C) at different depths.

3. X-ray burst light curves are characterized by a fast rise to peak followed by a power law-like decay, occassionally with double or triple peaks. Deviations from the general pattern are produced by a suite of different phenomena, including disturbances driven by the accretion flow, nonspherical emission, or the extent of the rp-process.

4. Oscillations with frequencies ranging between 11 and 600 Hz have been identified in the X-ray burst light curves, in about 25% of all bursting sources. While oscillations during the rising phase to peak luminosity have been linked to the spreading of a hot spot on a rotating neutron star, those observed during decline from peak have yet a controversial origin.

5. Quasi-periodic oscillations at 3–9 mHz have also been observed prior to a burst. They may result from marginally stable nuclear burning on the neutron star surface.

6. Observations suggest that a tiny fraction of the accreted envelope (1%, at most) is ejected from the neutron star by radiation-driven winds.

7. One-dimensional simulations have proved successful in reproducing the main observational features of type I X-ray bursts (i.e., light curve shapes, recurrence periods).

8. The burning front likely propagates subsonically (i.e., a deflagration). The outburst is likely quenched by fuel consumption (rather than by envelope expansion), and is driven by a suite of different nuclear processes. The most complex nuclear path is achieved for mixed H/He bursts, which are driven by the 3α-reaction, the αp-process (a sequence of (α, p) and (p, γ) reactions), and the rp-process (a series of rapid proton-captures and β^+-decays that play a key role in powering the light curve tail).

Box II. Mysteries, Unsolved Problems, and Challenges

- What causes the appearance of double and triple peaks in the X-ray burst light curves?

- Attempts to resolve gravitationally-redshifted atomic absorption lines from X-ray burst spectra through high-resolution spectroscopy have proved unsuccessful to date (mostly hampered by rotational broadening). To this end, efforts aimed at improving sensitivity in X-ray spectroscopy are highly advisible.

- What is the maximum extent of the rp-process in X-ray burst nucleosynthesis?

- From a nuclear physics viewpoint, efforts should be made to better constrain the rates of key reactions affected by nuclear uncertainties. Since most of the reactions of interest involve short-lived species, direct measurements will be extremely challenging (if not impossible!), and, therefore, improvements should mostly rely on indirect measurements and theoretical approaches.

- Do X-ray bursts contribute to the Galactic abundances?

- Further efforts are needed to clarify the existing controversy between theoretical and observationally inferred values of the mass-accretion rate at the transition between stable burning and bursting regimes.

- Multidimensional simulations of X-ray bursts are limited to date to models of flame propagation on the envelopes accreted onto neutron stars, as well as convection-in-a-box studies aimed at characterizing convective transport during the stages prior to ignition. Such efforts must be extended to 3D simulations of a full burst (and in a longer term, to series of bursts), taking advantage of state-of-the-art parallel computers.

- How localized is ignition in the different X-ray burst types?

- Is carbon the real fuel that powers superbursts? If so, how is it produced, and what heats the ignition layer to explain the shorter than predicted recurrence times?

6.5   Exercises

P1. Estimate the characteristic photon energies emitted by an accreting 1.4 $M_\odot$ (R = 10 km) neutron star. Assume that the temperature of the radiation emitted ranges between the blackbody value, T_{bb}, and the temperature obtained from full conversion of the gravitational energy released during infall into thermal energy, T_{th}. Adopt $T_{bb} = (L/4\pi R_{ns}^2 \sigma)^{1/4}$, where L is the accretion luminosity, and $T_{th} \sim GMm/kR_{ns}$.

P2. Estimate the energy released per gram and per nucleon during the incineration of a pure He envelope into Fe-peak nuclei. Use a simple α-chain network composed exclusively of α-capture reactions, beginning with the triple-α reaction, and extending all the way up to ^{56}Ni. Use Q-values from a standard compilation (e.g., http://starlib.physics.unc.edu/RateLib.php).

P3. A neutron star accretes solar composition material at a rate of 10^{-8} $M_\odot$ yr^{-1}. After $\sim$ 4 hr, a type I X-ray burst results.

   a) Determine the mass piled up on top of the neutron star.

   b) Estimate the mean value of the nuclear energy released during the burst required to power mass ejection from the neutron star surface. Compare the result with the mean value inferred from Figure 6.8 (middle right panel).

P4. One of the first long X-ray bursts ever recorded was detected in 1975 by the instruments on board the OSO-8 satellite (see Figure 6.5 [1752]). The maximum flux detected from the source was 6×10^{-8} erg s^{-1}. The X-ray spectra was best fitted by a blackbody with kT ranging between 0.87 keV and 2.3 keV. The distance inferred to the source was about 1 kpc. Estimate the size of the compact object hosting the explosion.

P5. Derive the expression for the Eddington luminosity. Assume that the pressure exerted by the radiation impinging on the material infalling onto an accreting 1.4 $M_\odot$ (R = 10 km) neutron star translates into a force that balances the gravitational pull of the compact object.

P6. ^{15}O(α, γ) is one of the main channels for CNO breakout, whenever $\tau_\alpha(^{15}\text{O})$ becomes shorter than $\tau_{\beta+}(^{15}\text{O})$ ($T_{1/2} = 122$ s; laboratory value).

   a) Demonstrate that T $-\rho$ conditions for which $\tau_\alpha(^{15}\text{O}) = \tau_{\beta+}(^{15}\text{O})$ satisfy

$$\rho = \frac{\ln 2}{T_{1/2}(^{15}\text{O})\,(Y/M_{\text{He}})\,N_{\text{Av}}\langle\sigma v\rangle_{^{15}\text{O}(\alpha,\gamma)}}.$$

   b) Consider the $N_{\text{Av}}\langle\sigma v\rangle_{^{15}\text{O}(\alpha,\gamma)}$ rate given in Table 6.4 (Iliadis et al. [882]) and prove that this is an important reaction for type I X-ray bursts but not for classical nova outbursts [414, 415]. How critical is the choice for the decay rate (i.e., laboratory vs. stellar)?

TABLE 6.4
The $^{15}O(\alpha, \gamma)$ (Recommended)
Reaction Rate

$\mathbf{T}(10^8 \mathbf{K})$	$N_{Av}\langle\sigma v\rangle_{^{15}O(\alpha,\gamma)}$
0.10	9.198×10^{-68}
0.15	8.496×10^{-58}
0.20	1.660×10^{-51}
0.25	4.932×10^{-47}
0.30	1.277×10^{-43}
0.40	1.213×10^{-38}
0.50	4.172×10^{-35}
0.60	2.083×10^{-32}
0.70	2.994×10^{-30}
0.80	1.788×10^{-28}
0.90	6.530×10^{-27}
1.00	4.481×10^{-25}
1.10	5.998×10^{-23}
1.20	4.382×10^{-21}
1.30	1.668×10^{-19}
1.40	3.731×10^{-18}
1.50	5.482×10^{-17}
1.60	5.729×10^{-16}
1.80	2.823×10^{-14}
2.00	6.283×10^{-14}
2.50	1.582×10^{-10}
3.00	5.995×10^{-9}
3.50	7.767×10^{-8}
4.00	5.163×10^{-7}
4.50	2.223×10^{-6}
5.00	7.127×10^{-6}

7

Core-Collapse Supernovae

"—Then if it's a supernova—which I think it may be, even though it shouldn't—it's just getting going.
—What do you mean, *shouldn't*? [...]
—Alpha Centauri is a double-star system—Wilmer said, in his relaxed drawl.—Double stars can become a Type Ia supernova, but only if one of the two stars is a white dwarf. Alpha centauri doesn't qualify.
—Then I guess Alpha Centauri doesn't know that—said Zoe.—It's not a good day to be an astronomical theorist. [...]
—Course, when the gas shell of the supernova expands, a big slug of gamma rays will break out. We have no idea which direction they'll emerge. But we have enough shielding to handle that, too. The big problem is going to be the high-energy particle flux. That will carry a lot more energy than the visible light or the gamma rays. It'll be an absolute killer.
Zoe came bolt upright.—And you say we're not in danger!
—We're not. The light and gammas travel at light speed, but the particles are much slower—five to ten percent of light speed. It will take them fifty years to get here."

Charles Sheffield, *Aftermath* (1998)

Stellar explosions are cosmic factories of nuclear species, and as such they rule most of the chemical evolution of the universe. On a human scale, supernovae have been particularly influential, sowing the interstellar medium with seeds that are essential for the emergence of life—e.g., oxygen, calcium, iron.

The emergence and evolution of life on Earth, with earliest undisputed evidence dating, at least, from 3.5 Gyr ago, has been challenged by a number of major extinction episodes[1]. The nature of such cataclysmic events remains far from clear, and different *suspects*, such as volcanoes, massive meteorites, and even exploding stars, have been invoked[2]. The potential threat posed by a nearby supernova has actually been stressed by several authors. Indeed, Iosif Shklovskii and Carl Sagan[3] suggested in 1966 that some of the major extinction events may have been triggered by a close supernova blast [1631]. The scenario was revisited by Terry and Tucker [1786], who estimated the probability that a nearby supernova explosion occurred in Earth's history and the possible biological impact of the radiation released in the event. The study concluded that appreciable doses of radiation, in the form of cosmic rays, from an explosion within a few 100 light-years, could hit the Earth every ~ 50 Myr.

The work by Terry and Tucker was shortly after criticized by Laster [1054], on the basis of too optimistic estimates of the cosmic-ray arrival time. Indeed, Laster stressed that the Milky Way's magnetic fields would diffuse the cosmic radiation emitted by a supernova,

---

[1] Although the Cretaceous–Tertiary event that wiped out dinosaurs from Earth is the most famous, a more severe extinction episode occurred at the end of the Permian period, when 96% of all existing species perished. These are two of the "Big Five" mass extinction events recorded in Earth's history, together with those in the Ordovician–Silurian, the Devonian, and the Triassic–Jurassic. All in all, more than $\sim 90\%$ of all organisms that ever lived on Earth are now extinct.

[2] See van den Bergh [1844] for a review on astronomical catastrophes in Earth's history.

[3] See also references [1017, 1140, 1584] for other early suggestions of the possible link between major extinctions and supernovae.

smoothing the expected effects on Earth over a timescale of years, rather than days. This, in turn, would reduce the area of influence to those supernovae exploding within a few light-years from Earth. A reanalysis performed by Tucker and Terry themselves [1831] unveiled that a burst of neutral, high-energy γ-rays is released in the event, together with charged cosmic-ray radiation. On impact, γ-rays may initiate air showers of secondary particles, raising the radiation levels on surface between 10 and 1000 rads, for an explosion at 100 light-years (see also reference [1546]).

The most catastrophic effect of a nearby supernova explosion was first unveiled by Ruderman [1539]. The impact of large amounts of ionizing radiation—i.e., cosmic rays and γ-rays—with Earth's atmosphere could produce free nitrogen atoms in an excited state, N*, which in turn can oxidize forming nitric oxides[4] (mainly NO and NO_2),

$$N^* + O_2 \rightarrow NO + O \tag{7.1}$$

$$NO + O_3 \rightarrow NO_2 + O_2 \tag{7.2}$$

$$NO_2 + O \rightarrow NO + O_2. \tag{7.3}$$

A small concentration of such oxides, at a level of few parts per billion (ppb), may reduce the ozone (O_3) atmospheric content by a factor of 2, leading to an increase of the surface UV radiation, with dramatic consequences for many life forms (e.g., large increase of skin carcinomas in humans)[5]. Indeed, the study suggested that more than 90% of the O_3 shield may have disappeared for periods ranging from a few years to a century, every few hundred million years[6]. Estimates of the level of atmospheric ozone depletion and recovery times, as well as of the expected effects on the biosphere[7], have been reported by different authors along the last two decades (see references [146, 408, 1226] for reviews). Current estimates yield lethal radiation doses for supernova explosions occurring every ~ 0.1–1 Gyr within 8 to 10 pc from Earth [620].

It is finally worth noting that, although a handful of nearby stars may explode as supernovae in the foreseeable future[8], no one is close enough to pose a serious threat to humans. In fact, the highest threat for humans seem to be, once more, humans themselves.

---

[4]Rood et al. [1516] were the first to suggest that a number of NO_3^- spikes observed in South Pole ice cores might correlate with the known historical supernovae. More recently, Motizuki et al. [1265] reported similar findings in ice samples from Antarctica. The corresponding NO_3^- spikes were dated from the 10th to the 11th century, somewhat coincident within uncertainties with two historical supernovae, SN 1006 and SN 1054 (see Chapter 5).

[5]Other sources of ionizing radiation, aside from nearby supernovae, include Galactic γ-ray bursts (within 1–10 kpc) and energetic solar-particle events (or *proton storms*) [88].

[6]See also references [338, 1931] for different estimates of the level of ozone depletion, based on a reevaluation of the frequency of nearby supernova explosions. See also reference [491] for the predicted effect on phytoplankton, reef communities, and marine life in general, which may even trigger a greenhouse episode.

[7]Another possible effect driven by a nearby supernova has been reported on the basis of the prominent blue light emission (400 nm) observed in the optical spectra of SN 2006gy. A short-time exposure to low levels of short wavelength optical light (below 450 nm) can strongly affect the endocrine system in humans and other mammals, inducing melatonin supresion and, eventually, cancer [1800].

[8]The list includes ρ Cassiopeiae, η Carinae, VY Canis Majoris, Betelgeuse, Antares, Spica, and a number of Wolf–Rayet precursors, such as WR 11 (γ^2 Velorum) and WR 104 [146, 1670, 1800]. The nearest candidate is IK Pegasi, a binary system that may eventually explode as a type Ia supernova, which is located at a distance of ~ 50 pc. Betelgeuse, at ~ 200 pc, is the closest type II supernova candidate [146]. Note that the detection of the radioisotope ^{60}Fe in the deep-sea, FeMn–rich terrestrial crust (see Section 3.4.2.5) [994, 995], as well as on lunar samples [370, 529], provides smoking-gun evidence of a past, nearby supernova explosion in the Earth's neighborhood.

7.1 SN 1987A

On February 23/24, 1987, neutrinos and light from the closest supernova since Kepler's SN 1604 hit the Earth. The event, known as SN 1987A, was spotted optically in the *suburbs* of the Tarantula nebula, within the Large Magellanic Cloud, a satellite of the Milky Way galaxy located about 168,000 light-years away (Figure 7.1). The exploding star was independently discovered by Ian Shelton and Oscar Duhalde at Las Campanas Observatory (Chile) and by Albert Jones in New Zealand[9]. It achieved a maximum apparent brightness of $\sim 3^{\mathrm{mag}}$, about three months from discovery—therefore, it was visible with the unaided eye. A series of photometric studies soon revealed that SN 1987A was not like any other discovered supernova (see Figure 5.11 and Sections 5.3 and 7.2).

The true interest in SN 1987A, besides its proximity and photometric peculiarities, was actually raised by the detection of a burst of neutrinos, about two to three hours before the optical discovery, at three different observatories. Indeed, the two water Cherenkov detectors Kamiokande II (Japan) and IMB Irvine–Michigan–Brookhaven (Ohio, US) registered 11 and 8 antineutrinos, respectively[10] [189, 806]; the Baksan liquid scintillation telescope (Russia) detected in turn another 5 antineutrinos[11] [14]. Observations were consistent with theoretical models of core-collapse supernovae, which predict that most of the energy released during the collapse of a massive star is radiated away in the form of neutrinos[12], preceding the emission of visible light [68]. This makes SN 1987A the only cosmic object outside the Solar System from which neutrinos have ever been detected[13].

About four days after discovery, a tentative progenitor was identified in a number of photographic plates and CCD images taken prior to the explosion. Evidence pointed toward a blue supergiant star known as Sanduleak $-69°$ 202. This actually came as a surprise, since such stars were not expected to explode as type II supernovae at the time (see, e.g., reference [993]). Indeed, the fact that SN 1987A achieved a peak luminosity of about 1/10th of a canonical type II supernova likely suggested a peculiar progenitor [68, 1220, 1391, 1436]. Since SN 1987A first became visible, astronomers have been earnestly searching for the neutron star that likely formed in the event, so far with negative results[14]. Several explanations have been proposed, including the possibility that the compact star could be shielded by a dense, dusty cloud[15], or that it further collapsed into a black hole when the amount of material piled up exceeded the Tolman–Oppenheimer–Volkoff limit (see Chapter 6). Evidence of the interaction between the supernova ejecta and circumstellar material was found, a few years later, in the form of mysterious, mirror-imaged rings enshrouding the debris of SN 1987A (Figure 7.2) [275, 390, 553]. A possible explanation assumes that the rings correspond to material previously ejected by the progenitor star through winds, subsequently ionized by the ultraviolet radiation emitted in the event [269, 1839]. All in

---

[9] A full account of the discovery and early observations of SN 1987A is reported in the IAU Circs. #4316, #4317, and #4330. See also references [1002, 1923].

[10] The total number of antineutrinos detected by Kamiokande II varies between 11 and 16, depending on the adopted time window [1863].

[11] Approximately three hours earlier, another burst of five neutrinos was detected by a liquid scintillator detector at the Mont Blanc. However, it is now widely accepted that such neutrino burst was not associated with SN 1987A.

[12] Theoretical estimates yield 5×10^{52} erg in the form of electron antineutrinos [1863].

[13] The IceCube Neutrino Observatory has also reported the discovery of high-energy neutrinos of unknown origin (likely extragalactic) [1].

[14] The discovery of the neutron star in SN 1987A has been tentatively announced several times. See, e.g., reference [1024].

[15] Dust condensation in the ejecta of SN 1987A was spectroscopically inferred about 500 days after the explosion [1206].

FIGURE 7.1

SN 1987A, the closest supernova detected since the invention of the telescope. The object corresponds to the very bright star in the middle right of the Tarantula Nebula, in the Large Magellanic Cloud. At the time of this picture, SN 1987A was visible with the naked eye. A color version of this image is available at `http://fisica.upc.edu/ca/users/jjose/CRC-Downloads`. Credit: ESO. Source: `http://www.eso.org/public/images/eso0708a/`; released into public domain by ESO.

FIGURE 7.2
Glowing gas rings surrounding SN 1987A, as seen by the Hubble Space Telescope in February 1994. A color version of this image is available at `http://fisica.upc.edu/ca/users/jjose/CRC-Downloads`. Credit: P. Challis, Harvard–Smithsonian Center for Astrophysics. Source: `http://hubblesite.org/newscenter/archive/releases/1998/08/image/g/`; released into public domain by NASA.

all, SN 1987A has provided astrophysicists with smoking-gun evidence for theories of core-collapse supernovae, revealing clues on the progenitor star, neutrino fluxes, nucleosynthesis, dust formation, and the interaction between the ejecta and circumstellar material.

7.2 Observations of Core-Collapse Supernovae: Spectra and Light Curves

Supernovae are mainly classified by a number of spectroscopic features. For instance, types I and II are distinguished by the absence or presence of prominent hydrogen lines in the spectra, respectively. Different type II subclasses have been further established on the basis of the strength and profile of the H_α line (Figure 7.3).

Following Branch [232], during the first months after a supernova explosion, the ejecta is optically thick. Therefore, early-time spectra reflect the complex interplay between photon emissions, absorptions, and scatterings that operate in such dense stellar plasmas. Clues on the composition of the ejecta and on the nature of the progenitor star are gradually revealed as the photosphere recedes and inner layers become progressively exposed, since the spectrum forms at and above the photosphere. At this early *photospheric* stage, the spectrum is characterized by broad lines (due to the high velocities of the ejecta, mostly with P-Cygni profiles) superimposed on a continuum. Hundreds of days after the explosion, the ejecta becomes optically thin and the supernova enters the *nebular* phase. At this stage, the spectrum forms mostly deep inside the ejecta, at dense layers containing pristine information of the nuclear processes that occurred during the explosion of the star. However, most of the (optical) spectra of type II supernovae are dominated by a strong H_α emission from the outer layers.

Type II supernovae are also characterized by a wide dispersion in absolute magnitudes[16]. Even though some energetic explosions may overlap with the dimmest thermonuclear type Ia supernovae, they are, in general 1.5^{mag} fainter [989]. Supernova light curves near peak luminosity also provide clues on the energy source and on the size, mass, and composition of the exploding star. On the other hand, decline from peak luminosity yields information on the composition of the interior, evidence of the synthesis of radioactive species and of the nuclear processing during the explosion, and constraints on the possible existence of an underlying neutron star [989]. Much of the light curve is, in fact, ruled by radioactive decays (see reference [953] for details). The maximum luminosity achieved by a supernova, for instance, reflects the overall amount of ^{56}Ni synthesized. The decline from peak luminosity is powered by the energy released in the decay chains $^{56}Ni \rightarrow {}^{56}Co$ ($T_{1/2} = 6.1$ days) and $^{56}Co \rightarrow {}^{56}Fe$ ($T_{1/2} = 77.3$ days), as first confirmed for type II supernova explosions by Uomoto and Kirshner [1840], through observations of SN 1980K. Once all ^{56}Co has decayed, the light curve may still be powered by the radioactive decay of ^{44}Ti ($T_{1/2} = 59$ yr).

Two major type II supernova classes have been established on the basis of their early-time light curves ($t \leq 100$ days) [122, 448]: While II-L (linear) supernovae are characterized by a steep decline from peak luminosity followed by an exponential tail, somehow resembling a typical type Ia event, II-P (plateau) supernovae remain at nearly constant luminosity, with variations $< 1^{mag}$) after maximum, for an extended period of time. Other subclasses (i.e., IIb and IIn) have been subsequently introduced based on additional spectroscopic peculiarities[17].

7.2.1   SN II-P

Type II-P supernovae exhibit a characteristic postmaximum plateau in the optical light curve, where they remain at nearly constant luminosity, with variations $< 1^{mag}$ for an extended period of time (typically, for about 100 days). The existence of such plateau has been explained as due to a change in the opacity of the outer layers in a massive star with a large hydrogen envelope. The arrival of the inner shock front, generated during core-collapse, deposits energy in the envelope. As temperature rises, hydrogen becomes ionized, driving a significant increase in the overall opacity. Following the expansion of the ejecta, these layers eventually cool down, allowing hydrogen to recombine. As a result, the opacity drops, and radiation is again efficiently emitted, causing the light curve to resume its steep decline [514].

Peak absolute magnitudes of type II-P supernovae show a very wide dispersion [1590, 2008], attributed to differences in size of the progenitor stars[18].

Most II-P supernovae exhibit a relatively well-defined spectral development. The early-

---

[16]Some events, tentatively classified as type II supernovae, such as SN 1961V and SN 2010dn, may not constitute true supernova explosions but rather super-outbursts of luminous blue variables, such as η Carinae. These events are known as *supernova impostors*.

[17]According to Faran [515], out of the 1154 type II supernovae discovered between 2000 and 2013, 697 have not been assigned a subtype, 228 have been classified as II-P, 158 are IIn, 68 are IIb, and only 3 are II-L. See also reference [1667] for type II supernova statistics based on different samples.

[18]The light curve of SN 1987A, although peculiar, basically followed the characteristic template of a type II-P supernova. The lower peak luminosity reported from SN 19877A has been interpreted as due to its peculiar progenitor, a blue supergiant, much smaller than a red supergiant (see Arnett et al. [68]

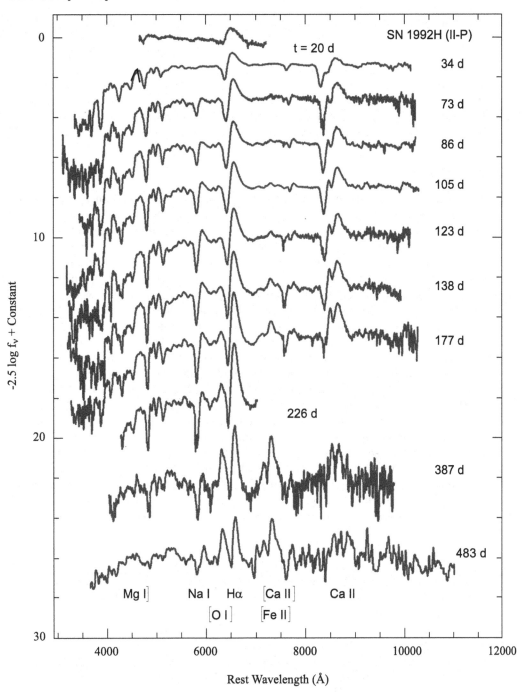

FIGURE 7.3

Spectral evolution of SN 1992H, a type II-P supernova. Figure from Filippenko [524]; reproduced with permission.

and references therein). See also Smartt [1667] and Poznanski [1442] for recent compilations of supernova progenitor masses.

time spectrum is nearly featureless and can be fitted by a blackbody temperature $> 10,000$ K. Very weak hydrogen (Balmer) lines and He I $\lambda 5876$ with a P-Cygni profile are sometimes visible. The tiny decline in the light curve, a few weeks after peak luminosity, is accompanied by a decrease in temperature, down to 5000 K, which remains almost constant along the plateau phase [525]. At this stage, strong Balmer and Ca II H and K lines with clear P-Cygni profiles appear. Weaker lines of Fe II, Sc II, and other iron-group elements also show up. Finally, as the light curve evolves into the late-time tail, the continuum fades. The H_α line becomes very strong, and prominent emission lines of [O I], [Ca II], and Ca II also emerge.

7.2.2   SN II-L

Type II-L supernovae are more scarcely detected than II-P, and, accordingly, they are not so well understood. On average, they are brighter than type II-P supernovae, and exhibit a bimodal distribution of absolute peak magnitudes [638,696]. The most relevant feature of a type II-L supernova is its steep, linear decline in luminosity, by $> 0.5^{\mathrm{mag}}$ during the first 50 days. Photometrically, type II-L supernovae are similar to type IIb (see Section 7.2.3) [515].

It is not yet clear whether types II-L and II-P constitute distinct spectroscopic subclasses [696]. Near peak luminosity, type II-L supernovae show almost featureless spectra, with weaker hydrogen absorption lines and shallower P-Cygni profiles than type II-P supernovae [696]. A week later, H_α emission is more easily discernible, together with Na I, H_β, and Fe II lines. By $t \sim 1$ month, H_α emission becomes stronger, while the absorption component is almost missing. In contrast, other lines show characteristic P-Cygni profiles.

It has been suggested that the photometric and spectroscopic properties of type II-L supernovae can be explained by massive progenitors with smaller mass H envelopes than their II-P counterparts, thus powering a light curve devoid of any plateau [122,198]. Evidence of the interaction between the ejecta and circumstellar, presupernova material has been inferred from the late-time spectra of a number of supernovae, such as SN 1979C, SN 1980K, and SN 1986E (see, e.g., reference [885]). Observational clues on the nature of type II-L supernova progenitors are, unfortunately, also scarce. A 10–15 $M_\odot$, blue giant star has been suggested as the star hosting SN 1980K [1667], while a 17–18 $M_\odot$, red supergiant has been claimed as the likely progenitor of SN 1979C [1846]. Other examples include SN 2009kr, for which several progenitors of different masses have been proposed [490,554].

7.2.3   SN IIb

Type IIb is just another rare variety of core-collapse supernovae, halfway between a type II and a type Ib. This class, inaugurated[19] with the discovery of SN 1987K [523], is characterized by an early-time spectrum with weak, high-velocity hydrogen lines, which unambiguously reveal their true type II identity. However, this characteristic H emission fades away at later times, becoming undetectable. This suggests that a type IIb supernova discovered well after peak luminosity may be misclassified as a type Ib.

A possible scenario for a type IIb explosion involves a giant star that has lost most of its H-rich envelope in previous interactions with a companion star within a binary system. As the ejecta expands, following the supernova explosion, the outer H-rich layers soon become optically thin, while inner layers of the star become progressively exposed[20]. The interaction with previously ejected H-rich winds may also yield late H_α emission during the nebular stage. Such scenario has been successfully applied to SN 1993J, for which a

---

[19]The concept, however, had been advanced theoretically by Ensman and Woosley [497].

[20]Two possible subclasses of type IIb supernovae have been proposed, involving compact and extended progenitor stars, with a dividing line given by a 0.1 $M_\odot$ H-rich envelope [323].

12–15 M$_\odot$, main sequence star progenitor, stripped of most of its H-rich envelope, has been proposed [1329, 1842]. Models naturally account for the late emergence of the He lines reported from SN 1993J between days 24 and 30, as driven by the recession of the photosphere, which progressively expose the He-rich interior of the star. It has also been claimed that Cassiopeia A could be the remnant leftover by a type IIb explosion [1022].

7.2.4 SN IIn

Schlegel [1585] was the first to recognize that a small subset of type II supernovae are characterized by narrow emission lines, and proposed for them the separate subclass IIn, the "n" denoting *narrow*. They are quite luminous compared to other core-collapse supernovae. Their light curves are characterized by a relatively long rise time (> 20 days), followed by diverse decline rates, ranging from slow to rapidly decaying events [984]. Observations suggest a significant interaction of the ejecta with circumstellar gas [333, 334, 1837], even at early times (unlike type II-L supernovae).

Type IIn supernova spectra often display relatively narrow and strong emission lines (most notably H) with little or nonexisting P-Cygni absorption component. He I emission lines, together with Fe II, Ca II, O I, or Na I absorption lines, are also present in the first few spectra. Recent studies support the association between type IIn supernovae and massive, luminous blue variable (LBV) progenitor stars[21] (with > 80 M$_\odot$), reinforced by the direct detection of the progenitor of the type IIn SN 2005gl [583].

7.2.5 Type I Core-Collapses: Supernovae Ib and Ic

Type I supernovae, and their subclasses Ia, Ib, and Ic, are defined by the absence of hydrogen in their spectra (see Chapter 5). There is, however, little in common between the type Ia and the types Ib–Ic classes, aside from such spectroscopic feature. In fact, the latter bear a much closer resemblance to the type II class, since they involve massive stars rather than white dwarfs, and are driven by the same physical mechanism: Core-collapse [1928]. Further observational evidence of the link between types Ib–Ic supernovae and massive stars comes from the fact that such explosions are only seen in spiral galaxies, near regions of recent star formation.

The absence of hydrogen[22] can naturally be explained assuming that their H-rich envelopes have been stripped in previous evolutionary stages[23], via stellar winds [1753, 1981] or through mass-transfer onto a stellar companion in a binary system [1333, 1982].

Bertola and collaborators [168, 169] were among the first to report that some type I supernovae, such as SN 1962L and 1964L, did not exhibit the deep absorption trough around 6150 Å, due to blueshifted Si II $\lambda\lambda$6347, 6371, that characterize type Ia supernovae. Accordingly, they were flagged as "peculiar" type I supernovae, or SN Ip. Their current denomination as type Ib was suggested by Elias and collaborators in an attempt to distinguish such anomalous events from normal type Ia supernovae [489]. As more data was gathered, it became crystal clear that type Ib supernovae constituted a rather heterogeneous

---

[21]Luminous blue variables (LBV) are a very rare group of massive, evolved stars that experience unpredictable spectroscopic and photometric variations. Examples of LBV stars include η Carinae and S Doradus.

[22]There have been reports of a weak hydrogen emission from a number of type Ic supernovae (e.g., 1987M, 1988L, and 1991A). Several explanations have been suggested, including the interaction between the ejecta and a previously expelled H-rich wind or contamination with superposed HII regions. See discussion in Filippenko [524, 526].

[23]Unlike type Ia supernovae, types Ib–Ic show radio emission, attributed to shocks produced in the interaction between the ejecta and circumstellar material likely ejected during the presupernova stage (see discussion in Kirschner [989]).

class of objects, with huge spectroscopic variations in the strength of He I. Wheeler and Harkness [727, 1926] suggested a further subdivision of type I supernovae into two distinct subclasses: those showing strong He I absorption lines in their early spectra, in particular He I λ5876 (type Ib), and those where He I is hardly detectable (type Ic).

The late-time spectra (> 4 months) of types Ib and Ic is dominated by several, relatively unblended lines of intermediate-mass elements[24], such as O and Ca [526]. The O I λ7774 line is frequently stronger in type Ic supernovae. Moreover, emission lines in type Ic appear to be broader than in type Ib at late times. This suggests either a larger amount of energy released in the explosion or the presence of smaller envelopes in type Ic supernovae [526, 1203].

It is finally worth noting that the amount of ejected nickel/iron in types Ib–Ic supernovae is not well constrained. First, crude estimates yielded, on average, ~ 0.15 $M_\odot$, based purely on the observational evidence that types Ib–Ic are roughly four times fainter than type Ia supernovae, which, on average, eject 0.6 $M_\odot$ of iron [1928]. More recent estimates have raised this contribution to about 0.3 $M_\odot$ of iron per type Ib–Ic explosion. Larger contributions, about 1.1 $M_\odot$, are even expected from the most energetic type Ic supernovae [1166, 1768] (i.e., those linked with γ-ray bursts; see Section 7.6).

7.2.6   Cosmology and Type II Supernovae

Type II supernovae are explosions of stars above $\simeq 10$ $M_\odot$. The large range of possible masses involved in such explosions translates into a broader dispersion of absolute magnitudes than for type Ia supernovae, that preferentially explode when the underlying white dwarf approaches the Chandrasekhar mass. A priori, type Ia supernovae are the obvious choice for precision cosmology, since type II supernovae clearly cannot be used as standard candles. Despite this limitation, different methods that use individual type II supernova as distance indicators have been proposed.

The *expanding photosphere method* [719, 990, 1586] derives from the classical Baade–Wesselink technique [99, 642, 1922], originally applied to determine sizes and distances to Cepheid variables. The method is based on theoretical models and is independent of the extragalactic distance scale [434, 479]. It relies on accurate determinations of the effective temperature and velocity of the star. This technique has proved particularly useful for type II-P supernovae, since during their characteristic plateau, the effective temperature of the star is approximately constant. In essence, the expanding photospheric method operates as follows (see Filippenko [525] for further details). First, the expansion speed of the supernova, v_e, is inferred from Doppler shift measurements using specific spectral lines. This allows the determination of the radius of the star at any time, $R(t)$, from the relation $R = v_e t$. Assuming a blackbody emission[25], one can relate the effective temperature and the radius through the intrinsic luminosity of the star, in the form, $L = 4\pi R^2 \sigma T^4$. Finally, a comparison with the apparent brightness of the supernova yields the distance to the object (see Problem 6, at the end of this chapter, for an application of the expanding photosphere method to the determination of the distance to SN 1980K).

The *standardized candle method*, introduced by Hamuy and Pinto [718], is an empirical technique that relies on the observational correlation[26] between the expansion velocity and the luminosity of a supernova during the plateau phase, in the form $L \propto v_e^n$.

Both techniques have proved successful for moderate redshifts. The distance accuracy achieved by any of these methods is about 10%–14% [1348], while the distance accuracy

---

[24]In contrast, type Ia supernova spectra are clearly dominated by blended Fe and Co lines at late times. Spectroscopically, type II supernovae resemble types Ib and Ic, but with narrower emission lines [989].

[25]See, e.g., reference [525] for dilution correction factors used to account for deviations from a blackbody law.

[26]See also Kasen and Woosley [953] for analytic correlations derived from models of type II supernovae.

obtained with type Ia supernovae is about 7%. First attempts to derive cosmological parameters from type II-P supernovae were performed by Nugent and collaborators [1336], who showed the enormous potential offered by core-collapse supernovae as cosmological probes.

7.3 Core-Collapse Supernovae: Evolution beyond Core Silicon Burning and Explosion

As described in Chapter 2, single, massive stars ($M \gtrsim 10$ M$_\odot$; Figure 7.4) undergo a handful of fusion stages, beginning with core H-burning and ending up with Si-burning[27]. After Si consumption, the stellar core is mainly composed of ^{56}Fe. Surrounding such core, there are inert and (thin) active nuclear-burning regions of different composition, in a multilayered structure that resembles an onion (see Figures 7.5 and 7.6).

Si-burning puts an end to the series of successive fusion stages that a star can undergo, since no further nuclear energy can be obtained through fusion (recall that Fe-peak nuclei resulting from Si-burning are the most tightly bound species; see Figure 2.3). And with the termination of the nuclear energy supply that helps supporting the star against gravitational collapse, its dramatic fate gets sealed.

7.3.1 Early Models: Prompt Shocks and Neutrino Transport

The establishment of two distinct supernova classes, I and II, on a purely observational basis, quickly sparked a number of speculative suggestions on the nature of such cosmic events. Walter Baade and Fritz Zwicky were the first to propose that a type II supernova *"represents the transition of an ordinary star into a body of considerable smaller mass[28]"*. By 1960, it was widely accepted that the central regions of a massive star, once its nuclear fuel has been fully transformed into tightly bound species (i.e., Fe-peak nuclei), undergo a gravitational collapse to some form of compact object, such as a neutron star. The collapse is actually driven by the continuous supply of mass from the overlying silicon burning shell

---

[27]Very massive stars, $M \gtrsim 100$ M$_\odot$, undergo different evolutionary paths (see Figure 7.4). These stars are characterized by high central temperatures, such that very energetic photons from the strong radiation field can actually undergo $e^+ - e^-$ pair production before core O-burning sets in. Radiation pressure plays an important role in supporting massive stars against gravity. Accordingly, a reduction of the number of photons in the stellar plasma through pair production leads to a pressure drop that can destabilize the star (see references [127, 190, 547, 550, 1467] for seminal work on this subject). A partial collapse, followed by a *pair-instability supernova*, is their likely fate.

In stars with initial masses ranging between 100 M$_\odot$ and 150 M$_\odot$, the collapse of the oxygen core is followed by explosive O- and Si-burning. The amount of nuclear energy released can actually halt and reverse the collapse into an explosion accompanied by moderate mass ejection. Afterward, the star relaxes back to hydrostatic conditions on a thermal timescale, and a second pair production episode may result if the remaining stellar mass is high enough. After several such episodes, known as pulsational pair-instabilities, the star can no longer support pair creation and ultimately collapses to a black hole [758, 760]. For stars in the mass range 150 M$_\odot$–260 M$_\odot$, the nuclear energy released during explosive burning is sufficient to blow off the star completely, powering a true pair-instability supernova. No remnant, neither a neutron star nor a black hole, is leftover. Finally, stars above 260 M$_\odot$ are expected to collapse to black holes after a single pair-instability episode. Observationally, a number of pair-instability supernova candidates, such as SN 2007bi [585], SN 2213-1745, and SN 1000+0216 [371], have been discovered in recent years. See references [758, 760] for further details.

[28]The idea, however, was somewhat put aside by Fred Hoyle and William Fowler who, decades later, still advocated for a common thermonuclear origin for both supernova types [547, 845]. Note, however, that Kushnir and Katz have recently revisited the collapse-induced thermonuclear explosion mechanism in massive stars [1035].

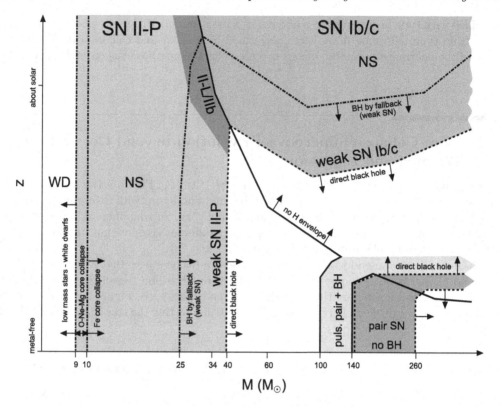

FIGURE 7.4

Supernova types resulting from the evolution of massive, single, nonrotating stars as a function of the initial mass and metallicity. A thick, diagonal line separates stars that retain their H-rich envelopes (i.e., type II supernovae; left and lower right regions of the diagram) from those that lost them in previous evolutionary stages (types Ib/c supernovae—upper right—and pulsational and nonpulsational pair-instability supernovae—a small region at the bottom of the figure, corresponding to stars within a mass range of 100–260 $M_\odot$). Other lines delimit regimes leading to direct and fall-back formation of black holes (BH), neutron stars (NS), or white dwarfs (WD). Figure adapted from Heger et al. [758]; reproduced with permission.

to the core, which is supported by electron degeneracy pressure. When the core exceeds the Chandrasekhar mass, it becomes unstable to gravitational collapse[29]. But how exactly the implosion turns into an explosion, with the subsequent ejection of a significant fraction of the mass of the star, was (and still is!) a matter of debate.

Two, somewhat competing models were proposed in the early years by Stirling Colgate and collaborators as possible mechanisms driving a type II supernova[30]. The first, known as the *prompt-shock* or *direct mechanism*, assumes that at some point during the collapse[31],

---

[29]The specific value at which the collapse of an iron core occurs requires a detailed account of a number of corrections (i.e., general relativity, Coulomb corrections, partial degeneracy) that have to be added to the traditional Chandrasekhar mass formula, $M_{\rm Ch} = 5.83 Y_e^2$ (e.g., 1.457 $M_\odot$ for a plasma with $Y_e = 0.5$). All in all, the mass of the iron core than can collapse ranges between 1.25 and 2.05 $M_\odot$, with 1.3–1.6 $M_\odot$ being the most typical values [1987].

[30]See references [174, 274, 769, 1268, 1622, 1975, 1987] for a historical insight.

[31]During the early stages, the collapse gets accelerated by two effects. On one hand, electron captures onto

Preexplosion

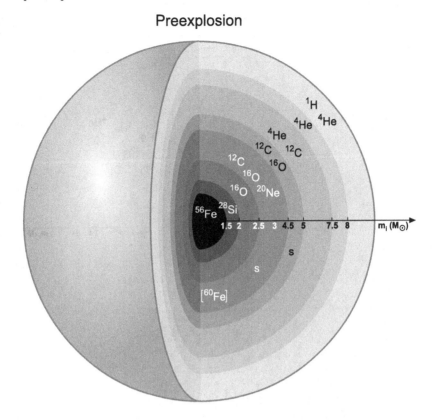

FIGURE 7.5

Structure of a ~ 25 $M_\odot$ star of solar composition, displayed as a function of the interior mass, m_i, at the end of central Si burning [1110, 1112]. Shortly before core-collapse ensues, the star presents a noticeable chemical stratification, with a layered structure that resembles an onion. ^{56}Fe, the ashes of core and shell Si-burning, is the most abundant species at the innermost regions of the star, up to $m_i \sim 1.5$ $M_\odot$. This region is surrounded by a ^{28}Si-rich layer (up to $m_i \sim 2$ $M_\odot$), resulting from shell O-burning, and, subsequently, by concentric layers mostly made of ^{16}O ($m_i \sim 2.5$ $M_\odot$; shell Ne-burning), ^{20}Ne-^{16}O-^{12}C ($m_i \sim 4.5$ $M_\odot$; shell C-burning), ^{16}O-^{12}C ($m_i \sim 5$ $M_\odot$; core He-burning), ^{12}C-^{4}He ($m_i \sim 7.5$ $M_\odot$; shell He-burning), ^{4}He ($m_i \sim 8$ $M_\odot$; shell H-burning), and an outer ^{1}H-^{4}He-rich layer resulting from core H-burning. A weak s-process (see Section 7.4.1) is also predicted during core He-burning ($4.5 \leq m_i(M_\odot) \leq 5$) and shell C-burning ($2.5 \leq m_i(M_\odot) \leq 4.5$). Note that the γ-ray emitter ^{60}Fe is also synthesized during shell C-burning [1111].

when the density exceeds the nuclear density[32], nuclei and free nucleons begin to experience the repulsive character of the strong nuclear force at short distances. Accordingly, the equation of state stiffens and the plasma becomes incompressible (i.e., it essentially behaves like a rigid wall). Collision with the overlying material, still infalling, generates pressure waves

---

nuclei, (e^-, ν_e), efficiently reduce the electron degeneracy pressure that supports the star against gravity. In turn, such electron captures give rise to an important burst of electron neutrinos. On the other hand, at the very high temperatures achieved in the stellar plasma, Fe-peak nuclei are efficiently disintegrated by high-energy photons. This reduces the radiation pressure, favoring in turn the implosion of the star.

[32]The nuclear density in a stellar plasma made of nucleons with equal number of neutrons and protons amounts to $\rho_{nuc} \sim 2.7 \times 10^{14}$ g cm^{-3}.

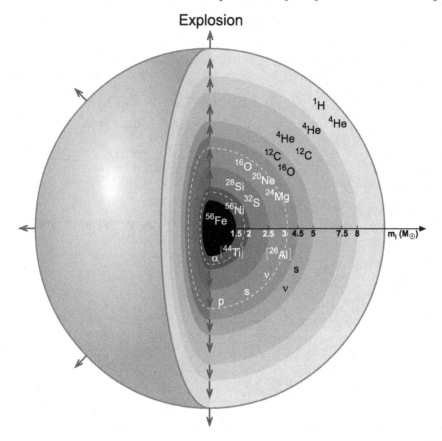

FIGURE 7.6
Same as Figure 7.5, right after core-collapse. The advance of the shock wave generated in the outer core subsequently gives rise to a suite of explosive burning stages and nucleosynthesis processes in the overlying layers: Explosive Si-burning and an α-rich freezout transform the first layer encountered by the shock into a ^{56}Ni-rich plasma (note that ^{44}Ti is also synthesized in this region). The next layer undergoes explosive O-burning, becoming ^{28}Si-^{32}S-rich. Finally, explosive Ne-C (together with a weak s-process, a p-process, and a ν-process) is triggered by the passage of the shock wave throughout the next layer, which becomes ^{16}O-^{20}Ne-^{24}Mg-rich. Note that ^{26}Al is also synthesized in this layer. Compositional changes driven by the shock wave are actually limited to layers below the outer dashed line shown in the figure. The inner dashed line locates the approximate boundary between ejected and fall-back matter, or *mass cut*.

that propagate outward, powering a shock that may eventually lead to mass ejection [364]. Unfortunately, most simulations revealed that even though a shock sets in, it rapidly degrades as most of its energy is invested in dissociating Fe-peak nuclei into nucleons. All in all, the shock stalls and fails to eject mass[33].

A few years later, Colgate and White [366] proposed the *neutrino transport model*, in an attempt to overcome the drawbacks of the prompt-shock mechanism. In this model, a fraction of the gravitational energy released during the collapse is converted into neutrinos, which may interact with, and ultimately unbind, the stellar mantle (see also reference [58]).

---

[33]See reference [378] for an overview on the direct mechanism.

Neutrinos carry more than 90% of the gravitational energy released during the collapse. In fact, the optical display that accompanies a supernova explosion, the *fireworks*, is just a residual side effect of the event, of little relevance in terms of the overall energy budget. Indeed, only a small fraction of the energy carried away by neutrinos is actually needed to power the explosion. Unfortunately, detailed simulations that included a more realistic treatment of neutrino transport seriously questioned the efficiency of this mechanism [59, 61, 1942], beginning an era in which the enthusiasm unleashed by promising developments was quickly neutralized by failed explosions in nearly all variants of numerical simulations. As an example, the publication of larger scattering cross-sections for neutrino-matter interaction, resulting from the inclusion of weak neutral currents [556], revitalized the interest in the neutrino-aided explosion model. However, while this may potentially increase the amount of energy (and momentum) deposited by neutrinos into the stellar mantle [258, 1592, 1943], detailed numerical simulations deceptively revealed that a large fraction of the neutrinos released were actually trapped inside the core[34] [1046, 1212, 1213, 1566, 1925], proving once more unsuccessful to power an explosion[35] [225, 259, 1945].

With neutrino transport models looking like a dead end, interest shifted once more toward core bounce and prompt-shock formation[36] [176, 1214]. Estimates suggested that while the energy carried by the shock can naturally account for the energetics of a normal type II supernova, the shock itself was found to degrade during its passage through the yet infalling plasma[37]. All in all, simulations revealed that a certain level of fine-tunning was needed for a successful explosion, such that slight differences in the input physics, for instance, may yield a totally different outcome. Indeed, while some researchers were getting explosions under specific conditions, others were not [65, 131, 132, 226, 272, 379, 795, 796, 801, 1946].

One of the few successful explosions modeled in the 1980s under realistic conditions was reported by Wilson [178, 1944], who found that several hundred milliseconds after the shock stalled, a weak explosion, with a net energy release of 4×10^{50} erg, developed (see Figure 7.7). The analysis of the intriguing delayed explosion pointed toward a late neutrino heating diffused out from the newly formed neutron star as the responsible for the revival of the shock. However, a number of concerns have been raised on different aspects of Wilson's model, such as the effect of the averaging of neutrino properties imposed by the assumption of spherical symmetry [797, 906]. A large number of increasingly sophisticated models have been developed since then. As confirmed by SN 1987A, neutrinos are indeed released during a type II supernova (see Section 7.1). However, the specific way in which the star makes use of such weakly-interacting particles to turn the stellar collapse into an explosion still remains unknown after more than 50 years of intensive research. The solution may involve neutrinos but also rotation, magnetic fields, and a number of fluid instabilities emerging from the true multidimensional nature of the problem[38] [1267, 1268, 1755, 1959].

While a fully self-consistent explosion model awaits to be developed, different techniques have been adopted to mimic the dynamics of a supernova. To this end, early models induced the shock wave by assuming a violent stimulus in a small region of the star[39]. In practice, this can be implemented in different ways [91]. One mechanism, although not tradition-

---

[34]The most important neutrino interactions during the collapse are (neutral current) elastic scattering on nuclei, (ν_e, ν'_e), electron-neutrino scattering, $e^- (\nu_e, \nu'_e) e'^-$, inverse β-decay, (ν_e, e^-), and inelastic scattering on nuclei, (ν_e, ν'_e) [261]. When the density reaches $\sim 10^{12}$ g cm^{-3}, the neutrino diffusion time becomes larger than the collapse time, and the neutrinos get trapped in the so-called *neutrino sphere* [174].

[35]See references [274, 1210] for overviews on neutrino transport models and their shortcomings.

[36]See also references [63, 64, 126].

[37]Energy losses are mostly due to the emission of neutrinos and to photodisitegration of Fe nuclei into free nucleons.

[38]See also references [260, 498, 1123, 1666] for studies based on the role of convection in type II supernovae.

[39]This technique is still implemented in spherically-symmetric type II supernova models.

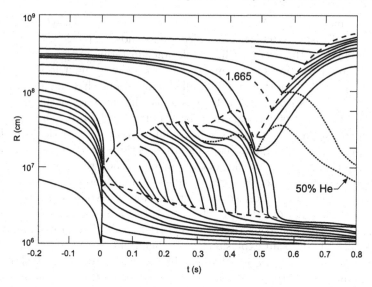

FIGURE 7.7

Trajectories of various mass points vs. time from core bounce, corresponding to the delayed explosion model of Wilson [178, 1944]. The upper, dashed line shows the location of the shock front. Note that the stalled shock gets revitalized by neutrino heating about 0.5 s after bounce. Figure from Bethe and Wilson [178]; reproduced with permission.

ally used in type II supernova simulations, is the *kinetic energy bomb*, in which a certain region of the stellar core is given outward velocities ($\sim 2 \times 10^9$ cm s^{-1}). The collision between such outgoing layers and the infalling material overlying the core ultimately induces a strong shock wave. A variation of the mechanism outlined above is the *internal energy bomb*, in which the temperature (rather than the velocity) of a region of the star is suddenly raised [733, 1629, 1795]. This, in turn, increases the pressure and subsequently pushes the surrounding stellar plasma outward. The free, adjustable parameters of the *bomb* mechanism are the amount, location, and time during the collapse when the energy deposition (temperature rise) is implemented. The last method relies on a *piston* to accelerate the plasma above a certain depth [1962, 1984, 1986]. The adjustable parameters are now the mass enclosed by the piston, the initial velocity of the piston, and a factor introduced to mimic the lifting effect driven by the existing pressure gradients accross the star. A drawback faced by the piston mechanism is the lack of a priori knowledge of the energy deposited by the shock. Nevertheless, this is a well-constrained quantity, since the total kinetic energy of the explosion must agree with the value inferred observationally from the light curve ($\sim (1-2) \times 10^{51}$). Therefore, the parameters of the model must be readjusted to match the required total kinetic energy. The location of the initial energy deposition, which defines the boundary between ejected and fall-back matter, or *mass cut*, is also constrained by a number of observables: Nucleosynthesis, for instance, reveals an anomalous overproduction of neutron-rich, Fe-group nuclei if the energy is initially deposited inside the iron core; on the other hand, a location beyond the O-burning shell would yield a too-large neutron star mass, after fall-back. Accordingly, most simulations place the *mass cut* within the Si-rich shell (see the inner, dashed line in Figure 7.6).

7.3.2 Shock Waves and Explosive Burning Regimes

The shock driven by core bounce propagates inside out, through layers of different composition, corresponding to the ashes of the different fusion episodes experienced by the star in previous evolutionary stages.

The nuclear activity induced by the passage of the shock wave will be here sketched in the framework of a representative 25 $M_\odot$ presupernova star of initial solar composition. The first layer encountered by the shock is mainly made of ^{28}Si, and as such undergoes complete explosive Si-burning (Section 2.4.3.4). At the typical temperatures achieved in this layer, $T \geq 5$ GK, the nuclear activity is characterized by equilibrium between strong and electromagnetic interactions. The abundances of all species present in this layer correspond to nuclear statistical equilibrium, and are given by the temperature, density, and neutron excess of the stellar plasma (see Section 2.4.3.5). In a typical 25 $M_\odot$ model, explosive Si-burning occurs in a layer characterized by a small neutron excess, $\eta \sim 0.003$. Since nuclear statistical equilibrium favors production of species with the largest binding energy and with individual neutron excesses close to η (i.e., $N \sim Z$, in this case), ^{56}Ni and, to a much lower extent, ^{57}Ni, ^{55}Co, and ^{54}Fe, become the main endproducts of explosive Si-burning (see Chapter 2 for a discussion on the largest binding energy of all Fe-peak nuclides).

As the incinerated layer expands and cools down, nuclear interactions begin to abandon equilibrium conditions at a characteristic *freeze-out* temperature [1969]. The first reaction to drop out of equilibrium is the triple-α. If the density (or the expansion time) when this reaction achieves freeze-out temperature is sufficiently large, nuclear statistical equilibrium predicts a deficiency of light particles (i.e., p, n, α) in the stellar plasma. This gives rise to the so-called *normal freeze-out*, during which capture reactions involving the very few light particles leftover do not alter the nuclear statistical equilibrium abundances (dominated by ^{56}Ni and the other Fe-peak nuclides). On the other hand, a sufficiently small density (or expansion time) at freeze-out results in a large excess of α-particles, whose subsequent captures by nuclei will dramatically alter the nuclear statistical equilibrium composition, in what is known as *α-rich freeze-out*. Calculations show that, for a stellar plasma characterized by $\eta \sim 0$, the most abundant nucleus is again ^{56}Ni, but other important species, such as the γ-ray emitter ^{44}Ti, are also produced in smaller amounts (see Section 7.5.1).

As the shock wave propagates outward, it continues to cross ^{28}Si-rich layers, which become progressively heated to temperatures in the range of $T \sim 4$–5 GK. Following the drop in temperature, nuclear statistical equilibrium breaks down in two quasi-equilibrium clusters, one around ^{28}Si and extending up to $A \sim 40$, and a second forming around the Fe-peak nuclei and involving species with $A \gtrsim 50$. A number of nuclear interactions, not in equilibrium, connect both clusters, progressively transforming silicon into Fe-group nuclei (Section 2.4.3.4). The abundance of any nuclide in those quasi-equilibrium clusters depends upon four main parameters (aside from nuclear properties, such as binding energies, masses, and spins): Temperature, density, neutron excess, and the abundance of ^{28}Si in the stellar plasma [208]. Subsequent expansion, following the passage of the shock wave, causes freeze-out. Since a significant amount of ^{28}Si still remains, the process is referred to as *incomplete silicon burning* [812], and the final composition in these layers is dominated by the presence of Fe-peak nuclei, ^{28}Si, and intermediate-mass species.

After freeze-out from nuclear statistical equilibrium and incomplete Si-burning, the next relevant burning regime induced by the shock wave is explosive O-burning ($T \sim 3$–4 GK) in the overlying ^{16}O-rich layers. As in incomplete Si-burning, the fuel (^{16}O) gets dissociated, giving rise to the onset of two quasi-equilibrium clusters centered around silicon and iron-peak nuclei, respectively. The lower temperatures achieved in the plasma translate into a much lower production of Fe-peak nuclei from silicon. Indeed, after freeze-out, the most abundant species in these layers are actually ^{28}Si, ^{32}S, ^{36}Ar, and ^{40}Ca.

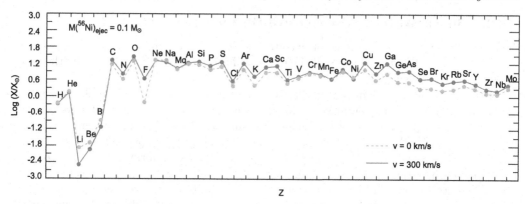

FIGURE 7.8
Integrated overproduction factors relative to solar abundances, for models of rotating (dark gray, solid line) and nonrotating (light gray, dashed line) massive stars (13 $M_\odot$–120 $M_\odot$), leading to core-collapse supernovae. The mass cut, that separates the ejecta from material which remains gravitationally bound, has been fixed in order to get 0.1 $M_\odot$ of ^{56}Ni in the ejecta of all models. Figure adapted from Chieffi and Limongi [324]; reproduced with permission.

Finally, the last burning stage takes place in the layers composed mostly of ^{16}O, ^{20}Ne, and ^{12}C, which become heated by the shock wave to $T \sim 2$–3 GK. As a result, ^{20}Ne and, to a lower extent, ^{12}C undergo explosive burning. Because of the lower temperatures attained in the plasma, reactions operate far from equilibrium (i.e., no quasi-equilibrium clusters are formed). Therefore, the abundance of any species is not anymore determined by a handful of parameters, such as temperature, density, and neutron excess, through nuclear statistical equilibrium calculations. Now, the final yields become sensitive to the initial composition of these layers and to the rates of the different nuclear reactions involved. Freeze-out is now characterized by the presence of ^{16}O, ^{20}Ne, ^{24}Mg, and ^{28}Si. Another important γ-ray emitter, ^{26}Al, is also produced during explosive Ne/C-burning (see Section 7.5.1).

All the other layers encountered by the shock become heated to peak temperatures below $\lesssim 2$ GK for very short times and, accordingly, do not experience significant nuclear processing. These layers, located beyond the outer dashed line in Figure 7.6, will be eventually ejected with an unaltered composition, resulting almost exclusively from the handful of presupernova burning stages. All in all, nucleosynthesis driven by the passage of the shock wave in core-collapse supernovae (Figure 7.8) proceeds through several burning regimes: normal or α-rich freeze-out from nuclear statistical equilibrium in the innermost regions, and incomplete Si-, O-, and C/Ne-burning in the overlying layers [1796, 1798, 1961]. About one hour since the onset of core collapse, the shock, traveling at an average speed of several thousand kilometers per second, hits the stellar surface.

7.4   Nucleosynthesis

The evolution of a massive star ($M \gtrsim 10 \ M_\odot$) constitutes one of the most complex stellar phenomena from a nuclear physics viewpoint. During its early, hydrostatic evolution, the star undergoes a suite of distinct nuclear fusion stages, from H to Si burning. Nucleosynthesis beyond Si also occurs (e.g., the weak s-process, which accounts for the creation of elements

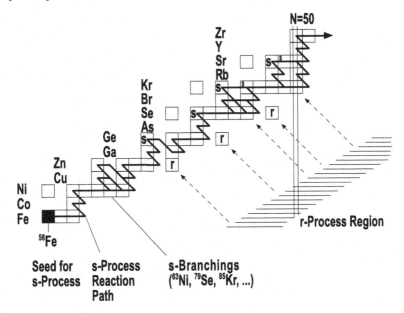

FIGURE 7.9
Main neutron-capture s- and r-process paths involved in the formation of nuclei beyond Fe. Figure adapted from Käppeler et al. [947]; reproduced with permission.

between iron and strontium, i.e., $A \lesssim 90$; see below), first during core He burning, and later on, during shell C burning. The rich variety of nuclear processes continues during the explosive stage and includes the classical r- and p-processes, together with an α-rich freeze-out and neutrino interactions—the ν and νp-processes. All these nucleosynthesis processes will be briefly addressed along the following sections.

7.4.1 The s-Process

The (slow) s-process occurs in stellar plasmas at relatively low neutron densities ($n_n \sim 10^7 - 10^{12}$ cm^{-3}) and moderate temperatures [947]. Under such conditions, nucleosynthesis is driven by a series of slow neutron captures on stable, seed nuclei, until an unstable species is reached. Most frequently, the neutron capture rates onto such unstable nuclei are slower than the corresponding β^--decay rates (hence the name) and, accordingly, the nuclide decays. In the process, a neutron transforms into a proton, increasing the atomic number of the synthesized species by one unit. Once a stable nuclei is reached, after one or several decays, another chain of neutron captures initiates. All in all, the process synthesizes stable isotopes by moving along the chart of nuclides in zigzag, close to the valley of stability (i.e., horizontally during n-captures and along diagonal lines during β-decays; Figure 7.9). Note, however, that at specific points on the chart of nuclides, certain unstable nuclei exhibit similar decay and neutron-capture probabilities, such that the main s-process path splits in two branches[40].

The abundances of the species synthesized through the s-process depend on the corresponding neutron-capture cross-sections and on the overall neutron flux (i.e., number of neutrons per unit time and area, frequently expressed in s^{-1} cm^{-2}). Nuclides characterized

---

[40]The study of the abundance pattern in such branchings provides important constraints on the physical conditions under which the s-process operates (i.e., temperature, neutron density) as well as on the stellar site where this process actually occurs [945, 946].

by very small neutron-capture cross-sections, like those with neutron-magic numbers (i.e., $N = 50$, 82, and 126), will not be dramatically affected by (n, γ) reactions and, therefore, will substantially pile up relative to those with larger cross sections. This accounts for the presence of narrow peaks at masses $A \sim 84$, 138, and 208 (which correspond to neutron-magic numbers) in the Solar System abundance distribution[41] (see Figure 2.2).

The analysis of the Solar System s-process abundance distribution has actually revealed the existence of three distinct components, referred to as *main* (for $A \sim$ 90–205), *weak* ($A \lesssim 90$), and *strong* ($A \sim 208$), which are characterized by the capture of about 10, 3, and 140 neutrons per seed nucleus, respectively [948]. Low-mass, thermally pulsing AGB stars, with $M \lesssim 4$ $M_\odot$, are considered the likely site of the main s-process component [594, 948]. Low-mass, metal-poor, thermally pulsing AGB stars are the main contributors to the strong s-process component [592, 1819], while core-helium burning and the subsequent carbon-shell burning in massive stars, as mentioned before, account for the weak s-process component [1419, 1446, 1466]. Simulations also suggest that the dominant neutron sources that trigger the weak and strong s-process components correspond to ^{22}Ne(α, n)^{25}Mg, and ^{13}C(α, n)^{16}O, respectively. With regard to the main s-process component, it is mostly driven by ^{13}C(α, n)^{16}O, with a marginal contribution from ^{22}Ne(α, n)^{25}Mg during thermal pulses.

Reliable determinations of s-process yields in the astrophysical scenarios depicted above require precise knowledge of the relevant nuclear reactions involved, at energies ranging from $kT \sim 8$ keV for low-mass AGB stars to 90 keV during shell C-burning in massive stars. The list includes neutron captures for nuclei up to $A \sim 210$, and all interactions with species that act as neutron sources (^{13}C(α, n)^{16}O and ^{22}Ne(α, n)^{25}Mg) or sinks (e.g., the neutron poisons ^{12}C, ^{14}N, ^{16}O, ^{22}Ne, and ^{25}Mg). The required uncertainties, smaller than a few percent, have been achieved for most nuclei. Interested readers can find further information on s-process nucleosynthesis and related topics in, e.g., references [277, 945, 947, 949, 1159, 1161, 1419, 1480].

7.4.2   The r-Process: Mergers or Blasts?

The (rapid) r-process occurs in stellar plasmas for more extreme neutron densities ($n_n \sim 10^{20} - 10^{28}$ cm^{-3}) and temperatures (~ 1 GK) than in the s-process [77, 1021, 1480]. It is driven by a series of rapid neutron captures, which proceed at a faster rate than the corresponding β-decays, for radioactive nuclides. The fast neutron captures drive the abundance flow far from the valley of stability, toward the neutron-drip line[42].

As in the s-process, r-process nucleosynthesis favors the production of species with (neutron) magic numbers. Once neutron irradiation terminates, such neutron-magic nuclei undergo a series of β-decays until the most neutron-rich stable (or long-lived) species is reached. Through this mechanism, about half of the nuclides beyond the iron peak, including the precious metals gold, silver, and platinum, and the nuclear cosmochronometers uranium and thorium (i.e., ^{235}U, ^{238}U, and ^{232}Th), are believed to be synthesized.

Subtraction of the (theoretical) main s-process component [76, 667] from the bulk of measured Solar System abundances[43] yields two prominent peaks in the residual distribution, at mass numbers near $A = 130$ and 195 (which correspond to neutron-magic numbers at $N = 82$ and 126). The presence of such r-process peaks in the abrupt topology of the chart of nuclides along the neutron-rich side, and of the long-lived radioisotopes ^{232}Th

---

[41]Several s-process signatures have been unambiguously identified in presolar meteoritic grains (see Chapter 3), giving clues on the nature of their stellar parent bodies.

[42]The neutron-drip line is the physical boundary for neutron-rich, bound nuclei. Conversely, the most proton-rich nuclei define the proton-drip line [503]. Nuclei beyond drip lines are particle-unbound and, therefore, are considered not to exist.

[43]Note, however, that whether the s-process component is better known than the r-process component is currently a matter of debate.

$(T_{1/2} = 1.4 \times 10^{10}$ yr$)$, ^{235}U $(T_{1/2} = 7.0 \times 10^8$ yr$)$, and ^{238}U $(T_{1/2} = 4.5 \times 10^9$ yr$)$, located beyond the s-process nucleosynthesis endpoint, provides strong evidence for the existence of the r-process.

Additional clues on the way r-process nucleosynthesis occurs have been obtained from spectroscopy of primitive, metal-poor, Galactic halo stars [1671] (Figure 7.10). Indeed, elemental abundances inferred for these stars (e.g., the remarkable CS 22892-052 [1672]) agree fairly well with the r-process abundances in the Solar System for $Z \gtrsim 56$–76 (i.e., between barium and osmium), but not so well for the range $Z \sim 38$–56 (from strontium to barium[44]). This may suggest different synthesis mechanisms for the light and heavy r-process nuclei[45]. Moreover, the presence of r-process nuclei in such metal-poor (and therefore old) stars suggests a synthesis mechanism that operated early in the history of the Galaxy[46].

A phenomenological account of r-process nucleosynthesis was already outlined in the seminal work of Seeger, Fowler, and Clayton [1603]. Fast neutron captures progressively drive the main r-process path toward the neutron drip line, until equilibrium between (n, γ) reactions and reverse (γ, n) photodisintegration reactions is reached. Such equilibrium translates into an accumulation of the abundance flow at the so-called *waiting points*, whose location depends upon the specific neutron density and plasma temperature, aside of the *neutron separation energies* (or reaction Q-values) of the different n-capture reactions[47]. Remarkably, this equilibrium is independent of the corresponding (n, γ) cross-sections. In fact, the final r-process abundances, once neutron captures cease, depend only on the β-decay half-lifes of the waiting-point nuclei.

First attempts to quantify the r-process contribution to the Solar System abundances relied on the *classical or canonical r-process model* [268, 1603]. It assumes constant temperature and neutron density, an instantaneous termination of the neutron irradiation after a characteristic time, τ_n, and (n, γ) – (γ, n) equilibrium at the waiting points [76, 77]. Unfortunately, no single combination of the parameters of the model (i.e., $T - n_n - \tau_n$) can match the overall distribution of r-nuclei abundances in the Solar System [1018]. Indeed, three different sets of values, in the range $T \sim 1.2 - 1.35$ GK, $n_n \sim 3 \times 10^{20} - 3 \times 10^{22}$ cm^{-3}, and $\tau_n \sim 1.5 - 2.5$ s, are at least required. A less restrictive parametric approach, known as the *multievent r-process* [222, 668], assumes a suite of different thermodynamic conditions in a single object to account for the bulk of r-process nuclei. The model adopts the same basic assumptions of the classical model, except the equilibrium between (n, γ) and (γ, n) reactions during the whole neutron irradiation process. Nevertheless, in a realistic scenario one would expect that both the temperature and the neutron density would change with time, such that the r-process actually occurs under dynamic conditions[48] [425, 798, 1565, 1591]. Pioneering dynamical r-process models relied on a hot stellar plasma[49] in nuclear statistical equilibrium, which subsequently expanded and cooled down in a characteristic timescale. Even though no specific astrophysical scenario was directly invoked, such dynamical models provided important clues on the conditions under which the r-process actually occurs, namely, the entropy and neutron excess $(\sum_i (N_i - Z_i) Y_i)$ of the stellar plasma and the charac-

---

[44]See, however, Roederer et al. [1512] for recent measurements below Ba and on the agreement between $Z = 52$ (Te) measurements and theoretical predictions.

[45]Several mechanisms have also been invoked to account for the synthesis of the light r-process nuclei, including a weak r-process [1021], a light element primary process (LEPP) [1262, 1818], and an α-process [1462].

[46]See Arnould and Goriely [76] and Roederer et al. [1510] for critical discussions on the universality of the r-process.

[47]Mathematically, the equilibrium is described by means of the Saha equation. See Equation 2.119 and references [344, 359, 878, 1513].

[48]See references [76, 77, 1020] for discussions on the limitations of the classical r-process model.

[49]Shinya Wanajo, however, has also presented a cold r-process scenario that can match the solar r-process abundance pattern when equilibrium between (n, γ) and (γ, n) reactions is not imposed [1886].

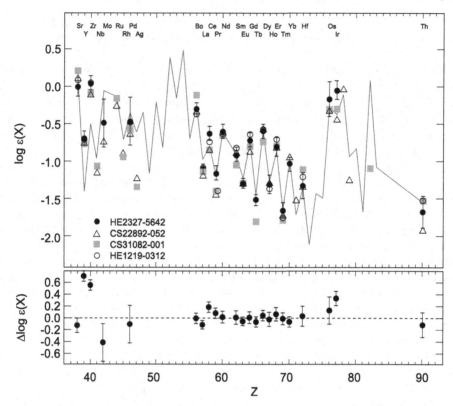

FIGURE 7.10

(Top) Heavy-element abundances of the low-metallicity halo stars HE1219-0312, HE2327-5642, CS22892-052, and CS31082-001, compared to the Solar System r-process abundances (solid line) scaled to match Ba–Hf. Abundances are given in standard spectroscopic notation, where $\log \varepsilon(i) \equiv \log_{10}(N_i/N_H) + 12.0$, and N_i denotes abundance by number. They have been normalized to the value derived for Eu in HE 2327-5642. (Bottom) Difference in $\log \varepsilon$ between HE 2327-5642 and Solar System r-process abundances (see also references [555, 1511], for different Pb r-process abundances). Figure adapted from Mashonkina et al. [1202]; reproduced with permission.

teristic timescale during which favorable conditions for the r-process are sustained. Furthermore, these efforts paved the road to identify two, somewhat competing scenarios, where the r-process can, in principle, occur: neutrino-driven winds accompanying core-collapse supernovae and neutron star mergers (Figure 7.11).

Neutrino-driven winds are the most popular site invoked for r-process nucleosynthesis. During core collapse, the stellar plasma is ruled by a suite of different nuclear processes that include β-decays, electron captures, and photodisintegration reactions. This gives rise to a burst of neutrinos that interact with the outermost layers of the newborn, proto-neutron star, as described in Section 7.3.1. These layers become progressively heated and ultimately ejected in a continuous wind. Early simulations suggested that about 10 s after the stellar core bounces, high entropy values are attained in the emitted neutrino-driven wind [1978, 1990], allowing the occurrence of a successful r-process for a wide range of neutron excesses and dynamical timescales. The expections, however, were shortly after lowered by additional simulations that yielded much lower entropies[50] [1771]. Subsequent studies confirmed that neutrino-driven winds can host r-process nucleosynthesis but only under unrealistic conditions [1463]. Indeed, by the time the wind evolves to high entropy, both the characteristic timescale and the neutron excess get too high for allowing a successful r-process [1802]. And moreover, the predicted wind mass-loss rates at those late times are found to be too small to significantly pollute the Galaxy in r-process nuclei [1463,1771]. More recent hydrodynamic simulations support the idea that the entropy needs can be attained during the late stages of the collapse[51] [57,1381] but already when the physical conditions required to power an r-process no longer hold [1034]. It is finally worth noting that, according to several core-collapse simulations [533,851], the resulting neutrino-driven wind can actually be proton-rich (rather than neutron-rich) for many seconds, a situation clearly at odds with r-process requirements[52].

Another, alternative r-process site is decompressed neutron star matter, ejected in the coalescence of a neutron star and a black hole in a compact binary system [1057]. The scenario was later extended to the case of double neutron star mergers[53] [487,558,1754]. Hydrodynamic simulations have proved that a substantial amount of matter enriched in r-process nuclei [558,1235] can be ejected during such stellar events[54] (see also refs. [905, 1342,1537]). The scenario has been frequently revisited [55]. Recent simulations of double neutron stars and neutron star-black hole mergers have reported a robust production of heavy r-process nuclei with $A > 130$–140 [142,669,1001,1506,1535]. A latitude-dependent production of r-process nuclei has also been reported in the framework of 3D simulations of neutrino-driven winds expelled from the remnant of a binary neutron star merger. The polar ejecta is enriched in $A \approx 80$–130 species, while contributions from lower latitudes include very neutron-rich nuclei, up to the third r-process peak around $A \approx 195$ [1407]. Successful

---

[50]Roberts et al. attributed the unrealistically high entropies reported to problems in the adopted equation of state [1507].

[51]The likelihood of a prompt, rather than late, r-process, occurring just 1–2 s after core bounce, has also been scrutinized [1802]. Only combinations of moderate entropies, short dynamical timescales, and high neutron excesses could power a successful r-process reaching the $A = 195$ peak. While the high wind mass-loss rates expected at such early times may have an impact on the Galactic chemical abundances, models request very restrictive conditions on the proto-neutron star.

[52]Neutrino flavor oscillations may also play a role in core-collapse supernova explosions, since they may even decrease further the resulting r-process yields [470,471].

[53]See also reference [1748] for r-process nucleosynthesis in hot accretion disk flows from black hole-neutron star mergers.

[54]The contribution of neutron star mergers to the Galactic inventory of chemical species includes three different processes: (i) direct ejection, (ii) neutrino-driven winds, and (iii) disk disintegration. Lattimer and Schramm [1057] provided the first estimates of the fraction of the neutron star that becomes unbound after a neutron star-black hole merger. Their calculations yield a mean ejected mass per merger of about 0.05 M_{ns}, or equivalently, an accumulated value of 2×10^{-4} $M_\odot$ pc^{-2}, which is approximately equal to the mass fraction of r-process elements in the Solar System. More realistic simulations [1342,1530,1536] yield a mean ejected mass of 3×10^{-3} $M_\odot$, with a mean electron mole fraction in the range of $0.01 < Y_e < 0.5$. Simulations suggest that neutrino-driven winds give rise to the ejection of about $\sim 10^{-4}$ $M_\odot$ of matter with $Y_e \sim 0.1$–0.2 per event [435]. Finally, the disk resulting from the merging episode may give rise to the ejection of neutron-rich material, either through viscous heating in an advective disk or through the energy produced during recombination of nucleons into nuclei [317,1236]. It has been claimed that about 0.03 $M_\odot$ of moderately neutron-rich material ($0.1 < Y_e < 0.5$) may be ejected from the disk [1237].

[55]Other pioneering simulations of double neutron star and neutron star-black hole mergers were performed by Oohara and Nakamura [1349] in the framework of Newtonian gravity and a simplified equation of state (see also references [1063,1470,2023]). Several improvements, aimed at providing a more realistic description compatible with general relativity include the use of pseudo-relativistic potentials [1062,1530], as well as post-Newtonian corrections [95,511]. The first fully general relativistic calculations of such mergers were performed by Shibata and Uryū [1628]. More details on the different methods adopted in the modeling of double neutron star and neutron star-black hole mergers can be found in reference [1532].

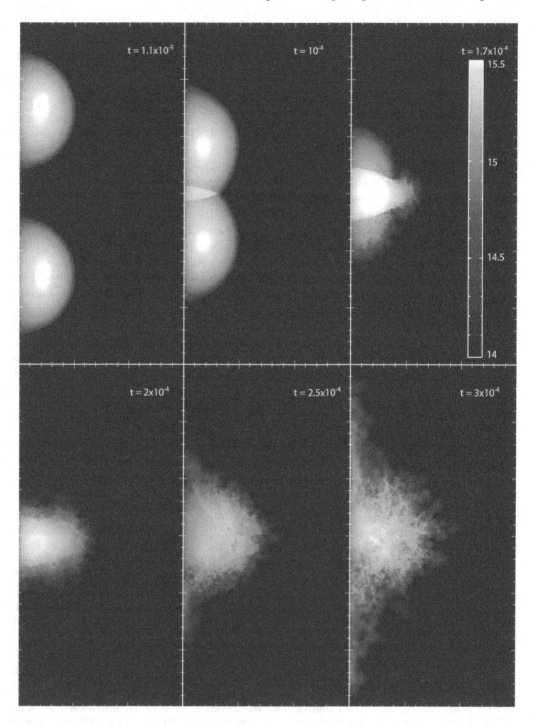

FIGURE 7.11
Time evolution of the density during the head-on collision of two neutron stars. Calculations were performed in 2D with an SPH code. A color version of this image is available at http://fisica.upc.edu/ca/users/jjose/CRC-Downloads. Figure courtesy of R. Cabezón [280].

production of r-process nuclei for $A > 90$ has also been recently reported[56] in the combined ejecta from compact binary mergers (i.e., double neutron stars and neutron star-black hole binaries) and in the neutrino and viscously-driven outflows of the relic black hole-accretion torus systems [938]. To date about ten stellar binaries likely containing double neutron stars have been identified [1153]. Of these, about half are expected to merge over the age of the universe. This is roughly in agreement with the predicted frequency, between 40 and 700 Myr^{-1} [148, 940], for a galaxy like the Milky Way[57].

All in all, an increasing number of simulations is supporting the feasibility of neutron star mergers as a likely r-process site. The specific abundance pattern is determined from detailed postprocessing calculations, that take into account not only the complex interplay between different nuclear interactions but also a set of thermodynamic conditions directly extracted from hydrodynamic models. In this regard, the physical conditions required to power a roburst r-process are fully consistent with the proposed scenario.

r-process nucleosynthesis calculations require huge reaction networks comprising about 5000 nuclear species (up to $Z \gtrsim 110$), lying between the valley of stability and the neutron drip line. Nuclear physics input needs include nuclear masses, β-decay half-lifs, and branching ratios for β-delayed neutron decay. If $(n, \gamma) - (\gamma, n)$ equilibrium conditions are not explicitly imposed, neutron capture rates and photodisintegration reaction rates are also required. Moreover, some detailed dynamical r-process calculations may also require rates for charged-particle and neutron-induced processes, such as (n, α) and (n, p). In general, these inputs almost exclusively rely on theoretical estimates (e.g., statistical Hauser–Feshbach models), since experimental information is very scarce. For additional information on topics related to the r-process, see references [76, 77, 1020, 1462, 1480, 1791].

7.4.3 The p-Process

Studies of element production through n-capture reactions soon identified about 30 proton-rich (neutron-deficient) nuclides, between ^{74}Se and ^{196}Hg, that are systematically bypassed by the s- and r-processes and, hence, require a different synthesis mechanism. Such *p-nuclei* are actually the rarest stable nuclides (no single p-nuclei is indeed the most abundant isotope of any element).

The origin of these proton-rich nuclei may, in principle, result from processing of already existing r- and s-seed nuclei, either by proton captures, (p, γ), or through photo-neutron emission reactions, (γ, n). The synthesis mechanism, known as the *p-process*, was originally envisaged as driven by the passage of a shock wave through the H-rich stellar envelope during a supernova explosion [268, 552, 895]. In this framework, the heaviest p-nuclei would be built up by photodisintegrations while light p-nuclei would result from p-captures. Efficient neutron removal through (γ, n) reactions requires temperatures of a few GK. But at such temperatures, proton-capture reactions have a low transmission probability through the huge Coulomb barriers exerted by nuclei beyond Fe. All in all, the necessary conditions for the synthesis of both light and heavy p-nuclei cannot be simultaneously achieved in the proposed scenario [90].

Core-collapse supernovae were revisited, with the main focus on the O-Ne burning layers [1447, 1479, 1979]. In this framework, the synthesis of p-nuclei could be driven by the photodisintegration of heavy seed nuclei, in a proton-exhausted, hot photon environment characterized by peak temperatures about $T \sim 2$–3 GK, and short timescales. In the pro-

---

[56]See also reference [1889] for the synthesis of r-process nuclei ($A \sim 90$–240) in the framework of fully general-relativistic simulations of double neutron star mergers with approximate neutrino transport.

[57]The frequency of neutron star-black hole mergers is highly uncertain, with estimates varying by orders of magnitude [148, 175]. Nevertheless, with one of the stars being a black hole, those systems are expected to play a minor role in the chemical evolution of the Galaxy.

cess, a number of seed nuclei are destroyed by (γ, n) reactions until a nucleus with a larger probability of occurrence of a (γ, p) or (γ, α) interaction over a (γ, n) is reached. This causes a branching in the abundance flow. In fact, the nuclear flow tends to accumulate at these branching points (or *waiting points*), particularly at nuclides characterized by closed neutron or proton shell configurations. These waiting points directly correspond to light p-nuclei or decay to heavy p-nuclei after cooling, expansion, and ejection of the stellar plasma. The suggested scenario prevents full processing into iron-peak nuclei through photodisinte-grations, because of the short timescale for expansion and cooling that characterize stellar explosions.

Numerical simulations of core-collapse supernovae reproduce the Solar System abundance of most p-nuclei within a factor of $2 - 3$, with the exception of the light p-isotopes of Mo and Ru, and the rare species ^{113}In, ^{115}In, and ^{138}La, which are systematically underproduced [1478, 1479]. A similar pattern has been found in other proposed sites, such as type Ia supernovae (see Chapter 5) and X-ray bursts (Chapter 6).

With few exceptions, most of the relevant reactions for p-process nucleosynthesis ($\gtrsim$ 10,000), such as (γ, n), (γ, p), and (γ, α) photodisintegrations, purely rely on theoretical grounds [58]. To this end, the statistical Hauser–Feshbach model is frequently adopted[59]. For more information on the p-process, including a detailed description of other proposed sites, see references [75, 1474, 1692].

7.4.4 Neutrino-Driven Nucleosynthesis: The ν- and νp-Processes

The hectic nuclear activity that characterizes core collapse during a type II supernova explosion gives rise to an important burst of neutrinos. Although cross-sections that define neutrino-matter interaction are very small (Figure 7.12), the huge number of neutrinos released during this stage can actually leave a number of characteristic nucleosynthesis imprints in the stellar plasma.

Neutrino-induced nucleosynthesis, or ν-*process*, traces back to the 1970s, when Domogatskii, Eramzhyan, and Nadyozhin first underlined the role played by neutrino-matter interactions in the synthesis of light nuclei (e.g., ^{7}Li, ^{9}Be, ^{11}B) [450–454] and of some intermediate-mass species[60] (such as ^{19}F [1974] and ^{26}Al [455]).

The mechanisms by which neutrinos interact with a stellar plasma include inelastic, neutral-current neutrino scattering of nuclei, as well as charged-current interactions, such as (ν_e, e^-) and $(\bar{\nu}_e, e^+)$ [261, 589]. All these interactions[61] can populate excited nuclear levels that subsequently decay via emission of light particles (e.g., n, p, α). As described in previous stages of the evolution of massive stars, the suite of light particles released participate as well in the complex pattern of nuclear interactions, contributing to the synthesis of certain

---

[58]See reference [1469] for a sensitivity study of p-process nucleosynthesis to the adopted rates for the different nuclear interactions.

[59]Reverse photodisintegration rates are frequently inferred from the experimentally-measured forward rates by means of the reciprocity theorem.

[60]See also Woosley [1960] for early work on neutrino-induced synthesis of p-nuclei, and Epstein, Colgate, and Haxton [499] for neutrino-driven production of neutrons during r-process nucleosynthesis.

[61]Neutrino cross-sections are proportional to the square of their energy. Therefore, neutrino-induced nucleosynthesis is mostly driven by μ and τ neutrinos emitted from the proto-neutron star since they have, on average, larger energies than electron neutrinos.

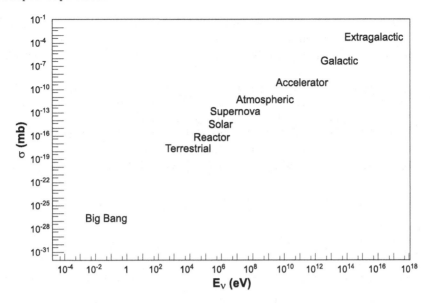

FIGURE 7.12
Neutrino cross-sections across relevant energy scales. Figure adapted from Formaggio and Zeller [545]; reproduced with permission.

species[62]. ν-process nucleosynthesis models[63] are actually hampered by uncertainties in cross-sections, average neutrino energies, and by details of the adopted explosion models.

A related nucleosynthesis process has been unveiled by recent simulations with energy-dependent neutrino transport, revealing that the neutrino-driven winds accompanying core-collapse supernovae can be proton-rich ($Y_e > 0.5$) for some seconds [267, 532, 851]. Indeed, the presence of protons in the hot neutrino-driven winds ($T > 10$ GK) gives rise to a distinct nucleosynthesis mechanism known as the νp-process [565, 1457, 1885].

At $T \sim 10$–5 GK, following the expansion and cooling of the ejected wind, all neutrons reassemble with protons, giving rise to a plasma made of α-particles and leftover protons. When the temperature decreases to $T \sim 5$–3 GK, α-particles combine to form mostly the $N = Z$ nuclides ^{56}Ni, ^{60}Zn, and ^{64}Ge. Strong photodisintegrations inhibit the further extension of the nuclear path toward heavier species. Once the plasma temperature has dropped to $T \sim 3$–1.5 GK, an important burst of neutrons (typically 10^{14} cm^{-3} for several seconds) is released in the interaction p + $\bar{\nu}_e \to$ n + e$^+$. This allows fast (n, p) reactions, followed by p-captures, onto the abundant nuclei ^{56}Ni, ^{60}Zn, and ^{64}Ge, extending the nuclear activity into the neutron-deficient (proton-rich) side of the valley of stability. Finally, when $T \lesssim 1.5$ GK, charged-particle interactions freeze out, and (n, p) reactions and β^+-decays drive the abundance flow toward the valley of stability. Accordingly, the νp-process is halted by cooling of the ejecta to temperatures below $T \sim 1$ GK, rather than by fuel (i.e., protons) exhaustion.

The recent interest on the νp-process has been sparked by claims of its contribution to

---

[62]Cross-sections for neutrino-nucleus interactions have been evaluated for a wide range of species. See, e.g., http://adg.llnl.gov/Research/RRSN/nu_csbr/neu_rate.html for tables of neutral- and charged-current cross-sections per nucleon, averaged over a normalized Fermi–Dirac distribution for low-energy neutrinos (i.e., 4, 6, and 8 MeV) [1973]. See also Formaggio and Zeller [545] for a recent review on neutrino cross-sections.

[63]See references [759, 2004, 2005] for other ν-process nucleosynthesis studies performed in the last decade.

the solar abundances of light p-nuclides up to ^{108}Cd, including the elusive ^{92}Mo, ^{94}Mo, ^{96}Ru, and ^{98}Ru [1888], systematically underproduced in other proposed scenarios[64] (see Section 7.4.3). This, however, has not been confirmed by further studies (see references [535, 1457]) and, therefore, the issue is currently a matter of debate.

The νp-process is sensitive to details of the explosion, the mass, and possibly the rotation rate of the proto-neutron star. Nuclear physics uncertainties play also an important role in the predicted nucleosynthesis[65]. For example, production of p-nuclei in the $A = 100$–110 mass region is influenced by the adopted rate of the slow triple-α reaction (Sec. 2.4.2), and by the rates of the (n, p) reactions on waiting point nuclei, that govern the path toward heavier species, particularly ^{56}Ni(n, p)^{56}Co and ^{60}Zn(n, p)^{60}Cu.

7.5   Observational Constraints

7.5.1 γ-Ray Observations

Gamma-rays can be released by different processes occurring in a core-collapse supernova [1749]: (i) when the shock wave breaks through the stellar surface (and during the subsequent interaction of the ejecta with the circumstellar material previously pulled off by radiation-driven winds); (ii) in the decay of freshly synthesized radioactive nuclei; (iii) and when the ejecta becomes thin enough for the radiation emitted from the underlying neutron star to penetrate this material. Unfortunately, the amount of radioactive material synthesized in a core-collapse supernova is relatively small, such that the expected γ-ray fluxes are an order of magnitude lower than those associated, for instance, with a thermonuclear, type Ia supernova. Therefore, only close events may provide observational constraints to current explosion models.

γ-rays released through the decay chains ^{56}Ni $\rightarrow$ ^{56}Co $\rightarrow$ ^{56}Fe and ^{57}Co $\rightarrow$ ^{57}Fe in a Galactic core-collapse supernova were already discussed by Clayton [342, 347]. If detectable, they may provide complementary insights into the explosion at different times, due to the different half-lifes of the nuclides involved (e.g., $T_{1/2}[^{57}$Co$] = 271.7$ days, $T_{1/2}[^{56}$Co$] = 77.2$ days, $T_{1/2}[^{56}$Ni$] = 6.1$ days).

Prospects for observations of γ-ray lines from individual supernovae[66] continued to be reported, based on numerical simulations with increasing level of sophistication [622]. However, the amount of radioactive ^{56}Ni synthesized in a core-collapse explosion remained quite uncertain, and values in the range 0.1–0.4 M$_\odot$ were considered plausible [1909]. Best expectations were met by the 847 keV line from ^{56}Co decay, with a maximum flux achieved about 400–600 days past explosion [622]. Remarkably, three lines from the radioactive chain ^{44}Ti $\rightarrow$ ^{44}Sc $\rightarrow$ ^{44}Ca were predicted for dedicated γ-ray spectroscopic observations of the supernova remnant Cassiopeia A (Figure 7.13). Such expectations could be soon validated with the discovery of SN 1987A, an event that marks the dawn of supernova γ-ray astrophysics. SN 1987A is, by far, the best-studied supernova to date. Specifically, it is the only witnessed core-collapse supernova from which γ-rays (and neutrinos) have been detected. A number of characteristic γ-ray lines at energies 0.847 MeV and 1.238 MeV (^{56}Co $\rightarrow$ ^{56}Fe),

---

[64]See, however, reference [516], for the synthesis of light p-, s-, and r-process isotopes in models of high-entropy wind in core-collapse supernovae.

[65]The limited impact of nuclear mass uncertainties on νp-process nucleosynthesis has been analyzed by Weber et al. [1913].

[66]Efforts were mostly devoted to type Ia supernovae, since the expected detectability of γ-ray signals from core-collapse supernovae was considered to be much poorer.

FIGURE 7.13

Composite image of the supernova remnant Cassiopeia A (Cas A), based on observations from Hubble and Spitzer telescopes as well as Chandra X-ray Observatory. While Spitzer has mapped the distribution of warm dust (~ 100 K) in the outer shells, Hubble has revealed filamentary structures of hot plasma (10,000 K), and Chandra has imaged the distribution of very hot plasma ($T \sim 10^7$ K) created in the collision between the ejecta and the surrounding gas and dust. Moreover, Chandra has also spotted the neutron star located near the center of Cassiopeia A's remnant. A color version of this image is available at `http://fisica.upc.edu/ca/users/jjose/CRC-Downloads`. Credit: NASA/JPL–Caltech/O. Krause (Steward Observatory). Source: `http://www.spitzer.caltech.edu/images/1445-ssc2005-14c-Cassiopeia-A-Death-Becomes-Her`; released into public domain by NASA.

and 0.122 MeV and 0.136 MeV (^{57}Co $\rightarrow$ ^{57}Fe) were detected from this event [443, 1209], yielding smoking-gun evidence to the proposed mechanism powering the late-time light curve[67] [621, 1033, 1423, 1972] (see Figure 5.32 for the corresponding decay schemes). About ~ 0.07 M$_\odot$ of ^{56}Co have been inferred in SN 1987A from observations[68] [1071, 1423].

Another nuclide synthesized in core-collapse supernovae[69] is ^{44}Ti ($T_{1/2} = 59$ yr [736]). γ-ray lines at 67.9 and 78.4 keV corresponding to the decay chain ^{44}Ti $\rightarrow$ ^{44}Sc $\rightarrow$ ^{44}Ca (Figure 7.14) have been detected with INTEGRAL's IBIS/ISGRI instrument in the su-

---

[67]See also reference [720] for optical photometry performed on SN 1987A, during the first 177 days past explosion.

[68]See also Seitenzahl et al. [1605] for a recent update on the masses of radioactive nuclides synthesized by SN 1987A.

[69]See reference [1000] for a report on the recent detection of ^{31}P in Cas A. The abundance ratio of ^{31}P to the major nucleosynthetic product, ^{56}Fe, in this supernova remnant is up to 100 times the average ratio in the Milky Way, confirming that phosphorus is synthesized in core-collapse supernovae.

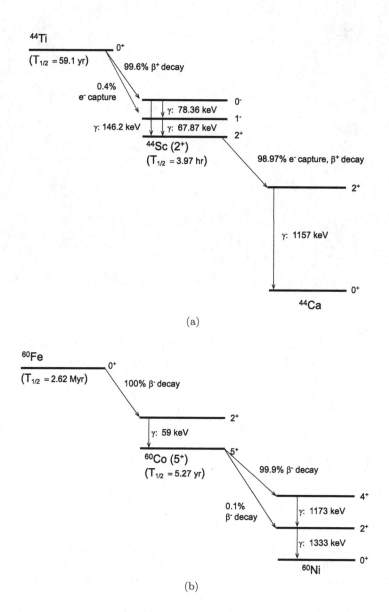

(a)

(b)

FIGURE 7.14
Simplified decay schemes for the radioactive chains ^{44}Ti → ^{44}Sc → ^{44}Ca (panel a), and ^{60}Fe → ^{60}Co → ^{60}Ni (panel b), showing the most important energy levels and gamma transitions. See http://www.nndc.bnl.gov/nudat2/reColor.jsp?newColor=dm, for details.

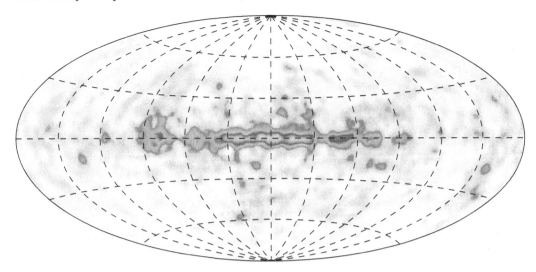

FIGURE 7.15
Maximum entropy, all-sky map of the Galactic ^{26}Al 1.809 MeV emission observed with the COMPTEL instrument on board CGRO satellite over 9 years. A color version of this plot is available at http://fisica.upc.edu/ca/users/jjose/CRC-Downloads. Figure adapted from Plüschke et al. [1433]; reproduced with permission.

pernova remnant Cassiopeia A [1487], located 11,000 light-years away[70]. Detection of the same γ-ray lines have also been reported from SN 1987A, from which $\sim (3.1 \pm 0.8) \times 10^{-4}$ $M_\odot$ of ^{44}Ti were inferred[71] [680]. Such discovery confirmed a key nucleosynthesis prediction of the α-rich freeze-out occurring during explosive Si-burning in core-collapse supernovae. However, why ^{44}Ti is not commonly detected in most supernova remnants[72] remains to be clarified [1787]. It has been proposed that ^{44}Ti synthesis at detectable levels may require special conditions, such as an asymmetric explosion[73] [213, 1282, 1649].

Models predict as well that other long-lived radioactive species, such as ^{26}Al ($T_{1/2} = 7.17 \times 10^5$ yr) and ^{60}Fe ($T_{1/2} = 2.62 \times 10^6$ yr), are also synthesized in core-collapse supernovae[74]. These species have long half-lifes compared to the characteristic frequency of ocurrence of Galactic (core-collapse) supernovae. Accordingly, γ-ray signatures associated with the decay of these species appear as diffuse rather than localized, tracing the cumulative emission from hundreds or thousands of sources (Figure 7.15). ^{60}Fe/^{26}Al γ-ray line flux ratios have actually been determined by the RHESSI instrument [1668] and the SPI spectrometer onboard INTEGRAL [1900]. Model predictions [1111, 1808] are consistent with the observed ratio[75] of ~ 0.15, when theoretical yields from single, massive stars are folded with the expected distribution of stellar masses or *initial mass function*. Such

---

[70]Light from this supernova explosion likely impinged our planet around 1680.

[71]See, however, Boggs et al. for improved estimates of ^{44}Ti synthesis in SN 1987A based on NuSTAR (Nuclear Spectroscopic Telescope Array) observations [213].

[72]A couple of candidates have been reported at low significance, including GRO J0852-4642 in the Vela region [902] and the Per OB2 association [476]. See discussion in references [443, 1859].

[73]See also Young [2006] for alternative explanations.

[74]See references [488, 1111, 1445, 1909, 1977] for studies of ^{26}Al synthesis during explosive Ne/C burning in massive stars.

[75]The γ-ray emitter ^{26}Al has been previously detected in the Galactic interstellar medium [441, 442, 1184, 1185, 1444].

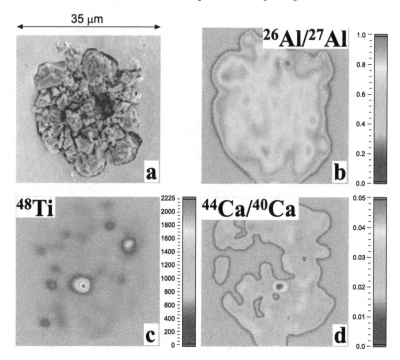

FIGURE 7.16
SEM image (panel a) and NanoSIMS isotopic maps of ^{26}Al/^{27}Al, ^{48}Ti, and ^{44}Ca/^{40}Ca (panels b, c, and d) of an unusually large (~ 30 μm) SiC grain of type X named *Bonanza*. The SEM image suggests that the grain is actually an aggregate of smaller grains. The anomalous size of Bonanza has allowed the determination of isotopic ratios for many different elements, including Li, B, C, N, Al-Mg, Si, S, Ca, Ti, Fe, and Ni [2027, 2028]. Isotopic analyses yield ^{12}C/^{13}C = 190, ^{14}N/^{15}N = 28, δ^{29}Si/^{28}Si = -282‰, and δ^{30}Si/^{28}Si = -442‰. Mg is completely dominated by the presence of radiogenic ^{26}Mg, from the decay of ^{26}Al (the ratio ^{26}Al/^{27}Al inferred in the grain lays in the range ~ 0.4–0.9). The region with the largest ^{44}Ca/^{40}Ca ratio (0.05), corresponds to a titanium subgrain, which exhibits as well the largest concentration of ^{48}Ti. This suggests that the ^{44}Ca excess corresponds to in situ decay of ^{44}Ti. The Ti, Fe, and Ni isotopic patterns are typical of other X grains and can be reproduced by mixing different layers in models of core-collapse supernovae. Color versions of these plots are available at http://fisica.upc.edu/ca/users/jjose/CRC-Downloads. Figure adapted from Zinner et al. [2028]; reproduced with permission.

estimates, however, are conditioned by nuclear uncertainties affecting a number of relevant interactions [819, 879, 1182, 1788, 1835].

7.5.2   Presolar Supernova Grains

The first isotopic anomalies identified in chondritic meteorites revealed injection of freshly synthesized material into the Solar System nebula shortly before condensation began[76]. Indeed, a number of studies suggested that dust spherules originally condensed in the ejecta

---

[76]The alternative in situ production attributed to an intense activity of the early Sun was soon disregarded. See, e.g., reference [1060].

of a nearby type II supernova could have survived the formation of the Solar System itself, contributing to the isotopic diversity evidenced in some meteoritic samples [191, 289, 343, 357, 514, 1058, 1060, 1434, 1549]. Today, core-collapse supernovae, together with AGB stars, are known as the most prolific dust producers in the universe[77] [475].

Since the discovery of presolar grains, successfully isolated from meteorites in the 1980s, the search for putative supernova grains has been guided by a number of characteristic isotopic features predicted from models, in particular, high $^{12}C/^{13}C$ ratios, low $^{14}N/^{15}N$, and excesses in ^{28}Si, ^{26}Mg (from ^{26}Al decay), ^{41}K (^{41}Ca), ^{44}Ca (^{44}Ti), and ^{49}Ti (^{49}V). From these, only ^{28}Si, ^{44}Ti and ^{49}V are specifically synthesized in supernova explosions. A number of supernova grains (e.g., SiC X and C grains, graphites, oxides) have been unambiguously identified within the rich inventory of presolar grains[78]. Supernova X grains (Figure 7.16) were, in fact, the next SiC grains isolated from meteorites after mainstream grains. They amount to $\sim 2\%$ of all SiC grains. About two-thirds of all graphite grains, all silicon nitrate grains, and $\sim 10\%$ of the presolar oxide and silicate grains have also a likely supernova origin [417]. Confrontation between isotopic abundances determined in the lab with theoretical predictions has provided a wealth of information about nucleosynthesis and mixing in supernovae, and their interplay with the chemical evolution of the universe[79].

Early work on dust formation in the supernova ejecta was performed in the 1970s, in the framework of equilibrium condensation sequences ruled by the formation of CO molecules [1059] (see Chapter 3). A major challenge, encountered both by nature and by theorists in these explosive environments, is the presence of a strong radiation field. This likely forces nucleation and condensation to proceed under nonequilibrium conditions. Homogeneous nucleation and grain growth were subsequently tackled by Kozasa, Hasegawa, and Nomoto in the late 1980s [1011,1012], who showed that the inclusion of the radiation emitted from the stellar photosphere increases the characteristic timescale for grain formation. Such studies also revealed that the expected mineralogy, potentially ranging from graphite to several types of oxides, depends critically on a number of assumptions adopted in the treatment of mixing between different supernova shells. In this context, Don Clayton and collaborators [350] have suggested that graphite dust can form even in environments characterized by O > C, driven by the efficient destruction of the newly formed CO molecules by high-energy photons. Whether this also applies to SiC dust has, however, been questioned. Indeed, Ebel and Grossman [480] have shown, on the basis of equilibrium sequences, that TiC (rather than SiC) likely condenses if Ti + Si $\gg$ C + O, even in a C < O environment. And even more remarkably, high-temperature sequences, aimed at mimicking the role played by radiation in dissociating CO molecules, yielded graphite dust, but neither TiC nor SiC.

More recently[80], Sarangi and Cherchneff [1563, 1564] have reported new studies of dust formation in the ejecta of type II-P supernovae, along the nucleation and condensation phases. The analysis reveals formation of three main types of dust: silicates, carbon, and alumina. The prevalence of a given dust type depends critically on the composition of the stellar progenitor at the presupernova stage (i.e., low-mass progenitors yield C-rich dust, while high-mass progenitors likely form silicates and alumina).

---

[77]SN 1987A provided first evidence of dust condensation in the ejecta of a core-collapse supernova through a strong decrease in the optical flux ~ 500 days after explosion, and a simultaneous increase of the mid-infrared flux [406,1157]. Similar effects have been reported from an increasingly large number of core-collapse supernovae, such as the type II-P SN 1999em [493], the type Ib/c SN 1990I [492], or, more recently, SN 2003gd, SN 2004dj, SN 2004et, SN 2006jc, and SN 2007od (see Szalai et al. [1756] and references therein).

[78]Isotopic determinations in putative supernova grains reflect however less extreme signatures than expected from models. A possible explanation [352] suggests that the passage of freshly condensed grains at high speeds through the supernova ejecta may lead to implantation of atoms from the environment and, therefore, may result in chemical dilution in the grains.

[79]See, e.g., reference [1810] for a study of dust formation in primordial supernovae.

[80]See also references [321, 322, 2010] for kinetic models of condensation of C-rich dust.

7.5.2.1   Silicon Carbide Grains of Types X and C

X grains are characterized by large ^{28}Si excesses (i.e., low ^{29}Si/^{28}Si and ^{30}Si/^{28}Si ratios), high ^{12}C/^{13}C, relatively low ^{14}N/^{15}N, and high inferred ^{26}Al/^{27}Al ratios[81] [28, 834, 835]. Some X grains exhibit as well ^{44}Ca excesses, likely incorporated into the grains as radioactive ^{44}Ti, a clear supernova imprint [1302]. Other titanium isotopes, such as ^{46,49,50}Ti, have also been isolated from X grains. Another evidence of their likely supernova origin comes from the presence of extinct ^{49}V [28, 829, 1302], whose relatively short half-life ($T_{1/2} \sim 330$ days) suggests grain formation within a few months of the explosion[82]. Isotopic ratios of Zr, Mo, and Ba have also been measured in X grains using resonance ionization mass spectrometry (RIMS) [1401, 1402] (see Chapter 3). Measurements in some X grains revealed excesses in ^{95}Mo and ^{97}Mo, but not in the r-process nuclide ^{100}Mo. This came as a surprise since, provided that the r-process occurs in core-collapse supernovae, one would expect to find characteristic imprints in those r-process elements isolated from supernova grains. Brad Meyer et al. [1238] have suggested that a rapid release of neutrons (on a time scale of seconds) in the He-rich layers heated by the passage of a shock front can yield a lower neutron flux than that required for the classical r-process. This can account for the ^{95}Mo and ^{97}Mo excesses, as well as the Zr and Ba isotopic ratios, found in these grains.

Attempts to reproduce the composition of X grains by core-collapse nucleosynthesis models require contributions from different ejected layers (from the core to the outer shells that experienced H- and incomplete He-burning in the early stages of the evolution), revealing extensive and heterogeneous mixing [835]. However, neither astronomical observations nor theory can account for the levels of mixing required by grain data[83].

A certain class of SiC grains known as C grains, has been recently identified on the basis of very prominent ^{29,30}Si and ^{32}S excesses, aside from other isotopic peculiarities [389, 701, 830, 832, 1421, 2035]. Their origin has been attributed to core-collapse supernovae, specifically in the C-rich layers exposed to lower shock temperatures than the more common type X grains [1421]. Other attemps to explain the ^{29}Si excesses include modifications of the canonical ^{26}Mg(α, n)^{29}Si reaction rate due to nuclear uncertainties (see discussion in reference [825]).

7.5.2.2   Graphite Grains

Presolar graphite grains exhibit a wide range of densities, with low-density grains ($\rho < 2.15$ g cm^{-3}) having somewhat higher trace-element concentrations than denser grains [37]. The isotopic signatures of low-density grains are in many ways similar to those observed in SiC-type X grains and Si_3N_4 grains (see Section 7.5.2.4), i.e., high ^{12}C/^{13}C, high ^{26}Al/^{27}Al, and Si-isotopic anomalies (except that graphite grains contain detectable concentrations of oxygen). Therefore, they likely share the same stellar paternity (i.e., core-collapse supernovae) [36]. On the other hand, high-density grains may have originated from different stellar environments, including AGB stars, supernovae, and novae. All in all, about 60% of all presolar graphite grains have a likely supernova origin.

A widely used feature for diagnosis of supernova grains is an ^{18}O excess. This nuclide is expected to be synthesized in the He-burning layers of massive stars (see Chapter 2). Unfortunately, in order to account for additional isotopic signatures found in supernova graphite grains, several regions of the star need to be invoked. For instance, the relatively high ^{26}Al/^{27}Al ratios (~ 0.1) inferred in some grains correspond to the values predicted

---

[81]Type X grains have been recently subdivided into types X0, X1, and X2, based on distinct isotopic features. See reference [1115] for details.

[82]See, however, reference [1115].

[83]Moreover, the agreement between grain data and theoretical yields presents some gaps, particularly in ^{15}N, much larger in X grains than predicted by models [172, 835].

for the outermost layers of a massive star, after H-burning. In contrast, ^{28}Si excesses may require material from the inner Si–S-rich layers (or from the He-C zone during explosive nucleosynthesis), while the presence of extinct ^{44}Ti—another strong supernova imprint—in the form of ^{44}Ca excesses, points toward the innermost ^{56}Ni-rich layers. All in all, mixing of different supernova layers are required to quantitatively match grain data, as no single zone can account for the distinct isotopic features found in supernova graphites. Travaglio et al. [1817] have performed mixing calculations using different zones from supernova models [1989]. A good match is obtained for ^{12}C/^{13}C, ^{18}O/^{16}O, and ^{30}Si/^{28}Si ratios, as well as for the inferred ^{41}Ca/^{40}Ca and ^{44}Ti/^{48}Ti ratios, provided that jets of material from the inner Si-rich zone penetrated the overlying O-rich zones and ultimately mixed with material from the outer C-rich layers. Unfortunately, models cannot account for the large concentrations of ^{15}N and ^{29}Si required to explain grain data [1817]. It is worth noting that while some success in reproducing grain data has been achieved through this approach, it is far from clear whether mixing of inner and outer layers can actually occur in the supernova ejecta. More recently, Pignatari et al. [1420], have reported new models of explosive nucleosynthesis in a core-collapse supernova. At $T > 3.5 \times 10^8$ K, the bottom of the He-C-rich zone becomes ^{12}C- and ^{28}Si-rich, and, therefore, mixing between the inner Si–S zone and the outer layers is no longer required.

7.5.2.3 Nanodiamonds

Nanodiamonds, the first type of presolar grain identified, are characterized by close-to-solar C and N isotopic ratios. Isotopic anomalies are only seen for minor elements, such as the two lightest, stable xenon isotopes, ^{124,126}Xe, attributed to the p-process, and the two heaviest isotopes ^{134,136}Xe synthesized in the r-process. Other anomalies include additional r-process nuclides, such as ^{110}Pd and ^{128,130}Te, and the r- and s-process nuclide ^{137}Ba (see Section 3.3.1).

Because of their small size, isotopic determinations are performed only on ensembles of millions of grains and, therefore, the observed anomalies may be carried by a reduced number of grains. Accordingly, while the isotopic anomalies associated to s- and r-process elements suggest a supernova origin, it is possible that most nanodiamonds actually condensed within the Solar System, and only a small fraction have a truly explosive origin [402]. Nevertheless, the abundances determined in nanodiamonds for the p- and r-process Xe isotopes cannot be reproduced by standard p- and r-process models, for reasons not well understood[84]. A number of possible explanations are discussed in Lodders and Amari [1137], to which the reader is referred for a deeper insight.

7.5.2.4 Silicon Nitride Grains

Silicon nitride (Si_3N_4) grains exhibit similar isotopic abundances to type X SiC grains and, therefore, likely condensed in the ejecta of a core-collapse supernova [835, 1115, 1304].

7.5.2.5 Oxide and Silicate Grains

Presolar oxides (and silicates) are subdivided into four distinct groups [1301, 1306], some of which may have a supernova origin (see Chapter 3). This includes group III, characterized by moderate ^{16}O enhancements, and group IV, with large ^{18}O and moderate ^{17}O enrichments. While group III grains have been attributed to different stellar sources, including low-mass, low-metallicity AGB stars and supernovae, isotopic composition of group IV grains suggests a unique progenitor, in the form of a core-collapse supernova. The latter is based upon the

---

[84]See, however, reference [1019].

perfect match between grain data[85] on O, Mg, and Ca (with ^{44}Ca resulting from extinct ^{44}Ti) and the corresponding predictions for models of 15 $M_\odot$ stars (the He-C- and He-rich layers, in particular). All in all, about 10% of all presolar oxides and silicates may have a supernova origin [825].

7.6 Supernovae, Neutron Star Mergers, and the Origin of Gamma-Ray Bursts

Core-collapse supernovae are also likely connected with γ-ray bursts (GRBs), brief, intense flashes of electromagnetic radiation whose origin constitutes one of the most striking enigmas of modern astrophysics. GRBs were serendipitously discovered in the 1960s by researchers from Los Alamos National Lab, led by Ray Klebesadel, using data recorded by the Vela satellites. These US military satellites were originally designed to unveil nuclear weapon tests potentially carried by the former USSR, in a clear violation of the Nuclear Test Ban Treaty signed in 1963 by the USSR, the US, and the United Kingdom. To this end, they were equipped with γ-ray detectors capable of registering the aftermath of a nuclear test.

It was 1969, when such bursts of γ-radiation were first found, buried under huge piles of data collected by Vela 3 and 4. Ray Klebesadel and his colleague Roy Olson identified an event that triggered the Vela detectors on July 2, 1967. Unfortunately, the first series of Vela satellites did not have enough resolution to determine the direction of the incoming photons. As other Vela satellites were launched, additional bursts of radiation were detected. A careful analysis of the arrival times of some of these events, performed in 1972 by Klebesadel, Olson, and Ian Strong, revealed that such mysterious signals did not originate from Earth or from the Sun. All in all, the discovery of 16 GRBs of *cosmic origin*, observed by Vela 5 and 6 between July 1969 and July 1972, with energies ranging from 0.2 to 1.5 MeV, was finally announced in 1973 [992]. In retrospect, the July 2, 1967, event can be considered the first observation of a GRB[86] (see Figure 7.17).

GRBs are probably the most luminous electromagnetic events in the universe. They are characterized by a sudden flash of γ-radiation, with typical photon energies $\sim$ 100 keV, usually followed by a longer-lived *afterglow* at longer wavelengths (from X-rays to radio). They are frequently detected, typically one or more bursts from different sources per day[87]. GRBs impinge the Earth from unpredictable locations, isotropically distributed on the sky [624]. This suggests a likely extragalactic origin, otherwise the flattened shape of the Milky Way would favor a larger concentration of GRBs near the Galactic plane[88]. To date, no repeated burst has ever been recorded from the same source.

In sharp contrast with many other stellar variables and transients, whose light curves

---

[85]See references [413, 1866] for additional isotopic features involving Si and Cr isotopes.

[86]γ-ray bursts are named after their discovery date (e.g., GRB 121027A): The first two digits correspond to the year, and are subsequently followed by the month (two intermediate digits) and the day (two final digits). A letter is appended to indicate the order of discovery along the same day (A is first, B is second, etc.). For bursts discovered before 2010, a letter was only appended if more than one burst was registered in the same day.

[87]However, they constitute extremely rare phenomena, with only a few bursting events per galaxy every million years.

[88]A number of recent studies have analyzed the possible threat posed by a Galactic GRB [51, 146, 408, 587, 588, 1225, 1226, 1289, 1425, 1426, 1799, 1801]. While lethal radiation doses are predicted for supernova explosions occurring within 8 to 10 pc from Earth, similar effects are foreseen for GRBs at critical distances of 1–10 kpc [146, 408]. See also Annis [51] for a possible link between Galactic GRBs and the solution to Fermi's paradox.

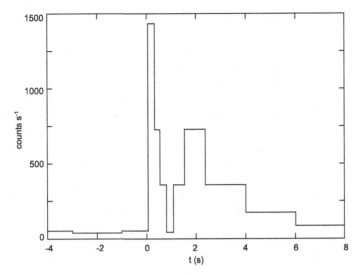

FIGURE 7.17
Histogram of the count rate of the first γ-ray burst ever observed, as registered by the Vela detectors on July 2, 1967. Figure from Bonnell and Klebesadel [218]; reproduced with permission.

frequently exhibit a rapid rise to peak luminosity followed by an exponential-like decay, GRBs display a suite of different and complex light curves. Some events are preceded by a weak *precursor* burst that anticipates the true bursting episode by seconds or minutes. The light curve itself spans for an overall duration that may range from ten milliseconds to several minutes, and is characterized by a single peak or by a handful of minipulses. The peaks themselves are sometimes symmetric or present huge differences between rise and fall (e.g., a fast rise and a very slow decline).

Despite the rich variety of light curves displayed, two broad categories[89] of GRBs have been established on the basis of a bimodal duration[90] distribution and differences in their spectral hardness: *short* (hard) γ-ray bursts, with a typical duration below 2 s, and *long* (soft) GRBs [155, 623]. Moreover, while short bursts predominantly emit, on average, at energies around 360 keV, maximum emission for long GRBs occurs, on average, at shorter energies, ~ 220 keV [712].

The nature of GRBs has puzzled theoreticians for decades. More than one hundred different models have been proposed to account for their main observational properties [1288]. Extensive searches for a counterpart at other wavelengths proved unsuccessful during years. In a few cases, for well-localized bursts, no bright object consistent with the position of the event was found, suggesting very faint stars or extremely distant galaxies as likely sites

---

[89]Other categories have been recently established. For instance, the class of ultra-long γ-ray bursts, that last hours rather than seconds. They have been proposed to form a separate class, resulting from the collapse of a blue supergiant star, or from the tidal disruption of a white dwarf in the presence of a supermassive black hole [634, 1080, 1726]. Other studies suggest, however, that such ultra-long events simply represent the tail of the distribution of long GRBs [1862]. Only a few ultra-long GRBs have been identified to date (e.g., GRB 091024A, GRB 101225A, GRB 111209A, GRB 121027A, and possibly GRB 130925A). See references [211, 676, 2020] for other studies aimed at determining whether such ultra-long events constitute a distinct class of GRBs.

[90]The duration of a γ-ray burst is frequently expressed in terms of $T90$, or time during which 90% of the burst's energy is actually emitted.

hosting such events[91]. A major observational breakthrough in the understanding of the physics powering GRBs and the nature of their progenitors was provided by the detection of marginal emission at longer wavelengths, the so-called *afterglows*, characterized by a progressive shift of the radiation toward lower energies: X-rays (after few hours), visible (days) and radio (weeks). The first afterglow was detected by the BeppoSAX satellite in X-rays, following the burst of γ rays emitted by the source GRB 970228. An optical afterglow was observed 20 hours later with the William Herschel telescope, in the Canary islands, which allowed the identification of a faint, distant host galaxy. GRB afterglows provide also a unique tool to infer the redshift to the source, either directly (from optical spectroscopy of the afterglow itself) or indirectly (via the host galaxy). Another BeppoSAX observation yielded the first accurate determination of a redshift (distance) to a γ-ray burst, GRB 970508. The value inferred, $z = 0.835$, placed the object at a distance of roughly 6 billion light-years, confirming the extragalactic nature of such events. This, in turn, opened Pandora's box: If GRB emission is assumed to be spherical, the overall energy output would be of the order of the rest-mass energy of the Sun. No known process in the universe can release such amount of energy in such a short time. Accordingly, early GRB models assumed a highly collimated (beamed) emission. Assuming emission angles between 2–20 degrees, a typical GRB may actually release $\sim 10^{51}$ erg, about 1/2000th of the rest-mass energy of the Sun, or about the energy released in a core-collapse supernova[92] [549].

The discovery of GRB 980425, in conjunction with SN 1998bw [586, 900], represents another breakthrough in the history of γ-ray bursts. This was followed by the discovery of several long γ-ray burst-supernova pairs, such as GRB 030329-SN 2003dh [1700] and GRB 031203-SN 2003lw [1187], involving in all cases type Ic supernovae. This raised the interest in the explosion of massive stars as possible long GRB progenitors. It is not clear, however, why supernova explosions are not always accompanied by GRBs (not even the Ic class). It has been proposed that GRBs result only from the most rapidly rotating and massive stars, which give rise to very energetic explosions popularly known as *hypernovae* [900, 1324, 1368]. In this framework, the collapse of the innermost layers of a massive star leading to a rapidly rotating black hole surrounded by a disk (i.e., the *collapsar* model [1173, 1280, 1964]) or to a highly magnetized, rotating neutron star (the *magnetar* model) have been suggested as possible engines of long GRBs [262, 469, 1167, 1234, 1676, 1841, 1929].

Statistically, about 30% of the γ-ray bursts detected by the BATSE instrument on board NASA's Compton Gamma-Ray Observatory (CGRO) belong to the short GRB class. However, there are reasons to believe that this could be an observational bias, since short GRBs release in general smaller amounts of energy than long bursts, and therefore face a more difficult detectability. In fact, it has been suggested that short GRBs may actually constitute the most frequent class of γ-ray bursts in the cosmos. Such short duration bursts are typically associated with distant, elliptical galaxies, suggesting old progenitors[93]. Proposed scenarios include the merger of binary neutron stars [199, 1366] (Figure 7.18) or a neutron star and a black hole[94]. [1283, 1367].

Nucleosynthesis in GRBs has been explored by different groups, particularly in the

---

[91]Most of the energy emitted by a GRB corresponds to γ-rays. However, some GRBs have revealed optical counterparts that can actually be briefly spotted from Earth with the unaided eye, despite of their distance (e.g., the optical counterpart of GRB 080319B peaked at a visible magnitude of $\sim 5.3^{\mathrm{mag}}$ [203, 1465]).

[92]This similarity did not receive much attention for decades. An exception, however, can be found in the seminal work of Colgate [362], who somehow predicted GRBs in the framework of relativistic shocks in supernovae.

[93]This rules out massive stars as likely progenitors, confirming that short GRBs are physically distinct from long GRBs. Indeed, no association between short GRBs and supernovae has ever been reported.

[94]A subset of the short GRB class may actually be powered by extragalactic giant flares from soft gamma-ray repeaters with long recurrence times [2, 3, 1343]. See Berger [155] for a recent review on short-duration GRBs.

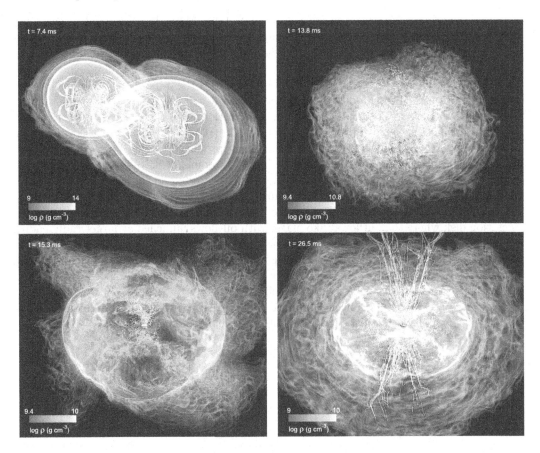

FIGURE 7.18
Evolution of the density and magnetic field during the merging of a binary neutron star system, at four different times (t = 7.4 ms, 13.8 ms, 15.3 ms, and 26.5 ms). The simulation yields a rapidly spinning black hole surrounded by a hot and highly magnetized torus. Magnetohydrodynamical instabilities can actually amplify an initially turbulent magnetic field along the black hole spin axis, naturally resulting in a relativistic jet, potentially powering a short GRB. A color version of this image is available at http://fisica.upc.edu/ca/users/jjose/CRC-Downloads. Figure from Rezzolla et al. [1494]; reproduced with permission.

framework of the collapsar model[95], distinguishing contributions from jet-induced explosions and element production in the disk[96]. Special emphasis has been devoted to the synthesis of ^{56}Ni, whose decay likely powers GRB light curves. Nomoto and collaborators [1177, 1332] have reported that a very energetic shock, driven by bipolar jets in hypernova explosions, may yield a peculiar abundance pattern resembling that inferred from some extremely metal-poor stars (in particular, C-rich metal-poor stars, and hyper metal-poor stars). The amount of ^{56}Ni synthesized turns out to be very sensitive to the energy (and mass) deposition rate near the black hole axis [1180, 1281, 1509].

The composition of the wind blown off from the disk orbiting the black hole in the

---

[95]See reference [119] for a discussion on type I and II collapsars and their associated nucleosynthesis.

[96]See reference [849] for recent studies on nucleosynthesis in the interstellar media during GRB afterglows.

collapsar model has also been investigated [1172, 1173, 1459]. It was found that matter ejected from the disk can exhibit a rather rich nucleosynthesis, depending on the specific mass-accretion rate adopted, the velocity and entropy of the outflow, and the location at which such matter is released [1747]. The corresponding outflow can actually be neutron- or proton-rich. In rapidly accreting disks ($\dot{M} \sim 10$ M$_\odot$ s^{-1}) the ouflow may exhibit r-process elements (Section 7.4.2). Moderate mass-accretion rates give rise to ^{56}Ni and a suite of light elements, such as Li or B. For lower mass-accretion rates ($\dot{M} \sim 1$ M$_\odot$ s^{-1}), the main products are likely ^{56}Ni, ^{49}Ti, ^{45}Sc, ^{64}Zn, and ^{92}Mo [1458, 1746, 1747]. In the very recent models[97] reported by Hu [848], ^{56}Ni is abundantly produced in the inner collapsar disk, for all the outflow cases investigated. The highest amount of ^{56}Ni in the ejecta reaches 0.463 M$_\odot$. The simulations support that the effective production of this and other heavy elements (e.g., ^{57,58}Ni, ^{59}Cu, ^{60}Zn) can occur in the outflows from the inner regions of the disk. The amount of heavy species produced actually depends on the eject-radius of the accretion disk where the outflows take place (with outflows launched at larger radii leading to larger amounts of heavy nuclei). Another study [991] has shown that disk models with mass-accretion rates near 1 M$_\odot$ s^{-1} may power a νp-process, for a certain combination of parameters (see Section 7.4.4). The time during which matter is exposed to neutrino interactions turns out to be critical to this end.

Reviews on the different, often speculative, mechanisms proposed as engines of GRBs, and their contribution to the Galactic inventory of chemical species, can be found, e.g., in references [1233, 1424, 1971].

---

[97]See also reference [904] for studies of nucleosynthesis in the disk, for different mass-accretion rates.

Box I. The Core-Collapse Supernova "Hall of Fame"

A selection of facts on type Ib/Ic and type II supernovae

1. Stars with initial masses $M \geq 10$ $M_\odot$ undergo a type II supernova explosion, accompanied by the formation of a neutron star or a black hole. Very massive stars, $M > 100$ $M_\odot$, explode as pair-instability supernovae, leaving behind a black hole or no remnant at all.

2. Type Ib and Ic supernovae are explosions driven by core-collapse of massive stars that have lost their H- and He-rich envelopes (i.e., Wolf–Rayet stars). Spectroscopically, they are characterized by the lack of hydrogen in their spectra, and are distinguished from type Ia supernovae because both Ib and Ic types lack as well the prominent Si absorption trough near 6150 Å. While type Ib show strong He I absorption lines in their early photospheric spectra, such lines are not easily discernible in type Ic supernovae.

3. Type II supernovae are spectroscopically distinguished from type I by the presence of prominent hydrogen lines. Four subclasses have been established on a purely photometric basis: SN II-P (plateau), which remain at nearly constant luminosity (with variations $< 1^{\text{mag}}$) after maximum, for an extended period of time; SN II-L (linear), characterized by a steep decline from peak luminosity, followed by a slower exponential tail; SN IIb, whose late-time spectra resemble those from SN Ib, with undetectable H emission lines; and the IIn (narrow) class, characterized by very narrow emission lines.

4. Nucleosynthesis in core-collapse supernovae is driven by a number of burning regimes, including normal and α-rich freeze-out from nuclear statistical equilibrium in the innermost regions, and explosive Si-, O-, and C/Ne-burning. Elements heavier than iron are mostly synthesized through the neutron-capture s- and r-processes in core-collapse supernovae (and neutron star mergers). Some extremely underabundant, proton-rich nuclei are formed through the p-process. Nucleosynthesis also proceeds through a number of additional channels, including the neutrino ν- and νp-processes.

5. A burst of (anti)neutrinos was detected from SN 1987A, about two to three hours before its optical discovery. Observations were consistent with theoretical models of core-collapse supernovae, that predict $\sim 10^{53}$ erg released in the form of electron antineutrinos.

6. Presolar supernova grains, identified by huge isotopic anomalies with respect to bulk Solar System abundances (in particular, high $^{12}C/^{13}C$, low $^{14}N/^{15}N$, and excesses in ^{28}Si, ^{26}Mg [from ^{26}Al decay], ^{44}Ca [^{44}Ti], and ^{49}Ti [^{49}V]), have been isolated from primitive meteorites. The inventory of supernova grains discovered so far includes silicon carbide grains of types X and C, graphites, nanodiamonds, silicon nitrides, and different oxides and silicates.

7. γ-ray emission at energies 0.847 MeV and 1.238 MeV ($^{56}Co \rightarrow ^{56}Fe$), and 0.122 MeV and 0.136 MeV ($^{57}Co \rightarrow ^{57}Fe$) was detected from SN 1987A, providing smoking-gun evidence of the radioactive-powered, late-time light curve. About 0.07 $M_\odot$ of ^{56}Co have been observationally inferred in SN 1987A.

8. γ-ray emission at 67.9 keV and 78.4 keV ($^{44}Ti \rightarrow ^{44}Sc$) has also been detected in the supernova remnant Cassiopeia A, as well as in SN 1987A.

9. Two broad categories of GRBs have been established on the basis of a bimodal duration distribution and differences in their spectral hardness: long (soft) GRBs with a duration above 2 s; and short (hard) GRBs, with a typical duration below 2 s. Even though the true nature of GRBs remains to be firmly established, the current view suggests that long GRBs may be triggered by the explosion of rapidly rotating, massive stars (hypernovae), while short GRB may result from the merging of binary neutron stars or a neutron star and a black hole.

10. Lethal radiation doses have been predicted to be released by supernova explosions occurring within 8–10 pc from Earth. However, there is no supernova candidate close enough to pose a serious threat to humans. In the unlikely event of a Galactic GRB, lethal radiation doses are expected from an explosion occurring within 1–10 kpc.

Box II. Mysteries, Unsolved Problems, and Challenges

- No self-consistent explosion model for core-collapse supernovae has been developed to date. A number of numerical artifacts (e.g., piston, energy deposition) have been adopted to mimic the dynamics of the supernova blast, and determine the accompanying nucleosynthesis. Efforts must be invested in identifying the mechanism by which the star turns the initial collapse into an explosion, which may involve neutrino-plasma interactions, rotation, magnetic fields and several hydrodynamic instabilities emerging from the true multididimensional nature of the problem.

- A handful of core-collapse supernova subclasses have been established on the basis on photometric and spectroscopic properties, but little is known about the physical reasons that cause such diversity.

- Where does the r-process actually occurs? Does it take place in a single astrophysical site or do both neutron stars mergers and neutrino-driven winds accompanying type II supernovae contribute?

- Where does the p-process occurs? Why the light p-isotopes of Mo, Ru, and some other rare species are systematically underproduced in most of the astrophysical scenarios proposed so far? Is this a nuclear physics problem?

- Need to reduce the uncertainties in cross-sections and average neutrino energies that currently affect ν- and νp-process nucleosynthesis predictions.

- Need to clarify why γ-rays from ^{44}Ti-decay are not commonly detected in supernova remnants. What makes Cassiopeia A special in this regard?

- Comparison between presolar grain data and theoretical predictions suggests extensive mixing between distinct supernova shells. It is important to assess whether the required levels of mixing can actually be self-consistently reproduced by current supernova models.

- What is the origin of GRBs? Do ultra-long events constitute a distinct GRB class or simply represent the tail of the distribution of long GRBs?

7.7 Exercises

P1. Estimate the total mass deposited on Earth by a 20 $M_\odot$ core-collapse supernova exploding 10 pc away. Consider the amount of material ejected in the explosion and the mass swept up by the ejecta while crossing the interstellar medium on its way to Earth. Compare the result with the mass of the impactor that caused the Cretaceous–Tertiary extinction event, 2.5×10^{17} g, as estimated by van den Bergh [1844].

P2. In September 2011, the OPERA experiment announced the results of a measurement in which neutrinos appeared to travel faster than light. Neutrinos were traveling from CERN accross 731 km, reaching the Grand Sasso National Laboratory (Italy) pressumably 60.7 ns earlier than expected for particles moving at the speed of light.

(a) Taking the OPERA results as a proxy for the neutrino speed, calculate the travel-time difference between neutrinos and light emitted from SN 1987A, the supernova that exploded in the Large Magellanic Cloud, some 168,000 light-years away.

(b) Compare the estimate with the travel-time difference of 3 hr between neutrinos and light reported from SN 1987A [807]. Prove that the faster-than-light claim for the OPERA neutrinos is hard to reconcile with SN 1987A.

(c) Discuss whether the detection of such neutrino burst, a few hours before the optical sighting of SN 1987A, is an indication of *slightly* faster-than-light neutrinos. Base your conclusions on current core-collapse supernova model predictions.

P3. Estimate the mean free path of a neutrino for the physical conditions at the center of the Sun ($\rho \sim 162$ g cm^{-3}), using a mean neutrino cross-section of $\sim 10^{-44}$ cm^2 per nucleus. Take Equation 1.30 and infer the opacity from the relation $\sigma = \kappa\mu$, where μ is the mean molecular weight. For simplicity, assume that the Sun's core is only made of ionized hydrogen. Express the result in astronomical units (i.e., mean distance between the Earth and the Sun, corresponding to 149.6×10^6 km).

P4. Prove that the collapse of a constant density stellar core is homologous (i.e., satisfies $v \propto r$).

P5. Compare the main neutron-capture s- and r-process paths involved in the formation of nuclei beyond iron. Starting from ^{56}Fe, draw separately the s- and r-process paths, adopting a characteristic neutron-capture timescale of 10 yr (s-process) and 0.1 s (r-process). Use β-decay times for the different radioactive species encountered along the path from, e.g., the *Nuclear Wallet Cards* [1832], and consider branching whenever the characteristic timescale for neutron-capture is comparable with the β-decay time. Determine which isotopes of Sr are created in the s- and r-process, and estimate the characteristic time required for the flow to reach Sr from Fe along the different paths.

P6. Consider the data on the core-collapse SN 1980K displayed in Table 7.1, where t is the time since the explosion, T the effective blackbody temperature of the star, and v the expansion velocity spectroscopically determined from Doppler shifts. Find the absolute magnitude of the supernova and the distance to SN 1980K using the expansion photosphere method.

P7. Prove that the radioactive-powered light curve of a core-collapse supernova during

TABLE 7.1
Derived Quantities for SN
1980K [1586]

t (days)	v (km s^{-1})	Temp (K)
34	7100	14,200
35	7100	14,200
36	7000	13,500
37	7000	13,200
39	6700	12,900
40	6600	12,100
41	6500	11,900
42	6500	11,400
49	6000	9200
63	4500	6900
64	4500	6700
65	4400	6500
66	4300	6400
69	4000	6100
70	3950	6100
71	3900	5900
85	3400	4700
88	3250	4700

decline has a slope given by

$$\frac{d \log L}{dt} = -0.43\lambda,$$
(7.4)

where λ is the corresponding decay constant, $\lambda = \ln 2/T_{1/2}$ (see Equation 2.7). Prove also that the absolute magnitude satisfies

$$dM/dt \propto \lambda$$
(7.5)

and determine the proportionality constant.

P8. Consider the following sample of GRBs, for which γ-ray fluxes, F_γ, and redshifts, z, have been determined: GRB 970828 [$F_\gamma = 96 \times 10^{-6}$ erg cm^{-2}; $z = 0.958$], GRB 990123 [$F_\gamma = 268 \times 10^{-6}$ erg cm^{-2}; $z = 1.600$], GRB 000131 [$F_\gamma = 41.8 \times 10^{-6}$ erg cm^{-2}; $z = 4.500$] (values adopted from Frail et al. [549]). Determine the overall energy emitted by the source under the assumption of:

(a) Isotropic emission.

(b) Beamed emission for several values of the opening angle of the jet: e.g., 1°, 4°, and 25°.

Appendix A

Henyey Method for Arbitrary Hydrodynamic Problems

A primer on spherically symmetric, hydrodynamic code development was presented in Chapter 1, in the framework of a simple problem: the free-fall collapse of a homogeneous sphere. This appendix extends the analysis to a general, 1D stellar evolution code. Let's start with the set of partial differential equations that describe a stellar plasma (i.e., conservation and transport equations; see Sections 1.2 and 1.4):

- Conservation of mass

$$\frac{1}{\rho} = \frac{4}{3}\pi\frac{\partial r^3}{\partial m}. \tag{A.1}$$

- Conservation of momentum

$$\frac{\partial u}{\partial t} + 4\pi r^2 \frac{\partial (P+q)}{\partial m} = -G\frac{m}{r^2}. \tag{A.2}$$

- Conservation of energy

$$\frac{\partial E}{\partial t} = \varepsilon - \frac{\partial L}{\partial m} + \frac{P+q}{\rho^2}\frac{\partial \rho}{\partial t}. \tag{A.3}$$

- Energy transport (by radiation and convection)

$$L = -256\sigma\pi^2 r^4 \frac{T^3}{3\kappa}\frac{\partial T}{\partial m} + L_{conv}. \tag{A.4}$$

- Lagrangian velocity

$$\frac{\partial r}{\partial t} = u. \tag{A.5}$$

Suitable expressions for this set of equations in finite difference form can be found in Section 1.4.1.

Following the same methodology applied to the free-fall collapse problem, three different regions of interest can be distinguished in a computational grid consisting of N concentric shells: the innermost (or central) shell, the $N-2$ intermediate shells, and the outermost (or surface) shell. The five equations that characterize the innermost shell can be rewritten in compact form as a function, C^j, that depends on eight unknowns:

$$C^j = C^j\left(\rho_{3/2}, T_{3/2}, r_2, L_2, u_2, \rho_{5/2}, T_{5/2}, r_3\right) = 0 \qquad (j = 1,5). \tag{A.6}$$

The same approach is applied to the $N-2$ intermediate shells,

$$F_i^j = F_i^j(r_i, L_i, \rho_{i+1/2}, T_{i+1/2}, r_{i+1}, L_{i+1}, u_{i+1}, \rho_{i+3/2}, T_{i+3/2}, r_{i+2}) = 0$$

$$(i = 2, N-1; j = 1,5), \tag{A.7}$$

and to the outermost shell,

$$S^j = S^j(r_N, L_N, \rho_{N+1/2}, T_{N+1/2}, r_{N+1}, L_{N+1}, u_{N+1}) = 0. \tag{A.8}$$

The system contains as many unknowns as equations, $5N$ (i.e., 5 for the innermost shell, $5(N-2)$ for the intermediate shells, and 5 for the surface shell).

As discussed in Section 1.6.6, one may think of a series of values of the unknowns, x^1, which satisfy the whole set of equations described above. The unknowns x^1 can be determined, through an iterative procedure, from guess values[1] and correction terms, in the form $x^1 = x^0 + \delta X$. Assuming small corrections, δX, the system of equations can be written in the form

$$C^j + \frac{\partial C^j}{\partial \rho_{3/2}}\delta\rho_{3/2} + \frac{\partial C^j}{\partial T_{3/2}}\delta T_{3/2} + \frac{\partial C^j}{\partial r_2}\delta r_2 + \frac{\partial C^j}{\partial L_2}\delta L_2 + \frac{\partial C^j}{\partial u_2}\delta u_2 +$$

$$+ \frac{\partial C^j}{\partial \rho_{5/2}}\delta\rho_{5/2} + \frac{\partial C^j}{\partial T_{5/2}}\delta T_{5/2} + \frac{\partial C^j}{\partial r_3}\delta r_3 = 0 \qquad (j = 1, 5) \tag{A.9}$$

$$F_i^j + \frac{\partial F_i^j}{\partial r_i}\delta r_i + \frac{\partial F_i^j}{\partial L_i}\delta L_i + \frac{\partial F_i^j}{\partial \rho_{i+1/2}}\delta\rho_{i+1/2} + \frac{\partial F_i^j}{\partial T_{i+1/2}}\delta T_{i+1/2} +$$

$$+ \frac{\partial F_i^j}{\partial r_{i+1}}\delta r_{i+1} + \frac{\partial F_i^j}{\partial L_{i+1}}\delta L_{i+1} + \frac{\partial F_i^j}{\partial u_{i+1}}\delta u_{i+1} + \frac{\partial F_i^j}{\partial \rho_{i+3/2}}\delta\rho_{i+3/2} +$$

$$+ \frac{\partial F_i^j}{\partial T_{i+3/2}}\delta T_{i+3/2} + \frac{\partial F_i^j}{\partial r_{i+2}}\delta r_{i+2} = 0 \qquad (i = 2, N-1; j = 1, 5) \tag{A.10}$$

$$S^j + \frac{\partial S^j}{\partial r_N}\delta r_N + \frac{\partial S^j}{\partial L_N}\delta L_N + \frac{\partial S^j}{\partial \rho_{N+1/2}}\delta\rho_{N+1/2} + \frac{\partial S^j}{\partial T_{N+1/2}}\delta T_{N+1/2} +$$

$$+ \frac{\partial S^j}{\partial r_{N+1}}\delta r_{N+1} + \frac{\partial S^j}{\partial L_{N+1}}\delta L_{N+1} + \frac{\partial S^j}{\partial u_{N+1}}\delta u_{N+1} = 0$$

$$(j = 1, 5). \tag{A.11}$$

Let's focus first on the system of equations corresponding to the innermost shell (Equations A.9): All terms that involve physical variables evaluated at the next shell, $i = 2$ (i.e., $\delta\rho_{5/2}, \delta T_{5/2}$, and δr_3), are moved to the right-hand side of the equation, in the form

$$\frac{\partial C^j}{\partial \rho_{3/2}}\delta\rho_{3/2} + \frac{\partial C^j}{\partial T_{3/2}}\delta T_{3/2} + \frac{\partial C^j}{\partial u_2}\delta u_2 + \frac{\partial C^j}{\partial L_2}\delta L_2 + \frac{\partial C^j}{\partial r_2}\delta r_2 =$$

$$-C^j - \frac{\partial C^j}{\partial \rho_{5/2}}\delta\rho_{5/2} - \frac{\partial C^j}{\partial T_{5/2}}\delta T_{5/2} - \frac{\partial C^j}{\partial r_3}\delta r_3 \qquad (j = 1, 5),$$

---

[1]Extrapolation from the last two converged models is frequently used to this end.

which can be rewritten in matrix form as

$$
\begin{pmatrix}
\dfrac{\partial C^1}{\partial \rho_{3/2}} & \dfrac{\partial C^1}{\partial T_{3/2}} & \dfrac{\partial C^1}{\partial u_2} & \dfrac{\partial C^1}{\partial L_2} & \dfrac{\partial C^1}{\partial r_2} \\[2ex]
\dfrac{\partial C^2}{\partial \rho_{3/2}} & \dfrac{\partial C^2}{\partial T_{3/2}} & \dfrac{\partial C^2}{\partial u_2} & \dfrac{\partial C^2}{\partial L_2} & \dfrac{\partial C^2}{\partial r_2} \\[2ex]
\dfrac{\partial C^3}{\partial \rho_{3/2}} & \dfrac{\partial C^3}{\partial T_{3/2}} & \dfrac{\partial C^3}{\partial u_2} & \dfrac{\partial C^3}{\partial L_2} & \dfrac{\partial C^3}{\partial r_2} \\[2ex]
\dfrac{\partial C^4}{\partial \rho_{3/2}} & \dfrac{\partial C^4}{\partial T_{3/2}} & \dfrac{\partial C^4}{\partial u_2} & \dfrac{\partial C^4}{\partial L_2} & \dfrac{\partial C^4}{\partial r_2} \\[2ex]
\dfrac{\partial C^5}{\partial \rho_{3/2}} & \dfrac{\partial C^5}{\partial T_{3/2}} & \dfrac{\partial C^5}{\partial u_2} & \dfrac{\partial C^5}{\partial L_2} & \dfrac{\partial C^5}{\partial r_2}
\end{pmatrix}
\begin{pmatrix}
\delta \rho_{3/2} \\[1ex] \delta T_{3/2} \\[1ex] \delta u_2 \\[1ex] \delta L_2 \\[1ex] \delta r_2
\end{pmatrix}
=
$$

$$
\begin{pmatrix}
-C^1 - \dfrac{\partial C^1}{\partial \rho_{5/2}} \delta \rho_{5/2} - \dfrac{\partial C^1}{\partial T_{5/2}} \delta T_{5/2} - \dfrac{\partial C^1}{\partial r_3} \delta r_3 \\[2ex]
-C^2 - \dfrac{\partial C^2}{\partial \rho_{5/2}} \delta \rho_{5/2} - \dfrac{\partial C^2}{\partial T_{5/2}} \delta T_{5/2} - \dfrac{\partial C^2}{\partial r_3} \delta r_3 \\[2ex]
-C^3 - \dfrac{\partial C^3}{\partial \rho_{5/2}} \delta \rho_{5/2} - \dfrac{\partial C^3}{\partial T_{5/2}} \delta T_{5/2} - \dfrac{\partial C^3}{\partial r_3} \delta r_3 \\[2ex]
-C^4 - \dfrac{\partial C^4}{\partial \rho_{5/2}} \delta \rho_{5/2} - \dfrac{\partial C^4}{\partial T_{5/2}} \delta T_{5/2} - \dfrac{\partial C^4}{\partial r_3} \delta r_3 \\[2ex]
-C^5 - \dfrac{\partial C^5}{\partial \rho_{5/2}} \delta \rho_{5/2} - \dfrac{\partial C^5}{\partial T_{5/2}} \delta T_{5/2} - \dfrac{\partial C^5}{\partial r_3} \delta r_3
\end{pmatrix} . \tag{A.12}
$$

Note that the matrix at the right-hand side of Equation A.12 has dimensions 5×1.

Corrections to the variables evaluated at the innermost shell (i.e., $\delta \rho_{3/2}$, $\delta T_{3/2}$, δr_2, δL_2, δu_2) can be conveniently expressed as a linear function of those variables evaluated at the next shell (i.e., $\delta \rho_{5/2}$, $\delta T_{5/2}$, δr_3), which are taken as parameters [824]

$$
\delta \rho_{3/2} = A_{12} \delta \rho_{5/2} + B_{12} \delta T_{5/2} + C_{12} \delta r_3 + D_{12}
$$

$$
\delta T_{3/2} = A_{22} \delta \rho_{5/2} + B_{22} \delta T_{5/2} + C_{22} \delta r_3 + D_{22}
$$

$$
\delta u_2 = A_{32} \delta \rho_{5/2} + B_{32} \delta T_{5/2} + C_{32} \delta r_3 + D_{32}
$$

$$
\delta L_2 = A_{42} \delta \rho_{5/2} + B_{42} \delta T_{5/2} + C_{42} \delta r_3 + D_{42}
$$

$$
\delta r_2 = A_{52} \delta \rho_{5/2} + B_{52} \delta T_{5/2} + C_{52} \delta r_3 + D_{52}. \tag{A.13}
$$

Equivalently,

$$
\begin{pmatrix}
\delta \rho_{3/2} \\ \delta T_{3/2} \\ \delta u_2 \\ \delta L_2 \\ \delta r_2
\end{pmatrix}
=
\begin{pmatrix}
A_{12} & B_{12} & C_{12} & D_{12} \\
A_{22} & B_{22} & C_{22} & D_{22} \\
A_{32} & B_{32} & C_{32} & D_{32} \\
A_{42} & B_{42} & C_{42} & D_{42} \\
A_{52} & B_{52} & C_{52} & D_{52}
\end{pmatrix}
\begin{pmatrix}
\delta \rho_{5/2} \\ \delta T_{5/2} \\ \delta r_3 \\ 1
\end{pmatrix} . \tag{A.14}
$$

Moreover, the matrix at the right-hand side of Equation A.12 can be rewritten as the product

$$
\begin{pmatrix}
-C^1 - \dfrac{\partial C^1}{\partial \rho_{5/2}}\delta\rho_{5/2} - \dfrac{\partial C^1}{\partial T_{5/2}}\delta T_{5/2} - \dfrac{\partial C^1}{\partial r_3}\delta r_3 \\[2mm]
-C^2 - \dfrac{\partial C^2}{\partial \rho_{5/2}}\delta\rho_{5/2} - \dfrac{\partial C^2}{\partial T_{5/2}}\delta T_{5/2} - \dfrac{\partial C^2}{\partial r_3}\delta r_3 \\[2mm]
-C^3 - \dfrac{\partial C^3}{\partial \rho_{5/2}}\delta\rho_{5/2} - \dfrac{\partial C^3}{\partial T_{5/2}}\delta T_{5/2} - \dfrac{\partial C^3}{\partial r_3}\delta r_3 \\[2mm]
-C^4 - \dfrac{\partial C^4}{\partial \rho_{5/2}}\delta\rho_{5/2} - \dfrac{\partial C^4}{\partial T_{5/2}}\delta T_{5/2} - \dfrac{\partial C^4}{\partial r_3}\delta r_3 \\[2mm]
-C^5 - \dfrac{\partial C^5}{\partial \rho_{5/2}}\delta\rho_{5/2} - \dfrac{\partial C^5}{\partial T_{5/2}}\delta T_{5/2} - \dfrac{\partial C^5}{\partial r_3}\delta r_3
\end{pmatrix} =
$$

$$
\begin{pmatrix}
-\dfrac{\partial C^1}{\partial \rho_{5/2}} & -\dfrac{\partial C^1}{\partial T_{5/2}} & -\dfrac{\partial C^1}{\partial r_3} & -C^1 \\[2mm]
-\dfrac{\partial C^2}{\partial \rho_{5/2}} & -\dfrac{\partial C^2}{\partial T_{5/2}} & -\dfrac{\partial C^2}{\partial r_3} & -C^2 \\[2mm]
-\dfrac{\partial C^3}{\partial \rho_{5/2}} & -\dfrac{\partial C^3}{\partial T_{5/2}} & -\dfrac{\partial C^3}{\partial r_3} & -C^3 \\[2mm]
-\dfrac{\partial C^4}{\partial \rho_{5/2}} & -\dfrac{\partial C^4}{\partial T_{5/2}} & -\dfrac{\partial C^4}{\partial r_3} & -C^4 \\[2mm]
-\dfrac{\partial C^5}{\partial \rho_{5/2}} & -\dfrac{\partial C^5}{\partial T_{5/2}} & -\dfrac{\partial C^5}{\partial r_3} & -C^5
\end{pmatrix}
\begin{pmatrix}
\delta\rho_{5/2} \\[2mm]
\delta T_{5/2} \\[2mm]
\delta r_3 \\[2mm]
1
\end{pmatrix} . \qquad \text{(A.15)}
$$

Plugging Equations A.14 and A.15 into Equation A.12 yields

$$
\begin{pmatrix}
\dfrac{\partial C^1}{\partial \rho_{3/2}} & \dfrac{\partial C^1}{\partial T_{3/2}} & \dfrac{\partial C^1}{\partial u_2} & \dfrac{\partial C^1}{\partial L_2} & \dfrac{\partial C^1}{\partial r_2} \\[2mm]
\dfrac{\partial C^2}{\partial \rho_{3/2}} & \dfrac{\partial C^2}{\partial T_{3/2}} & \dfrac{\partial C^2}{\partial u_2} & \dfrac{\partial C^2}{\partial L_2} & \dfrac{\partial C^2}{\partial r_2} \\[2mm]
\dfrac{\partial C^3}{\partial \rho_{3/2}} & \dfrac{\partial C^3}{\partial T_{3/2}} & \dfrac{\partial C^3}{\partial u_2} & \dfrac{\partial C^3}{\partial L_2} & \dfrac{\partial C^3}{\partial r_2} \\[2mm]
\dfrac{\partial C^4}{\partial \rho_{3/2}} & \dfrac{\partial C^4}{\partial T_{3/2}} & \dfrac{\partial C^4}{\partial u_2} & \dfrac{\partial C^4}{\partial L_2} & \dfrac{\partial C^4}{\partial r_2} \\[2mm]
\dfrac{\partial C^5}{\partial \rho_{3/2}} & \dfrac{\partial C^5}{\partial T_{3/2}} & \dfrac{\partial C^5}{\partial u_2} & \dfrac{\partial C^5}{\partial L_2} & \dfrac{\partial C^5}{\partial r_2}
\end{pmatrix}
\begin{pmatrix}
A_{12} & B_{12} & C_{12} & D_{12} \\
A_{22} & B_{22} & C_{22} & D_{22} \\
A_{32} & B_{32} & C_{32} & D_{32} \\
A_{42} & B_{42} & C_{42} & D_{42} \\
A_{52} & B_{52} & C_{52} & D_{52}
\end{pmatrix} =
$$

$$
\begin{pmatrix}
-\dfrac{\partial C^1}{\partial \rho_{5/2}} & -\dfrac{\partial C^1}{\partial T_{5/2}} & -\dfrac{\partial C^1}{\partial r_3} & -C^1 \\[2mm]
-\dfrac{\partial C^2}{\partial \rho_{5/2}} & -\dfrac{\partial C^2}{\partial T_{5/2}} & -\dfrac{\partial C^2}{\partial r_3} & -C^2 \\[2mm]
-\dfrac{\partial C^3}{\partial \rho_{5/2}} & -\dfrac{\partial C^3}{\partial T_{5/2}} & -\dfrac{\partial C^3}{\partial r_3} & -C^3 \\[2mm]
-\dfrac{\partial C^4}{\partial \rho_{5/2}} & -\dfrac{\partial C^4}{\partial T_{5/2}} & -\dfrac{\partial C^4}{\partial r_3} & -C^4 \\[2mm]
-\dfrac{\partial C^5}{\partial \rho_{5/2}} & -\dfrac{\partial C^5}{\partial T_{5/2}} & -\dfrac{\partial C^5}{\partial r_3} & -C^5
\end{pmatrix} , \qquad \text{(A.16)}
$$

from which constants A_{k2}, B_{k2}, C_{k2}, and D_{k2} $(k = 1, 5)$ can be obtained.

The outlined procedure is subsequently extended to the $N - 2$ intermediate shells. First, the linear dependence between variables evaluated at the first and second shells described above (Equation A.13) is generalized for shells $(i - 1)^{th}$ and i^{th} as follows:

$$\delta\rho_{i-1/2} = A_{1i}\delta\rho_{i+1/2} + B_{1i}\delta T_{i+1/2} + C_{1i}\delta r_{i+1} + D_{1i}$$

$$\delta T_{i-1/2} = A_{2i}\delta\rho_{i+1/2} + B_{2i}\delta T_{i+1/2} + C_{2i}\delta r_{i+1} + D_{2i}$$

$$\delta u_i = A_{3i}\delta\rho_{i+1/2} + B_{3i}\delta T_{i+1/2} + C_{3i}\delta r_{i+1} + D_{3i}$$

$$\delta L_i = A_{4i}\delta\rho_{i+1/2} + B_{4i}\delta T_{i+1/2} + C_{4i}\delta r_{i+1} + D_{4i}$$

$$\delta r_i = A_{5i}\delta\rho_{i+1/2} + B_{5i}\delta T_{i+1/2} + C_{5i}\delta r_{i+1} + D_{5i}. \tag{A.17}$$

Plugging the last two relations, for δL_i and δr_i, into Equation A.10 yields

$$\left[\frac{\partial F_i^j}{\partial \rho_{i+1/2}} + A_{4i}\frac{\partial F_i^j}{\partial L_i} + A_{5i}\frac{\partial F_i^j}{\partial r_i}\right]\delta\rho_{i+1/2} + \left[\frac{\partial F_i^j}{\partial T_{i+1/2}} + B_{4i}\right.$$

$$\left.\frac{\partial F_i^j}{\partial L_i} + B_{5i}\frac{\partial F_i^j}{\partial r_i}\right]\delta T_{i+1/2} + \frac{\partial F_i^j}{\partial u_{i+1}}\delta u_{i+1} + \frac{\partial F_i^j}{\partial L_{i+1}}\delta L_{i+1} + \left[\frac{\partial F_i^j}{\partial r_{i+1}} + C_{4i}\frac{\partial F_i^j}{\partial L_i} + \right.$$

$$\left. + C_{5i}\frac{\partial F_i^j}{\partial r_i}\right]\delta r_{i+1} + \frac{\partial F_i^j}{\partial \rho_{i+3/2}}\delta\rho_{i+3/2} + \frac{\partial F_i^j}{\partial T_{i+3/2}}\delta T_{i+3/2} + \frac{\partial F_i^j}{\partial r_{i+2}}\delta r_{i+2} =$$

$$-F_i^j - D_{4i}\frac{\partial F_i^j}{\partial L_i} + D_{5i}\frac{\partial F_i^j}{\partial r_i} \qquad (j = 1, 5), \tag{A.18}$$

which represents a system of 5 linearized equations with 8 unknowns. For convenience, three of the unknowns are, as before, taken as parameters and subsequently moved to the right-hand side of the equation, in the form

$$\alpha_i^j\delta\rho_{i+1/2} + \beta_i^j\delta T_{i+1/2} + \frac{\partial F_i^j}{\partial u_{i+1}}\delta u_{i+1} + \frac{\partial F_i^j}{\partial L_{i+1}}\delta L_{i+1} + \gamma_i^j\delta r_{i+1} + \frac{\partial F_i^j}{\partial \rho_{i+3/2}}\delta\rho_{i+3/2} +$$

$$+ \frac{\partial F_i^j}{\partial T_{i+3/2}}\delta T_{i+3/2} + \frac{\partial F_i^j}{\partial r_{i+2}}\delta r_{i+2} = -F_i^j - D_{4i}\frac{\partial F_i^j}{\partial L_i} -$$

$$- D_{5,i}\frac{\partial F_i^j}{\partial r_i} \qquad (j = 1, 5), \tag{A.19}$$

where

$$\alpha_i^j \equiv \frac{\partial F_i^j}{\partial \rho_{i+1/2}} + A_{4i}\frac{\partial F_i^j}{\partial L_i} + A_{5i}\frac{\partial F_i^j}{\partial r_i}$$

$$\beta_i^j \equiv \frac{\partial F_i^j}{\partial T_{i+1/2}} + B_{4i}\frac{\partial F_i^j}{\partial L_i} + B_{5i}\frac{\partial F_i^j}{\partial r_i}$$

$$\gamma_i^j \equiv \frac{\partial F_i^j}{\partial r_{i+1}} + C_{4i}\frac{\partial F_i^j}{\partial L_i} + C_{5i}\frac{\partial F_i^j}{\partial r_i}.$$

Equation A.19 can now be written in the matrix form

$$
\begin{pmatrix}
\alpha_i^1 & \beta_i^1 & \dfrac{\partial F_i^1}{\partial u_{i+1}} & \dfrac{\partial F_i^1}{\partial L_{i+1}} & \gamma_i^1 \\[2ex]
\alpha_i^2 & \beta_i^2 & \dfrac{\partial F_i^2}{\partial u_{i+1}} & \dfrac{\partial F_i^2}{\partial L_{i+1}} & \gamma_i^2 \\[2ex]
\alpha_i^3 & \beta_i^3 & \dfrac{\partial F_i^3}{\partial u_{i+1}} & \dfrac{\partial F_i^3}{\partial L_{i+1}} & \gamma_i^3 \\[2ex]
\alpha_i^4 & \beta_i^4 & \dfrac{\partial F_i^4}{\partial u_{i+1}} & \dfrac{\partial F_i^4}{\partial L_{i+1}} & \gamma_i^4 \\[2ex]
\alpha_i^5 & \beta_i^5 & \dfrac{\partial F_i^5}{\partial u_{i+1}} & \dfrac{\partial F_i^5}{\partial L_{i+1}} & \gamma_i^5
\end{pmatrix}
\begin{pmatrix}
\delta\rho_{i+1/2} \\[2ex]
\delta T_{i+1/2} \\[2ex]
\delta u_{i+1} \\[2ex]
\delta L_{i+1} \\[2ex]
\delta r_{i+1}
\end{pmatrix}
=
\begin{pmatrix}
\delta_i^1 \\[2ex]
\delta_i^2 \\[2ex]
\delta_i^3 \\[2ex]
\delta_i^4 \\[2ex]
\delta_i^5
\end{pmatrix},
\tag{A.20}
$$

where

$$
\delta_i^j \equiv -F_i^j - D_{4i}\frac{\partial F_i^j}{\partial L_i} - D_{5i}\frac{\partial F_i^j}{\partial r_i} - \frac{\partial F_i^j}{\partial \rho_{i+3/2}}\delta\rho_{i+3/2} -
$$

$$
- \frac{\partial F_i^j}{\partial T_{i+3/2}}\delta T_{i+3/2} - \frac{\partial F_i^j}{\partial r_{i+2}}\delta r_{i+2}.
$$

Expressing the 5 unknowns of the ith shell in terms of the variables $\delta\rho_{i+3/2}, \delta T_{i+3/2}$, and δr_{i+2},

$$
\delta\rho_{i+1/2} = A_{1,i+1}\delta\rho_{i+3/2} + B_{1,i+1}\delta T_{i+3/2} + C_{1,i+1}\delta r_{i+2} + D_{1,i+1}
$$

$$
\delta T_{i+1/2} = A_{2,i+1}\delta\rho_{i+3/2} + B_{2,i+1}\delta T_{i+3/2} + C_{2,i+1}\delta r_{i+2} + D_{2,i+1}
$$

$$
\delta u_{i+1} = A_{3,i+1}\delta\rho_{i+3/2} + B_{3,i+1}\delta T_{i+3/2} + C_{3,i+1}\delta r_{i+2} + D_{3,i+1}
$$

$$
\delta L_{i+1} = A_{4,i+1}\delta\rho_{i+3/2} + B_{4,i+1}\delta T_{i+3/2} + C_{4,i+1}\delta r_{i+2} + D_{4,i+1}
$$

$$
\delta r_{i+1} = A_{5,i+1}\delta\rho_{i+3/2} + B_{5,i+1}\delta T_{i+3/2} + C_{5,i+1}\delta r_{i+2} + D_{5,i+1},
$$

the constants $A_{k,i+l}, B_{k,i+l}, C_{k,i+l}$, and $D_{k,i+l}$ $(k = 1, 5;\ i = 2, N-1)$ can be determined by solving the matrix system

$$
\begin{pmatrix}
\alpha_i^1 & \beta_i^1 & \dfrac{\partial F_i^1}{\partial u_{i+1}} & \dfrac{\partial F_i^1}{\partial L_{i+1}} & \gamma_i^1 \\[2ex]
\alpha_i^2 & \beta_i^2 & \dfrac{\partial F_i^2}{\partial u_{i+1}} & \dfrac{\partial F_i^2}{\partial L_{i+1}} & \gamma_i^2 \\[2ex]
\alpha_i^3 & \beta_i^3 & \dfrac{\partial F_i^3}{\partial u_{i+1}} & \dfrac{\partial F_i^3}{\partial L_{i+1}} & \gamma_i^3 \\[2ex]
\alpha_i^4 & \beta_i^4 & \dfrac{\partial F_i^4}{\partial u_{i+1}} & \dfrac{\partial F_i^4}{\partial L_{i+1}} & \gamma_i^4 \\[2ex]
\alpha_i^5 & \beta_i^5 & \dfrac{\partial F_i^5}{\partial u_{i+1}} & \dfrac{\partial F_i^5}{\partial L_{i+1}} & \gamma_i^5
\end{pmatrix}
\begin{pmatrix}
A_{1,i+1} & B_{1,i+1} & C_{1,i+1} & D_{1,i+1} \\[2ex]
A_{2,i+1} & B_{2,i+1} & C_{2,i+1} & D_{2,i+1} \\[2ex]
A_{3,i+1} & B_{3,i+1} & C_{3,i+1} & D_{3,i+1} \\[2ex]
A_{4,i+1} & B_{4,i+1} & C_{4,i+1} & D_{4,i+1} \\[2ex]
A_{5,i+1} & B_{5,i+1} & C_{5,i+1} & D_{5,i+1}
\end{pmatrix}
=
$$

$$
\begin{pmatrix}
-\dfrac{\partial F_i^1}{\partial \rho_{i+3/2}} & -\dfrac{\partial F_i^1}{\partial T_{3+1/2}} & -\dfrac{\partial F_i^1}{\partial r_{i+2}} & \epsilon_i^1 \\[2mm]
-\dfrac{\partial F_i^2}{\partial \rho_{i+3/2}} & -\dfrac{\partial F_i^2}{\partial T_{3+1/2}} & -\dfrac{\partial F_i^2}{\partial r_{i+2}} & \epsilon_i^2 \\[2mm]
-\dfrac{\partial F_i^3}{\partial \rho_{i+3/2}} & -\dfrac{\partial F_i^3}{\partial T_{3+1/2}} & -\dfrac{\partial F_i^3}{\partial r_{i+2}} & \epsilon_i^3 \\[2mm]
-\dfrac{\partial F_i^4}{\partial \rho_{i+3/2}} & -\dfrac{\partial F_i^4}{\partial T_{3+1/2}} & -\dfrac{\partial F_i^4}{\partial r_{i+2}} & \epsilon_i^4 \\[2mm]
-\dfrac{\partial F_i^5}{\partial \rho_{i+3/2}} & -\dfrac{\partial F_i^5}{\partial T_{3+1/2}} & -\dfrac{\partial F_i^5}{\partial r_{i+2}} & \epsilon_i^5
\end{pmatrix},
\tag{A.21}
$$

where

$$
\epsilon_i^j \equiv -F_i^j - D_{4i}\frac{\partial F_i^j}{\partial L_i} - D_{5i}\frac{\partial F_i^j}{\partial r_i}.
$$

Finally, the same procedure is applied to the system of equations for the outermost shell. Expressing all variables in terms of quantities belonging to the Nth shell,

$$
\delta\rho_{N-1/2} = A_{1N}\delta\rho_{N+1/2} + B_{1N}\delta T_{N+1/2} + C_{1N}\delta r_{N+1} + D_{1N}
$$

$$
\delta T_{N-1/2} = A_{2N}\delta\rho_{N+1/2} + B_{2N}\delta T_{N+1/2} + C_{2N}\delta r_{N+1} + D_{2N}
$$

$$
\delta u_N = A_{3N}\delta\rho_{N+1/2} + B_{3N}\delta T_{N+1/2} + C_{3N}\delta r_{N+1} + D_{3N}
$$

$$
\delta L_N = A_{4N}\delta\rho_{N+1/2} + B_{4N}\delta T_{N+1/2} + C_{4N}\delta r_{N+1} + D_{4N}
$$

$$
\delta r_N = A_{5N}\delta\rho_{N+1/2} + B_{5N}\delta T_{N+1/2} + C_{5N}\delta r_{N+1} + D_{5N},
$$

and, subsequently, plugging these relations into the system of linearized equations for the surface layer, Equations A.11, yields

$$
\left[\frac{\partial S^j}{\partial \rho_{N+1/2}} + A_{4N}\frac{\partial S^j}{\partial L_N} + A_{5N}\frac{\partial S^j}{\partial r_N}\right]\delta\rho_{N+1/2} + \left[\frac{\partial S^j}{\partial T_{N+1/2}} + B_{4N}\frac{\partial S^j}{\partial L_N} + \right.
$$

$$
\left. + B_{5N}\frac{\partial S^j}{\partial r_N}\right]\delta T_{N+1/2} + \frac{\partial S^j}{\partial u_{N+1}}\delta u_{N+1} + \frac{\partial S^j}{\partial L_{N+1}}\delta L_{N+1} + \left[\frac{\partial S^j}{\partial r_{N+1}} + C_{4N}\frac{\partial S^j}{\partial L_N} + \right.
$$

$$
\left. + C_{5N}\frac{\partial S^j}{\partial r_N}\right]\delta r_{N+1} = -S^j - D_{4N}\frac{\partial S^j}{\partial L_N} + D_{5N}\frac{\partial S^j}{\partial r_N}
$$

$$
(j = 1, 5).
\tag{A.22}
$$

Equation A.22 can now be rewritten in the following matrix form

$$
\begin{pmatrix}
\alpha_{N+1}^1 & \beta_{N+1}^1 & \dfrac{\partial S^1}{\partial u_{N+1}} & \dfrac{\partial S^1}{\partial L_{N+1}} & \gamma_{N+1}^1 \\[2mm]
\alpha_{N+1}^2 & \beta_{N+1}^2 & \dfrac{\partial S^2}{\partial u_{N+1}} & \dfrac{\partial S^2}{\partial L_{N+1}} & \gamma_{N+1}^2 \\[2mm]
\alpha_{N+1}^3 & \beta_{N+1}^3 & \dfrac{\partial S^3}{\partial u_{N+1}} & \dfrac{\partial S^3}{\partial L_{N+1}} & \gamma_{N+1}^3 \\[2mm]
\alpha_{N+1}^4 & \beta_{N+1}^4 & \dfrac{\partial S^4}{\partial u_{N+1}} & \dfrac{\partial S^4}{\partial L_{N+1}} & \gamma_{N+1}^4 \\[2mm]
\alpha_{N+1}^5 & \beta_{N+1}^5 & \dfrac{\partial S^5}{\partial u_{N+1}} & \dfrac{\partial S^5}{\partial L_{N+1}} & \gamma_{N+1}^5
\end{pmatrix}
\begin{pmatrix}
\delta\rho_{N+1/2} \\[2mm]
\delta T_{N+1/2} \\[2mm]
\delta u_{N+1} \\[2mm]
\delta L_{N+1} \\[2mm]
\delta r_{N+1}
\end{pmatrix} =
$$

$$\begin{pmatrix} \delta^1_{N+1} \\[1mm] \delta^2_{N+1} \\[1mm] \delta^3_{N+1} \\[1mm] \delta^4_{N+1} \\[1mm] \delta^5_{N+1} \end{pmatrix}, \tag{A.23}$$

where

$$\alpha^j_{N+1} \equiv \frac{\partial S^j}{\partial \rho_{N+1/2}} + A_{4N}\frac{\partial S^j}{\partial L_N} + A_{5N}\frac{\partial S^j}{\partial r_N}$$

$$\beta^j_{N+1} \equiv \frac{\partial S^j}{\partial T_{N+1/2}} + B_{4N}\frac{\partial S^j}{\partial L_N} + B_{5N}\frac{\partial S^j}{\partial r_N}$$

$$\gamma^j_{N+1} \equiv \frac{\partial S^j}{\partial r_{N+1}} + C_{4N}\frac{\partial S^j}{\partial L_N} + C_{5N}\frac{\partial S^j}{\partial r_N}$$

$$\delta^j_{N+1} \equiv -S^j - D_{4N}\frac{\partial S^j}{\partial L_N} - D_{5N}\frac{\partial S^j}{\partial r_N}.$$

The solution of this matrix equation yields a first set of corrections for the variables belonging to the surface layer, $\delta\rho_{N+1/2}$, $\delta T_{N+1/2}$, δu_{N+1}, δL_{N+1}, and δr_{N+1}. From these values, and the stored auxiliar quantities A_{kl}, B_{kl}, C_{kl}, and D_{kl} ($k = 1,5; l = 2, N$), first corrections to the variables for the whole grid can also be obtained:

$$\delta\rho_{i-1/2} = A_{1,i}\delta\rho_{i+1/2} + B_{1,i}\delta T_{i+1/2} + C_{1,i}\delta r_{i+1} + D_{1,i}$$

$$\delta T_{i-1/2} = A_{2,i}\delta\rho_{i+1/2} + B_{2,i}\delta T_{i+1/2} + C_{2,i}\delta r_{i+1} + D_{2,i}$$

$$\delta u_i = A_{3,i}\delta\rho_{i+1/2} + B_{3,i}\delta T_{i+1/2} + C_{3,i}\delta r_{i+1} + D_{3,i}$$

$$\delta L_i = A_{4,i}\delta\rho_{i+1/2} + B_{4,i}\delta T_{i+1/2} + C_{4,i}\delta r_{i+1} + D_{4,i}$$

$$\delta r_i = A_{5,i}\delta\rho_{i+1/2} + B_{5,i}\delta T_{i+1/2} + C_{5,i}\delta r_{i+1} + D_{5,i}$$

$$(i = N, 2).$$

These corrections are finally applied to the first guess values, in the form

$$\rho^1_{i+1/2} = \rho^o_{i+1/2} + \delta\rho_{i+1/2}$$

$$T^1_{i+1/2} = T^o_{i+1/2} + \delta T_{i+1/2}$$

$$u^1_{i+1} = u^o_{i+1} + \delta u_{i+1}$$

$$L^1_{i+1} = L^o_{i+1} + \delta L_{i+1}$$

$$r^1_{i+1} = r^o_{i+1} + \delta r_{i+1} \qquad (i = 1, N).$$

The overall procedure is then iterated until a given accuracy criterion (i.e., corrections smaller than a given quantity) is satisfied.

Appendix B

Computer Program for the Free-Fall Collapse Problem

```
C  ***********************************************************************
C  ....................................................................
C            1D HYDRODYNAMIC SIMULATION OF THE FREE-FALL COLLAPSE
C            OF A HOMOGENEOUS SPHERE
C
C            (c) Jordi Jose, 2015
C            Stellar Explosions: Hydrodynamics and Nucleosynthesis
C            CRC/Taylor and Francis
C  ....................................................................
C  ***********************************************************************
      IMPLICIT REAL*8(A-H,O-Z)
C
      PARAMETER (N=100)
      COMMON/F0/F1(N+1),F2(N+1),F5(N+1)
      COMMON/F1/DF1W(N+1),DF1RI(N+1),DF1RIO(N+1)
      COMMON/F21/DF2UI(N+1),DF2RI(N+1)
      COMMON/F5/DF5UI(N+1),DF5RI(N+1)
      COMMON/CONST/PI,G
      COMMON/PHYSICS/U(N+1),UU(N+1),R(N+1),RR(N+1)
      DIMENSION UUU(N+1),RRR(N+1)
C
C  -------------------------------------------------------------------
C  ............N       = Number of shells of the computational domain.
C                        N is defined in Main Program and in subroutines
C                        HENYEY3 and ENERGYC
C  ........... F1      = Conservation of mass in finite differences
C  ........... F2      = Conservation of momentum in finite differences
C  ........... F5      = Lagrangian velocity in finite differences
C  ........... DF1W    = Partial derivative of function F1 with respect
C                        to Wi-1/2
C  ........... DF1RI   = Partial derivative of function F1 with respect
C                        to Ri
C  ........... DF1RIO  = Partial derivative of function F1 with respect
C                        to Ri-1
C  ........... DF2UI   = Partial derivative of function F2 with respect
C                        to Ui
C  ........... DF2RI   = Partial derivative of function F2 with respect
C                        to Ri
C  ........... DF5RI   = Partial derivative of function F5 with respect
C                        to Ri
C  ........... DF5UI   = Partial derivative of function F5 with respect
C                        to Ui
C  ........... G       = Gravitational constant       (dyn cm^2 g^-2)
C  ........... U       = Velocity                     (cm s^-1)
C                        [while U represents velocity at time t_n,
C                        UU is the velocity evaluated at time t_n-1 and
C                        UUU is the velocity at time t_n-2]
C  ........... R       = Radius                        (cm)
C                        [while R represents the radius at time t_n,
C                        RR is the radius at time t_n-1 and
C                        RRR is the radius at time t_n-2]
C  -------------------------------------------------------------------
C
      DIMENSION RM(N+1),RHO(N),V(N),W(N)
      DIMENSION RRM(N+1),VV(N),VVV(N),WW(N)
      DIMENSION DM(N+1),Q(N+1),S(N+1),AMM(N+1),QM(N+1)
      DIMENSION FF(N+1),F(N+1),DFRI(N+1)
      DIMENSION AX1(N+1),AX2(N+1)
      DIMENSION HW(N),HU(N+1),HR(N+1)
C
```

```
C -----------------------------------------------------------------------
C ............ RM      = Ln(R)
C                        [while RM is evaluated at time t_n,
C                        RRM corresponds to time t_n-1]
C ............ RHO     = Density (1/V)                    (g cm^-3)
C ............ V       = Specific volume (1/RHO)          (cm^3 g^-1)
C                        [while V is evaluated at time t_n,
C                        VV corresponds to time t_n-1, and VVV
C                        is evaluated at time t_n-2]
C ............ W       = Ln(V)
C                        [while W is evaluated at time t_n,
C                        WW corresponds to time t_n-1]
C ............ DM      = Mass interior to the i-th interface (g)
C ............ Q       = 1 - DM/Total Mass
C ............ S       = Ln(Q)
C ............ AMM     = DSQRT[DMi*DMi-1]                  (g)
C ............ QM      = 1 - AMM/Total Mass
C ............ F       = Auxiliar function F used in the equation of
C                        momentum conservation at time t_n
C ............ FF      = Auxiliar function F at time t_n-1
C ............ DFRI    = Partial derivative of function F with
C                        respect to Ri
C ............ AX1     = Auxiliar function for F1
C ............ AX2     = Auxiliar function for F2
C ............ HW      = Corrections obtained for W
C ............ HU      = Corrections obtained for U
C ............ HR      = Corrections obtained for R
C -----------------------------------------------------------------------
C
      OPEN(UNIT=5,FILE='FREE.INI',STATUS='OLD')
      READ(5,*) ICOI,ICOF,IWRT,DACUR
      READ(5,*) TIMEI,TIMEF,DTIME,EPS
      READ(5,*) BETA,RINI,RHOI
      OPEN(UNIT=7,FILE='OUTFIL7.OUT1',STATUS='NEW')
      OPEN(UNIT=10,FILE='PLOT.OUT1',STATUS='NEW')
      OPEN(UNIT=11,FILE='RADTIM.OUT1',STATUS='NEW')
      OPEN(UNIT=13,FILE='ANALYTIC.OUT1',STATUS='NEW')
C
C -----------------------------------------------------------------------
C ............ ICOI    = Initial model
C ............ ICOF    = Final model
C ............ IWRT    = Label that controls the frequency
C                        of data output
C ............ DACUR   = Precision requirement for the
C                        iterative procedure
C ............ TIMEI   = Initial time                     (s)
C ............ TIMEF   = Final time                       (s)
C ............ DTIME   = Initial time-step                (s)
C ............ EPS     = Maximum variation allowance for
C                        physical variables R and V
C ............ BETA    = Parameter of the scheme
C             ...... 0 : Explicit scheme
C             ...... 1 : Implicit scheme
C ............ RINI    = Initial radius of the sphere     (cm)
C ............ RHOI    = Initial density of the sphere    (g cm^-3)
C
C -----------------------------------------------------------------------
C
      PI=4.D0*DATAN(1.D0)
      DMENV=4.D0*PI*RHOI*(RINI**3.D0)/3.D0
      G=6.67D-8
C
C -----------------------------------------------------------------------
C Physical constants
C ............ DMENV   = Total mass of the sphere         (g)
C ............ G       = Gravitational constant           (dyne cm^2 g^-2)
C -----------------------------------------------------------------------
C
C Discretization of the sphere in N concentric shells. Subroutine ZONING
C assigns a mass DMi+1 - DMi to each individual shell i
C
      CALL ZONING(N,DMENV,DM)
C
C Boundary conditions
C
      R(1)=0.D0
```

```
      U(1)=0.D0
C
C Assumption of hydrostatic equilibrium (Ui = 0) and homogeneity of the
C sphere (Vi = 1/RHOI).
C Determination of the radius of each intershell, Ri+1, from mass
C conservation, using the initial density RHOI and the mass distribution
C obtained in subroutine ZONING
C
      DO 90 I=1,N
      U(I+1)=0.D0
      V(I)=1.D0/RHOI
      R(I+1)=(R(I)**3.D0+(3.D0*(DM(I+1)-DM(I))/4.D0/PI/RHOI))
     &**(1.D0/3.D0)
  90  CONTINUE
C
C -----------------------------------------------------------------------
C Initialization of counters and variables
C ............ ICO  = Model number
C ........... INDY = Number of iterations needed for the current model
C                      (step)
C ............ IWW  = Output counter
C ............ TIME = Total computed time                 (s)
C -----------------------------------------------------------------------
C
      ICO=ICOI
      INDY=0
      IWW=1
      TIME=TIMEI
C
C Storage of input values and initial model in Unit 7
C
      IF(ICO.EQ.1) THEN
       WRITE(7,1800) ICOI,ICOF,IWRT,DACUR
       WRITE(7,1802) TIMEI,TIMEF,DTIME,EPS
       WRITE(7,1801) BETA,DMENV,RINI,RHOI
       WRITE(7,1900)
       WRITE(7,1902)
       DO 94 J=1,N
       WRITE(7,905) J,1.D0/V(J),R(J+1),U(J+1)
  94   CONTINUE
       WRITE(7,777)
      ENDIF
C
C -----------------------------------------------------------------------
C Analytical solution for the free-fall collapse problem
C [S.A. Colgate and R.H. White, The Astrophysical Journal (1966),
C vol. 143, p. 632].
C ........... RX   = Ratio of current total radius of the sphere R(N+1)
C                      over the initial radius RINI
C ........... FUNC = TIME*SQRT(8*PI*G*RHOI/3) = SQRT(1-RX)*SQRT(RX) +
C                      ASIN[SQRT(1-RX)]
C ........... FUNX = PI/2 - FUNC
C ........... DRX  = Step used to generate the output in RX
C -----------------------------------------------------------------------
C
      RX=1.0005D0
      DRX=0.0025D0
C
C Storage of RX and FUNX in Unit 13
C
      DO 756 I=1,400
      RX=RX-DRX
      FUNC=DASIN((1.D0-RX)**0.5D0)+((1.D0-RX)**0.5D0)*(RX**0.5D0)
      FUNX=PI/2.D0-FUNC
      WRITE(13,790) FUNX,RX
 756  CONTINUE
C
C For each variable Y, we identify Y=YY (i.e., previous value of Y at
C time t_n-1 is equal to the current value of Y at time t_n), since no
C earlier values for Y are available.
C
C
      RR(1)=R(1)
      UU(1)=U(1)
C
      DO 8 I=1,N
```

```
          RR(I+1)=R(I+1)
          RM(I+1)=DLOG(R(I+1))
          RRM(I+1)=DLOG(RR(I+1))
          VV(I)=V(I)
          UU(I+1)=U(I+1)
          W(I)=DLOG(V(I))
          WW(I)=DLOG(VV(I))
        8 CONTINUE
C
C Definition of several mass variables used in the calculations
C
          DO 4 I=1,N+1
          Q(I)=1.D0-DM(I)/DMENV
          S(I)=DLOG(Q(I))
        4 CONTINUE
C
          DO 101 I=2,N+1
          AMM(I)=DSQRT(DM(I)*DM(I-1))
          QM(I)=1.D0-AMM(I)/DMENV
      101 CONTINUE
C
C Special treatment for the innermost shell to avoid numerical problems
C when DM1=0 (i.e., simulation extending all the way down to the center
C of the sphere)
C
          IF (DM(1).EQ.0.D0) THEN
           AMM(2)=DM(2)/2.D0
           QM(2)=1.D0-AMM(2)/DMENV
          ENDIF
C
C Definition of auxiliar functions AX1 and AX2 that appear in the
C system of finite difference equations
C
          DO 907 I=2,N+1
          AX1(I)=4.D0*PI/(3.D0*DMENV*(Q(I)-Q(I-1)))
      907 CONTINUE
C
          DO 5 I=2,N
          AX2(I)=4.D0*PI/(DMENV*(QM(I+1)-QM(I)))
          FF(I)=-G*DM(I)/RR(I)/RR(I)
        5 CONTINUE
          AX2(N+1)=4.D0*PI/(DMENV*(Q(N+1)-QM(N+1)))
          FF(N+1)=-G*DMENV/RR(N+1)/RR(N+1)
C
C ------------------------------------------------------------------------
C Energy budget
C ............ EKIN  =  Kinetic energy                         (erg)
C ............ EGRAV =  Gravitational potential energy         (erg)
C ............ EOLD  =  Total initial energy                   (erg)
C ------------------------------------------------------------------------
C
          EKIN=0.D0
          EGRAV=0.D0
C
          DO 813 I=2,N+1
          EKIN=EKIN+DABS(UU(I)*UU(I-1))*(DM(I)-DM(I-1))/2.D0
          IF (RR(I-1).NE.0.D0) THEN
           EGRAV=EGRAV-G*(DM(I)-DM(I-1))*AMM(I)/DSQRT(RR(I)*RR(I-1))
          ELSE
           EGRAV=0.D0
          ENDIF
      813 CONTINUE
          EOLD=EKIN+EGRAV
          WRITE(7,793) EOLD
C
C ------------------------------------------------------------------------
C                     M A I N          L O O P
C ------------------------------------------------------------------------
C For each physical magnitud Y, the program determines the set of
C corrections HY that should be applied to the first guess value,
C by means of an iterative procedure
C
      309 CONTINUE
C
          INDY=INDY+1
C
```

```
C -------------------------------------------------------------------
C Innermost and Intermediate Shells
C -------------------------------------------------------------------
C All physical magnitudes, such as V (or W), that are evaluated at the
C middle of a shell, i+1/2, are written as:
C     V(I-1/2) --> V(I-1)
C     V(I+1/2) --> V(I),              I=1, N
C
      DO 160 I=2,N
C
C Function F1 and its derivatives
C
      F1(I)=AX1(I)*(R(I)*R(I)*R(I)-R(I-1)*R(I-1)*R(I-1))+V(I-1)
      DF1W(I)=V(I-1)
      DF1RI(I)=AX1(I)*3.D0*R(I)*R(I)*R(I)
      DF1RI0(I)=-AX1(I)*3.D0*R(I-1)*R(I-1)*R(I-1)
C
C Function F2 and its derivatives
C
      F(I)=-G*DM(I)/R(I)/R(I)
      F2(I)=(U(I)-UU(I))/DTIME-(1.D0-BETA)*FF(I)-BETA*F(I)
      DFRI(I)=2.D0*G*DM(I)/R(I)/R(I)/R(I)
      DF2UI(I)=1.D0/DTIME
      DF2RI(I)=-R(I)*BETA*DFRI(I)
C
C Function F5 and its derivatives
C
      F5(I)=(RM(I)-RRM(I))/DTIME-BETA*U(I)/R(I)-(1.D0-BETA)*UU(I)/
     &RR(I)
      DF5UI(I)=-BETA/R(I)
      DF5RI(I)=1.D0/DTIME+BETA*U(I)/R(I)
C
 160  CONTINUE
C
C -------------------------------------------------------------------
C Surface Shell
C -------------------------------------------------------------------
C
C Function F1 at the surface and its derivatives
C
      F1(N+1)=AX1(N+1)*(R(N+1)*R(N+1)*R(N+1)-R(N)*R(N)*R(N))+V(N)
      DF1W(N+1)=V(N)
      DF1RI(N+1)=AX1(N+1)*3.D0*R(N+1)*R(N+1)*R(N+1)
      DF1RI0(N+1)=-AX1(N+1)*3.D0*R(N)*R(N)*R(N)
C
C Function F2 at the surface and its derivatives
C
      F(N+1)=-G*DMENV/R(N+1)/R(N+1)
      F2(N+1)=(U(N+1)-UU(N+1))/DTIME-(1.D0-BETA)*FF(N+1)-BETA*F(N+1)
      DFRI(N+1)=2.D0*G*DMENV/R(N+1)/R(N+1)/R(N+1)
      DF2UI(N+1)=1.D0/DTIME
      DF2RI(N+1)=-R(N+1)*BETA*DFRI(N+1)
C
C Function F5 at the surface and its derivatives
C
      F5(N+1)=(RM(N+1)-RRM(N+1))/DTIME-BETA*U(N+1)/R(N+1)-
     &(1.D0-BETA)*UU(N+1)/RR(N+1)
      DF5UI(N+1)=-BETA/R(N+1)
      DF5RI(N+1)=1.D0/DTIME+BETA*U(N+1)/R(N+1)
C
C Solution of the system based on Henyey method. Corrections HY
C are applied to the first guess values of Y
C
      CALL HENYEY3(HW,HU,HR)
C
      DO 35 I=1,N
      W(I)=W(I)+HW(I)
      U(I+1)=U(I+1)+HU(I+1)
      RM(I+1)=RM(I+1)+HR(I+1)
   35 CONTINUE
C
      DO 41 I=1,N
      V(I)=EXP(W(I))
      R(I+1)=EXP(RM(I+1))
   41 CONTINUE
C
```

```
C ------------------------------------------------------------------------
C Checkpoint for the corrections. If HY > DACUR, convergence
C has not yet been reached and the procedure is iterated with
C the corrected guess values of Y, Y' = Y + HY.
C ............ ICONT  = Accuracy counter
C              ...... 0 : Accuracy criterion is satisfied
C              ...... 2 : Accuracy criterion in not satisfied and
C                         model is forced to another iteration
C ------------------------------------------------------------------------
C
      ICONT=0
      DO 260 K=1,N
      IF(DABS(HW(K)*2.D0/(W(K)+WW(K))).GT.DACUR) THEN
       ICONT=2
      ENDIF
      IF(DABS(HU(K+1)*2.D0/(U(K+1)+UU(K+1))).GT.DACUR) THEN
       ICONT=2
      ENDIF
      IF(DABS(HR(K+1)*2.D0/(RM(K+1)+RRM(K+1))).GT.DACUR) THEN
       ICONT=2
      ENDIF
  260 CONTINUE
C
C If convergence is not reached after 10 iterations, the adopted
C time-step is reduced by a factor of 10 and the overall iteration
C process is reinitialized for this model
C
      IF(INDY.GE.10) THEN
      DTIME=DTIME/10.D0
      ICONT=2
      ENDIF
C
      IF(ICONT.EQ.2) GOTO 309
C
C ------------------------------------------------------------------------
C Storage of results in Units 7, 10, and 11, whenever IWW = IWRT
C UNIT  7: output file containing density (1/V), radius (R), and velocity
C          (V) at t=TIME.
C UNIT 10: output file aimed at comparing the numerical and the
C          analytical solutions of the free-fall collapse problem
C          [see UNIT 13]. It also includes the density at the innermost
C          and outermost shells to check whether the sphere remains
C          homogeneous during collapse, as proved analytically.
C
C ............ DA   = Ratio of current total radius of the sphere
C                     over the initial radius [DA = RX]
C ............ DB   = PI/2 - TIME*SQRT(8*PI*G*RHOI/3) =
C                     PI/2 - SQRT(1-RX)*SQRT(RX)-ASIN[SQRT(1-RX)]
C                     [DB = FUNX]
C UNIT 11: output data listing the overall size of the sphere,
C          R(N+1), as a function of time (TIME-TIMEI).
C ------------------------------------------------------------------------
C
      WRITE(7,*) '********* ICO=',ICO
      WRITE(7,*) '........ INDY=',INDY
C
      IF(IWW.EQ.IWRT) THEN
      WRITE(7,1806) ICO,INDY,TIME,DTIME
      WRITE(7,1902)
      DO 92 J=1,N
      WRITE(7,905) J,1.D0/V(J),R(J+1),U(J+1)
   92 CONTINUE
C
      DA=R(N+1)/RINI
      DB=PI/2.D0-(TIME-TIMEI)*((8.D0*PI*G*RHOI/3.D0)**(1./2.))
      WRITE(10,990) DB,DA,1.D0/V(1),1.D0/V(N)
      WRITE(11,790) TIME-TIMEI,R(N+1)
C
C ------------------------------------------------------------------------
C Check for energy conservation
C ............ ENEW = Total energy at t=TIME                     (erg)
C ------------------------------------------------------------------------
C
      CALL ENERGYC(DM,AMM,EOLD,ENEW)
C
C Initialization of counter IWW
```

```
C
      IWW=0
      ENDIF
C
C Writing last configuration after completion of the simulation
C
      IF((ICO.GT.ICOF).OR.(TIME.GT.TIMEF)) GOTO 310
C
C -----------------------------------------------------------------------
C Time-step evaluation
C ........... DDTIME  = Current time-step                    (s)
C ........... DTIME   = First guess for the new time-step    (s)
C -----------------------------------------------------------------------
C
      DDTIME=DTIME
      CALL TSTEP(DDTIME,EPS,V,VV,R,RR,DTIME,N)
C
C Redefinition of variables at previous time and first guess;
C linear extrapolation of the new values at t(new)=t(old)+DTIME
C
      DO 78 I=1,N
      RRR(I+1)=RR(I+1)
      UUU(I+1)=UU(I+1)
      VVV(I)=VV(I)
   78 CONTINUE
C
      DO 76 I=1,N
      RR(I+1)=R(I+1)
      RRM(I+1)=RM(I+1)
      UU(I+1)=U(I+1)
      VV(I)=V(I)
      WW(I)=W(I)
      FF(I+1)=F(I+1)
   76 CONTINUE
C
      DO 77 J=1,N
      R(J+1)=RR(J+1)+((RR(J+1)-RRR(J+1))*DTIME/DDTIME)
      U(J+1)=UU(J+1)+((UU(J+1)-UUU(J+1))*DTIME/DDTIME)
      V(J)=VV(J)+((VV(J)-VVV(J))*DTIME/DDTIME)
   77 CONTINUE
C
C Update on total time and model counters
C
      TIME=TIME+DTIME
      ICO=ICO+1
      IWW=IWW+1
      INDY=0
C
C -----------------------------------------------------------------------
C Density control
C ........... INEG  = check counter
C               ....... 0 : All shells have positive densities
C               ....... 5 : At least one shell has a negative density,
C                           and execution is aborted
C -----------------------------------------------------------------------
C
      INEG=0
      DO 3934 K=1,N
      IF(V(K).LT.0.D0) INEG=5
 3934 CONTINUE
C
      IF(INEG.EQ.5) THEN
      WRITE(7,*) 'Numerical problem (RHO<0). Execution aborted'
      GOTO 310
      ENDIF
C
      GOTO 309
C
  310 CONTINUE
C
C     Formats
C
  777 FORMAT(70('*'))
  790 FORMAT(2(2X,1PD12.5))
  793 FORMAT(2X,'INITIAL ENERGY=',1PD11.4)
  905 FORMAT(2X,I8,3(2X,1PD12.5))
```

```
 990 FORMAT(2(2X,1PD12.5),2(2X,1PD14.7))
1800 FORMAT(2X,'INITIAL MODEL =',I8,2X,'FINAL MODEL =',I8,/,2X,'PRINT
    &STEP =',I8,2X,'PRECISION =',1PD10.3)
1801 FORMAT(2X,'BETA =',1PD10.3,2X,'MASS =',1PD10.3,/,2X,'RADIUS =',
    &1PD10.3,2X,'DENSITY =',1PD10.3)
1802 FORMAT(2X,'INITIAL TIME =',1PD10.3,2X,'FINAL TIME =',1PD10.3,/,
    &2X,'TIME-STEP =',1PD10.3,2X,'TIME-STEP VARIATION =',1PD10.3)
1806 FORMAT(2X,'ICO = ',I8,2X,'ITERATIONS/MODEL =',I8,2X,
    &/,'TIME =',1PD12.5,2X,'DTIME =',1PD12.5)
1900 FORMAT(2X,'INITIAL MODEL:')
1902 FORMAT(2X,'     SHELL     DENSITY      RADIUS       VELOCITY')
4780 FORMAT(2X,I5,3(2X,1PD10.3))
C
      STOP
      END
C -------------------------------------------------------------------------
C -------------------------------------------------------------------------
C                          SUBROUTINES
C -------------------------------------------------------------------------
C -------------------------------------------------------------------------
C ****************************************************************
      SUBROUTINE ZONING(N,DMV,DM)
C
C Discretization of the sphere in N concentric shells, with N+1
C intershells. For simplicity, all shells are assumed to contain
C the same mass. Note that for other applications, it may be
C convenient to use an unequal mass grid (i.e., lower-mass shells
C close to the center or at the outer edge, where more resolution
C is often needed).
C ...........N    = Number of shells of the computational
C                   domain
C ......... DMV   = Total mass of the sphere             (g)
C ......... DMS   = Mass per shell                       (g)
C ......... DM    = Mass interior to the i-th intershell (g)
C ****************************************************************
C
      IMPLICIT REAL*8(A-H,O-Z)
      DIMENSION DM(N+1)
C
      DMS=DMV/DFLOAT(N)
C
      DM(1)=0.D0
      DO 401 L=1,N
      DM(L+1)=DFLOAT(L)*DMS
 401  CONTINUE
C
      RETURN
      END
C
C ****************************************************************
      SUBROUTINE SCALING(XC,DFAC,ND)
C
C Re-scaling of matrix X to improve accuracy
C ......... XC    = Input matrix
C ......... ND    = Dimension of matrix XC [NDxND]
C ......... DFAC = Re-scaling factor per row
C ****************************************************************
C
      IMPLICIT DOUBLE PRECISION (A-H,O-Z)
      DIMENSION DFAC(3),XC(3,3)
C
      DO 13 I=1,ND
      DFAC(I)=DABS(XC(I,1))
      DO 14 K=1,ND
      IF(DABS(XC(I,K)).GE.DFAC(I)) DFAC(I)=DABS(XC(I,K))
  14  CONTINUE
  13  CONTINUE
C
      SUP=DFAC(1)
C
      DO 15 I=1,ND
      IF (DFAC(I).GE.SUP) SUP=DFAC(I)
  15  CONTINUE
C
      DO 16 I=1,ND
      IF (DFAC(I).NE.0.D0) THEN
```

```
            DFAC(I)=SUP/DFAC(I)
             ELSE
            DFAC(I)=1.D0
            ENDIF
      16  CONTINUE
C
            RETURN
            END
C
C ***********************************************************************
            SUBROUTINE ENERGYC(DM,AMM,EOLD,ENEW)
C
C This subroutine checks whether energy is conserved
C .......... N      = Number of shells of the computational
C                     domain
C .......... ENEW   = Total energy at t=TIME                (erg)
C .......... EKIN   = Kinetic energy                        (erg)
C .......... EGRAV  = Gravitational potential energy        (erg)
C .......... EOLD   = Total initial energy                  (erg)
C .......... ERATIO = Relative error (in %) in energy
C                     conservation
C .......... G      = Gravitational constant                (dyn cm^2 g-2)
C .......... U      = Velocity                              (cm s^-1)
C                     [while U represents velocity at time
C                     t_n, UU is the velocity at time t_n-1]
C .......... R      = Radius                                (cm)
C                     [R represents radius at time t_n;
C                     RR is the radius at time t_n-1]
C .......... DM     = Mass interior to the i-th intershell  (g)
C .......... AMM    = DSQRT[DMi*DMi-1]                       (g)
C ***********************************************************************
C
            PARAMETER (N=100)
            IMPLICIT DOUBLE PRECISION (A-H,O-Z)
C
            COMMON/CONST/PI,G
            COMMON/PHYSICS/U(N+1),UU(N+1),R(N+1),RR(N+1)
            DIMENSION AMM(N+1),DM(N+1)
C
C Initialization of kinetic and gravitational potential
C energy terms
C
            EKIN=0.D0
            EGRAV=0.D0
C
            DO 10 I=2,N+1
            EKIN=EKIN+DABS(U(I)*U(I-1))*(DM(I)-DM(I-1))/2.D0
            IF (RR(I-1).NE.0.D0) THEN
             EGRAV=EGRAV-G*(DM(I)-DM(I-1))*AMM(I)/DSQRT(R(I)*R(I-1))
            ELSE
             EGRAV=0.D0
            ENDIF
      10  CONTINUE
C
            ENEW=EKIN+EGRAV
C
            ERATIO=(ENEW-EOLD)*100.D0/EOLD
            WRITE(7,*) '*********** ERATIO (%)=',ERATIO
C
            RETURN
            END
C
C ***********************************************************************
            SUBROUTINE HENYEY3(DW,DU,DR)
C
C Subroutine that solves the matrix equation A*B = C based on Henyey
C method [L.G. Henyey, J.E. Forbes and N.L. Gould, The Astrophysical
C Journal (1964), vol. 139, p. 306; L.G. Henyey, L. Wilets, K.H. Bohm,
C R. LeLevier and R.D. Levee, The Astrophysical Journal (1959), vol. 129,
C p. 628].
C
C .......... N      = Number of shells of the computational domain
C .......... DW     = Corrections obtained for W
C .......... DU     = Corrections obtained for U
C .......... DR     = Corrections obtained for R
C .......... F1     = Conservation of mass in finite differences
```

```
C .......... F2     = Conservation of momentum in finite differences
C .......... F5     = Lagrangian velocity in finite differences
C .......... DF1W   = Partial derivative of function F1 with respect to Wi-1/2
C .......... DF1RI  = Partial derivative of function F1 with respect to Ri
C .......... DF1RIO = Partial derivative of function F1 with respect to Ri-1
C .......... DF2UI  = Partial derivative of function F2 with respect to Ui
C .......... DF2RI  = Partial derivative of function F2 with respect to Ri
C .......... DF5RI  = Partial derivative of function F5 with respect to Ri
C .......... DF5UI  = Partial derivative of function F5 with respect to Ui
C .......... X      = Matrix that contains the partial derivatives of F1, F2,
C                     and F5, for the innermost shell
C .......... Z      = Matrix that contains the values of F1, F2, and F5, for
C                     the innermost shell
C .......... Y      = Matrix of unknowns DW, DU, and DR, for the innermost
C                     shell [ X * Y = Z ]
C .......... XBUR   = Matrix X after rescaling
C .......... ZBUR   = Matrix Z after rescaling
C .......... W      = Matrix that contains the partial derivatives of F1, F2,
C                     and F5, for the N-2 intermediate shells
C .......... S      = Matrix that contains the values of F1, F2, and F5, for
C                     the intermediate shells
C .......... T      = Matrix of unknowns DW, DU, and DR, for the intermediate
C                     shells [ W * T = S ]
C .......... WBUR   = Matrix W after rescaling
C .......... SBUR   = Matrix S after rescaling
C .......... F      = Matrix that contains the partial derivatives of F1, F2,
C                     and F5, for the outermost shell
C .......... G      = Matrix that contains the values of F1, F2, and F5, for
C                     the outermost shell
C .......... V      = Matrix of unknowns DW, DU, and DR, for the outermost
C                     shell [ F * V = G ]
C .......... FBUR   = Matrix F after rescaling
C .......... GBUR   = Matrix G after rescaling
C .......... DFC    = Rescaling factor for matrices XBUR and ZBUR
C .......... DFC2   = Rescaling factor for matrices WBUR and SBUR
C .......... DFC3   = Rescaling factor for matrices FBUR and GBUR
C ****************************************************************************
C
      PARAMETER (N=100)
      IMPLICIT DOUBLE PRECISION (A-H,O-Z)
C
      COMMON/FO/F1(N+1),F2(N+1),F5(N+1)
      COMMON/F1/DF1W(N+1),DF1RI(N+1),DF1RIO(N+1)
      COMMON/F21/DF2UI(N+1),DF2RI(N+1)
      COMMON/F5/DF5UI(N+1),DF5RI(N+1)
      DIMENSION X(3,3),Z(3),Y(3)
      DIMENSION W(3,3),S(3),T(3)
      DIMENSION F(3,3),G(3),V(3)
      DIMENSION DW(N),DR(N+1),DU(N+1)
      DIMENSION XBUR(3,3),ZBUR(3),DFC(3)
      DIMENSION WBUR(3,3),SBUR(3),DFC2(3)
      DIMENSION FBUR(3,3),GBUR(3),DFC3(3)
C
C Matrix form of the linearized system of equations for the innermost
C shell of the sphere [ X * Y = Z ]
C
      X(1,1)=DF1W(2)
      X(1,2)=0.D0
      X(1,3)=DF1RI(2)
      X(2,1)=0.D0
      X(2,2)=DF2UI(2)
      X(2,3)=DF2RI(2)
      X(3,1)=0.D0
      X(3,2)=DF5UI(2)
      X(3,3)=DF5RI(2)
C
      Z(1)=-F1(2)
      Z(2)=-F2(2)
      Z(3)=-F5(2)
C
C Rescaling of matrices X and Z to improve accuracy
C
      CALL SCALING(X,DFC,3)
      DO 55 J=1,3
      DO 56 I=1,3
      XBUR(J,I)=X(J,I)*DFC(J)
```

```
      56    CONTINUE
      55    CONTINUE
            DO 57 K=1,3
            ZBUR(K)=Z(K)*DFC(K)
      57    CONTINUE
C
C Solution of the matrix system XBUR * Y = ZBUR
C through Gaussian elimination
C
            CALL GAUSS(XBUR,ZBUR,Y,3)
C
            DW(1)=Y(1)
            DU(2)=Y(2)
            DR(2)=Y(3)
C
C Matrix form of the linearized system of equations for the N-2
C intermediate shells of the sphere [ W * T = S ]. Correcting terms
C obtained for the innermost shell are moved to matrix S
C
            DO 20 I=3,N
            W(1,1)=DF1W(I)
            W(1,2)=0.D0
            W(1,3)=DF1RI(I)
            W(2,1)=0.D0
            W(2,2)=DF2UI(I)
            W(2,3)=DF2RI(I)
            W(3,1)=0.D0
            W(3,2)=DF5UI(I)
            W(3,3)=DF5RI(I)
C
            S(1)=-F1(I)-DF1RIO(I)*DR(I-1)
            S(2)=-F2(I)
            S(3)=-F5(I)
C
C Rescaling of matrices W and S to improve accuracy
C
            CALL SCALING(W,DFC2,3)
            DO 855 J=1,3
            DO 856 IL=1,3
            WBUR(J,IL)=W(J,IL)*DFC2(J)
      856 CONTINUE
      855 CONTINUE
            DO 857 K=1,3
            SBUR(K)=S(K)*DFC2(K)
      857 CONTINUE
C
C Solution of the matrix system WBUR * T = SBUR
C through Gaussian elimination
C
            CALL GAUSS(WBUR,SBUR,T,3)
C
            DW(I-1)=T(1)
            DU(I)=T(2)
            DR(I)=T(3)
      20    CONTINUE
C
C Matrix form of the linearized system of equations for the outermost
C shell of the sphere [ F * V = G ]
C
            F(1,1)=DF1W(N+1)
            F(1,2)=0.D0
            F(1,3)=DF1RI(N+1)
            F(2,1)=0.D0
            F(2,2)=DF2UI(N+1)
            F(2,3)=DF2RI(N+1)
            F(3,1)=0.D0
            F(3,2)=DF5UI(N+1)
            F(3,3)=DF5RI(N+1)
C
            G(1)=-F1(N+1)-DF1RIO(N+1)*DR(N)
            G(2)=-F2(N+1)
            G(3)=-F5(N+1)
C
C Rescaling of matrices F and G to improve accuracy
C
            CALL SCALING(F,DFC3,3)
```

```
          DO 955 J=1,3
          DO 136 I=1,3
          FBUR(J,I)=F(J,I)*DFC3(J)
  136 CONTINUE
  955 CONTINUE
          DO 255 J=1,3
          GBUR(J)=G(J)*DFC3(J)
  255 CONTINUE
C
C Solution of the matrix system FBUR * V = GBUR
C through Gaussian elimination
C
          CALL GAUSS(FBUR,GBUR,V,3)
C
          DW(N)=V(1)
          DU(N+1)=V(2)
          DR(N+1)=V(3)
C
          RETURN
          END
C
C ************************************************************************
          SUBROUTINE TSTEP(DTOLD,EPS,V,VV,R,RR,DTIME,N)
C
C .......... N       = Number of shells of the computational
C                      domain
C .......... DTOLD   = Current time-step                     (s)
C .......... DTIME   = First guess for the new time-step     (s)
C .......... V       = Specific volume (1/RHO)               (cm^3 g^-1)
C                      [while V is evaluated at time tn,
C                      VV corresponds to time tn-1]
C .......... R       = Radius                                (cm)
C                      [while R represents radius at time t_n,
C                      RR is the radius at time t_n-1]
C .......... EPS     = Maximum variation allowance for variables
C                      R and V
C .......... AM2     = Auxiliar function used in the
C                      determination of the new time-step
C .......... AM4     = Auxiliar function used in the
C                      determination of the new time-step
C .......... DTMIN   = Auxiliar function used in the
C                      determination of the new time-step
C ************************************************************************
C
          IMPLICIT DOUBLE PRECISION (A-H,O-Z)
C
          DIMENSION AM2(N),AM4(N),DTF(N)
          DIMENSION V(N),VV(N),R(N+1),RR(N+1)
C
          DTIME=DTOLD
C
C First guess of the new time-step based on the maximum allowed
C variation of the physical variables V and R (U is not considered
C since it dramatically slows down the calculation. For other
C applications however, it may be wise to add an additional
C control based on the variation of the Lagrangian fluid velocity).
C From the current variation of a variable Y, DY/Dt, the maximum
C allowed variation of Y, EPS*Y, yields an estimate of the new
C time-step
C
      DO 290 JK=1,N
      IF ((V(JK)-VV(JK)).NE.0.D0) THEN
        AM2(JK)=DABS(EPS*V(JK)*DTIME/(V(JK)-VV(JK)))
      ELSE
        AM2(JK)=1.D40
      ENDIF
  290 CONTINUE
C
      DO 490 JK=1,N
      IF ((R(JK+1)-RR(JK+1)).NE.0.D0) THEN
        AM4(JK)=DABS(EPS*R(JK+1)*DTIME/(R(JK+1)-RR(JK+1)))
      ELSE
        AM4(JK)=1.D40
      ENDIF
  490 CONTINUE
C
```

```
C Determination of the minimum time-step
C
      DO 300 I=1,N
      DTF(I)=DMIN1(AM2(I),AM4(I))
  300 CONTINUE
C
      DTMIN=DTF(1)
      DO 350 I=2,N
      IF (DTF(I).LT.DTMIN) DTMIN=DTF(I)
  350 CONTINUE
C
C DTMIN is the new time-step, provided that it does not
C exceed the previous time-step by more than 50%
C
      DTM=1.5D0*DTIME
      IF (DTMIN.GT.DTM) THEN
      DTIME=DTM
      ELSE
      DTIME=DTMIN
      ENDIF
C
      RETURN
      END
C
C *********************************************************************
      SUBROUTINE GAUSS(X,Z,Y,N)
C
C Solution of the matrix system of equations X(N,N) * Y(N) = Z(N)
C through Gaussian elimination.
C
C .......... X     = Input matrix containing the partial derivatives
C                    of F1, F2, and F5
C .......... Z     = Input matrix containing the values of F1, F2, and
C                    F5
C .......... Y     = Output matrix of unknowns DW, DU, and DR
C .......... DFAC  = Rescaling factor for matrix X, per row
C *********************************************************************
C
      IMPLICIT DOUBLE PRECISION (A-H,O-Z)
C
      DIMENSION X(N,N),Z(N),DFAC(N),Y(N),IJJ(N)
C
C Determination of rescaling factors for matrix X
C
      DO 20 I=1,N
      SUP=0.D0
      DO 30 J=1,N
      SUP=DMAX1(SUP,DABS(X(I,J)))
   30 CONTINUE
      DFAC(I)=SUP
   20 CONTINUE
C
C Identification of the largest pivoting element. For each column,
C the largest rescaled value of X(i,j)/DFAC(i) is stored as DLPE and its
C location is registered by counter L
C
      DO 10 K=1,N
      IJJ(K)=K
   10 CONTINUE

      DO 40 J=1,N-1
      XPI=0.D0
C
      DO 50 I= J,N
      DLPE=DABS(X(IJJ(I),J))/DFAC(IJJ(I))
      IF (DLPE.GT.XPI) THEN
       XPI=DLPE
       L=I
      ELSE
      ENDIF
   50 CONTINUE
C
C Row exchange through IJJ counter
C
      IAUX=IJJ(J)
      IJJ(J)=IJJ(L)
```

```
      IJJ(L)=IAUX
C
      DO 60 I=J+1,N
      XPO=X(IJJ(I),J)/X(IJJ(J),J)
      X(IJJ(I),J)=XPO
C
      DO 70 M=J+1,N
      X(IJJ(I),M)=X(IJJ(I),M)-XPO*X(IJJ(J),M)
   70 CONTINUE
   60 CONTINUE
C
   40 CONTINUE
C
      DO 80 I=1,N-1
      DO 90 J= I+1,N
      Z(IJJ(J))=Z(IJJ(J))-X(IJJ(J),I)*Z(IJJ(I))
   90 CONTINUE
   80 CONTINUE
C
      Y(N)=Z(IJJ(N))/X(IJJ(N),N)
      DO 100 I=N-1,1,-1
      Y(I)= Z(IJJ(I))
      DO 110 J=I+1,N
      Y(I)=Y(I)-X(IJJ(I),J)*Y(J)
  110 CONTINUE
      Y(I)=Y(I)/X(IJJ(I),I)
  100 CONTINUE
C
      RETURN
      END
C
```

Bibliography

[1] M. G. Aartsen, K. Abraham, M. Ackermann, et al. *Phys. Rev. Lett.*, 115:081102 (7 pp), 2015.

[2] J. Abadie, B. P. Abbott, T. D. Abbott, et al. *Astrophys. J.*, 755:2 (8 pp), 2012.

[3] B. Abbott, R. Abbott, R. Adhikari, et al. *Astrophys. J.*, 681:1419–1430, 2008.

[4] A. A. Abdo, M. Ackermann, M. Ajello, et al. *Science*, 329:817–821, 2010.

[5] T. Accadia, F. Acernese, F. Antonucci, et al. *Class. Quantum Grav.*, 28:114002 (10 pp), 2011.

[6] I. A. Acharova, Y. N. Mishurov, and V. V. Kovtyukh. *Mon. Not. R. Astron. Soc.*, 420:1590–1605, 2012.

[7] M. Ackermann, M. Ajello, A. Albert, et al. *Science*, 345:554–558, 2014.

[8] P. A. R. Ade, N. Aghanim, M. I. R. Alves, et al. *Astron. Astrophys.*, 571:A1 (48 pp), 2014.

[9] A. S. Adekola, D. W. Bardayan, J. C. Blackmon, et al. *Phys. Rev. C*, 83:052801 (5 pp), 2011.

[10] E. G. Adelberger, S. M. Austin, J. N. Bahcall, et al. *Rev. Mod. Phys.*, 70:1265–1291, 1998.

[11] O. Agertz, B. Moore, J. Stadel, et al. *Mon. Not. R. Astron. Soc.*, 380:963–978, 2007.

[12] D. K. Aitken, C. H. Smith, S. D. James, et al. *Mon. Not. R. Astron. Soc.*, 231:7P–14P, 1988.

[13] T. Akahori, Y. Funaki, and K. Yabana. *ArXiv e-prints*, 2014.

[14] E. N. Alekseev, L. N. Alekseeva, V. I. Volchenko, and I. V. Krivosheina. *JETP Lett.*, 45:589–592, 1987.

[15] A. Alexakis, A. C. Calder, L. J. Dursi, et al. *Phys. Fluids*, 16:3256–3268, 2004.

[16] A. Alexakis, A. C. Calder, A. Heger, et al. *Astrophys. J.*, 602:931–937, 2004.

[17] A. Alibés, J. Labay, and R. Canal. *Astrophys. J.*, 571:326–333, 2002.

[18] C. J. Allegre, J. L. Birck, S. Fourcade, and M. P. Semet. *Science*, 187:436–438, 1975.

[19] D. A. Allen, J. P. Swings, and P. M. Harvey. *Astron. Astrophys.*, 20:333–336, 1972.

[20] L. H. Aller. *Physics of Thermal Gaseous Nebulae*. Reidel, Dordrecht, The Netherlands, 1984.

[21] D. Altamirano, L. Keek, A. Cumming, et al. *Mon. Not. R. Astron. Soc.*, 426:927–934, 2012.

[22] D. Altamirano, M. van der Klis, R. Wijnands, and A. Cumming. *Astrophys. J. Lett.*, 673:L35–L38, 2008.

[23] L. G. Althaus, A. H. Córsico, J. Isern, and E. García-Berro. *Astron. Astrophys. Rev.*, 18:471–566, 2010.

[24] S. Amari. *New Astron. Rev.*, 46:519–524, 2002.

[25] S. Amari. *New Astron. Rev.*, 50:578–581, 2006.

[26] S. Amari, A. Anders, A. Virag, and E. Zinner. *Nature*, 345:238–240, 1990.

[27] S. Amari, X. Gao, L. R. Nittler, et al. *Astrophys. J.*, 551:1065–1072, 2001.

[28] S. Amari, P. Hoppe, E. Zinner, and R. S. Lewis. *Astrophys. J. Lett.*, 394:L43–L46, 1992.

[29] S. Amari, P. Hoppe, E. Zinner, and R. S. Lewis. *Nature*, 365:806–809, 1993.

[30] S. Amari, P. Hoppe, E. Zinner, and R. S. Lewis. *Meteoritics*, 30:679–693, 1995.

[31] S. Amari, R. S. Lewis, and E. Anders. *Geochim. Cosmochim. Acta*, 58:459–470, 1994.

[32] S. Amari, R. S. Lewis, and E. Anders. *Geochim. Cosmochim. Acta*, 59:1411–1426, 1995.

[33] S. Amari and K. Lodders. In *Highlights of Astronomy, Vol. 14*, K. A. van der Hucht (editor). Cambridge Univ. Press, Cambridge, UK, 2007, pp 349–352.

[34] S. Amari, L. R. Nittler, E. Zinner, et al. *Astrophys. J.*, 546:248–266, 2001.

[35] S. Amari, L. R. Nittler, E. Zinner, K. Lodders, and R. S. Lewis. *Astrophys. J.*, 559:463–483, 2001.

[36] S. Amari and E. Zinner. In *Astrophysical Implications of the Laboratory Study of Presolar Materials*, T. J. Bernatowicz and E. Zinner (editor). American Inst. Phys., Woodbury (New York), 1997, pp 287–305.

[37] S. Amari, E. Zinner, and R. Gallino. In *Origin of Matter and Evolution of Galaxies 2013*, S. Jeong, N. Imai, H. Miyatake, and T. Kajino (editor). American Inst. Phys., Melville (New York), 2014, pp 307–312.

[38] S. Amari, E. Zinner, and R. Gallino. *Geochim. Cosmochim. Acta*, 133:479–522, 2014.

[39] S. Amari, E. Zinner, and R. S. Lewis. *Astrophys. J. Lett.*, 447:L147–L150, 1995.

[40] S. Amari, E. Zinner, and R. S. Lewis. *Astrophys. J. Lett.*, 470:L101–L104, 1996.

[41] S. Amari, E. Zinner, and R. S. Lewis. *Meteorit. Planet. Sci.*, 35:997–1014, 2000.

[42] P. Amaro-Seoane, S. Aoudia, S. Babak, et al. *Class. Quantum Grav.*, 29:124016 (20 pp), 2012.

[43] V. A. Ambartsumyan and G. S. Saakyan. *Sov. Astron.*, 4:187–201, 1960.

[44] K. Ambwani and P. Sutherland. *Astrophys. J.*, 325:820–827, 1988.

[45] M. A. Amthor, D. Galaviz, A. Heger, et al. *Proc. Science, PoS(NIC-IX)* 068 (6 pp), 2006.

[46] E. Anders. In *Meteorites and the Early Solar System*, J. F. Kerridge and M. S. Matthews (editor). Univ. Arizona Press, Tucson (Arizona), 1988, pp 927–955.

[47] E. Anders and N. Grevesse. *Geochim. Cosmochim. Acta*, 53:197–214, 1989.

[48] A. C. Andersen. In *Why Galaxies Care about AGB Stars II: Shining Examples and Common Inhabitants*, F. Kerschbaum, T. Lebzelter, and R. F. Wing (editor). Astron. Soc. Pac. Conf. Series, San Francisco (California), 2011, pp 215–225.

[49] J. Andreä, H. Drechsel, and S. Starrfield. *Astron. Astrophys.*, 291:869–889, 1994.

[50] C. Angulo, M. Arnould, M. Rayet, et al. *Nucl. Phys. A*, 656:3–183, 1999.

[51] J. Annis. *J. British Interplanetary Soc.*, 52:19–22, 1999.

[52] J. Antoniadis, P. C. C. Freire, N. Wex, et al. *Science*, 340:1233232 (9 pp), 2013.

[53] V. M. Antonov, V. P. Bashurin, A. I. Golubev, et al. *J. Appl. Mech. Technol. Phys.*, 26:757–763, 1985.

[54] F. Antonucci, M. Armano, H. Audley, et al. *Class. Quantum Grav.*, 29:124014 (10 pp), 2012.

[55] G. C. Anupama. In *Classical Nova Explosions*, M. Hernanz and J. José (editor). American Inst. Phys., Melville (New York), 2002, pp 32–41.

[56] K. M. V. Apparao and S. M. Chitre. *Astrophys. Space Sci.*, 63:125–129, 1979.

[57] A. Arcones, H.-T. Janka, and L. Scheck. *Astron. Astrophys.*, 467:1227–1248, 2007.

[58] W. D. Arnett. *Can. J. Phys.*, 44:2553–2594, 1966.

[59] W. D. Arnett. *Can. J. Phys.*, 45:1621–1641, 1967.

[60] W. D. Arnett. *Nature*, 219:1344–1346, 1968.

[61] W. D. Arnett. *Astrophys. J.*, 153:341–348, 1968.

[62] W. D. Arnett. *Astrophys. Space Sci.*, 5:180–212, 1969.

[63] W. D. Arnett. *Astrophys. J.*, 194:373–383, 1974.

[64] W. D. Arnett. *Astrophys. J.*, 195:727–733, 1975.

[65] W. D. Arnett. *Astrophys. J. Lett.*, 263:L55–L57, 1982.

[66] W. D. Arnett. *Astrophys. J.*, 253:785–797, 1982.

[67] W. D. Arnett. *Supernovae and Mucleosynthesis. An Investigation of the History of Matter from the Big Bang to the Present*. Princeton Univ. Press, Princeton (New Jersey), 1996.

[68] W. D. Arnett, J. N. Bahcall, R. P. Kirshner, and S. E. Woosley. *Annu. Rev. Astron. Astr.*, 27:629–700, 1989.

[69] W. D. Arnett, B. Fryxell, and E. Müller. *Astrophys. J. Lett.*, 341:L63–L66, 1989.

[70] W. D. Arnett, C. Meakin, and P. A. Young. *Astrophys. J.*, 690:1715–1729, 2009.

[71] W. D. Arnett, C. Meakin, and P. A. Young. *Astrophys. J.*, 710:1619–1626, 2010.

[72] W. D. Arnett and F.-K. Thielemann. *Astrophys. J.*, 295:589–619, 1985.

[73] W. D. Arnett and J. W. Truran. *Astrophys. J.*, 157:339–365, 1969.

[74] W. D. Arnett, J. W. Truran, and S. E. Woosley. *Astrophys. J.*, 165:87–103, 1971.

[75] M. Arnould and S. Goriely. *Phys. Rep.*, 384:1–84, 2003.

[76] M. Arnould and S. Goriely. In *Astrophysics*, I. Kucuk (editor). InTech, Rijeka (Croatia), 2012, pp 61–88.

[77] M. Arnould, S. Goriely, and K. Takahashi. *Phys. Reps.*, 450:97–213, 2007.

[78] M. Arnould and H. Norgaard. *Astron. Astrophys.*, 42:55–70, 1975.

[79] M. Arnould and M. Rayet. *Ann. Phys.*, 15:183–254, 1990.

[80] C. Arpesella, E. Bellotti, C. Broggini, et al. *Phys. Lett. B*, 389:452–456, 1996.

[81] N. W. Ashcroft and N. D. Mermin. *Solid State Physics.* Saunders College Pub., Forth Worth (Texas), 1976.

[82] A. J. Aspden, J. B. Bell, M. S. Day, S. E. Woosley, and M. Zingale. *Astrophys. J.*, 689:1173–1185, 2008.

[83] M. Asplund. *Annu. Rev. Astron. Astr.*, 43:481–530, 2005.

[84] M. Asplund, N. Grevesse, A. J. Sauval, and P. Scott. *Annu. Rev. Astron. Astr.*, 47:481–522, 2009.

[85] F.W. Aston. *Phil. Mag.*, 39:611–625, 1920.

[86] R. d'E. Atkinson. *Astrophys. J.*, 84:73–84, 1936.

[87] R. d'E. Atkinson and F. G. Houtermans. *Z. Phys.*, 54:656–665, 1929.

[88] D. Atri and A. L. Melott. *Astropart. Phys.*, 53:186–190, 2014.

[89] G. Audi, K. F. G., W. M., et al. *Chinese Phys. C*, 36:1157–1286, 2012.

[90] J. Audouze and J. W. Truran. *Astrophys. J.*, 202:204–213, 1975.

[91] M. B. Aufderheide, E. Baron, and F.-K. Thielemann. *Astrophys. J.*, 370:630–642, 1991.

[92] S. J. Austin, R. M. Wagner, S. Starrfield, et al. *Astron. J.*, 111:869–898, 1996.

[93] A. A. Avdyeva, Y. P. Zakharov, V. V. Maksimov, et al. *J. Appl. Mech. Technol. Phys.*, 30:892–895, 1989.

[94] J. N. Ávila, T. R. Ireland, M. Lugaro, et al. 43rd *Lunar and Planetary Science Conf.*, Abstract #2709, 2012.

[95] S. Ayal, T. Piran, R. Oechslin, M. B. Davies, and S. Rosswog. *Astrophys. J.*, 550:846–859, 2001.

[96] S. Ayasli and P. C. Joss. *Astrophys. J.*, 256:637–665, 1982.

[97] G. Aznar-Siguán, E. García-Berro, P. Lorén-Aguilar, J. José, and J. Isern. *Mon. Not. R. Astron. Soc.*, 434:2539–2555, 2013.

[98] G. Aznar-Siguán, E. García-Berro, M. Magnien, and P. Lorén-Aguilar. *Mon. Not. R. Astron. Soc.*, 443:2372–2383, 2014.

[99] W. Baade. *Astron. Nachr.*, 228:359–362, 1926.

[100] W. Baade. *Pub. Astron. Soc. Pac.*, 48:226–229, 1936.

[101] W. Baade. *Astrophys. J.*, 97:119–127, 1943.

[102] W. Baade, G. R. Burbidge, F. Hoyle, et al. *Pub. Astron. Soc. Pac.*, 68:296–300, 1956.

[103] W. Baade and F. Zwicky. *P. Natl. Acad. Sci. USA*, 20:254–259, 1934.

[104] W. Baade and F. Zwicky. *Phys. Rev.*, 46:76–77, 1934.

[105] W. A. Baan. *Astrophys. J.*, 214:245–250, 1977.

[106] O. P. Babushkina, M. I. Kudriavtsev, A. S. Melioranskii, et al. *Sov. Astron. Lett.*, 1:32–34, 1975.

[107] G. Bader and P. Deuflhard. *Numer. Math.*, 41:373–398, 1983.

[108] J. N. Bahcall. *Neutrino Astrophysics*. Cambridge Univ. Press, Cambridge, UK, 1989.

[109] J. N. Bahcall, S. Basu, M. Pinsonneault, and A. M. Serenelli. *Astrophys. J.*, 618:1049–1056, 2005.

[110] J. N. Bahcall and R. M. May. *Astrophys. J.*, 155:501–510, 1969.

[111] J. N. Bahcall, M. H. Pinsonneault, and S. Basu. *Astrophys. J.*, 555:990–1012, 2001.

[112] J. N. Bahcall and A. M. Serenelli. *Astrophys. J.*, 626:530–542, 2005.

[113] J. N. Bahcall, A. M. Serenelli, and S. Basu. *Astrophys. J. Suppl. S.*, 165:400–431, 2006.

[114] J. N. Bahcall and R. K. Ulrich. *Rev. Mod. Phys.*, 60:297–372, 1988.

[115] S. Bailey, G. Aldering, P. Antilogus, et al. *Astron. Astrophys.*, 500:L17–L20, 2009.

[116] M. J. Baines and I. P. Williams. *Nature*, 205:59–60, 1965.

[117] M. J. Baines and I. P. Williams. *Nature*, 208:1191–1193, 1965.

[118] J. E. Baldwin and D. O. Edge. *Observatory*, 77:139–143, 1957.

[119] I. Banerjee and B. Mukhopadhyay. *Astrophys. J.*, 778:8 (12 pp), 2013.

[120] K. Barbary, G. Aldering, R. Amanullah, et al. *Astrophys. J.*, 745:32 (28 pp), 2012.

[121] R. Barbon, M. Capaccioli, and F. Ciatti. *Astron. Astrophys.*, 44:267–271, 1975.

[122] R. Barbon, F. Ciatti, and L. Rosino. *Astron. Astrophys.*, 72:287–292, 1979.

[123] D. W. Bardayan, J. C. Batchelder, J. C. Blackmon, et al. *Phys. Rev. Lett.*, 89:262501 (4 pp), 2002.

[124] D. W. Bardayan, J. C. Blackmon, C. R. Brune, et al. *Phys. Rev. C*, 62:055804 (14 pp), 2000.

[125] D. W. Bardayan, J. C. Blackmon, K. Y. Chae, et al. *Phys. Rev. C*, 81:065802 (4 pp), 2010.

[126] Z. Barkat, G. Rakavy, Y. Reiss, and J. R. Wilson. *Astrophys. J.*, 196:633–638, 1975.

[127] Z. Barkat, G. Rakavy, and N. Sack. *Phys. Rev. Lett.*, 18:379–381, 1967.

[128] F. C. Barker. *Nucl. Phys. A*, 575:361–373, 1994.

[129] F. C. Barker. *Nucl. Phys. A*, 637:576–582, 1998.

[130] J. Barnes and P. Hut. *Nature*, 324:446–449, 1986.

[131] E. Baron, J. Cooperstein, and S. Kahana. *Nucl. Phys. A*, 440:744–754, 1985.

[132] E. Baron, J. Cooperstein, and S. Kahana. *Phys. Rev. Lett.*, 55:126–129, 1985.

[133] E. Baron, D. J. Jeffery, D. Branch, et al. *Astrophys. J.*, 672:1038–1042, 2008.

[134] R. L. Barone-Nugent, C. Lidman, J. S. B. Wyithe, et al. *Mon. Not. R. Astron. Soc.*, 425:1007–1012, 2012.

[135] M. Barranco, J. R. Buchler, and M. Livio. *Astrophys. J.*, 242:1226–1231, 1980.

[136] J. A. Barrat, B. Zanda, F. Moynier, et al. *Geochim. Cosmochim. Acta*, 83:79–92, 2012.

[137] J. G. Barzyk, M. R. Savina, A. M. Davis, et al. *Meteoritics and Planet. Sci.*, 42:1103–1119, 2007.

[138] J. G. Barzyk, M. R. Savina, A. M. Davis, et al. *New Astron. Rev.*, 50:587–590, 2006.

[139] E. M. Basinska, W. H. G. Lewin, M. Sztajno, L. R. Cominsky, and F. J. Marshall. *Astrophys. J.*, 281:337–353, 1984.

[140] S. Basu and H. M. Antia. *Phys. Reps.*, 457:217–283, 2008.

[141] A. H. Batten. *Rep. Prog. Phys.*, 58:885–928, 1995.

[142] A. Bauswein, S. Goriely, and H.-T. Janka. *Astrophys. J.*, 773:78 (21 pp), 2013.

[143] G. Bazán, D. S. P. Dearborn, D. D. Dossa, et al. In *3D Stellar Evolution*, S. Turcotte, S. C. Keller, and R. M. Cavallo (editor). Astron. Soc. Pac. Conf. Series, San Francisco (California), 2003, pp 1–14.

[144] D. Bazin, F. Montes, A. Becerril, et al. *Phys. Rev. Lett.*, 101:252501 (4 pp), 2008.

[145] R. Becker and W. Döring. *Ann. Phys.*, 24:719–752, 1935.

[146] M. Beech. *Astrophys. Space Sci.*, 336:287–302, 2011.

[147] C. E. Beer, A. M. Laird, A. S. J. Murphy, et al. *Phys. Rev. C*, 83:042801 (4 pp), 2011.

[148] K. Belczynski, R. E. Taam, V. Kalogera, F. A. Rasio, and T. Bulik. *Astrophys. J.*, 662:504–511, 2007.

[149] R. D. Belian, J. P. Conner, and W. D. Evans. *Astrophys. J. Lett.*, 171:L87–L90, 1972.

[150] R. D. Belian, J. P. Conner, and W. D. Evans. *Astrophys. J. Lett.*, 206:L135–L138, 1976.

[151] A. M. Beloborodov. *Mon. Not. R. Astron. Soc.*, 438:169–176, 2014.

[152] A. Benninghoven, F.G. Rüdenauer, and H.W. Werner. *Secondary Ion Mass Spectrometry Basic. Concepts, Instrumental Aspects, Applications, and Trends.* John Wiley and Sons, New York, 1987.

[153] W. Benz. In *Late Stages of Stellar Evolution. Computational Methods in Astrophysical Hydrodynamics*, C. B. De Loore (editor). 1991, pp 259–312.

[154] W. Benz. In *Thermonuclear Supernovae*, P. Ruiz-Lapuente, R. Canal, and J. Isern (editor). Kluwer Acad. Publ., Dordrecht, The Netherlands, 1997, pp 457–474.

[155] E. Berger. *Annu. Rev. Astron. Astr.*, 52:43–105, 2014.

[156] M. J. Berger and P. Colella. *J. Comput. Phys.*, 82:64–84, 1989.

[157] M. J. Berger and J. Oliger. *J. Comput. Phys.*, 53:484–512, 1984.

[158] T. E. Berger, G. Slater, N. Hurlburt, et al. *Astrophys. J.*, 716:1288–1307, 2010.

[159] J. Beringer, J.-F. Arguin, R. M. Barnett, et al. *Phys. Rev. D*, 86:010001 (1526 pp), 2012.

[160] R. G. Berkhout and Y. Levin. *Mon. Not. R. Astron. Soc.*, 385:1029–1035, 2008.

[161] T. J. Bernatowicz, S. Amari, and R. S. Lewis. 23rd *Lunar and Planetary Science Conf.*, Abstract #91, 1992.

[162] T. J. Bernatowicz, S. Amari, E. K. Zinner, and R. S. Lewis. *Astrophys. J. Lett.*, 373:L73–L76, 1991.

[163] T. J. Bernatowicz and R. Cowsik. In *Astrophysical Implications of the Laboratory Study of Presolar Materials*, T. J. Bernatowicz and E. Zinner (editor). American Inst. Phys., Woodbury (New York), 1997, pp 451–474.

[164] T. J. Bernatowicz, R. Cowsik, P. C. Gibbons, et al. *Astrophys. J.*, 472:760–782, 1996.

[165] T. J. Bernatowicz, G. Fraundorf, T. Ming, et al. *Nature*, 330:728–730, 1987.

[166] T. J. Bernatowicz, P. C. Gibbons, and R. S. Lewis. *Astrophys. J.*, 359:246–255, 1990.

[167] T. J. Bernatowicz, S. Messenger, O. Pravdivtseva, P. Swan, and R. M. Walker. *Geochim. Cosmochim. Acta*, 67:4679–4691, 2003.

[168] F. Bertola. *Ann. Astrophys.*, 27:319–326, 1964.

[169] F. Bertola, A. Mammano, and M. Perinotto. *Contrib. Osserv. Astrofis. Univ. Padova in Asiago*, 174:51–61, 1965.

[170] G. F. Bertsch. In *Fifty Years of Nuclear BCS: Pairing in Finite Systems*, R. A. Broglia and V. Zelevinsky (editor). World Scientific, Singapore, 2013, pp 26–39.

[171] C. A. Bertulani and A. Gade. *Phys. Rep.*, 485:195–259, 2010.

[172] A. Besmehn and P. Hoppe. *Geochim. Cosmochim. Ac.*, 67:4693–4703, 2003.

[173] H. A. Bethe. *Phys. Rev.*, 55:434–456, 1939.

[174] H. A. Bethe. *Rev. Mod. Phys.*, 62:801–866, 1990.

[175] H. A. Bethe and G. E. Brown. *Astrophys. J.*, 506:780–789, 1998.

[176] H. A. Bethe, G. E. Brown, J. Applegate, and J. M. Lattimer. *Nucl. Phys. A*, 324:487–533, 1979.

[177] H. A. Bethe and C. L. Critchfield. *Phys. Rev.*, 54:248–254, 1938.

[178] H. A. Bethe and J. R. Wilson. *Astrophys. J.*, 295:14–23, 1985.

[179] S. Bhattacharyya, M. C. Miller, and D. K. Galloway. *Mon. Not. R. Astron. Soc.*, 401:2–6, 2010.

[180] F. B. Bianco, D. A. Howell, M. Sullivan, et al. *Astrophys. J.*, 741:20 (12 pp), 2011.

[181] M. F. Bietenholz. *Astrophys. J.*, 645:1180–1187, 2006.

[182] L. Bildsten. In *The Many Faces of Neutron Stars*, R. Buccheri, J. van Paradijs, and A. Alpar (editor). Kluwer Acad. Publ., Dordrecht, The Netherlands, 1998, pp 419–449.

[183] L. Bildsten. In *Cosmic Explosions: Tenth Astrophysics Conf.*, S. S. Holt and W. W. Zhang (editor). American Inst. Phys., Melville (New York), 2000, pp 359–369.

[184] L. Bildsten, P. Chang, and F. Paerels. *Astrophys. J. Lett.*, 591:L29–L32, 2003.

[185] L. Bildsten and A. Cumming. *Astrophys. J.*, 506:842–862, 1998.

[186] L. Bildsten and C. Cutler. *Astrophys. J.*, 449:800–812, 1995.

[187] L. Bildsten and D. M. Hall. *Astrophys. J. Lett.*, 549:L219–L223, 2001.

[188] L. Bildsten, K. J. Shen, N. N. Weinberg, and G. Nelemans. *Astrophys. J. Lett.*, 662:L95–L98, 2007.

[189] R. M. Bionta, G. Blewitt, C. B. Bratton, D. Casper, and A. Ciocio. *Phys. Rev. Lett.*, 58:1494–1496, 1987.

[190] G. S. Bisnovatyi-Kogan and Y. M. Kazhdan. *Sov. Astron.*, 10:604–612, 1967.

[191] D. C. Black. *Geochim. Cosmochim. Ac.*, 36:377–394, 1972.

[192] D. C. Black and R. O. Pepin. *Earth Planet. Sc. Lett.*, 6:395–405, 1969.

[193] J. C. Blackmon, D. W. Bardayan, W. Bradfield-Smith, et al. *Nucl. Phys. A*, 718:127c–130c, 2003.

[194] W. P. Blair, P. Ghavamian, K. S. Long, et al. *Astrophys. J.*, 662:998–1013, 2007.

[195] A. Blanco, G. Falcicchia, and F. Merico. *Astrophys. Space Sci.*, 89:163–168, 1983.

[196] J. M. Blatt and V. F. Weisskopf. *Theoretical Nuclear Physics*. John Wiley and Sons, New York, 1952.

[197] K. Blaum, S. Eliseev, T. Eronen, and Y. Litvinov. *J. Phys. Conf. Ser.*, 381:012013 (8 pp), 2012.

[198] S. I. Blinnikov and O. S. Bartunov. *Astron. Astrophys.*, 273:106–122, 1993.

[199] S. I. Blinnikov, I. D. Novikov, T. V. Perevodchikova, and A. G. Polnarev. *Sov. Astron. Lett.*, 10:177–179, 1984.

[200] S. Blondin, L. Dessart, D. J. Hillier, and A. M. Khokhlov. *Mon. Not. R. Astron. Soc.*, 429:2127–2142, 2013.

[201] S. Blondin, T. Matheson, R. P. Kirshner, et al. *Astron. J.*, 143:126 (33 p), 2012.

[202] J. S. Bloom, D. Kasen, K. J. Shen, et al. *Astrophys. J. Lett.*, 744:L17 (5 pp), 2012.

[203] J. S. Bloom, D. A. Perley, W. Li, et al. *Astrophys. J.*, 691:723–737, 2009.

[204] D. Bodansky, D. D. Clayton, and W. A. Fowler. *Astrophys. J. Suppl. S.*, 16:299–371, 1968.

[205] M. F. Bode, T. J. O'Brien, J. P. Osborne, et al. *Astrophys. J.*, 652:629–635, 2006.

[206] P. Bodenheimer. *Astrophys. J.*, 142:451–461, 1965.

[207] P. Bodenheimer. *Astrophys. J.*, 144:103–107, 1966.

[208] P. Bodenheimer. *Astrophys. J.*, 153:483–494, 1968.

[209] P. Bodenheimer, G. P. Laughlin, M. Różyczka, and H. W. Yorke. *Numerical Methods in Astrophysics: An Introduction*. Taylor & Francis/CRC, Boca Raton (Florida), 2007.

[210] P. Bodenheimer and A. Sweigart. *Astrophys. J.*, 152:515–522, 1968.

[211] M. Boër, B. Gendre, and G. Stratta. *Astrophys. J.*, 800:16 (6 pp), 2015.

[212] H. M. J. Boffin, G. Paulus, M. Arnould, and N. Mowlavi. *Astron. Astrophys.*, 279:173–178, 1993.

[213] S. E. Boggs, F. A. Harrison, H. Miyasaka, et al. *Science*, 348:670–671, 2015.

[214] E. Böhm-Vitense. *Z. Astrophys.*, 46:108–143, 1958.

[215] N. Bohr. *Nature*, 137:344–348, 1936.

[216] J. R. Boisseau, J. C. Wheeler, E. S. Oran, and A. M. Khokhlov. *Astrophys. J. Lett.*, 471:L99–L102, 1996.

[217] I. Bonaparte, F. Matteucci, S. Recchi, et al. *Mon. Not. R. Astron. Soc.*, 435:2460–2473, 2013.

[218] J. T. Bonnell and R. W. Klebesadel. In *Gamma-Ray Bursts: 3rd Huntsville Symposium*, C. Kouveliotou, M. F. Briggs, and G. J. Fishman (editor). American Inst. Phys., Woodbury (New York), 1996, pp 977–980.

[219] G. Börner, L. Qibin, and B. Aschenbach. In *Mining the Sky*, A. J. Banday, S. Zaroubi, and M. Bartelmann (editor). Springer-Verlag, Berlin, Germany, 2001, pp 649–655.

[220] J. E. Borovsky, M. B. Pongratz, R. A. Roussel-Dupre, and T.-H. Tan. *Astrophys. J.*, 280:802–808, 1984.

[221] M. Bose, C. Floss, and F. J. Stadermann. *Astrophys. J.*, 714:1624–1636, 2010.

[222] V. Bouquelle, N. Cerf, M. Arnould, T. Tachibana, and S. Goriely. *Astron. Astrophys.*, 305:1005–1018, 1996.

[223] S. Boutloukos, M. C. Miller, and F. K. Lamb. *Astrophys. J. Lett.*, 720:L15–L19, 2010.

[224] R. L. Bowers and T. Deeming. *Astrophysics. Vols. I and II.* Jones and Bartlett Publ., Boston (Massachusetts), 1984.

[225] R. L. Bowers and J. R. Wilson. *Space Sci. Rev.*, 27:537–543, 1980.

[226] R. L. Bowers and J. R. Wilson. *Astrophys. J.*, 263:366–376, 1982.

[227] R. L. Bowers and J. R. Wilson. *Numerical Modeling in Applied Physics and Astrophysics.* Jones and Bartlett Publ., Boston (Massachusetts), 1991.

[228] R. N. Boyd. *An Introduction to Nuclear Astrophysics.* Univ. Chicago Press, Chicago (Illinois), 2008.

[229] F. Brachwitz, D. J. Dean, W. R. Hix, et al. *Astrophys. J.*, 536:934–947, 2000.

[230] C. F. Bradshaw, E. B. Fomalont, and B. J. Geldzahler. *Astrophys. J. Lett.*, 512:L121–L124, 1999.

[231] D. Branch. *Astrophys. J.*, 258:35–40, 1982.

[232] D. Branch. In *Supernovae*, A. G. Petschek (editor). Springer-Verlag, Berlin, Germany, 1990, pp 30–58.

[233] D. Branch. *Annu. Rev. Astron. Astr.*, 36:17–56, 1998.

[234] D. Branch. In *Cosmic Explosions in Three Dimensions: Asymmetries in Supernovae and Gamma-Ray Bursts*, P. Höflich, P. Kumar, and J. C. Wheeler (editor). Cambridge Univ. Press, Cambridge, UK, 2004, pp 132–141.

[235] D. Branch, E. Baron, N. Hall, M. Melakayil, and J. Parrent. *Pub. Astron. Soc. Pac.*, 117:545–552, 2005.

[236] D. Branch and C. Bettis. *Astron. J.*, 83:224–227, 1978.

[237] D. Branch, L. C. Dang, and E. Baron. *Pub. Astron. Soc. Pac.*, 121:238–247, 2009.

[238] D. Branch, L. C. Dang, N. Hall, et al. *Pub. Astron. Soc. Pac.*, 118:560–571, 2006.

[239] D. Branch, D. J. Jeffery, J. Parrent, et al. *Pub. Astron. Soc. Pac.*, 120:135–149, 2008.

[240] D. Branch and B. Patchett. *Mon. Not. R. Astron. Soc.*, 161:71–83, 1973.

[241] D. Branch, M. A. Troxel, D. J. Jeffery, et al. *Pub. Astron. Soc. Pac.*, 119:709–721, 2007.

[242] E. Bravo and D. García-Senz. *Mon. Not. R. Astron. Soc.*, 307:984–992, 1999.

[243] E. Bravo and D. García-Senz. *Astrophys. J. Lett.*, 642:L157–L160, 2006.

[244] E. Bravo and D. García-Senz. *Astron. Astrophys.*, 478:843–853, 2008.

[245] E. Bravo and D. García-Senz. *Astrophys. J.*, 695:1244–1256, 2009.

[246] E. Bravo, D. García-Senz, R. M. Cabezón, and I. Domínguez. *Astrophys. J.*, 695:1257–1272, 2009.

[247] E. Bravo, J. Isern, R. Canal, and J. Labay. *Astron. Astrophys.*, 257:534–538, 1992.

[248] E. Bravo, L. Piersanti, I. Domínguez, et al. *Astron. Astrophys.*, 535:A114 (15 pp), 2011.

[249] A. Breen and D. McCarthy. *Vista Ast. S.*, 39:363–379, 1995.

[250] S. C. Brenner and L. R. Scott. *The Mathematical Theory of Finite Element Methods.* Springer-Verlag, New York, 2008.

[251] M. Brouillette. *Annu. Rev. Fluid Mech.*, 34:445–468, 2002.

[252] B. A. Brown, R. R. Clement, H. Schatz, A. Volya, and W. A. Richter. *Phys. Rev. C*, 65:045802 (12 pp), 2002.

[253] E. F. Brown. *Astrophys. J. Lett.*, 614:L57–L60, 2004.

[254] E. F. Brown and L. Bildsten. *Astrophys. J.*, 496:915–933, 1998.

[255] P. J. Brown, K. S. Dawson, M. de Pasquale, et al. *Astrophys. J.*, 753:22 (9 pp), 2012.

[256] D. Brownlee. *Annu. Rev. Earth Pl. Sc.*, 42:179–205, 2014.

[257] S. W. Bruenn. *Astrophys. J. Suppl. S.*, 24:283–318, 1972.

[258] S. W. Bruenn. *Ann. NY Acad. Sci.*, 262:80–94, 1975.

[259] S. W. Bruenn, W. D. Arnett, and D. N. Schramm. *Astrophys. J.*, 213:213–224, 1977.

[260] S. W. Bruenn, J. R. Buchler, and M. Livio. *Astrophys. J. Lett.*, 234:L183–L186, 1979.

[261] S. W. Bruenn and W. C. Haxton. *Astrophys. J.*, 376:678–700, 1991.

[262] N. Bucciantini. In *Death of Massive Stars: Supernovae and Gamma-Ray Bursts*, P. Roming, N. Kawai, and E. Pian (editor). Cambridge Univ. Press, Cambridge, UK, 2012, pp 289–296.

[263] J. R. Buchler, M. Barranco, and M. Livio. *Space Sci. Rev.*, 27:585–589, 1980.

[264] J.-R. Buchler and T. J. Mazurek. *Mem. Soc. R. Sci. Liege*, 8:435–445, 1975.

[265] J. R. Buchler, T. J. Mazurek, and J. W. Truran. *Comm. Astrophys. Space Phys.*, 6:45–55, 1974.

[266] L. R. Buchmann and C. A. Barnes. *Nucl. Phys. A*, 777:254–290, 2006.

[267] R. Buras, M. Rampp, H.-T. Janka, and K. Kifonidis. *Astron. Astrophys.*, 447:1049–1092, 2006.

[268] E. M. Burbidge, G. R. Burbidge, W. A. Fowler, and F. Hoyle. *Rev. Mod. Phys.*, 29:547–650, 1957.

[269] L. Burderi and A. R. King. *Mon. Not. R. Astron. Soc.*, 276:1141–1147, 1995.

[270] P. G. Burke and W. D. Robb. *Adv. Atom. Mol. Phys.*, 11:143–214, 1976.

[271] D. S. Burnett. *Meteorit. Planet. Sci.*, 48:2351–2370, 2013.

[272] A. Burrows and J. M. Lattimer. *Astrophys. J. Lett.*, 299:L19–L22, 1985.

[273] A. Burrows and L.-S. The. *Astrophys. J.*, 360:626–638, 1990.

[274] A. S. Burrows. In *Supernovae*, A. G. Petschek (editor). Springer-Verlag, Berlin, Germany, 1990, pp 143–181.

[275] C. J. Burrows, J. Krist, J. J. Hester, et al. *Astrophys. J.*, 452:680–684, 1995.

[276] W. Buscombe and G. de Vaucouleurs. *Observatory*, 75:170–175, 1955.

[277] M. Busso, R. Gallino, and G. J. Wasserburg. *Annu. Rev. Astron. Astr.*, 37:239–309, 1999.

[278] G.D. Byrne and A.C. Hindmarsh. *ACM Trans. Math. Soft.*, 1:71–96, 1975.

[279] C. B. De Loore (editor). *Late Stages of Stellar Evolution. Computational Methods in Astrophysical Hydrodynamics.* Springer-Verlag, Berlin, Germany, 1991.

[280] R. M. Cabezón. *Hydrodynamical simulations of DNS systems: Gravitational emission and equation of state.* PhD thesis, Tech. Univ. Catalonia (UPC), Barcelona (Spain), 2010.

[281] R. Cadonau, G. A. Tammann, and A. Sandage. In *Supernovae as Distance Indicators*, N. Bartel (editor). Springer-Verlag, Berlin, Germany, 1985, pp 151–165.

[282] A. C. Calder, B. C. Curtis, L. J. Dursi, et al. *Proc. SC2000: High Performance Networking and Computing Conf.* (16 pp), 2000.

[283] A. C. Calder, B. Fryxell, T. Plewa, et al. *Astrophys. J. Suppl. S.*, 143:201–229, 2002.

[284] A. G. W. Cameron. *Astrophys. J.*, 121:144–160, 1955.

[285] A. G. W. Cameron. *Astrophys. J.*, 130:916–940, 1959.

[286] A. G. W. Cameron. *Astron. J.*, 65:485, 1960.

[287] A. G. W. Cameron. In *Interstellar Dust and Related Topics*, J. M. Greenberg and H. C. van de Hulst (editor). Reidel, Dordrecht, The Netherlands, 1973, pp 545–547.

[288] A. G. W. Cameron. *Stellar Evolution, Nuclear Astrophysics, and Nucleogenesis.* 2nd Ed., Dover Publ., Mineola (New York), 2013.

[289] A. G. W. Cameron and J. W. Truran. *Icarus*, 30:447–461, 1977.

[290] F. Camilo, A. G. Lyne, R. N. Manchester, et al. *Astrophys. J. Lett.*, 548:L187–L191, 2001.

[291] W. W. Campbell. *Astron. Nachr.*, 131:201–206, 1892.

[292] W. W. Campbell. *Astrophys. J.*, 1:49–51, 1895.

[293] W. W. Campbell. *Astrophys. J.*, 5:233–242, 1897.

[294] R. Canal, D. Garcia, J. Isern, and J. Labay. *Astrophys. J. Lett.*, 356:L51–L53, 1990.

[295] E. Cappellaro and M. Turatto. In *The Influence of Binaries on Stellar Population Studies*, D. Vanbeveren (editor). Kluwer Acad. Publ., Dordrecht, The Netherlands, 2001, pp 199–213.

[296] B. W. Carroll and D. A. Ostlie. *An Introduction to Modern Astrophysics and Cosmology*. 2nd Ed., Addison-Wesley, San Francisco (California), 2006.

[297] S. M. Carroll. *Spacetime and Geometry. An Introduction to General Relativity*. Addison-Wesley, San Francisco (California), 2004.

[298] S. M. Carroll, W. H. Press, and E. L. Turner. *Annu. Rev. Astron. Astr.*, 30:499–542, 1992.

[299] J. Casanova, J. José, E. García-Berro, A. Calder, and S. N. Shore. *Astron. Astrophys.*, 513:L5 (4 pp), 2010.

[300] J. Casanova, J. José, E. García-Berro, A. Calder, and S. N. Shore. *Astron. Astrophys.*, 527:A5 (7 pp), 2011.

[301] J. Casanova, J. José, E. García-Berro, S. N. Shore, and A. C. Calder. *Nature*, 478:490–492, 2011.

[302] C. Casella, H. Costantini, A. Lemut, et al. *Nucl. Phys. A*, 706:203–216, 2002.

[303] S. Cassisi, I. Iben, Jr., and A. Tornambe. *Astrophys. J.*, 496:376–385, 1998.

[304] G. R. Caughlan and W. A. Fowler. *Atom. Data Nucl. Data*, 40:283–334, 1988.

[305] Y. Cavecchi, A. L. Watts, J. Braithwaite, and Y. Levin. *Mon. Not. R. Astron. Soc.*, 434:3526–3541, 2013.

[306] Y. Cavecchi, A. L. Watts, Y. Levin, and J. Braithwaite. *Mon. Not. R. Astron. Soc.*, 448:445–455, 2015.

[307] J. Chadwick. *Nature*, 129:312, 1932.

[308] A. E. Champagne and M. Wiescher. *Annu. Rev. Nucl. Part. S.*, 42:39–76, 1992.

[309] D. A. Chamulak, E. F. Brown, and F. X. Timmes. *Astrophys. J. Lett.*, 655:L93–L96, 2007.

[310] D. A. Chamulak, E. F. Brown, F. X. Timmes, and K. Dupczak. *Astrophys. J.*, 677:160–168, 2008.

[311] P. Chang, L. Bildsten, and I. Wasserman. *Astrophys. J.*, 629:998–1007, 2005.

[312] P. Chang, S. Morsink, L. Bildsten, and I. Wasserman. *Astrophys. J. Lett.*, 636:L117–L120, 2006.

[313] D. L. Chapman. *Phil. Mag.*, 47:90–104, 1899.

[314] A. A. Chen, R. Lewis, K. B. Swartz, D. W. Visser, and P. D. Parker. *Phys. Rev. C*, 63:065807 (11 pp), 2001.

[315] M. C. Chen, F. Herwig, P. A. Denissenkov, and B. Paxton. *Mon. Not. R. Astron. Soc.*, 440:1274–1280, 2014.

[316] W.-C. Chen and X.-D. Li. *Astrophys. J.*, 702:686–691, 2009.

[317] W.-X. Chen and A. M. Beloborodov. *Astrophys. J.*, 657:383–399, 2007.

[318] X. Chen, C. S. Jeffery, X. Zhang, and Z. Han. *Astrophys. J. Lett.*, 755:L9 (5 pp), 2012.

[319] I. Cherchneff. In *Hot and Cool: Bridging Gaps in Massive Star Evolution*, C. Leitherer, P. D. Bennett, P. W. Morris, and J. T. Van Loon (editor). Astron. Soc. Pac. Conf. Series, San Francisco (California), 2010, pp 237–246.

[320] I. Cherchneff. *EAS Publications*, 60:175–184, 2013.

[321] I. Cherchneff and E. Dwek. *Astrophys. J.*, 703:642–661, 2009.

[322] I. Cherchneff and E. Dwek. *Astrophys. J.*, 713:1–24, 2010.

[323] R. A. Chevalier and A. M. Soderberg. *Astrophys. J. Lett.*, 711:L40–L43, 2010.

[324] A. Chieffi and M. Limongi. *Astrophys. J.*, 764:21 (36 pp), 2013.

[325] A. Chieffi, M. Limongi, and O. Straniero. *Astrophys. J.*, 502:737–762, 1998.

[326] Y.-N. Chin and Y.-L. Huang. *Nature*, 371:398–399, 1994.

[327] K. A. Chipps, D. W. Bardayan, J. C. Blackmon, et al. *Phys. Rev. Lett.*, 102:152502 (4 pp), 2009.

[328] B.-G. Choi, G. R. Huss, G. J. Wasserburg, and R. Gallino. *Science*, 282:1284–1289, 1998.

[329] B.-G. Choi, G. J. Wasserburg, and G. R. Huss. *Astrophys. J. Lett.*, 522:L133–L136, 1999.

[330] L. Chomiuk, A. M. Soderberg, M. Moe, et al. *Astrophys. J.*, 750:164 (9 pp), 2012.

[331] C. J. Christensen, A. Nielsen, A. Bahnsen, W. K. Brown, and B. M. Rustad. *Phys. Rev. D*, 5:1628–1640, 1972.

[332] R. F. Christy. *Rev. Mod. Phys.*, 36:555–571, 1964.

[333] N. N. Chugai. *Mon. Not. R. Astron. Soc.*, 250:513–518, 1991.

[334] N. N. Chugai and I. J. Danziger. *Mon. Not. R. Astron. Soc.*, 268:173–180, 1994.

[335] E. Churazov, R. Sunyaev, J. Isern, et al. *Nature*, 512:406–408, 2014.

[336] F. Ciaraldi-Schoolmann, I. R. Seitenzahl, and F. K. Röpke. *Astron. Astrophys.*, 559:A117 (10 pp), 2013.

[337] J. S. W. Claeys, O. R. Pols, R. G. Izzard, J. Vink, and F. W. M. Verbunt. *Astron. Astrophys.*, 563:A83 (24 pp), 2014.

[338] D. H. Clark, W. H. McCrea, and F. R. Stephenson. *Nature*, 265:318–319, 1977.

[339] D. H. Clark and F. R. Stephenson. *Observatory*, 95:190–195, 1975.

[340] D. H. Clark and F. R. Stephenson. *The Historical Supernovae*. Pergamon Press, New York, 1977.

[341] D. Clarke. *Stellar Polarimetry*. Wiley-VCH, Weinheim, Germany, 2010.

[342] D. D. Clayton. *Astrophys. J.*, 188:155–158, 1974.

[343] D. D. Clayton. *Nature*, 257:36–37, 1975.

[344] D. D. Clayton. *Principles of Stellar Evolution and Nucleosynthesis*. Univ. Chicago Press, Chicago (Illinois), 1983.

[345] D. D. Clayton. *Astrophys. J.*, 340:613–619, 1989.

[346] D. D. Clayton. In *Astronomy with Radioactivities*, R. Diehl, D. H. Hartmann, and N. Prantzos (editor). Springer-Verlag, Berlin, Germany, 2011, pp 25–82.

[347] D. D. Clayton, S. A. Colgate, and G. J. Fishman. *Astrophys. J.*, 155:75–82, 1969.

[348] D. D. Clayton and F. Hoyle. *Astrophys. J. Lett.*, 187:L101–L103, 1974.

[349] D. D. Clayton and F. Hoyle. *Astrophys. J.*, 203:490–496, 1976.

[350] D. D. Clayton, W. Liu, and A. Dalgarno. *Science*, 283:1290, 1999.

[351] D. D. Clayton, B. S. Meyer, C. I. Sanderson, S. S. Russell, and C. T. Pillinger. *Astrophys. J.*, 447:894–905, 1995.

[352] D. D. Clayton, B. S. Meyer, L.-S. The, and M. F. El Eid. *Astrophys. J. Lett.*, 578:L83–L86, 2002.

[353] D. D. Clayton and L. R. Nittler. *Annu. Rev. Astron. Astr.*, 42:39–78, 2004.

[354] D. D. Clayton and N. C. Wickramasinghe. *Astrophys. Space Sci.*, 42:463–475, 1976.

[355] G. C. Clayton, T. R. Geballe, F. Herwig, C. Fryer, and M. Asplund. *Astrophys. J.*, 662:1220–1230, 2007.

[356] R. N. Clayton. *Ann. Rev. Earth Planet. Sci.*, 21:115–149, 1993.

[357] R. N. Clayton, N. Onuma, L. Grossman, and T. K. Mayeda. *Earth Planet. Sc. Lett.*, 34:209–224, 1977.

[358] A. M. Clerke. *Problems in Astrophysics*. A. & C. Black, London, 1903.

[359] F. E. Clifford and R. J. Tayler. *Mem. R. Astron. Soc.*, 69:21–81, 1965.

[360] A. Coc, M. Hernanz, J. José, and J.-P. Thibaud. *Astron. Astrophys.*, 357:561–571, 2000.

[361] A. Coc, R. Mochkovitch, Y. Oberto, J.-P. Thibaud, and E. Vangioni-Flam. *Astron. Astrophys.*, 299:479–492, 1995.

[362] S. A. Colgate. *Can. J. Phys.*, 46:S476–S480, 1968.

[363] S. A. Colgate. *Astrophys. J.*, 232:404–408, 1979.

[364] S. A. Colgate and M. H. Johnson. *Phys. Rev. Lett.*, pp 235–238, 1960.

[365] S. A. Colgate and C. McKee. *Astrophys. J.*, 157:623–643, 1969.

[366] S. A. Colgate and R. H. White. *Astrophys. J.*, 143:626–681, 1966.

[367] G. W. Collins, W. P. Claspy, and J. C. Martin. *Pub. Astron. Soc. Pac.*, 111:871–880, 1999.

[368] J. M. Comella, H. D. Craft, R. V. E. Lovelace, and J. M. Sutton. *Nature*, 221:453–454, 1969.

[369] A. Conley, J. Guy, M. Sullivan, et al. *Astrophys. J. Suppl. S.*, 192:1 (29 pp), 2011.

[370] D. L. Cook, E. Berger, T. Faestermann, et al. 40th *Lunar and Planetary Science Conf.*, Abstract #1129, 2009.

[371] J. Cooke, M. Sullivan, A. Gal-Yam, et al. *Nature*, 491:228–231, 2012.

[372] R. L. Cooper, B. Mukhopadhyay, D. Steeghs, and R. Narayan. *Astrophys. J.*, 642:443–454, 2006.

[373] R. L. Cooper and R. Narayan. *Astrophys. J.*, 629:422–437, 2005.

[374] R. L. Cooper and R. Narayan. *Astrophys. J.*, 652:584–596, 2006.

[375] R. L. Cooper and R. Narayan. *Astrophys. J. Lett.*, 648:L123–L126, 2006.

[376] R. L. Cooper and R. Narayan. *Astrophys. J.*, 661:468–476, 2007.

[377] R. L. Cooper, A. W. Steiner, and E. F. Brown. *Astrophys. J.*, 702:660–671, 2009.

[378] J. Cooperstein and E. A. Baron. In *Supernovae*, A. G. Petschek (editor). Springer-Verlag, Berlin, Germany, 1990, pp 213–266.

[379] J. Cooperstein, H. A. Bethe, and G. E. Brown. *Nucl. Phys. A*, 429:527–555, 1984.

[380] R. Cornelisse, J. Heise, E. Kuulkers, F. Verbunt, and J. J. M. in't Zand. *Astron. Astrophys.*, 357:L21–L24, 2000.

[381] J. Cottam, F. Paerels, and M. Mendez. *Nature*, 420:51–54, 2002.

[382] J. Cottam, F. Paerels, M. Méndez, et al. *Astrophys. J.*, 672:504–509, 2008.

[383] R. G. Couch and W. D. Arnett. *Astrophys. J.*, 196:791–803, 1975.

[384] J.P. Cox and R.T. Giuli. *Principles of Stellar Structure, Vols. I & II.* Gordon and Breech, New York, 1968.

[385] T. K. Croat, T. Berg, T. Bernatowicz, E. Groopman, and M. Jadhav. *Meteorit. Planet. Sci.*, 48:686–699, 2013.

[386] T. K. Croat, T. Bernatowicz, S. Amari, S. Messenger, and F. J. Stadermann. *Geochim. Cosmochim. Acta*, 67:4705–4725, 2003.

[387] T. K. Croat, F. J. Stadermann, and T. J. Bernatowicz. *Astrophys. J.*, 631:976–987, 2005.

[388] T. K. Croat, F. J. Stadermann, and T. J. Bernatowicz. *Meteorit. Planet. Sci.*, 43:1497–1516, 2008.

[389] T. K. Croat, F. J. Stadermann, and T. J. Bernatowicz. *Astron. J.*, 139:2159–2169, 2010.

[390] A. P. S. Crotts, W. E. Kunkel, and P. J. McCarthy. *Astrophys. J. Lett.*, 347:L61–L64, 1989.

[391] P. A. Crowther. *Astrophys. Space Sci.*, 285:677–685, 2003.

[392] J. Cruz, H. Luís, M. Fonseca, and A. P. Jesus. *J. Phys. Conf. Ser.*, 337:012062 (3 pp), 2012.

[393] A. Cumming. *Astrophys. J.*, 595:1077–1085, 2003.

[394] A. Cumming. *Astrophys. J.*, 630:441–453, 2005.

[395] A. Cumming. *Nucl. Phys. A*, 758:439–446, 2005.

[396] A. Cumming and L. Bildsten. *Astrophys. J.*, 544:453–474, 2000.

[397] A. Cumming and L. Bildsten. *Astrophys. J. Lett.*, 559:L127–L130, 2001.

[398] A. Cumming and J. Macbeth. *Astrophys. J. Lett.*, 603:L37–L40, 2004.

[399] A. Cumming, J. Macbeth, J. J. M. in 't Zand, and D. Page. *Astrophys. J.*, 646:429–451, 2006.

[400] R. H. Cyburt, A. M. Amthor, R. Ferguson, et al. *Astrophys. J. Suppl. S.*, 189:240–252, 2010.

[401] M. Czerny and M. Jaroszynski. *Acta Astronom.*, 30:157–166, 1980.

[402] Z. R. Dai, J. P. Bradley, D. J. Joswiak, et al. *Nature*, 418:157–159, 2002.

[403] E. Damen, E. Magnier, W. H. G. Lewin, et al. *Astron. Astrophys.*, 237:103–109, 1990.

[404] M. Dan, S. Rosswog, M. Brüggen, and P. Podsiadlowski. *Mon. Not. R. Astron. Soc.*, 438:14–34, 2014.

[405] M. Dan, S. Rosswog, J. Guillochon, and E. Ramirez-Ruiz. *Mon. Not. R. Astron. Soc.*, 422:2417–2428, 2012.

[406] I. J. Danziger, P. Bouchet, C. Gouiffes, and L. B. Lucy. *Ann. NY Acad. Sci.*, 647:42–51, 1991.

[407] M. J. Darnley, S. C. Williams, M. F. Bode, et al. *Astron. Astrophys.*, 563:L9 (4 pp), 2014.

[408] L. R. Dartnell. *Astrobiol.*, 11:551–582, 2011.

[409] T. L. Daulton, T. J. Bernatowicz, R. S. Lewis, et al. *Science*, 296:1852–1855, 2002.

[410] T. L. Daulton, T. J. Bernatowicz, R. S. Lewis, et al. *Geochim. Cosmochim. Acta*, 67:4743–4767, 2003.

[411] T. L. Daulton, D. D. Eisenhour, T. J. Bernatowicz, R. S. Lewis, and P. R. Buseck. *Geochim. Cosmochim. Acta*, 60:4853–4872, 1996.

[412] N. Dauphas, T. Rauscher, B. Marty, and L. Reisberg. *Nucl. Phys. A*, 719:c287–c295, 2003.

[413] N. Dauphas, L. Remusat, J. H. Chen, et al. *Astrophys. J.*, 720:1577–1591, 2010.

[414] B. Davids, R. H. Cyburt, J. José, and S. Mythili. *Astrophys. J.*, 735:40 (10 pp), 2011.

[415] B. Davids, A. M. van den Berg, P. Dendooven, et al. *Phys. Rev. C*, 67:012801 (4 pp), 2003.

[416] B. Davids, A. M. van den Berg, P. Dendooven, et al. *Phys. Rev. C*, 67:065808 (8 pp), 2003.

[417] A. M. Davis. *P. Natl. Acad. Sci. USA*, 108:19142–19146, 2011.

[418] T. M. Davis, E. Mörtsell, J. Sollerman, et al. *Astrophys. J.*, 666:716–725, 2007.

[419] N. de Séréville, C. Angulo, A. Coc, et al. *Phys. Rev. C*, 79:015801 (7 pp), 2009.

[420] G. de Vaucouleurs and H. G. Corwin, Jr. *Astrophys. J.*, 295:287–304, 1985.

[421] D. S. P. Dearborn, J. R. Wilson, and G. J. Mathews. *Astrophys. J.*, 630:309–320, 2005.

[422] A. Decourchelle, J. L. Sauvageot, M. Audard, et al. *Astron. Astrophys.*, 365:L218–L224, 2001.

[423] N. Degenaar and R. Wijnands. *Mon. Not. R. Astron. Soc.*, 422:581–589, 2012.

[424] J. C. del Toro Iniesta. *Introduction to Spectropolarimetry*. Cambridge Univ. Press, Cambridge, UK, 2003.

[425] M. D. Delano and A. G. W. Cameron. *Astrophys. Space Sci.*, 10:203–226, 1971.

[426] T. Delbar, W. Galster, P. Leleux, et al. *Phys. Rev. C*, 48:3088–3096, 1993.

[427] M. Della Valle and R. Gilmozzi. *Science*, 296:1275, 2002.

[428] M. Della Valle and M. Livio. *Astrophys. J.*, 452:704–709, 1995.

[429] M. Della Valle, L. Pasquini, D. Daou, and R. E. Williams. *Astron. Astrophys.*, 390:155–166, 2002.

[430] C. J. Deloye, R. E. Taam, C. Winisdoerffer, and G. Chabrier. *Mon. Not. R. Astron. Soc.*, 381:525–542, 2007.

[431] P. B. Demorest, T. Pennucci, S. M. Ransom, M. S. E. Roberts, and J. W. T. Hessels. *Nature*, 467:1081–1083, 2010.

[432] P. A. Denissenkov, J. W. Truran, M. Pignatari, et al. *Mon. Not. R. Astron. Soc.*, 442:2058–2074, 2014.

[433] P. Descouvemont and D. Baye. *Rep. Prog. Phys.*, 73:036301 (44 pp), 2010.

[434] L. Dessart and D. J. Hillier. *Astron. Astrophys.*, 439:671–685, 2005.

[435] L. Dessart, C. D. Ott, A. Burrows, S. Rosswog, and E. Livne. *Astrophys. J.*, 690:1681–1705, 2009.

[436] H. E. Dewitt, H. C. Graboske, and M. S. Cooper. *Astrophys. J.*, 181:439–456, 1973.

[437] R. Di Stefano. *Astrophys. J.*, 719:474–482, 2010.

[438] R. Di Stefano, R. Voss, and J. Claeys. In *Binary Paths to Type Ia Supernovae Explosions*, R. Di Stefano, M. Orio, and M. Moe (editor). Cambridge Univ. Press, Cambridge, UK, 2013, pp 64–67.

[439] M. P. Diaz and A. Bruch. *Astron. Astrophys.*, 322:807–816, 1997.

[440] R. Diehl. *Rep. Prog. Phys.*, 76:026301 (30 pp), 2013.

[441] R. Diehl, K. Bennett, H. Bloemen, et al. *Astron. Astrophys.*, 298:L25–L28, 1995.

[442] R. Diehl, H. Halloin, K. Kretschmer, et al. *Nature*, 439:45–47, 2006.

[443] R. Diehl, N. Prantzos, and P. von Ballmoos. *Nucl. Phys. A*, 777:70–97, 2006.

[444] R. Diehl, T. Siegert, W. Hillebrandt, et al. *Science*, 345:1162–1165, 2014.

[445] F. S. Dietrich. *Lawrence Livermore Nat. Lab., UCRL-TR-201718* (11 pp), 2004.

[446] B. Dilday, D. A. Howell, S. B. Cenko, et al. *Science*, 337:942–945, 2012.

[447] G. Dimonte, P. Ramaprabhu, D. L. Youngs, M. J. Andrews, and R. Rosner. *Phys. Plasmas*, 12:056301 (6 pp), 2005.

[448] J. B. Doggett and D. Branch. *Astron. J.*, 90:2303–2311, 1985.

[449] C. L. Doherty, L. Siess, J. C. Lattanzio, and P. Gil-Pons. *Mon. Not. R. Astron. Soc.*, 401:1453–1464, 2010.

[450] G. V. Domogatskii, R. A. Eramzhian, and D. K. Nadyozhin. *Astrophys. Space Sci.*, 58:273–299, 1978.

[451] G. V. Domogatskii and V. S. Imshenik. *Sov. Astron. Letters*, 8:190–193, 1982.

[452] G. V. Domogatskii and D. K. Nadyozhin. *Mon. Not. R. Astron. Soc.*, 178:33P–36P, 1977.

[453] G. V. Domogatskii and D. K. Nadyozhin. *Sov. Astron.*, 22:297–305, 1978.

[454] G. V. Domogatskii and D. K. Nadyozhin. *Astrophys. Space Sci.*, 70:33–53, 1980.

[455] G. V. Domogatskii and D. K. Nadyozhin. *Sov. Astron. Letters*, 6:127–130, 1980.

[456] S. Dong, B. Katz, D. Kushnir, and J. L. Prieto. *ArXiv e-prints*, 2014.

[457] B. Donn, J. Hecht, R. Khanna, et al. *Surf. Sci.*, 106:576–581, 1981.

[458] B. Donn, N. C. Wickramasinghe, J. Hudson, and T. P. Stecher. *Astrophys. J.*, 153:451–464, 1968.

[459] W. Döring. *Ann. Phys.*, 43:421–436, 1943.

[460] A. Dotter and B. Paxton. *Astron. Astrophys.*, 507:1617–1619, 2009.

[461] L. N. Downen, C. Iliadis, J. José, and S. Starrfield. *Astrophys. J.*, 762:105 (11 pp), 2013.

[462] A. Drago, G. Pagliara, and J. Schaffner-Bielich. *J. Phys. G Nucl. Partic.*, 35:014052 (6 pp), 2008.

[463] B. T. Draine. In *The Cold Universe*, A. W. Blain, F. Combes, B. T. Draine, D. Pfenniger, and Y. Revaz (editor). Springer-Verlag, Berlin, Germany, 2004, pp 213–305.

[464] B. T. Draine and H. M. Lee. *Astrophys. J.*, 285:89–108, 1984.

[465] B. T. Draine and E. E. Salpeter. *Astrophys. J.*, 231:77–94, 1979.

[466] R. P. Drake. *J. Geophys. Res.*, 104:14505–14516, 1999.

[467] R. P. Drake, J. J. Carroll, III, K. Estabrook, et al. *Astrophys. J. Lett.*, 500:L157–L161, 1998.

[468] R. P. Drake, S. G. Glendinning, K. Estabrook, et al. *Phys. Rev. Lett.*, 81:2068–2071, 1998.

[469] G. Drenkhahn and H. C. Spruit. *Astron. Astrophys.*, 391:1141–1153, 2002.

[470] H. Duan and A. Friedland. *Phys. Rev. Lett.*, 106:091101 (4 pp), 2011.

[471] H. Duan, A. Friedland, G. C. McLaughlin, and R. Surman. *J. Phys. G Nucl. Partic.*, 38:035201 (18 pp), 2011.

[472] H. W. Duerbeck. In *Classical Novae*, M. F. Bode and A. Evans (editor). 2nd Ed., Cambridge Univ. Press, Cambridge, UK, 2008, pp 1–15.

[473] H. W. Duerbeck. *Astron. Nachr.*, 330:568–573, 2009.

[474] D. N. Dunbar, R. E. Pixley, W. A. Wenzel, and W. Whaling. *Phys. Rev.*, 92:649–650, 1953.

[475] L. Dunne, S. Eales, R. Ivison, H. Morgan, and M. Edmunds. *Nature*, 424:285–287, 2003.

[476] C. Dupraz, H. Bloemen, K. Bennett, et al. *Astron. Astrophys.*, 324:683–689, 1997.

[477] R. H. Durisen. *Astrophys. J.*, 213:145–156, 1977.

[478] J. J. L. Duyvendak. *Pub. Astron. Soc. Pac.*, 54:91–94, 1942.

[479] R. G. Eastman, B. P. Schmidt, and R. Kirshner. *Astrophys. J.*, 466:911–937, 1996.

[480] D. S. Ebel and L. Grossman. *Geochim. Cosmochim. Acta*, 65:469–477, 2001.

[481] T. Ebisuzaki, T. Hanawa, and D. Sugimoto. *Pub. Astron. Soc. Jpn.*, 35:17–32, 1983.

[482] A. S. Eddington. *Observatory*, 43:341–358, 1920.

[483] R. T. Edwards and M. Bailes. *Astrophys. J. Lett.*, 547:L37–L40, 2001.

[484] Z. I. Edwards, A. Pagnotta, and B. E. Schaefer. *Astrophys. J. Lett.*, 747:L19 (5 pp), 2012.

[485] G. Efstathiou. *Mon. Not. R. Astron. Soc.*, 440:1138–1152, 2014.

[486] P. P. Eggleton. *Astrophys. J.*, 268:368–369, 1983.

[487] D. Eichler, M. Livio, T. Piran, and D. N. Schramm. *Nature*, 340:126–128, 1989.

[488] M. F. El Eid, L.-S. The, and B. S. Meyer. *Space Sci. Rev.*, 147:1–29, 2009.

[489] J. H. Elias, K. Matthews, G. Neugebauer, and S. E. Persson. *Astrophys. J.*, 296:379–389, 1985.

[490] N. Elias-Rosa, S. D. Van Dyk, W. Li, et al. *Astrophys. J. Lett.*, 714:L254–L259, 2010.

[491] J. Ellis and D. N. Schramm. *P. Natl. Acad. Sci. USA*, 92:235–238, 1995.

[492] A. Elmhamdi, I. J. Danziger, E. Cappellaro, et al. *Astron. Astrophys.*, 426:963–977, 2004.

[493] A. Elmhamdi, I. J. Danziger, N. Chugai, et al. *Mon. Not. R. Astron. Soc.*, 338:939–956, 2003.

[494] V.-V. Elomaa, G. K. Vorobjev, A. Kankainen, et al. *Phys. Rev. Lett.*, 102:252501 (4 pp), 2009.

[495] A. F. Emery. *J. Comput. Phys.*, 2:306–331, 1968.

[496] J. Emsley. *The Elements*. 3rd Ed., Clarendon Press, Oxford, UK, 1998.

[497] L. Ensman and S. E. Woosley. *Bull. American Astron. Soc.*, 19:757, 1987.

[498] R. I. Epstein. *Mon. Not. R. Astron. Soc.*, 188:305–325, 1979.

[499] R. I. Epstein, S. A. Colgate, and W. C. Haxton. *Phys. Rev. Lett.*, 61:2038–2041, 1988.

[500] E. V. Ergma and A. V. Tutukov. *Acta Astronom.*, 26:69–76, 1976.

[501] E. V. Ergma and A. V. Tutukov. *Astron. Astrophys.*, 84:123–127, 1980.

[502] L. Erikson, C. Ruiz, F. Ames, et al. *Phys. Rev. C*, 81:045808 (12 pp), 2010.

[503] J. Erler, N. Birge, M. Kortelainen, et al. *Nature*, 486:509–512, 2012.

[504] A. Evans. In *Physics of Classical Novae*, A. Cassatella and R. Viotti (editor). Springer-Verlag, Berlin, Germany, 1990, pp 253–263.

[505] A. Evans and M. C. Rawlings. In *Classical Novae*, M. F. Bode and A. Evans (editor). 2nd Ed., Cambridge Univ. Press, Cambridge, UK, 2008, pp 308–334.

[506] A. Evans, C. E. Woodward, L. A. Helton, et al. *Astrophys. J. Lett.*, 671:L157–L160, 2007.

[507] R. Eymard, T. Gallouet, and R. Herbin. In *Handbook of Numerical Analysis, Vol. VII*, D. S. Lauretta and H. Y. McSween, Jr. (editor). North-Holland Pub., Amsterdam, The Netherlands, 2000, pp 713–1020.

[508] D. Ezer and A. G. W. Cameron. *Can. J. Phys.*, 43:1497–1517, 1965.

[509] D. Ezer and A. G. W. Cameron. *Can. J. Phys.*, 45:3429–3460, 1967.

[510] D. Ezer and G. W. Cameron. *Icarus*, 1:422–441, 1963.

[511] J. A. Faber and F. A. Rasio. *Phys. Rev. D*, 62:064012 (23 pp), 2000.

[512] M. Falanga, J. Chenevez, A. Cumming, et al. *Astron. Astrophys.*, 484:43–50, 2008.

[513] M. Falanga, A. Cumming, E. Bozzo, and J. Chenevez. *Astron. Astrophys.*, 496:333–342, 2009.

[514] S. W. Falk, J. M. Lattimer, and S. H. Margolis. *Nature*, 270:700–701, 1977.

[515] T. Faran, D. Poznanski, A. V. Filippenko, et al. *Mon. Not. R. Astron. Soc.*, 445:554–569, 2014.

[516] K. Farouqi, K.-L. Kratz, and B. Pfeiffer. *Pub. Astron. Soc. Aust.*, 26:194–202, 2009.

[517] J. Feder, K. C. Russell, J. Lothe, and G. M. Pound. *Adv. Phys.*, 15:111–178, 1966.

[518] J. Feige, A. Wallner, L. K. Fifield, et al. *EPJ Web of Conferences*, 63:03003 (5 pp), 2013.

[519] J. Feige, A. Wallner, S. R. Winkler, et al. *Pub. Astron. Soc. Aust.*, 29:109–114, 2012.

[520] G. J. Ferland, K. T. Korista, D. A. Verner, et al. *Pub. Astron. Soc. Pac.*, 110:761–778, 1998.

[521] G. J. Ferland, R. L. Porter, P. A. M. van Hoof, et al. *Rev. Mex. Astron. Astr.*, 49:137–163, 2013.

[522] W. Fickett and W. C. Davis. *Detonation*. Univ. California Press, Berkeley (California), 1979.

[523] A. V. Filippenko. *Astron. J.*, 96:1941–1948, 1988.

[524] A. V. Filippenko. *Annu. Rev. Astron. Astr.*, 35:309–355, 1997.

[525] A. V. Filippenko. In *Cosmic Explosions: Tenth Astrophysics Conf.*, S. S. Holt and W. W. Zhang (editor). American Inst. Phys., Melville (New York), 2000, pp 123–140.

[526] A. V. Filippenko. In *The Fate of the Most Massive Stars*, R. Humphreys and K. Stanek (editor). Astron. Soc. Pac. Conf. Series, San Francisco (California), 2005, pp 33–43.

[527] A. V. Filippenko and D. C. Leonard. In *Cosmic Explosions in Three Dimensions: Asymmetries in Supernovae and Gamma-Ray Bursts*, P. Höflich, P. Kumar, and J. C. Wheeler (editor). Cambridge Univ. Press, Cambridge, UK, 2004, pp 30–42.

[528] A. V. Filippenko, M. W. Richmond, T. Matheson, et al. *Astrophys. J. Lett.*, 384:L15–L18, 1992.

[529] L. Fimiani, D. L. Cook, T. Faestermann, et al. 43rd *Lunar and Planetary Science Conf.*, Abstract #1279, 2012.

[530] M. H. Finger, R. B. Wilson, B. A. Harmon, K. Hagedon, and T. A. Prince. *IAU Circ.* #6285, 1996.

[531] M. Fink, F. K. Röpke, W. Hillebrandt, et al. *Astron. Astrophys.*, 514:A53 (10 pp), 2010.

[532] T. Fischer, S. C. Whitehouse, A. Mezzacappa, F.-K. Thielemann, and M. Liebendörfer. *Astron. Astrophys.*, 499:1–15, 2009.

[533] T. Fischer, S. C. Whitehouse, A. Mezzacappa, F.-K. Thielemann, and M. Liebendörfer. *Astron. Astrophys.*, 517:A80 (25 pp), 2010.

[534] J. L. Fisker, J. Görres, M. Wiescher, and B. Davids. *Astrophys. J.*, 650:332–337, 2006.

[535] J. L. Fisker, R. D. Hoffman, and J. Pruet. *Astrophys. J. Lett.*, 690:L135–L139, 2009.

[536] J. L. Fisker, H. Schatz, and F.-K. Thielemann. *Astrophys. J. Suppl. S.*, 174:261–276, 2008.

[537] J. L. Fisker, W. Tan, J. Görres, M. Wiescher, and R. L. Cooper. *Astrophys. J.*, 665:637–641, 2007.

[538] J. L. Fisker, F.-K. Thielemann, and M. Wiescher. *Astrophys. J. Lett.*, 608:L61–L64, 2004.

[539] C. Floss and F. Stadermann. *Geochim. Cosmochim. Acta*, 73:2415–2440, 2009.

[540] C. Floss and F. J. Stadermann. *Meteorit. Planet. Sci.*, 47:992–1009, 2012.

[541] C. Floss, F. J. Stadermann, J. P. Bradley, et al. *Geochim. Cosmochim. Acta*, 70:2371–2399, 2006.

[542] G. Folatelli, M. M. Phillips, N. Morrell, et al. *Astrophys. J.*, 745:74 (17 pp), 2012.

[543] R. Forcada, D. Garcia-Senz, and J. José. *Proc. Science, PoS(NIC-IX)* 096 (7 pp), 2006.

[544] M. Forestini, S. Goriely, A. Jorissen, and M. Arnould. *Astron. Astrophys.*, 261:157–163, 1992.

[545] J. A. Formaggio and G. P. Zeller. *Rev. Mod. Phys.*, 84:1307–1341, 2012.

[546] W. A. Fowler, G. R. Caughlan, and B. A. Zimmerman. *Annu. Rev. Astron. Astr.*, 5:525–570, 1967.

[547] W. A. Fowler and F. Hoyle. *Astrophys. J. Suppl. S.*, 9:201–319, 1964.

[548] C. Fox, C. Iliadis, A. E. Champagne, et al. *Phys. Rev. Lett.*, 93:081102 (4 pp), 2004.

[549] D. A. Frail, S. R. Kulkarni, R. Sari, et al. *Astrophys. J. Lett.*, 562:L55–L58, 2001.

[550] G. S. Fraley. *Astrophys. J. Suppl. S.*, 2:96–114, 1968.

[551] J. Frank, A. King, and D. J. Raine. *Accretion Power in Astrophysics*. 3rd Ed., Cambridge Univ. Press, Cambridge, UK, 2002.

[552] D. A. Frank-Kamenetskii. *Sov. Astron.*, 5:66–71, 1961.

[553] C. Fransson, A. Cassatella, R. Gilmozzi, et al. *Astrophys. J.*, 336:429–441, 1989.

[554] M. Fraser, K. Takáts, A. Pastorello, et al. *Astrophys. J. Lett.*, 714:L280–L284, 2010.

[555] A. Frebel and K.-L. Kratz. In *The Ages of Stars*, E. E. Mamajek, D. R. Soderblom, and R. F. G. Wyse (editor). Cambridge Univ. Press, Cambridge, UK, 2009, pp 449–456.

[556] D. Z. Freedman. *Phys. Rev. D*, 9:1389–1392, 1974.

[557] W. L. Freedman and B. F. Madore. *Annu. Rev. Astron. Astr.*, 48:673–710, 2010.

[558] C. Freiburghaus, S. Rosswog, and F.-K. Thielemann. *Astrophys. J. Lett.*, 525:L121–L124, 1999.

[559] J. Frenkel. *J. Chem. Phys.*, 7:538–547, 1939.

[560] M. Friedjung. *Astron. Astrophys.*, 77:357–358, 1979.

[561] A. Friedmann. *Z. Phys.*, 10:377–386, 1922.

[562] A. Friedmann. *Z. Phys.*, 21:326–332, 1924.

[563] A. Friedmann. *Gen. Relat. Gravit.*, 31:1991–2000, 1999.

[564] A. Friedmann. *Gen. Relat. Gravit.*, 31:2001–2008, 1999.

[565] C. Fröhlich, G. Martínez-Pinedo, M. Liebendörfer, et al. *Phys. Rev. Lett.*, 96:142502 (4 pp), 2006.

[566] B. A. Fryxell, E. Müller, and W. D. Arnett. *Hydrodynamics and Nuclear Burning.* Rep. 449, Max Planck Inst. Astrophysics, Garching, Germany, 1989.

[567] B. A. Fryxell, K. Olson, P. Ricker, et al. *Astrophys. J. Suppl. S.*, 131:273–334, 2000.

[568] B. A. Fryxell and S. E. Woosley. *Astrophys. J.*, 258:733–739, 1982.

[569] B. A. Fryxell and S. E. Woosley. *Astrophys. J.*, 261:332–336, 1982.

[570] M. Fujimoto and I. Iben, Jr. *Astrophys. J.*, 399:646–655, 1992.

[571] M. Y. Fujimoto. *Astrophys. J.*, 257:767–779, 1982.

[572] M. Y. Fujimoto. *Astrophys. J.*, 257:752–766, 1982.

[573] M. Y. Fujimoto, T. Hanawa, and S. Miyaji. *Astrophys. J.*, 246:267–278, 1981.

[574] M. Y. Fujimoto and D. Sugimoto. *Astrophys. J.*, 257:291–302, 1982.

[575] W. Fujiya, P. Hoppe, E. Zinner, M. Pignatari, and F. Herwig. *Astrophys. J. Lett.*, 776:L29 (6 pp), 2013.

[576] G. M. Fuller, W. A. Fowler, and M. J. Newman. *Astrophys. J. Suppl. S.*, 42:447–473, 1980.

[577] G. M. Fuller, W. A. Fowler, and M. J. Newman. *Astrophys. J.*, 252:715–740, 1982.

[578] G. M. Fuller, W. A. Fowler, and M. J. Newman. *Astrophys. J. Suppl. S.*, 48:279–319, 1982.

[579] I. Fushiki and D. Q. Lamb. *Astrophys. J. Lett.*, 323:L55–L60, 1987.

[580] H.-P. Gail, R. Keller, and E. Sedlmayr. *Astron. Astrophys.*, 133:320–332, 1984.

[581] H.-P. Gail and E. Sedlmayr. *Astron. Astrophys.*, 166:225–236, 1986.

[582] H.-P. Gail and E. Sedlmayr. *Astron. Astrophys.*, 206:153–168, 1988.

[583] A. Gal-Yam and D. C. Leonard. *Nature*, 458:865–867, 2009.

[584] A. Gal-Yam and D. Maoz. *Mon. Not. R. Astron. Soc.*, 347:942–950, 2004.

[585] A. Gal-Yam, P. Mazzali, E. O. Ofek, et al. *Nature*, 462:624–627, 2009.

[586] T. J. Galama, P. M. Vreeswijk, J. van Paradijs, et al. *Nature*, 395:670–672, 1998.

[587] D. Galante and J. E. Horvath. *Int. J. Astrobiol.*, 6:19–26, 2007.

[588] D. Galante and J. E. Horvath. *Int. J. Mod. Phys. D*, 16:509–514, 2007.

[589] H. Gallagher, G. Garvey, and G. P. Zeller. *Annu. Rev. Nucl. Part. S.*, 61:355–378, 2011.

[590] J. S. Gallagher. *Astron. J.*, 82:209–215, 1977.

[591] J. S. Gallagher and A. D. Code. *Astrophys. J.*, 189:303–314, 1974.

[592] R. Gallino, C. Arlandini, M. Busso, et al. *Astrophys. J*, 497:388–403, 1998.

[593] R. Gallino, M. Busso, G. Picchio, and C. M. Raiteri. *Nature*, 348:298–302, 1990.

[594] R. Gallino, C. M. Raiteri, and M. Busso. *Astrophys. J.*, 410:400–411, 1993.

[595] R. Gallino, C. M. Raiteri, M. Busso, and F. Matteucci. *Astrophys. J.*, 430:858–869, 1994.

[596] D. K. Galloway and A. Cumming. *Astrophys. J.*, 652:559–568, 2006.

[597] D. K. Galloway, A. Cumming, E. Kuulkers, et al. *Astrophys. J.*, 601:466–473, 2004.

[598] D. K. Galloway, M. P. Muno, J. M. Hartman, D. Psaltis, and D. Chakrabarty. *Astrophys. J. Suppl. S.*, 179:360–422, 2008.

[599] V. N. Gamezo, A. M. Khokhlov, and E. S. Oran. *Astrophys. J.*, 623:337–346, 2005.

[600] V. N. Gamezo, A. M. Khokhlov, E. S. Oran, A. Y. Chtchelkanova, and R. O. Rosenberg. *Science*, 299:77–81, 2003.

[601] V. N. Gamezo, J. C. Wheeler, A. M. Khokhlov, and E. S. Oran. *Astrophys. J.*, 512:827–842, 1999.

[602] G. Gamow and E. Teller. *Phys. Rev.*, 53:608–609, 1938.

[603] M. Ganeshalingam, W. Li, and A. V. Filippenko. *Mon. Not. R. Astron. Soc.*, 416:2607–2622, 2011.

[604] M. Ganeshalingam, W. Li, and A. V. Filippenko. *Mon. Not. R. Astron. Soc.*, 433:2240–2258, 2013.

[605] X. Gao and L. R. Nittler. 28th *Lunar and Planetary Science Conf.*, Abstract #1769, 1997.

[606] G. Garavini, G. Aldering, A. Amadon, et al. *Astron. J.*, 130:2278–2292, 2005.

[607] E. García-Berro, P. Lorén-Aguilar, G. Aznar-Siguán, et al. *Astrophys. J.*, 749:25 (5 pp), 2012.

[608] E. García-Berro, P. Lorén-Aguilar, A. G. Pedemonte, et al. *Astrophys. J. Lett.*, 661:L179–L182, 2007.

[609] E. García-Berro, A. G. Pedemonte, D. García-Senz, et al. *J. Physics Con. Ser.*, 66(1):012040 (10 pp), 2007.

[610] D. García-Senz, C. Badenes, and N. Serichol. *Astrophys, J.*, 745:75 (14 pp), 2012.

[611] D. García-Senz and E. Bravo. *Astron. Astrophys.*, 430:585–602, 2005.

[612] D. García-Senz, E. Bravo, R. M. Cabezón, and S. E. Woosley. *Astrophys. J.*, 660:509–515, 2007.

[613] D. García-Senz, E. Bravo, and N. Serichol. *Astrophys. J. Suppl. S.*, 115:119–139, 1998.

[614] D. García-Senz, E. Bravo, and S. E. Woosley. *Astron. Astrophys.*, 349:177–188, 1999.

[615] D. García-Senz, R. M. Cabezón, A. Arcones, A. Relaño, and F. K. Thielemann. *Mon. Not. R. Astron. Soc.*, 436:3413–3429, 2013.

[616] D. García-Senz, A. Relaño, R. M. Cabezón, and E. Bravo. *Mon. Not. R. Astron. Soc.*, 392:346–360, 2009.

[617] F. F. Gardner and D. K. Milne. *Astron. J.*, 70:754, 1965.

[618] L. R. Gasques, E. F. Brown, A. Chieffi, et al. *Phys. Rev. C*, 76:035802 (10 pp), 2007.

[619] C.W. Gear. *Commun. ACM*, 14:176–179, 1971.

[620] N. Gehrels, C. M. Laird, C. H. Jackman, et al. *Astrophys. J.*, 585:1169–1176, 2003.

[621] N. Gehrels, M. Leventhal, and C. J. MacCallum. *Astrophys. J.*, 322:215–233, 1987.

[622] N. Gehrels, C. J. MacCallum, and M. Leventhal. *Astrophys. J. Lett.*, 320:L19–L22, 1987.

[623] N. Gehrels and P. Mészáros. *Science*, 337:932–936, 2012.

[624] N. Gehrels, E. Ramirez-Ruiz, and D. B. Fox. *Annu. Rev. Astron. Astr.*, 47:567–617, 2009.

[625] R. D. Gehrz. *Annu. Rev. Astron. Astr.*, 26:377–412, 1988.

[626] R. D. Gehrz. *Phys. Rep.*, 311:405–418, 1999.

[627] R. D. Gehrz. In *Classical Nova Explosions*, M. Hernanz and J. José (editor). American Inst. Phys., Melville (New York), 2002, pp 198–207.

[628] R. D. Gehrz. In *Classical Novae*, M. F. Bode and A. Evans (editor). 2nd Ed., Cambridge Univ. Press, Cambridge, UK, 2008, pp 167–193.

[629] R. D. Gehrz, G. L. Grasdalen, and J. A. Hackwell. *Astrophys. J. Lett.*, 298:L47–L50 (Erratum: Astrophys. J. Lett. 306:L49), 1985.

[630] R. D. Gehrz and E. P. Ney. *P. Natl. Acad. Sci. USA*, 84:6961–6964, 1987.

[631] R. D. Gehrz and E. P. Ney. *P. Natl. Acad. Sci. USA*, 87:4354–4357, 1990.

[632] R. D. Gehrz, J. W. Truran, R. E. Williams, and S. Starrfield. *Pub. Astron. Soc. Pac.*, 110:3–26, 1998.

[633] J. Geiss and G. Gloeckler. *Space Sci. Rev.*, 84:239–250, 1998.

[634] B. Gendre, G. Stratta, J. L. Atteia, et al. *Astrophys. J.*, 766:30 (9 pp), 2013.

[635] R. Georgii, S. Plüschke, R. Diehl, et al. *Astron. Astrophys.*, 394:517–523, 2002.

[636] C. L. Gerardy, W. P. S. Meikle, R. Kotak, et al. *Astrophys. J.*, 661:995–1012, 2007.

[637] N. Gershenfeld. *The Nature of Mathematical Modeling*. Cambridge Univ. Press, Cambridge, UK, 1998.

[638] S. Gezari, J. P. Halpern, D. Grupe, et al. *Astrophys. J.*, 690:1313–1321, 2009.

[639] R. Giacconi, H. Gursky, E. Kellogg, E. Schreier, and H. Tananbaum. *Astrophys. J. Lett.*, 167:L67–L73, 1971.

[640] R. Giacconi, H. Gursky, F. R. Paolini, and B. B. Rossi. *Phys. Rev. Lett.*, 9:439–443, 1962.

[641] P. Giannone and A. Weigert. *Z. Astrophys.*, 67:41–63, 1967.

[642] W. Gieren, J. Storm, N. Nardetto, et al. In *Advancing the Physics of Cosmic Distances*, R. de Grijs (editor). Cambridge Univ. Press, Cambridge, UK, 2013, pp 138–144.

[643] P. Gil-Pons, E. García-Berro, J. José, M. Hernanz, and J. W. Truran. *Astron. Astrophys.*, 407:1021–1028, 2003.

[644] M. Gilfanov and Á. Bogdán. *Nature*, 463:924–925, 2010.

[645] R. C. Gilman. *Astrophys. J. Suppl. S.*, 28:397–403, 1974.

[646] R. A. Gingold and J. J. Monaghan. *Mon. Not. R. Astron. Soc.*, 181:375–389, 1977.

[647] R. A. Gingold and J. J. Monaghan. *J. Comput. Phys.*, 46:429–453, 1982.

[648] S. A. Glasner and E. Livne. *Astrophys. J. Lett.*, 445:L149–L151, 1995.

[649] S. A. Glasner, E. Livne, and J. W. Truran. *Astrophys. J.*, 475:754–762, 1997.

[650] S. A. Glasner, E. Livne, and J. W. Truran. *Astrophys. J.*, 625:347–350, 2005.

[651] S. A. Glasner and J. W. Truran. *Astrophys. J. Lett.*, 692:L58–L61, 2009.

[652] T. Gold. *Nature*, 218:731–732, 1968.

[653] G. Goldhaber, S. Deustua, S. Gabi, et al. In *Thermonuclear Supernovae*, P. Ruiz-Lapuente, R. Canal, and J. Isern (editor). Kluwer Acad. Publ., Dordrecht, The Netherlands, 1997, pp 777–784.

[654] G. Goldhaber, D. E. Groom, A. Kim, et al. *Astrophys. J.*, 558:359–368, 2001.

[655] B. R. Goldstein. *Astron. J.*, 70:105–114, 1965.

[656] B. R. Goldstein and H. Peng Yoke. *Astron. J.*, 70:748–753, 1965.

[657] J. Goldstein, D.E. Newbury, D.C. Joy, et al. *Scanning Electron Microscopy and X-Ray Microanalysis*. 3rd. Ed., Springer, New York, 2003.

[658] I. Golombek and J. C. Niemeyer. *Astron. Astrophys.*, 438:611–616, 2005.

[659] J. Gómez del Campo, A. Galindo-Uribarri, J. R. Beene, et al. *Phys. Rev. Lett.*, 86:43–46, 2001.

[660] J. Gómez-Gomar, M. Hernanz, J. José, and J. Isern. *Mon. Not. R. Astron. Soc.*, 296:913–920, 1998.

[661] J. Gómez-Gomar, J. Isern, and P. Jean. *Mon. Not. R. Astron. Soc.*, 295:1–9, 1998.

[662] J. I. González Hernández, P. Ruiz-Lapuente, A. V. Filippenko, et al. *Astrophys. J.*, 691:1–15, 2009.

[663] J. I. González Hernández, P. Ruiz-Lapuente, H. M. Tabernero, et al. *Nature*, 489:533–536, 2012.

[664] R. Gonzalez-Riestra, M. Orio, and J. Gallagher. *Astron. Astrophys. Suppl.*, 129:23–30, 1998.

[665] A. Goobar and B. Leibundgut. *Annu. Rev. Nucl. Part. S.*, 61:251–279, 2011.

[666] R. W. Goodrich. *Pub. Astron. Soc. Pac.*, 103:1314–1322, 1991.

[667] S. Goriely. *Astron. Astrophys.*, 327:845–853, 1997.

[668] S. Goriely and M. Arnould. *Astron. Astrophys.*, 312:327–337, 1996.

[669] S. Goriely, A. Bauswein, and H.-T. Janka. *Astrophys. J. Lett.*, 738:L32 (6 pp), 2011.

[670] S. Goriely, D. García-Senz, E. Bravo, and J. José. *Astron. Astrophys.*, 444:L1–L4, 2005.

[671] S. Goriely, J. José, M. Hernanz, M. Rayet, and M. Arnould. *Astron. Astrophys.*, 383:L27–L30, 2002.

[672] J. Görres, M. Wiescher, and F.-K. Thielemann. *Phys. Rev. C*, 51:392–400, 1995.

[673] N. B. Gove and M. J. Martin. *Atom. Data Nucl. Data*, 10:205–219, 1971.

[674] H. C. Graboske, H. E. Dewitt, A. S. Grossman, and M. S. Cooper. *Astrophys. J.*, 181:457–474, 1973.

[675] G. A. Graham, N. E. Teslich, A. T. Kearsley, et al. *Meteorit. Planet. Sci.*, 43:561–569, 2008.

[676] J. F. Graham and A. S. Fruchter. *Astrophys. J.*, 774:119 (23 pp), 2013.

[677] M. A. Granada. In *1604-2004: Supernovae as Cosmological Lighthouses*, M. Turatto, S. Benetti, L. Zampieri, and W. Shea (editor). Astron. Soc. Pac. Conf. Series, San Francisco (California), 2005, pp 30–42.

[678] O. Graur and D. Maoz. *Mon. Not. R. Astron. Soc.*, 430:1746–1763, 2013.

[679] O. Graur, D. Poznanski, D. Maoz, et al. *Mon. Not. R. Astron. Soc.*, 417:916–940, 2011.

[680] S. A. Grebenev, A. A. Lutovinov, S. S. Tsygankov, and C. Winkler. *Nature*, 490:373–375, 2012.

[681] D. A. Green. *Mon. Not. R. Astron. Soc.*, 209:449–478, 1984.

[682] N. Grevesse and A. J. Sauval. *Space Sci. Rev.*, 85:161–174, 1998.

[683] N. Grevesse, P. Scott, M. Asplund, and A. J. Sauval. *Astron. Astrophys.*, 573:A27 (23 pp), 2015.

[684] D. J. Griffiths. *Introduction to quantum mechanics*. Prentice Hall, Englewood Cliffs (New Jersey), 1995.

[685] J. Grindlay and H. Gursky. *Astrophys. J. Lett.*, 205:L131–L133, 1976.

[686] J. Grindlay, H. Gursky, H. Schnopper, et al. *Astrophys. J. Lett.*, 205:L127–L130, 1976.

[687] J. E. Grindlay. *Astrophys. J.*, 221:234–257, 1978.

[688] J. E. Grindlay, J. E. McClintock, C. R. Canizares, et al. *Nature*, 274:567–568, 1978.

[689] E. Groopman, L. R. Nittler, T. Bernatowicz, and E. Zinner. *Astrophys. J.*, 790:9 (13 pp), 2014.

[690] J. Grun, J. Stamper, C. Manka, J. Resnick, and R. Burris. *Phys. Rev. Lett.*, 66:2738–2741, 1991.

[691] J. Guerrero, E. García-Berro, and J. Isern. *Astron. Astrophys.*, 413:257–272, 2004.

[692] R. Guerriero, D. W. Fox, J. Kommers, et al. *Mon. Not. R. Astron. Soc.*, 307:179–189, 1999.

[693] J. Guillochon, M. Dan, E. Ramirez-Ruiz, and S. Rosswog. *Astrophys. J. Lett.*, 709:L64–L69, 2010.

[694] L. E. Gurevitch and A. I. Lebedinsky. In *Non-Stable Stars*, G. H. Herbig (editor). Cambridge Univ. Press, Cambridge, UK, 1957, pp 77–82.

[695] H. Gursky. In *Structure and Evolution of Close Binary Systems*, P. Eggleton, S. Mitton, and J. Whelan (editor). Kluwer Acad. Publ., Dordrecht, The Netherlands, 1976, pp 19–25.

[696] C. P. Gutiérrez, J. P. Anderson, M. Hamuy, et al. *Rev. Mex. Ast. Astr.*, 44:111, 2014.

[697] J. Gutierrez, E. Garcia-Berro, I. Iben, Jr., et al. *Astrophys. J.*, 459:701–705, 1996.

[698] T. Güver, F. Özel, and D. Psaltis. *Astrophys. J.*, 747:77 (12 pp), 2012.

[699] T. Güver, D. Psaltis, and F. Özel. *Astrophys. J.*, 747:76 (16 pp), 2012.

[700] F. Gyngard, L. Nittler, E. Zinner, and J. José. *Proc. Science, PoS(NIC-XI)* 141 (5 pp), 2010.

[701] F. Gyngard, L. R. Nittler, and E. Zinner. *Meteorit. Planet. Sci. Suppl.*, 73:Abstract #5242, 2010.

[702] F. Gyngard, L. R. Nittler, E. Zinner, J. José, and S. Cristallo. 42nd *Lunar and Planetary Science Conf.*, Abstract #2675, 2011.

[703] F. Gyngard, E. Zinner, L. R. Nittler, et al. *Astrophys. J.*, 717:107–120, 2010.

[704] F. Haberl, L. Stella, N. E. White, M. Gottwald, and W. C. Priedhorsky. *Astrophys. J.*, 314:266–271, 1987.

[705] I. Hachisu, M. Kato, and G. J. M. Luna. *Astrophys. J. Lett.*, 659:L153–L156, 2007.

[706] I. Hachisu, M. Kato, and K. Nomoto. *Astrophys. J.*, 522:487–503, 1999.

[707] I. Hachisu, M. Kato, and K. Nomoto. *Astrophys. J. Lett.*, 756:L4 (5 pp), 2012.

[708] J. A. Hackwell, R. D. Gehrz, and G. L. Grasdalen. *Astrophys. J.*, 234:133–139, 1979.

[709] J. A. Hackwell, G. L. Grasdalen, R. D. Gehrz, et al. *Astrophys. J. Lett.*, 233:L115–L119, 1979.

[710] P. Haenecour, X. Zhao, C. Floss, Y. Lin, and E. Zinner. *Astrophys. J.*, 768:L17 (5 pp), 2013.

[711] K. I. Hahn, A. García, E. G. Adelberger, et al. *Phys. Rev. C*, 54:1999–2013, 1996.

[712] J. Hakkila, D. J. Haglin, G. N. Pendleton, et al. *Astrophys. J.*, 538:165–180, 2000.

[713] A. S. Hamers, O. R. Pols, J. S. W. Claeys, and G. Nelemans. *Mon. Not. R. Astron. Soc.*, 430:2262–2280, 2013.

[714] A. J. S. Hamilton, C. L. Sarazin, and A. E. Szymkowiak. *Astrophys. J.*, 300:698–712, 1986.

[715] A. J. S. Hamilton, C. L. Sarazin, and A. E. Szymkowiak. *Astrophys. J.*, 300:713–721, 1986.

[716] M. Hamuy, M. M. Phillips, J. Maza, et al. *Astron. J.*, 108:2226–2232, 1994.

[717] M. Hamuy, M. M. Phillips, N. B. Suntzeff, et al. *Nature*, 424:651–654, 2003.

[718] M. Hamuy and P. A. Pinto. *Astrophys. J. Lett.*, 566:L63–L65, 2002.

[719] M. Hamuy, P. A. Pinto, J. Maza, et al. *Astrophys. J.*, 558:615–642, 2001.

[720] M. Hamuy, N. B. Suntzeff, R. Gonzalez, and G. Martin. *Astron. J.*, 95:63–83, 1988.

[721] Z. Han. *Astrophys. J. Lett.*, 677:L109–L112, 2008.

[722] T. Hanawa and M. Y. Fujimoto. *Pub. Astron. Soc. Jpn.*, 36:199–214, 1984.

[723] T. Hanawa, D. Sugimoto, and M.-A. Hashimoto. *Pub. Astron. Soc. Jpn.*, 35:491–506, 1983.

[724] R. Hanbury-Brown and C. Hazard. *Nature*, 170:364–365, 1952.

[725] P. J. Hancock, B. M. Gaensler, and T. Murphy. *Astrophys. J. Lett.*, 735:L35 (5 pp), 2011.

[726] C. J. Hansen and H. M. van Horn. *Astrophys. J.*, 195:735–741, 1975.

[727] R. P. Harkness and J. C. Wheeler. In *Supernovae*, A. G. Petschek (editor). Springer-Verlag, Berlin, Germany, 1990, pp 1–29.

[728] M. J. Harris, M. D. Leising, and G. H. Share. *Astrophys. J.*, 375:216–220, 1991.

[729] M. J. Harris, J. E. Naya, B. J. Teegarden, et al. *Astrophys. J.*, 522:424–432, 1999.

[730] M. J. Harris, B. J. Teegarden, G. Weidenspointner, et al. *Astrophys. J.*, 563:950–957, 2001.

[731] J. B. Hartle. *Gravity. An Introduction to Einstein's General Relativity.* Addison-Wesley, San Francisco (California), 2003.

[732] J. Hartmann. *Astron. Nachr.*, 226:63, 1925.

[733] M. Hashimoto, K. Nomoto, and T. Shigeyama. *Astron. Astrophys.*, 210:L5–L8, 1989.

[734] M.-A. Hashimoto, T. Hanawa, and D. Sugimoto. *Pub. Astron. Soc. Jpn.*, 35:1–15, 1983.

[735] M.-A. Hashimoto, K.-I. Nomoto, K. Arai, and K. Kaminisi. *Astrophys. J.*, 307:687–693, 1986.

[736] T. Hashimoto, K. Nakai, Y. Wakasaya, et al. *Nucl. Phys. A*, 686:591–599, 2001.

[737] P. H. Hauschildt. In *Classical Novae*, M. F. Bode and A. Evans (editor). 2nd Ed., Cambridge Univ. Press, Cambridge, UK, 2008, pp 102–120.

[738] P. H. Hauschildt, S. N. Shore, G. J. Schwarz, et al. *Astrophys. J.*, 490:803–818, 1997.

[739] P. H. Hauschildt, R. Wehrse, S. Starrfield, and G. Shaviv. *Astrophys. J.*, 393:307–328, 1992.

[740] W. Hauser and H. Feshbach. *Phys. Rev.*, 87:366–373, 1952.

[741] J. F. Hawley, L. L. Smarr, and J. R. Wilson. *Astrophys. J.*, 277:296–311, 1984.

[742] J. F. Hawley, L. L. Smarr, and J. R. Wilson. *Astrophys. J. Suppl. S.*, 55:211–246, 1984.

[743] S. Hayakawa. *Space Sci. Rev.*, 29:221–290, 1981.

[744] C. Hayashi. *Pub. Astron. Soc. Jpn.*, 13:450–452, 1961.

[745] C. Hayashi. *Annu. Rev. Astron. Astr.*, 4:171–192, 1966.

[746] C. Hayashi, R. Hōshi, and D. Sugimoto. *Prog. Theor. Phys. Supp.*, 22:1–183, 1962.

[747] C. Hayashi, R. Hōshi, and D. Sugimoto. *Prog. Theor. Phys.*, 34:885–911, 1965.

[748] C. Hayashi and T. Nakano. *Prog. Theor. Phys.*, 34:754–775, 1965.

[749] B. T. Hayden, P. M. Garnavich, D. Kasen, et al. *Astrophys. J.*, 722:1691–1698, 2010.

[750] J. J. He, J. Hu, S. W. Xu, et al. *Eur. Phys. J. A*, 47:67 (8 pp), 2011.

[751] J. J. He, P. J. Woods, T. Davinson, et al. *Phys. Rev. C*, 80:042801 (3 pp), 2009.

[752] J. J. He, L. Y. Zhang, A. Parikh, et al. *Phys. Rev. C*, 88:012801 (6 pp), 2013.

[753] N. C. Hearn, T. Plewa, R. P. Drake, and C. Kuranz. *Astrophys. Space Sci.*, 307:227–231, 2007.

[754] P. R. Heck, S. Amari, P. Hoppe, et al. *Astrophys. J.*, 701:1415–1425, 2009.

[755] R. P. Hedrosa, C. Abia, M. Busso, et al. *Astrophys. J. Lett.*, 768:L11 (5 pp), 2013.

[756] A. Heger, A. Cumming, D. K. Galloway, and S. E. Woosley. *Astrophys. J. Lett.*, 671:L141–L144, 2007.

[757] A. Heger, A. Cumming, and S. E. Woosley. *Astrophys. J.*, 665:1311–1320, 2007.

[758] A. Heger, C. L. Fryer, S. E. Woosley, N. Langer, and D. H. Hartmann. *Astrophys. J.*, 591:288–300, 2003.

[759] A. Heger, E. Kolbe, W. C. Haxton, et al. *Phys. Lett. B*, 606:258–264, 2005.

[760] A. Heger and S. E. Woosley. *Astrophys. J.*, 567:532–543, 2002.

[761] H. Hekiri and G. Emanuel. *Shock Waves*, 21:511–521, 2011.

[762] L. A. Helton, C. E. Woodward, F. M. Walter, et al. *Astron. J.*, 140:1347–1369, 2010.

[763] R. N. Henriksen. *Astrophys. J. Lett.*, 210:L19–L22, 1976.

[764] L. G. Henyey, J. E. Forbes, and N. L. Gould. *Astrophys. J.*, 139:306–317, 1964.

[765] L. G. Henyey and J. L'Ecuyer. *Astrophys. J.*, 156:549–558, 1969.

[766] L. G. Henyey, R. Lelevier, and R. D. Levée. *Pub. Astron. Soc. Pac.*, 67:154–160, 1955.

[767] M. Henze, W. Pietsch, F. Haberl, et al. *Astron. Astrophys.*, 523:A89 (16 pp), 2010.

[768] M. Henze, W. Pietsch, F. Haberl, et al. *Astron. Astrophys.*, 533:A52 (30 pp), 2011.

[769] M. Herant. *Space Sci. Rev.*, 74:335–341, 1995.

[770] M. Herant and W. Benz. *Astrophys. J. Lett.*, 370:L81–L84, 1991.

[771] M. Herant and W. Benz. *Astrophys. J.*, 387:294–308, 1992.

[772] M. Herant, W. Benz, W. R. Hix, C. L. Fryer, and S. A. Colgate. *Astrophys. J.*, 435:339–361, 1994.

[773] M. Hernandez, W. P. S. Meikle, A. Aparicio, et al. *Mon. Not. R. Astron. Soc.*, 319:223–234, 2000.

[774] M. Hernanz. In *Classical Novae*, M. F. Bode and A. Evans (editor). 2nd Ed., Cambridge Univ. Press, Cambridge, UK, 2008, pp 252–284.

[775] M. Hernanz. In *Stella Novae: Past and Future Decades*, P. A. Woudt and V. A. R. M. Ribeiro (editor). Astron. Soc. Pac. Conf. Series, San Francisco (California), 2014, pp 319–326.

[776] M. Hernanz, J. Isern, R. Canal, J. Labay, and R. Mochkovitch. *Astrophys. J.*, 324:331–344, 1988.

[777] M. Hernanz and J. José. *New Astron. Rev.*, 48:35–39, 2004.

[778] M. Hernanz and J. José. In *5th INTEGRAL Workshop on the INTEGRAL Universe*, V. Schoenfelder, G. Lichti, and C. Winkler (editor). ESA SP-552, Noordwijk, The Netherlands, 2004, pp 95–98.

[779] M. Hernanz and J. José. *Exp. Astron.*, 20:57–64, 2005.

[780] M. Hernanz and J. José. *Proc. Science, PoS(SUPERNOVA)* 014 (10 pp), 2008.

[781] M. Hernanz and J. José. *New Astron. Rev.*, 52:386–389, 2008.

[782] M. Hernanz, J. José, A. Coc, J. Gómez-Gomar, and J. Isern. *Astrophys. J. Lett.*, 526:L97–L100, 1999.

[783] M. Hernanz, J. José, A. Coc, and J. Isern. *Astrophys. J. Lett.*, 465:L27–L30, 1996.

[784] M. Hernanz and G. Sala. *Science*, 298:393–395, 2002.

[785] M. Hernanz and G. Sala. *Astron. Nachr.*, 331:169–174, 2010.

[786] M. Hernanz, M. Salaris, J. Isern, and J. José. In *Thermonuclear Supernovae*, P. Ruiz-Lapuente, R. Canal, and J. Isern (editor). Kluwer Acad. Publ., Dordrecht, The Netherlands, 1997, pp 167–176.

[787] M. Hernanz, D. M. Smith, J. Fishman, et al. In *Fifth Compton Symposium*, M. L. McConnell and J. M. Ryan (editor). American Inst. Phys., Melville (New York), 2000, pp 82–86.

[788] H. Herndl, J. Görres, M. Wiescher, B. A. Brown, and L. van Wormer. *Phys. Rev. C*, 52:1078–1094, 1995.

[789] L. Hernquist and N. Katz. *Astrophys. J. Suppl. S.*, 70:419–446, 1989.

[790] M. Herzog and F. K. Röpke. *Phys. Rev. D*, 84:083002 (13 pp), 2011.

[791] A. Hewish, S. J. Bell, J. D. H. Pilkington, P. F. Scott, and R. A. Collins. *Nature*, 217:709–713, 1968.

[792] J. S. Heyl. *Astrophys. J.*, 600:939–945, 2004.

[793] J. S. Heyl. *Mon. Not. R. Astron. Soc.*, 361:504–510, 2005.

[794] M. Hicken, P. M. Garnavich, J. L. Prieto, et al. *Astrophys. J. Lett.*, 669:L17–L20, 2007.

[795] W. Hillebrandt. *Astron. Astrophys.*, 110:L3–L6, 1982.

[796] W. Hillebrandt. In *Supernovae: A Survey of Current Research*, M. J. Rees and R. J. Stoneham (editor). Reidel, Dordrecht, The Netherlands, 1982, pp 123–155.

[797] W. Hillebrandt. In *Supernovae*, S. A. Bludman, R. Mochkovitch, and J. Zinn-Justin (editor). 1994, pp 251–300.

[798] W. Hillebrandt, T. Kodama, and K. Takahashi. *Astron. Astrophys.*, 52:63–68, 1976.

[799] W. Hillebrandt, M. Kromer, F. K. Röpke, and A. J. Ruiter. *Front. Phys.*, 8:116–143, 2013.

[800] W. Hillebrandt and J. C. Niemeyer. *Annu. Rev. Astron. Astr.*, 38:191–230, 2000.

[801] W. Hillebrandt, K. Nomoto, and R. G. Wolff. *Astron. Astrophys.*, 133:175–184, 1984.

[802] W. Hillebrandt, S. A. Sim, and F. K. Röpke. *Astron. Astrophys.*, 465:L17–L20, 2007.

[803] W. Hillebrandt and F.-K. Thielemann. *Astrophys. J.*, 255:617–623, 1982.

[804] Y. Hillman, D. Prialnik, A. Kovetz, and M. M. Shara. *Mon. Not. R. Astron. Soc.*, 446:1924–1930, 2015.

[805] Y. Hillman, D. Prialnik, A. Kovetz, M. M. Shara, and J. D. Neill. *Mon. Not. R. Astron. Soc.*, 437:1962–1975, 2014.

[806] K. Hirata, T. Kajita, M. Koshiba, M. Nakahata, and Y. Oyama. *Phys. Rev. Lett.*, 58:1490–1493, 1987.

[807] K. S. Hirata, T. Kajita, M. Koshiba, et al. *Phys. Rev. D*, 38:448–458, 1988.

[808] C. W. Hirt, A. A. Amsden, and J. L. Cook. *J. Comput. Phys.*, 14:227–253, 1974.

[809] W. R. Hix, S. T. Parete-Koon, C. Freiburghaus, and F.-K. Thielemann. *Astrophys. J.*, 667:476–488, 2007.

[810] W. R. Hix, M. S. Smith, A. Mezzacappa, S. Starrfield, and D. L. Smith. *NASA STI/Recon Technical Report N*, 2001.

[811] W. R. Hix and F.-K. Thielemann. *Astrophys. J.*, 460:869–894, 1996.

[812] W. R. Hix and F.-K. Thielemann. *Astrophys. J.*, 511:862–875, 1999.

[813] R. M. Hjellming and C. M. Wade. *Astrophys. J. Lett.*, 162:L1–L4, 1970.

[814] P.-Y. Ho, F. W. Paar, and P. W. Parsons. *Vista Ast. S.*, 13:1–13, 1972.

[815] D. C. Hoffman, F. O. Lawrence, J. L. Mewherter, and F. M. Rourke. *Nature*, 234:132–134, 1971.

[816] J. A. Hoffman, W. H. G. Lewin, and J. Doty. *Mon. Not. R. Astron. Soc.*, 179:57P–64P, 1977.

[817] J. A. Hoffman, W. H. G. Lewin, and J. Doty. *Astrophys. J. Lett.*, 217:L23–L28, 1977.

[818] J. A. Hoffman, H. L. Marshall, and W. H. G. Lewin. *Nature*, 271:630–633, 1978.

[819] R. D. Hoffman, S. E. Woosley, T. A. Weaver, T. Rauscher, and F.-K. Thielemann. *Astrophys. J.*, 521:735–752, 1999.

[820] P. Höflich and A. Khokhlov. *Astrophys. J.*, 457:500–528, 1996.

[821] P. Höflich, A. Khokhlov, and E. Müller. *Astrophys. J. Suppl. S.*, 92:501–504, 1994.

[822] P. Höflich, A. M. Khokhlov, and J. C. Wheeler. *Astrophys. J.*, 444:831–847, 1995.

[823] P. A. Höflich, A. Khokhlov, J. C. Wheeler, K. Nomoto, and F. K. Thielemann. In *Thermonuclear Supernovae*, P. Ruiz-Lapuente, R. Canal, and J. Isern (editor). Kluwer Acad. Publ., Dordrecht, The Netherlands, 1997, pp 659–679.

[824] E. Hofmeister, R. Kippenhahn, and A. Weigert. *Z. Astrophys.*, 59:215–241, 1964.

[825] P. Hoppe. *Proc. Science, PoS(NIC-XI)* 021 (10 pp), 2010.

[826] P. Hoppe, S. Amari, E. Zinner, T. Ireland, and R. S. Lewis. *Astrophys. J.*, 430:870–890, 1994.

[827] P. Hoppe, S. Amari, E. Zinner, and R. S. Lewis. *Geochim. Cosmochim. Acta*, 59:4029–4056, 1995.

[828] P. Hoppe, P. Annen, R. Strebel, et al. *Astrophys. J. Lett.*, 487:L101–L104, 1997.

[829] P. Hoppe and A. Besmehn. *Astrophys. J. Lett.*, 576:L69–L72, 2002.

[830] P. Hoppe and W. Fujiya. 42nd *Lunar and Planetary Science Conf.*, Abstract #1059, 2011.

[831] P. Hoppe, W. Fujiya, and E. Zinner. *Astrophys. J. Lett.*, 745:L26 (5 pp), 2012.

[832] P. Hoppe, J. Leitner, E. Gröner, et al. *Astrophys. J.*, 719:1370–1384, 2010.

[833] P. Hoppe, U. Ott, and G. W. Lugmair. *New Astron. Rev.*, 48:171–176, 2004.

[834] P. Hoppe, R. Strebel, P. Eberhardt, S. Amari, and R. S. Lewis. *Science*, 272:1314–1316, 1996.

[835] P. Hoppe, R. Strebel, P. Eberhardt, S. Amari, and R. S. Lewis. *Meteorit. Planet. Sci.*, 35:1157–1176, 2000.

[836] A. Horesh, S. R. Kulkarni, D. B. Fox, et al. *Astrophys. J.*, 746:21 (8 pp), 2012.

[837] R. Horiuchi, T. Kadonaga, and A. Tomimatsu. *Prog. Theor. Phys.*, 66:172–179, 1981.

[838] J. H. Hough, J. A. Bailey, M. F. Rouse, and D. C. B. Whittet. *Mon. Not. R. Astron. Soc.*, 227:1P–5P, 1987.

[839] D. A. Howell. *Nature Comm.*, 2:350 (10 pp), 2011.

[840] D. A. Howell, P. Höflich, L. Wang, and J. C. Wheeler. *Astrophys. J.*, 556:302–321, 2001.

[841] D. A. Howell, M. Sullivan, P. E. Nugent, et al. *Nature*, 443:308–311, 2006.

[842] F. Hoyle. *Mon. Not. R. Astron. Soc.*, 106:343–383, 1946.

[843] F. Hoyle. *Astrophys. J. Suppl. S.*, 1:121–146, 1954.

[844] F. Hoyle, D. N. Dunbar, W. A. Wenzel, and W. Whaling. *Phys. Rev.*, 92:1095, 1953.

[845] F. Hoyle and W. A. Fowler. *Astrophys. J.*, 132:565–590, 1960.

[846] F. Hoyle and N. C. Wickramasinghe. *Mon. Not. R. Astron. Soc.*, 124:417–433, 1962.

[847] J. Hu, J. J. He, A. Parikh, et al. *Phys. Rev. C*, 90:025803 (7 pp), 2014.

[848] T. Hu. *Astron. Astrophys.*, 578:A132 (37 pp), 2015.

[849] T. Hu and M. Wang. *Mon. Not. R. Astron. Soc.*, 438:3443–3455, 2014.

[850] E. Hubble. *P. Natl. Acad. Sci. USA*, 15:168–173, 1929.

[851] L. Hüdepohl, B. Müller, H.-T. Janka, A. Marek, and G. G. Raffelt. *Phys. Rev. Lett.*, 104:251101 (4 pp), 2010.

[852] W. Huggins and W. A. Miller. *R. Soc. London Proc.*, 15:146–149, 1866.

[853] D. W. Hughes. *Nature*, 285:132–133, 1980.

[854] H. Hugoniot. *J. École Polytech.*, 57:3–97, 1887.

[855] H. Hugoniot. *J. École Polytech.*, 58:1–125, 1889.

[856] M. L. Humason. *Pub. Astron. Soc. Pac.*, 48:110–113, 1936.

[857] M. L. Humason and R. Minkowski. *Pub. Astron. Soc. Pac.*, 53:130–132, 1941.

[858] G.S. Hurst and M.G. Payne. *Principles and Applications of Resonance Ionisation Spectroscopy.* Taylor & Francis/CRC, London, 1988.

[859] G. R. Huss, A. J. Fahey, R. Gallino, and G. J. Wasserburg. *Astrophys. J. Lett.*, 430:L81–L84, 1994.

[860] G. R. Huss, I. D. Hutcheon, and G. J. Wasserburg. *Geochim. Cosmochim. Acta*, 61:5117–5148, 1997.

[861] G. R. Huss, I. D. Hutcheon, G. J. Wasserburg, and J. Stone. *Lunar Planet. Sci. Contrib.*, 781:29–33, 1992.

[862] G. R. Huss and R. S. Lewis. *Meteoritics*, 29:791–810, 1994.

[863] G. R. Huss and R. S. Lewis. *Geochim. Cosmochim. Acta*, 59:115–160, 1995.

[864] M. Y. Hussaini and T. A. Zang. *Annu. Rev. Fluid Mech.*, 19:339–367, 1987.

[865] I. D. Hutcheon, G. R. Huss, A. J. Fahey, and G. J. Wasserburg. *Astrophys. J. Lett.*, 425:L97–L100, 1994.

[866] K. M. Hynes, T. K. Croat, S. Amari, A. F. Mertz, and T. J. Bernatowicz. *Meteorit. Planet. Sci.*, 45:596–614, 2010.

[867] K. M. Hynes and F. Gyngard. 40th *Lunar and Planetary Science Conf.*, Abstract #1198, 2009.

[868] L. Iapichino and P. Lesaffre. *Astron. Astrophys.*, 512:A27 (7 pp), 2010.

[869] I. Iben, Jr. *Astrophys. J.*, 141:993–1018, 1965.

[870] I. Iben, Jr. *Astrophys. J.*, 259:244–266, 1982.

[871] I. Iben, Jr. *Astrophys. J. Suppl. S.*, 76:55–114, 1991.

[872] I. Iben, Jr. *Stellar Evolution Physics. Vol. 1: Physical Processes in Stellar Interiors.* Cambridge Univ. Press, Cambridge, UK, 2013.

[873] I. Iben, Jr., M. Y. Fujimoto, and J. MacDonald. *Astrophys. J. Lett.*, 375:L27–L29, 1991.

[874] I. Iben, Jr., M. Y. Fujimoto, and J. MacDonald. *Astrophys. J.*, 388:521–540, 1992.

[875] I. Iben, Jr. and A. V. Tutukov. *Astrophys. J. Suppl. S.*, 54:335–372, 1984.

[876] I. Iben, Jr. and A. V. Tutukov. *Astrophys. J. Suppl. S.*, 105:145–180, 1996.

[877] C. Iliadis. *Nucl. Phys. A*, 618:166–175, 1997.

[878] C. Iliadis. *Nuclear Physics of Stars.* 2nd Ed., Wiley-VCH Verlag, Weinheim, Germany, 2015.

[879] C. Iliadis, A. Champagne, A. Chieffi, and M. Limongi. *Astrophys. J. Suppl. S.*, 193:16 (23 pp), 2011.

[880] C. Iliadis, A. Champagne, J. José, S. Starrfield, and P. Tupper. *Astrophys. J. Suppl. S.*, 142:105–137, 2002.

[881] C. Iliadis, P. M. Endt, N. Prantzos, and W. J. Thompson. *Astrophys. J.*, 524:434–453, 1999.

[882] C. Iliadis, R. Longland, A. E. Champagne, A. Coc, and R. Fitzgerald. *Nucl. Phys. A*, 841:31–250, 2010.

[883] S. Immler, P. J. Brown, P. Milne, et al. *Astrophys. J. Lett.*, 648:L119–L122, 2006.

[884] J. J. M. in 't Zand, D. Altamirano, D. R. Ballantyne, et al. *ArXiv e-prints*, 2015.

[885] C. Inserra, A. Pastorello, M. Turatto, et al. *Astron. Astrophys.*, 555:A142 (26 pp), 2013.

[886] J. J. M. in't Zand, A. Cumming, M. V. van der Sluys, F. Verbunt, and O. R. Pols. *Astron. Astrophys.*, 441:675–684, 2005.

[887] J. J. M. in't Zand, D. K. Galloway, H. L. Marshall, et al. *Astron. Astrophys.*, 553:A83 (11 pp), 2013.

[888] J. J. M. in't Zand, L. Keek, and Y. Cavecchi. *Astron. Astrophys.*, 568:A69 (11 p), 2014.

[889] J. J. M. in't Zand, E. Kuulkers, F. Verbunt, J. Heise, and R. Cornelisse. *Astron. Astrophys.*, 411:L487–L491, 2003.

[890] J. J. M. in't Zand and N. N. Weinberg. *Astron. Astrophys.*, 520:A81 (13 pp), 2010.

[891] J. Isern, M. Hernanz, and J. José. In *Astronomy with Radioactivities*, R. Diehl, D. H. Hartmann, and N. Prantzos (editor). Springer-Verlag, Berlin, Germany, 2011, pp 233–308.

[892] J. Isern, M. Hernanz, M. Salaris, et al. In *Thermonuclear Supernovae*, P. Ruiz-Lapuente, R. Canal, and J. Isern (editor). Kluwer Acad. Publ., Dordrecht, The Netherlands, 1997, pp 127–146.

[893] J. Isern, P. Jean, E. Bravo, et al. *Astron. Astrophys.*, 552:A97 (9 pp), 2013.

[894] S. Ishikawa. *Phys. Rev. C*, 87:055804 (9 pp), 2013.

[895] K. Ito. *Prog. Theor. Phys.*, 26:990–1004, 1961.

[896] N. Itoh, F. Kuwashima, and H. Munakata. *Astrophys. J.*, 362:620–623, 1990.

[897] L. N. Ivanova, V. S. Imshennik, and V. M. Chechetkin. *Astrophys. Space Sci.*, 31:497–514, 1974.

[898] N. Ivanova, S. Justham, X. Chen, et al. *Astron. Astrophys. Rev.*, 21:59 (73 pp), 2013.

[899] K. Iwamoto, F. Brachwitz, K. Nomoto, et al. *Astrophys. J. Suppl. S.*, 125:439–462, 1999.

[900] K. Iwamoto, P. A. Mazzali, K. Nomoto, et al. *Nature*, 395:672–674, 1998.

[901] A. F. Iyudin, K. Bennett, H. Bloemen, et al. *Astron. Astrophys.*, 300:422–428, 1995.

[902] A. F. Iyudin, V. Schönfelder, K. Bennett, et al. *Nature*, 396:142–144, 1998.

[903] M. Jadhav, E. Zinner, S. Amari, et al. *Geochim. Cosmochim. Acta*, 113:193–224, 2013.

[904] A. Janiuk. *Astron. Astrophys.*, 568:A105 (7 pp), 2014.

[905] H.-T. Janka, T. Eberl, M. Ruffert, and C. L. Fryer. *Astrophys. J. Lett.*, 527:L39–L42, 1999.

[906] H.-T. Janka and W. Hillebrandt. *Astron. Astrophys. Sup.*, 78:375–397, 1989.

[907] P. Jean, M. Hernanz, J. Gómez-Gomar, and J. José. *Mon. Not. R. Astron. Soc.*, 319:350–364, 2000.

[908] J. H. Jeans. *Phil. Trans. R. Soc. Lond.*, 199:1–53, 1902.

[909] S. Jha, D. Branch, R. Chornock, et al. *Astron. J.*, 132:189–196, 2006.

[910] S. Jha, A. G. Riess, and R. P. Kirshner. *Astrophys. J.*, 659:122–148, 2007.

[911] S. Ji, R. T. Fisher, E. García-Berro, et al. *Astrophys. J.*, 773:136 (14 pp), 2013.

[912] C. Johnson. *Numerical Solution of Partial Differential Equations by the Finite Element Method*. Cambridge Univ. Press, Cambridge, UK, 1987.

[913] A. P. Jones. In *From Stardust to Planetesimals*, Y. J. Pendleton (editor). Astron. Soc. Pac. Conf. Series, San Francisco (California), 1997, pp 97–106.

[914] A. P. Jones and J. A. Nuth. *Astron. Astrophys.*, 530:A44 (12 pp), 2011.

[915] G. C. Jordan, IV, H. B. Perets, R. T. Fisher, and D. R. van Rossum. *Astrophys. J. Lett.*, 761:L23 (5 pp), 2012.

[916] J. José, A. Coc, and M. Hernanz. *Astrophys. J.*, 520:347–360, 1999.

[917] J. José, A. Coc, and M. Hernanz. *Astrophys. J.*, 560:897–906, 2001.

[918] J. José, E. García-Berro, M. Hernanz, and P. Gil-Pons. *Astrophys. J. Lett.*, 662:L103–L106, 2007.

[919] J. José and M. Hernanz. *Astrophys. J.*, 494:680–690, 1998.

[920] J. José and M. Hernanz. *J. Phys. G Nucl. Partic.*, 34:R431–R458, 2007.

[921] J. José and M. Hernanz. *Meteorit. Planet. Sci.*, 42:1135–1143, 2007.

[922] J. José, M. Hernanz, S. Amari, K. Lodders, and E. Zinner. *Astrophys. J.*, 612:414–428, 2004.

[923] J. José, M. Hernanz, and A. Coc. *Astrophys. J. Lett.*, 479:L55–L58, 1997.

[924] J. José, M. Hernanz, and C. Iliadis. *Nucl. Phys. A*, 777:550–578, 2006.

[925] J. José, M. Hernanz, and J. Isern. *Astron. Astrophys.*, 269:291–300, 1993.

[926] J. José and F. Moreno. *Proc. Science, PoS(NIC-IX)* 123 (6 pp), 2006.

[927] J. José, F. Moreno, A. Parikh, and C. Iliadis. *Astrophys. J. Suppl. S.*, 189:204–239, 2010.

[928] J. José and S. N. Shore. In *Classical Novae*, M. F. Bode and A. Evans (editor). 2nd Ed., Cambridge Univ. Press, Cambridge, UK, 2008, pp 121–140.

[929] P. C. Joss. *Nature*, 270:310–314, 1977.

[930] P. C. Joss. *Astrophys. J. Lett.*, 225:L123–L127, 1978.

[931] P. C. Joss and F. K. Li. *Astrophys. J.*, 238:287–295, 1980.

[932] P. C. Joss and F. Melia. *Astrophys. J.*, 312:700–710, 1987.

[933] P. C. Joss and S. Rappaport. *Nature*, 265:222–224, 1977.

[934] J.C.E. Jouguet. *J. Math. Pure Appl.*, 1:347–425, 1905.

[935] J.C.E. Jouguet. *J. Math. Pure Appl.*, 2:5–85, 1906.

[936] A. H. Joy. *Astrophys. J.*, 120:377–383, 1954.

[937] M. Junker, A. D'alessandro, S. Zavatarelli, et al. *Phys. Rev. C*, 57:2700–2710, 1998.

[938] O. Just, A. Bauswein, R. A. Pulpillo, S. Goriely, and H.-T. Janka. *Mon. Not. R. Astron. Soc.*, 448:541–567, 2015.

[939] P. Kahabka and E. P. J. van den Heuvel. *Annu. Rev. Astron. Astr.*, 35:69–100, 1997.

[940] V. Kalogera, C. Kim, D. R. Lorimer, et al. *Astrophys. J. Lett.*, 601:L179–L182, 2004.

[941] G. Kanbach and L. Nittler. In *Astronomy with Radioactivities*, R. Diehl, D. H. Hartmann, and N. Prantzos (editor). Springer-Verlag, Berlin, Germany, 2011, pp 490–518.

[942] J. Kane, D. Arnett, B. A. Remington, et al. *Astrophys. J. Suppl. S.*, 127:365–369, 2000.

[943] J. Kane, D. Arnett, B. A. Remington, et al. *Astrophys. J. Lett.*, 478:L75–L78, 1997.

[944] J. O. Kane, H. F. Robey, B. A. Remington, et al. *Phys. Rev. E*, 63:055401 (4 pp), 2001.

[945] F. Käppeler. *Prog. Part. Nucl. Phys.*, 43:419–483, 1999.

[946] F. Käppeler, S. Bisterzo, R. Gallino, et al. *Pub. Astron. Soc. Aust.*, 26:209–216, 2009.

[947] F. Käppeler, R. Gallino, S. Bisterzo, and W. Aoki. *Rev. Mod. Phys.*, 83:157–194, 2011.

[948] F. Käppeler, R. Gallino, M. Busso, G. Picchio, and C. M. Raiteri. *Astrophys. J.*, 354:630–643, 1990.

[949] A. I. Karakas and J. C. Lattanzio. *Pub. Astron. Soc. Aust.*, 31:30 (62 pp), 2014.

[950] D. Kasen. *Astrophys. J.*, 708:1025–1031, 2010.

[951] D. Kasen, P. Nugent, R. C. Thomas, and L. Wang. *Astrophys. J.*, 610:876–887, 2004.

[952] D. Kasen, F. K. Röpke, and S. E. Woosley. *Nature*, 460:869–872, 2009.

[953] D. Kasen and S. E. Woosley. *Astrophys. J.*, 703:2205–2216, 2009.

[954] K. Kashiyama, K. Ioka, and P. Mészáros. *Astrophys. J. Lett.*, 776:L39 (4 pp), 2013.

[955] M. M. Kasliwal, S. B. Cenko, S. R. Kulkarni, et al. *Astrophys. J.*, 735:94 (12 pp), 2011.

[956] M. Kato. *Pub. Astron. Soc. Jpn.*, 35:33–46, 1983.

[957] M. Kato and I. Hachisu. *Astrophys. J.*, 437:802–826, 1994.

[958] K. Kawabe (for the LIGO Collaboration). *J. Phys. Conf. Ser.*, 120:032003 (7 pp), 2008.

[959] M. Kawai, A. K. Kerman, and K. W. McVoy. *Ann. Phys.*, 75:156–170, 1973.

[960] Y. Kawai, H. Saio, and K. Nomoto. *Astrophys. J.*, 315:229–233, 1987.

[961] Y. Kawai, H. Saio, and K. Nomoto. *Astrophys. J.*, 328:207–212, 1988.

[962] L. Keek. *Astrophys. J.*, 756:130 (8 pp), 2012.

[963] L. Keek, R. H. Cyburt, and A. Heger. *Astrophys. J.*, 787:101 (11 pp), 2014.

[964] L. Keek, D. K. Galloway, J. J. M. in't Zand, and A. Heger. *Astrophys. J.*, 718:292–305, 2010.

[965] L. Keek and A. Heger. *Astrophys. J.*, 743:189 (15 pp), 2011.

[966] L. Keek, A. Heger, and J. J. M. in't Zand. *Astrophys. J.*, 752:150 (13 pp), 2012.

[967] L. Keek and J. J. M. in't Zand. *Proc. Science, PoS(Integral08)* 032 (11 pp), 2008.

[968] L. Keek, J. J. M. in't Zand, E. Kuulkers, et al. *Astron. Astrophys.*, 479:177–188, 2008.

[969] K. J. Kelly, C. Iliadis, L. Downen, J. José, and A. Champagne. *Astrophys. J.*, 777:130 (7 pp), 2013.

[970] S. J. Kenyon, M. Livio, J. Mikolajewska, and C. A. Tout. *Astrophys. J. Lett.*, 407:L81–L84, 1993.

[971] A. Kercek, W. Hillebrandt, and J. W. Truran. *Astron. Astrophys.*, 337:379–392, 1998.

[972] A. Kercek, W. Hillebrandt, and J. W. Truran. *Astron. Astrophys.*, 345:831–840, 1999.

[973] W. E. Kerzendorf. In *Binary Paths to Type Ia Supernovae Explosions*, R. Di Stefano, M. Orio, and M. Moe (editor). Cambridge Univ. Press, Cambridge, UK, 2013, pp 326–330.

[974] W. E. Kerzendorf, B. P. Schmidt, M. Asplund, et al. *Astrophys. J.*, 701:1665–1672, 2009.

[975] W. E. Kerzendorf, D. Yong, B. P. Schmidt, et al. *Astrophys. J.*, 774:99 (19 pp), 2013.

[976] R. Kessler, J. Guy, J. Marriner, et al. *Astrophys. J.*, 764:48 (18 pp), 2013.

[977] A. Khokhlov, E. Müller, and P. Höflich. *Astron. Astrophys.*, 270:223–248, 1993.

[978] A. M. Khokhlov. *Astron. Astrophys.*, 245:114–128, 1991.

[979] A. M. Khokhlov. *Astron. Astrophys.*, 245:L25–L28, 1991.

[980] A. M. Khokhlov. In *Supernovae and Gamma-Ray Bursts: The Greatest Explosions since the Big Bang*, M. Livio, N. Panagia, and K. Sahu (editor). Cambridge Univ. Press, Cambridge, UK, 2001, pp 239–249.

[981] A. M. Khokhlov and E. S. Oran. *Combust. Flame*, 119:400–416, 1999.

[982] A. M. Khokhlov, E. S. Oran, and G. O. Thomas. *Combust. Flame*, 117:323–339, 1999.

[983] R. E. Kidder and H. E. de Witt. *J. Nucl. Energy*, 2:218–223, 1961.

[984] M. Kiewe, A. Gal-Yam, I. Arcavi, et al. *Astrophys. J.*, 744:10 (19 pp), 2012.

[985] K. Kifonidis, T. Plewa, H.-T. Janka, and E. Müller. *Astron. Astrophys.*, 408:621–649, 2003.

[986] R. Kippenhahn and H.-C. Thomas. *Astron. Astrophys.*, 63:265–272, 1978.

[987] R. Kippenhahn, A. Weigert, and E. Hofmeister. *Meth. Comput. Phys.*, 7:129–190, 1967.

[988] R. Kippenhahn, A. Weigert, and A. Weiss. *Stellar Structure and Evolution.* 2nd Ed., Springer-Verlag, Berlin, Germany, 2012.

[989] R. P. Kirshner. In *Supernovae*, A. G. Petschek (editor). Springer-Verlag, Berlin, Germany, 1990, pp 59–75.

[990] R. P. Kirshner and J. Kwan. *Astrophys. J.*, 193:27–36, 1974.

[991] L.-T. Kizivat, G. Martínez-Pinedo, K. Langanke, R. Surman, and G. C. McLaughlin. *Phys. Rev. C*, 81:025802 (9 pp), 2010.

[992] R. W. Klebesadel, I. B. Strong, and R. A. Olson. *Astrophys. J. Lett.*, 182:L85–L88, 1973.

[993] I. K. W. Kleiser, D. Poznanski, D. Kasen, et al. *Mon. Not. R. Astron. Soc.*, 415:372–382, 2011.

[994] K. Knie, G. Korschinek, T. Faestermann, et al. *Phys. Rev. Lett.*, 93:171103 (4 pp), 2004.

[995] K. Knie, G. Korschinek, T. Faestermann, et al. *Phys. Rev. Lett.*, 83:18–21, 1999.

[996] P. Knupp and K. Salari. *Verification of Computer Codes in Computational Science and Engineering.* Chapman and Hall/CRC, London, UK, 2002.

[997] O. Koike, M.-A. Hashimoto, K. Arai, and S. Wanajo. *Astron. Astrophys.*, 342:464–473, 1999.

[998] O. Koike, M.-A. Hashimoto, R. Kuromizu, and S.-i. Fujimoto. *Astrophys. J.*, 603:242–251, 2004.

[999] A. K. H. Kong, J. M. Miller, M. Méndez, et al. *Astrophys. J. Lett.*, 670:L17–L20, 2007.

[1000] B.-C. Koo, Y.-H. Lee, D.-S. Moon, S.-C. Yoon, and J. C. Raymond. *Science*, 342:1346–1348, 2013.

[1001] O. Korobkin, S. Rosswog, A. Arcones, and C. Winteler. *Mon. Not. R. Astron. Soc.*, 426:1940–1949, 2012.

[1002] M. Koshiba. *Phys. Rep.*, 220:229–381, 1992.

[1003] R. Kotak, P. Meikle, M. Pozzo, et al. *Astrophys. J. Lett.*, 651:L117–L120, 2006.

[1004] R. Kotak, W. P. S. Meikle, A. Adamson, and S. K. Leggett. *Mon. Not. R. Astron. Soc.*, 354:L13–L17, 2004.

[1005] C. Kouveliotou, J. Kommers, W. H. G. Lewin, et al. *IAU Circ.* #6286, 1996.

[1006] A. Kovetz and D. Prialnik. *Astrophys. J.*, 291:812–821, 1985.

[1007] A. Kovetz and D. Prialnik. *Astrophys. J.*, 424:319–332, 1994.

[1008] A. Kovetz and D. Prialnik. *Astrophys. J.*, 477:356–367, 1997.

[1009] C. T. Kowal. *Astron. J.*, 73:1021–1024, 1968.

[1010] M. Kowalski, D. Rubin, G. Aldering, et al. *Astrophys. J.*, 686:749–778, 2008.

[1011] T. Kozasa, H. Hasegawa, and K. Nomoto. *Astrophys. J.*, 344:325–331, 1989.

[1012] T. Kozasa, H. Hasegawa, and K. Nomoto. *Astrophys. J. Lett*, 346:L81–L84, 1989.

[1013] C. Kozma, C. Fransson, W. Hillebrandt, et al. *Astron. Astrophys.*, 437:983–995, 2005.

[1014] R. P. Kraft. *Leaflet of the Astron. Soc. Pac.*, 9:137–144, 1964.

[1015] R. P. Kraft. *Astrophys. J.*, 139:457–475, 1964.

[1016] K. S. Krane. *Introductory Nuclear Physics.* 3rd Ed., John Wiley and Sons, New York, 1987.

[1017] V. I. Krasovski and I. Shklovsky. *Dokl. Akad. Nauk. SSR*, 116:197–199, 1957.

[1018] K.-L. Kratz, J.-P. Bitouzet, F.-K. Thielemann, P. Moeller, and B. Pfeiffer. *Astrophys. J.*, 403:216–238, 1993.

[1019] K. L. Kratz, K. Farouqi, O. Hallmann, B. Pfeiffer, and U. Ott. In *Seventh European Summer School on Experimental Nuclear Astrophysics*, C. Spitaleri, L. Lamia, and R. G. Pizzone (editor). American Inst. Phys., Melville (New York), 2014, pp 62–68.

[1020] K.-L. Kratz, K. Farouqi, and B. Pfeiffer. *Prog. Part. Nucl. Phys.*, 59:147–155, 2007.

[1021] K.-L. Kratz, K. Farouqi, B. Pfeiffer, et al. *Astrophys. J.*, 662:39–52, 2007.

[1022] O. Krause, S. M. Birkmann, T. Usuda, et al. *Science*, 320:1195–1197, 2008.

[1023] O. Krause, M. Tanaka, T. Usuda, et al. *Nature*, 456:617–619, 2008.

[1024] J. Kristian, C. R. Pennypacker, D. E. Morris, et al. *Nature*, 338:234–236, 1989.

[1025] M. Kromer, M. Fink, V. Stanishev, et al. *Mon. Not. R. Astron. Soc.*, 429:2287–2297, 2013.

[1026] M. Kromer, R. Pakmor, S. Taubenberger, et al. *Astrophys. J.*, 778:L18 (6 pp), 2013.

[1027] M. Kromer, S. A. Sim, M. Fink, et al. *Astrophys. J.*, 719:1067–1082, 2010.

[1028] A. N. Krot, K. Keil, E. R. D. Scott, C. A. Goodrich, and M. K. Weisberg. In *Meteorites, Comets and Planets: Treatise on Geochemistry, Vol. 1*, A. M. Davis (editor). 2nd Ed., Elsevier B. V., Amsterdam, The Netherlands, 2014, pp 1–63.

[1029] E. Krügel. *An Introduction to the Physics of Interstellar Dust*. Taylor & Francis/CRC, London, 2008.

[1030] M. J. Kuchner, R. P. Kirshner, P. A. Pinto, and B. Leibundgut. *Astrophys. J. Lett.*, 426:L89–L92, 1994.

[1031] P. V. Kulkarni, N. M. Ashok, K. M. V. Apparao, and S. M. Chitre. *Nature*, 280:819–820, 1979.

[1032] S. Kumagai, K. Iwabuchi, and K. Nomoto. *Adv. Space Res.*, 25:699–702, 2000.

[1033] S. Kumagai, T. Shigeyama, K. Nomoto, M. Itoh, and J. Nishimura. *Astron. Astrophys.*, 197:L7–L10, 1988.

[1034] T. Kuroda, S. Wanajo, and K. Nomoto. *Astrophys. J.*, 672:1068–1078, 2008.

[1035] D. Kushnir and B. Katz. *Astrophys. J.*, 811:97 (12 pp), 2015.

[1036] D. Kushnir, B. Katz, S. Dong, E. Livne, and R. Fernández. *Astrophys. J. Lett.*, 778:L37 (6 pp), 2013.

[1037] G. S. Kutter and W. M. Sparks. *Astrophys. J.*, 175:407–415, 1972.

[1038] G. S. Kutter and W. M. Sparks. *Astrophys. J.*, 321:386–393, 1987.

[1039] E. Kuulkers. *Nucl. Phys. B Proc. Sup.*, 132:466–475, 2004.

[1040] E. Kuulkers, P. R. den Hartog, J. J. M. in't Zand, et al. *Astron. Astrophys.*, 399:663–680, 2003.

[1041] E. Kuulkers, J. J. M. in't Zand, and J.-P. Lasota. *Astron. Astrophys.*, 503:889–897, 2009.

[1042] E. Kuulkers, J. J. M. in't Zand, M. H. van Kerkwijk, et al. *Astron. Astrophys.*, 382:503–512, 2002.

[1043] J. Lachner, I. Dillmann, T. Faestermann, et al. *Phys. Rev. C*, 85:015801 (6 pp), 2012.

[1044] D. Q. Lamb and F. K. Lamb. *Ann. NY Acad. Sci.*, 302:261–299, 1977.

[1045] D. Q. Lamb and F. K. Lamb. *Astrophys. J.*, 220:291–302, 1978.

[1046] D. Q. Lamb and C. J. Pethick. *Astrophys. J. Lett.*, 209:L77–L81, 1976.

[1047] F. K. Lamb, A. C. Fabian, J. E. Pringle, and D. Q. Lamb. *Astrophys. J.*, 217:197–212, 1977.

[1048] L.D. Landau and E.M. Lifshitz. *Fluid Mechanics. Course of Theoretical Physics, Vol. 6*. 2nd Ed., Butterworth-Heinemann, Oxford, UK, 1987.

[1049] A. M. Lane and R. G. Thomas. *Rev. Mod. Phys.*, 30:257–353, 1958.

[1050] K. Langanke and G. Martínez-Pinedo. *Nucl. Phys. A*, 673:481–508, 2000.

[1051] K. Langanke and G. Martínez-Pinedo. *Atom. Data Nucl. Data*, 79:1–46, 2001.

[1052] K. Langanke, M. Wiescher, W. A. Fowler, and J. Gorres. *Astrophys. J.*, 301:629–633, 1986.

[1053] R. B. Larson. *Mon. Not. R. Astron. Soc.*, 145:271–295, 1969.

[1054] H. Laster. *Science*, 160:1138, 1968.

[1055] J. M. Lattimer and M. Prakash. *Science*, 304:536–542, 2004.

[1056] J. M. Lattimer and M. Prakash. *Nucl. Phys. A*, 777:479–496, 2006.

[1057] J. M. Lattimer and D. N. Schramm. *Astrophys. J. Lett.*, 192:L145–L147, 1974.

[1058] J. M. Lattimer, D. N. Schramm, and L. Grossman. *Nature*, 269:116–118, 1977.

[1059] J. M. Lattimer, D. N. Schramm, and L. Grossman. *Astrophys. J.*, 219:230–249, 1978.

[1060] T. Lee, D. A. Papanastassiou, and G. J. Wasserburg. *Astrophys. J. Lett.*, 211:L107–L110, 1977.

[1061] U. Lee and T. E. Strohmayer. *Mon. Not. R. Astron. Soc.*, 361:659–672, 2005.

[1062] W. H. Lee and W. Kluźniak. *Mon. Not. R. Astron. Soc.*, 308:780–794, 1999.

[1063] W. H. Lee and W. Kluźniak. *Astrophys. J.*, 526:178–199, 1999.

[1064] B. Leibundgut. *Annu. Rev. Astron. Astr.*, 39:67–98, 2001.

[1065] B. Leibundgut. *Gen. Relat. Gravit.*, 40:221–248, 2008.

[1066] B. Leibundgut, R. P. Kirshner, M. M. Phillips, et al. *Astron. J.*, 105:301–313, 1993.

[1067] B. Leibundgut and P. A. Pinto. *Astrophys. J.*, 401:49–59, 1992.

[1068] B. Leibundgut, R. Schommer, M. Phillips, et al. *Astrophys. J. Lett.*, 466:L21–L24, 1996.

[1069] M. D. Leising. *Astron. Astrophys.*, 97:299–301, 1993.

[1070] M. D. Leising and D. D. Clayton. *Astrophys. J.*, 323:159–169, 1987.

[1071] M. D. Leising and G. H. Share. *Astrophys. J.*, 357:638–648, 1990.

[1072] M. D. Leising, G. H. Share, E. L. Chupp, and G. Kanbach. *Astrophys. J.*, 328:755–762, 1988.

[1073] G. Lemaître. *Ann. Soc. Sci. Brux.*, 47:49–59, 1927.

[1074] G. Lemaître. *Mon. Not. R. Astron. Soc.*, 91:483–490, 1931.

[1075] E. J. Lentz, D. Branch, and E. Baron. *Astrophys. J.*, 512:678–682, 1999.

[1076] D. C. Leonard. *Astrophys. J.*, 670:1275–1282, 2007.

[1077] D. C. Leonard, A. V. Filippenko, and T. Matheson. In *Cosmic Explosions: Tenth Astrophysics Conf.*, S. S. Holt and W. W. Zhang (editor). American Inst. Phys., Melville (New York), 2000, pp 165–168.

[1078] D. C. Leonard, W. Li, A. V. Filippenko, R. J. Foley, and R. Chornock. *Astrophys. J.*, 632:450–475, 2005.

[1079] V.S. Letokhov. *Laser Photoionization Spectroscopy.* Academic Press, Orlando (Florida), 1987.

[1080] A. J. Levan, N. R. Tanvir, R. L. C. Starling, et al. *Astrophys. J.*, 781:13 (22 pp), 2014.

[1081] R. D. Levee. *Astrophys. J.*, 117:200–210, 1953.

[1082] M. Leventhal, C. MacCallum, and A. Watts. *Astrophys. J.*, 216:491–502, 1977.

[1083] R.J. LeVeque, D. Mihalas, E.A. Dorfi, and E. Müller. *Computational Methods for Astrophysical Fluid Flows.* Springer-Verlag, Berlin, Germany, 1998.

[1084] W. H. G. Lewin. *Ann. NY Acad. Sci.*, 302:210–227, 1977.

[1085] W. H. G. Lewin, G. Clark, and J. Doty. *IAU Circ.* #2922, 1976.

[1086] W. H. G. Lewin, L. R. Cominsky, A. R. Walker, and B. S. C. Robertson. *Nature*, 287:27–28, 1980.

[1087] W. H. G. Lewin, J. Doty, G. W. Clark, et al. *Astrophys. J. Lett.*, 207:L95–L99, 1976.

[1088] W. H. G. Lewin and P. C. Joss. *Nature*, 270:211–216, 1977.

[1089] W. H. G. Lewin and P. C. Joss. *Space Sci. Rev.*, 28:3–87, 1981.

[1090] W. H. G. Lewin, W. D. Vacca, and E. M. Basinska. *Astrophys. J. Lett.*, 277:L57–L60, 1984.

[1091] W. H. G. Lewin, J. van Paradijs, and R. E. Taam. *Space Sci. Rev.*, 62:223–389, 1993.

[1092] W. H. G. Lewin, J. van Paradijs, and R. E. Taam. In *X-Ray Binaries*, W. H. G. Lewin, J. van Paradijs, and E. P. J. van den Heuvel (editor). Cambridge Univ. Press, Cambridge, UK, 1995, pp 175–232.

[1093] K. M. Lewis, M. Lugaro, B. K. Gibson, and K. Pilkington. *Astrophys. J. Lett.*, 768:L19 (5 pp), 2013.

[1094] R. S. Lewis, S. Amari, and E. Anders. *Nature*, 348:293–298, 1990.

[1095] R. S. Lewis, S. Amari, and E. Anders. *Geochim. Cosmochim. Acta*, 58:471–494, 1994.

[1096] R. S. Lewis, G. R. Huss, E. Anders, Y.-G. Liu, and R. A. Schmitt. *Meteoritics*, 26:363–364, 1991.

[1097] R. S. Lewis, G. R. Huss, and G. Lugmair. 22nd *Lunar and Planetary Science Conf.*, Abstract #807, 1991.

[1098] R. S. Lewis, T. Ming, J. F. Wacker, E. Anders, and E. Steel. *Nature*, 326:160–162, 1987.

[1099] W. Li, J. S. Bloom, P. Podsiadlowski, et al. *Nature*, 480:348–350, 2011.

[1100] W. Li, A. V. Filippenko, R. Chornock, et al. *Pub. Astron. Soc. Pac.*, 115:453–473, 2003.

[1101] W. Li, A. V. Filippenko, E. Gates, et al. *Pub. Astron. Soc. Pac.*, 113:1178–1204, 2001.

[1102] W. Li, J. Leaman, R. Chornock, et al. *Mon. Not. R. Astron. Soc.*, 412:1441–1472, 2011.

[1103] W. D. Li, A. V. Filippenko, A. G. Riess, et al. In *Cosmic Explosions: Tenth Astrophysics Conf.*, S. S. Holt and W. W. Zhang (editor). American Inst. Phys., Melville (New York), 2000, pp 91–94.

[1104] X.-D. Li and E. P. J. van den Heuvel. *Astron. Astrophys.*, 322:L9–L12, 1997.

[1105] A. Liñán and F. A. Williams. *Fundamental Aspects of Combustion*. Oxford Univ. Press, Oxford, UK, 1993.

[1106] E. P. T. Liang. *Astrophys. J. Lett.*, 211:L67–L70, 1977.

[1107] E. P. T. Liang. *Astrophys. J.*, 218:243–246, 1977.

[1108] G. G. Lichti, K. Bennett, J. W. den Herder, et al. *Astron. Astrophys.*, 292:569–579, 1994.

[1109] M. Liebendörfer, S. Rosswog, and F.-K. Thielemann. *Astrophys. J. Suppl. S.*, 141:229–246, 2002.

[1110] M. Limongi and A. Chieffi. *Astrophys. J.*, 592:404–433, 2003.

[1111] M. Limongi and A. Chieffi. *Astrophys. J.*, 647:483–500, 2006.

[1112] M. Limongi, O. Straniero, and A. Chieffi. *Astrophys. J. Suppl. S.*, 129:625–664, 2000.

[1113] M. Limongi and A. Tornambe. *Astrophys. J.*, 371:317–331, 1991.

[1114] D. J. Lin, A. Bayliss, and R. E. Taam. *Astrophys. J.*, 653:545–557, 2006.

[1115] Y. Lin, F. Gyngard, and E. Zinner. *Astrophys. J.*, 709:1157–1173, 2010.

[1116] M. Linares, D. Altamirano, D. Chakrabarty, A. Cumming, and L. Keek. *Astrophys. J.*, 748:82 (13 pp), 2012.

[1117] M. Linares, V. Connaughton, P. Jenke, et al. *Astrophys. J.*, 760:133 (11 pp), 2012.

[1118] M. Linares, A. L. Watts, R. Wijnands, et al. *Mon. Not. R. Astron. Soc.*, 392:L11–L15, 2009.

[1119] A. E. Litherland. *Annu. Rev. Nucl. Part. S.*, 30:437–473, 1980.

[1120] G. R. Liu and M. B. Liu. *Smoothed Particle Hydrodynamics: A Meshfree Particle Method*. World Scientific, Singapore, 2003.

[1121] Z.-W. Liu, R. Pakmor, F. K. Röpke, et al. *Astron. Astrophys.*, 554:A109 (12 pp), 2013.

[1122] M. Livio. In *Type Ia Supernovae, Theory and Cosmology*, J. C. Niemeyer and J. W. Truran (editor). Cambridge Univ. Press, Cambridge, UK, 2000, pp 33–48.

[1123] M. Livio, J. R. Buchler, and S. A. Colgate. *Astrophys. J. Lett.*, 238:L139–L143, 1980.

[1124] M. Livio, A. Mastichiadis, H. Ögelman, and J. W. Truran. *Astrophys. J.*, 394:217–220, 1992.

[1125] M. Livio, D. Prialnik, and O. Regev. *Astrophys. J.*, 341:299–305, 1989.

[1126] M. Livio and J. E. Pringle. *Astrophys. J. Lett.*, 740:L18 (4 pp), 2011.

[1127] M. Livio and M. M. Shara. *Astrophys. J.*, 319:819–826, 1987.

[1128] M. Livio and J. W. Truran. *Astrophys. J.*, 318:316–325, 1987.

[1129] M. Livio and J. W. Truran. *Astrophys. J.*, 425:797–801, 1994.

[1130] E. Livne. *Astrophys. J.*, 412:634–647, 1993.

[1131] E. Livne. *Astrophys. J. Lett.*, 527:L97–L100, 1999.

[1132] E. Livne and W. D. Arnett. *Astrophys. J.*, 452:62–74, 1995.

[1133] E. Livne and A. S. Glasner. *Astrophys. J.*, 370:272–281, 1991.

[1134] H. M. Lloyd, T. J. O'Brien, M. F. Bode, et al. *Nature*, 356:222–224, 1992.

[1135] K. Lodders. *Astrophys. J.*, 591:1220–1247, 2003.

[1136] K. Lodders. In *Principles and Perspectives in Cosmochemistry*, A. Goswami and B. E. Reddy (editor). Springer-Verlag, Berlin, Germany, 2010, pp 379–417.

[1137] K. Lodders and S. Amari. *Chem. Erde-Geochem.*, 65:93–166, 2005.

[1138] K. Lodders and B. Fegley, Jr. *Meteoritics*, 30:661–678, 1995.

[1139] K. Lodders, H. Palme, and H.-P. Gail. In *Landolt Börnstein, Vol. VI/4B*, J. E. Trümper (editor). Springer-Verlag, Berlin, Germany, 2009, pp 1–59.

[1140] A. R. Loeblich and H. Tappan. *Geol. Soc. Amer. Bull.*, 75:367–392, 1964.

[1141] R. Lohner. *Comput. Method. Appl. M.*, 61:323–338, 1987.

[1142] A. Lombardi. In *1604-2004: Supernovae as Cosmological Lighthouses*, M. Turatto, S. Benetti, L. Zampieri, and W. Shea (editor). Astron. Soc. Pac. Conf. Series, San Francisco (California), 2005, pp 21–29.

[1143] R. A. London, W. M. Howard, and R. E. Taam. *Astrophys. J. Lett.*, 287:L27–L30, 1984.

[1144] M. Long, G. C. Jordan, IV, D. R. van Rossum, et al. *Astrophys. J.*, 789:103 (22 pp), 2014.

[1145] M. S. Longair. *High-Energy Astrophysics*. 3rd Ed., Cambridge Univ. Press, Cambridge, UK, 2011.

[1146] R. Longland, P. Lorén-Aguilar, J. José, E. García-Berro, and L. G. Althaus. *Astron. Astrophys.*, 542:A117 (6 pp), 2012.

[1147] R. Longland, P. Lorén-Aguilar, J. José, et al. *Astrophys. J. Lett.*, 737:L34 (4 pp), 2011.

[1148] R. Longland, D. Martin, and J. José. *Astron. Astrophys.*, 563:A67 (13 pp), 2014.

[1149] F. D. Lora-Clavijo, J. P. Cruz-Perez, F. S. Guzman, and J. A. Gonzalez. *Rev. Mex. Fis. E*, 59:28–50, 2013.

[1150] P. Lorén-Aguilar, J. Guerrero, J. Isern, J. A. Lobo, and E. García-Berro. *Mon. Not. R. Astron. Soc.*, 356:627–636, 2005.

[1151] P. Lorén-Aguilar, J. Isern, and E. García-Berro. *Astron. Astrophys.*, 500:1193–1205, 2009.

[1152] P. Lorén-Aguilar, J. Isern, and E. García-Berro. *Mon. Not. R. Astron. Soc.*, 406:2749–2763, 2010.

[1153] D. R. Lorimer. *Living Rev. Relativ.*, 11:8 (90 pp), 2008.

[1154] F. J. Lu, Q. D. Wang, M. Y. Ge, et al. *Astrophys. J.*, 732:11 (6 pp), 2011.

[1155] G. Lü, C. Zhu, Z. Wang, and N. Wang. *Mon. Not. R. Astron. Soc.*, 396:1086–1095, 2009.

[1156] L. B. Lucy. *Astron. J.*, 82:1013–1024, 1977.

[1157] L. B. Lucy, I. J. Danziger, C. Gouiffes, and P. Bouchet. In *Structure and Dynamics of the Interstellar Medium*, G. Tenorio-Tagle, M. Moles, and J. Melnick (editor). Springer-Verlag, Berlin, Germany, 1989, pp 164–179.

[1158] M. Lugaro. *Stardust from Meteorites. An Introduction to Presolar Grains.* World Scientific, Singapore, 2005.

[1159] M. Lugaro and A. Chieffi. In *Astronomy with Radioactivities*, R. Diehl, D. H. Hartmann, and N. Prantzos (editor). Springer-Verlag, Berlin, Germany, 2011, pp 83–152.

[1160] M. Lugaro, A. M. Davis, R. Gallino, et al. *Astrophys. J.*, 593:486–508, 2003.

[1161] M. Lugaro, A.I. Karakas, and S.W. Campbell. *Proc. Science, PoS(NIC-XIII)* 008 (10 pp), 2014.

[1162] M. Lugaro, C. Ugalde, A. I. Karakas, et al. *Astrophys. J.*, 615:934–946, 2004.

[1163] G. W. Lugmair. *Meteoritics*, 9:369, 1974.

[1164] D. Lunney. *Proc. Science, PoS(NIC-IX)* 010 (14 pp), 2006.

[1165] D. Lunney, J. M. Pearson, and C. Thibault. *Rev. Mod. Phys.*, 75:1021–1082, 2003.

[1166] J. Lyman, D. Bersier, P. James, et al. *ArXiv e-prints*, 2014.

[1167] M. Lyutikov and E. G. Blackman. *Mon. Not. R. Astron. Soc.*, 321:177–186, 2001.

[1168] Z. Ma, R. N. Thompson, K. R. Lykke, M. J. Pellin, and A. M. Davis. *Rev. Sci. Instrum.*, 66:3168–3176, 1995.

[1169] R. Maas, R. D. Loss, K. J. R. Rosman, et al. *Meteorit. Planet. Sci.*, 36:849–858, 2001.

[1170] J. MacDonald. *Astrophys. J.*, 267:732–746, 1983.

[1171] J. MacDonald. *Astrophys. J.*, 273:289–298, 1983.

[1172] A. I. MacFadyen. In *From Twilight to Highlight: The Physics of Supernovae*, W. Hillebrandt and B. Leibundgut (editor). Springer, Berlin (Germany), 2003, pp 97–103.

[1173] A. I. MacFadyen and S. E. Woosley. *Astrophys. J.*, 524:262–289, 1999.

[1174] P. MacNeice, K. M. Olson, C. Mobarry, R. de Fainchtein, and C. Packer. *Comput. Phys. Commun.*, 126:330–354, 2000.

[1175] J. Madej, P. C. Joss, and A. Różańska. *Astrophys. J.*, 602:904–912, 2004.

[1176] K. Maeda, S. Benetti, M. Stritzinger, et al. *Nature*, 466:82–85, 2010.

[1177] K. Maeda and K. Nomoto. *Astrophys. J.*, 598:1163–1200, 2003.

[1178] K. Maeda, F. K. Röpke, M. Fink, et al. *Astrophys. J.*, 712:624–638, 2010.

[1179] K. Maeda, Y. Terada, D. Kasen, et al. *Astrophys. J.*, 760:54 (9 pp), 2012.

[1180] K. Maeda and N. Tominaga. *Mon. Not. R. Astron. Soc.*, 394:1317–1324, 2009.

[1181] Z. Magic, A. Serenelli, A. Weiss, and B. Chaboyer. *Astrophys. J.*, 718:1378–1387, 2010.

[1182] G. Magkotsios, F. X. Timmes, A. L. Hungerford, et al. *Astrophys. J. Suppl. S.*, 191:66–95, 2010.

[1183] P. V. Magnus, M. S. Smith, A. J. Howard, P. D. Parker, and A. E. Champagne. *Nucl. Phys. A*, 506:332–345, 1990.

[1184] W. A. Mahoney, J. C. Ling, A. S. Jacobson, and R. E. Lingenfelter. *Astrophys. J.*, 262:742–748, 1982.

[1185] W. A. Mahoney, J. C. Ling, W. A. Wheaton, and A. S. Jacobson. *Astrophys. J.*, 286:578–585, 1984.

[1186] A. Majczyna, J. Madej, P. C. Joss, and A. Różańska. *Astron. Astrophys.*, 430:643–654, 2005.

[1187] D. Malesani, G. Tagliaferri, G. Chincarini, et al. *Astrophys. J. Lett.*, 609:L5–L8, 2004.

[1188] C. M. Malone, A. Nonaka, A. S. Almgren, J. B. Bell, and M. Zingale. *Astrophys. J.*, 728:118 (18 pp), 2011.

[1189] C. M. Malone, M. Zingale, A. Nonaka, A. S. Almgren, and J. B. Bell. *Astrophys. J.*, 788:115 (12 pp), 2014.

[1190] F. Mannucci, M. Della Valle, and N. Panagia. *Mon. Not. R. Astron. Soc.*, 370:773–783, 2006.

[1191] D. Maoz and C. Badenes. *Mon. Not. R. Astron. Soc.*, 407:1314–1327, 2010.

[1192] D. Maoz and F. Mannucci. *Pub. Astron. Soc. Aust.*, 29:447–465, 2012.

[1193] D. Maoz, K. Sharon, and A. Gal-Yam. *Astrophys. J.*, 722:1879–1894, 2010.

[1194] L. Maraschi and A. Cavaliere. In *Highlights of Astronomy, Vol. 4*, E. A. Müller (editor). Reidel, Dordrecht, The Netherlands, 1977, pp 127–128.

[1195] K. K. Marhas, S. Amari, F. Gyngard, E. Zinner, and R. Gallino. *Astrophys. J.*, 689:622–645, 2008.

[1196] E. Marietta, A. Burrows, and B. Fryxell. *Astrophys. J. Suppl. S.*, 128:615–650, 2000.

[1197] G. H. Marion, P. Höflich, W. D. Vacca, and J. C. Wheeler. *Astrophys. J.*, 591:316–333, 2003.

[1198] G. H. Marion, P. Höflich, J. C. Wheeler, et al. *Astrophys. J.*, 645:1392–1401, 2006.

[1199] T. R. Marsh, G. Nelemans, and D. Steeghs. *Mon. Not. R. Astron. Soc.*, 350:113–128, 2004.

[1200] R. G. Martin, C. A. Tout, and P. Lesaffre. *Mon. Not. R. Astron. Soc.*, 373:263–270, 2006.

[1201] B. Marty, M. Chaussidon, R. C. Wiens, A. J. G. Jurewicz, and D. S. Burnett. *Science*, 332:1533–1536, 2011.

[1202] L. Mashonkina, N. Christlieb, P. S. Barklem, et al. *Astron. Astrophys.*, 516:A46 (22 pp), 2010.

[1203] T. Matheson, A. V. Filippenko, W. Li, D. C. Leonard, and J. C. Shields. *Astron. J.*, 121:1648–1675, 2001.

[1204] A. Matic, A. M. van den Berg, M. N. Harakeh, et al. *Phys. Rev. C*, 80:055804 (18 pp), 2009.

[1205] Y. Matsuo, H. Tsujimoto, T. Noda, et al. *Prog. Theor. Phys.*, 126:1177–1186, 2011.

[1206] M. Matsuura, E. Dwek, M. Meixner, et al. *Science*, 333:1258–1261, 2011.

[1207] F. Matteucci, A. Renda, A. Pipino, and M. Della Valle. *Astron. Astrophys.*, 405:23–30, 2003.

[1208] T. G. Matthews, D. M. Smith, and M. Hernanz. *Bull. American Astron. Soc.*, 38:344, 2006.

[1209] S. M. Matz, G. H. Share, M. D. Leising, E. L. Chupp, and W. T. Vestrand. *Nature*, 331:416–418, 1988.

[1210] R. W. Mayle. In *Supernovae*, A. G. Petschek (editor). Springer-Verlag, Berlin, Germany, 1990, pp 267–289.

[1211] J. Maza and S. van den Bergh. *Astrophys. J.*, 204:519–529, 1976.

[1212] T. J. Mazurek. *Nature*, 252:287–289, 1974.

[1213] T. J. Mazurek. *Astrophys. Space Sci.*, 35:117–135, 1975.

[1214] T. J. Mazurek, J. M. Lattimer, and G. E. Brown. *Astrophys. J.*, 229:713–727, 1979.

[1215] T. J. Mazurek, D. L. Meier, and J. C. Wheeler. *Astrophys. J.*, 213:518–526, 1977.

[1216] P. A. Mazzali and L. B. Lucy. *Mon. Not. R. Astron. Soc.*, 295:428–436, 1998.

[1217] M. L. McCall. *Mon. Not. R. Astron. Soc.*, 210:829–837, 1984.

[1218] M. L. McCall, N. Reid, M. S. Bessell, and D. Wickramasinghe. *Mon. Not. R. Astron. Soc.*, 210:839–843, 1984.

[1219] J. E. McClintock, C. R. Canizares, L. Cominsky, et al. *Nature*, 279:47–49, 1979.

[1220] R. McCray. *Annu. Rev. Astron. Astr.*, 31:175–216, 1993.

[1221] D. B. McLaughlin. *Pub. Astron. Soc. Pac.*, 75:133–148, 1963.

[1222] C. A. Meakin and W. D. Arnett. *Astrophys. J.*, 667:448–475, 2007.

[1223] M. M. M. Meier, P. R. Heck, S. Amari, H. Baur, and R. Wieler. *Geochim. Cosmochim. Acta*, 76:147–160, 2012.

[1224] W. P. S. Meikle, R. J. Cumming, T. R. Geballe, et al. *Mon. Not. R. Astron. Soc.*, 281:263–280, 1996.

[1225] A. L. Melott, B. S. Lieberman, C. M. Laird, et al. *Int. J. Astrobiol.*, 3:55–61, 2004.

[1226] A. L. Melott and B. C. Thomas. *Astrobiol.*, 11:343–361, 2011.

[1227] X. Meng, X. Chen, and Z. Han. *Mon. Not. R. Astron. Soc.*, 395:2103–2116, 2009.

[1228] N. Mennekens, D. Vanbeveren, J. P. De Greve, and E. De Donder. *Astron. Astrophys.*, 515:A89 (11 p), 2010.

[1229] P. W. Merrill. *Astrophys. J.*, 116:21–26, 1952.

[1230] E.E. Meshkov. *Fluid Dyn.*, 4:101–104, 1969.

[1231] S. Messenger, L. P. Keller, F. J. Stadermann, R. M. Walker, and E. Zinner. *Science*, 300:105–108, 2003.

[1232] A. Messiah. *Quantum Mechanics*. Dover Publ., New York, 1999.

[1233] P. Mészáros. *Annu. Rev. Astron. Astr.*, 40:137–169, 2002.

[1234] B. D. Metzger, D. Giannios, T. A. Thompson, N. Bucciantini, and E. Quataert. *Mon. Not. R. Astron. Soc.*, 413:2031–2056, 2011.

[1235] B. D. Metzger, G. Martínez-Pinedo, S. Darbha, et al. *Mon. Not. R. Astron. Soc.*, 406:2650–2662, 2010.

[1236] B. D. Metzger, A. L. Piro, and E. Quataert. *Mon. Not. R. Astron. Soc.*, 390:781–797, 2008.

[1237] B. D. Metzger, A. L. Piro, and E. Quataert. *Mon. Not. R. Astron. Soc.*, 396:304–314, 2009.

[1238] B. S. Meyer, D. D. Clayton, and L.-S. The. *Astrophys. J. Lett.*, 540:L49–L52, 2000.

[1239] B. S. Meyer and E. Zinner. In *Meteorites and the Early Solar System II*, D. S. Lauretta and H. Y. McSween, Jr. (editor). University of Arizona Press, Tucson (Arizona), 2006, pp 69–108.

[1240] G. Meynet and M. Arnould. In *Nuclei in the Cosmos II*, F. Käppeler and K. Wisshak (editor). Inst. Phys. Publ., Bristol (Philadelphia), 1993, pp 503–508.

[1241] G. Meynet and M. Arnould. *Astron. Astrophys.*, 355:176–180, 2000.

[1242] D. Mihalas and B. W. Mihalas. *Foundations of Radiation Hydrodynamics*. Oxford Univ. Press, New York, 1984.

[1243] J. Mikołajewska. *Balt. Astron.*, 21:5–12, 2012.

[1244] P. A. Milne, A. L. Hungerford, C. L. Fryer, et al. *Astrophys. J.*, 613:1101–1119, 2004.

[1245] R. Minkowski. *Astrophys. J.*, 89:156–217, 1939.

[1246] R. Minkowski. *Pub. Astron. Soc. Pac.*, 52:206–207, 1940.

[1247] R. Minkowski. *Pub. Astron. Soc. Pac.*, 53:224–225, 1941.

[1248] J. Miralda-Escude, B. Paczyński, and P. Haensel. *Astrophys. J.*, 362:572–583, 1990.

[1249] C. W. Misner, K. S. Thorne, and J. A. Wheeler. *Gravitation*. W.H. Freeman and Co., San Francisco (California), 1973.

[1250] S. Mitton. *The Crab Nebula*. Scribner, New York, 1978.

[1251] M. Mocák, C. Meakin, M. Viallet, and W. D. Arnett. *ArXiv e-prints*, 2014.

[1252] R. Mochkovitch and M. Livio. *Astron. Astrophys.*, 236:378–384, 1990.

[1253] P. Mohr, R. Longland, and C. Iliadis. *Phys. Rev. C*, 90:065806 (11 pp), 2014.

[1254] P. Mohr and A. Matic. *Phys. Rev. C*, 87:035801 (9 pp), 2013.

[1255] R. Moll and S. E. Woosley. *Astrophys. J.*, 774:137 (15 pp), 2013.

[1256] J. J. Monaghan. *Comput. Phys. Rep.*, 3:71–124, 1985.

[1257] J. J. Monaghan. *Annu. Rev. Astron. Astr.*, 30:543–574, 1992.

[1258] J. J. Monaghan. *Rep. Prog. Phys.*, 68:1703–1759, 2005.

[1259] J. J. Monaghan. *Annu. Rev. Fluid Mech.*, 44:323–346, 2012.

[1260] J. J. Monaghan and R. A. Gingold. *J. Comput. Phys.*, 52:374–389, 1983.

[1261] J. J. Monaghan and J. C. Lattanzio. *Astron. Astrophys.*, 149:135–143, 1985.

[1262] F. Montes, T. C. Beers, J. Cowan, et al. *Astrophys. J.*, 671:1685–1695, 2007.

[1263] C. Morisset and D. Pequignot. *Astron. Astrophys.*, 312:135–159, 1996.

[1264] S. Mostefaoui and P. Hoppe. *Astrophys. J. Lett.*, 613:L149–L152, 2004.

[1265] Y. Motizuki, Y. Naka, and K. Takahashi. *Highlights Astron.*, 15:630–631, 2010.

[1266] N. Mowlavi, A. Jorissen, and M. Arnould. *Astron. Astrrophys.*, 311:803–816, 1996.

[1267] E. Mueller and W. Hillebrandt. *Astron. Astrophys.*, 103:358–366, 1981.

[1268] E. Müller. In *Supernovae*, S. A. Bludman, R. Mochkovitch, and J. Zinn-Justin (editor). North Holland Pub., Amsterdam, The Netherlands, 1994, pp 393–488.

[1269] E. Müller, B. Fryxell, and W. D. Arnett. *Astron. Astrophys.*, 251:505–514, 1991.

[1270] M. P. Muno. In *X-Ray Timing 2003: Rossi and Beyond*, P. Kaaret, F. K. Lamb, and J. H. Swank (editor). American Inst. Phys., Melville (New York), 2004, pp 239–244.

[1271] M. P. Muno, D. Chakrabarty, D. K. Galloway, and D. Psaltis. *Astrophys. J.*, 580:1048–1059, 2002.

[1272] M. P. Muno, F. Özel, and D. Chakrabarty. *Astrophys. J.*, 581:550–561, 2002.

[1273] E. R. Mustel. *Vista Ast. S.*, 2:1486–1491, 1956.

[1274] E. R. Mustel. *Sov. Astron.*, 7:772–775, 1964.

[1275] É. R. Mustel. *Sov. Astron.*, 15:1–8, 1971.

[1276] E. R. Mustel and A. A. Boyarchuk. *Astrophys. Space Sci.*, 6:183–204, 1970.

[1277] E. G. Myers. *Int. J. Mass Spectrom.*, 349:107–122, 2013.

[1278] M. Nagasawa, T. Nakamura, and S. M. Miyama. *Pub. Astron. Soc. Jpn.*, 40:691–708, 1988.

[1279] K. Nagashima, A. N. Krot, and H. Yurimoto. *Nature*, 428:921–924, 2004.

[1280] S. Nagataki. *Proc. Science, PoS(GRB 2012)* 096 (8 pp), 2006.

[1281] S. Nagataki. In *Deciphering the Ancient Universe with Gamma-Ray Bursts*, N. Kawai and S. Nagataki (editor). American Inst. Phys., Melville (New York), 2010, pp 77–80.

[1282] S. Nagataki, M.-a. Hashimoto, K. Sato, S. Yamada, and Y. S. Mochizuki. *Astrophys. J.*, 492:L45–L48, 1998.

[1283] R. Narayan, T. Piran, and A. Shemi. *Astrophys. J. Lett.*, 379:L17–L20, 1991.

[1284] G. Nelemans. In *The Astrophysics of Cataclysmic Variables and Related Objects*, J.-M. Hameury and J. P. Lasota (editor). Astron. Soc. Pac. Conf. Series, San Francisco (California), 2005, pp 27–40.

[1285] G. Nelemans, S. F. Portegies Zwart, F. Verbunt, and L. R. Yungelson. *Astron. Astrophys.*, 368:939–949, 2001.

[1286] G. Nelemans, L. R. Yungelson, and S. F. Portegies Zwart. *Astron. Astrophys.*, 375:890–898, 2001.

[1287] G. Nelemans, L. R. Yungelson, S. F. Portegies Zwart, and F. Verbunt. *Astron. Astrophys.*, 365:491–507, 2001.

[1288] R. J. Nemiroff. *Comment. Astrophys.*, 17:189–205, 1994.

[1289] B. Neuenswander and A. Melott. *Adv. Space Res.*, 55:2946–2949, 2015.

[1290] J. R. Newton, C. Iliadis, A. E. Champagne, et al. *Phys. Rev. C*, 75:045801 (4 pp), 2007.

[1291] A. N. Nguyen and S. Messenger. *Astrophys. J.*, 784:149 (15 pp), 2014.

[1292] A. N. Nguyen, L. R. Nittler, F. J. Stadermann, R. M. Stroud, and C. M. O'D. Alexander. *Astrophys. J.*, 719:166–189, 2010.

[1293] A. N. Nguyen, F. J. Stadermann, E. Zinner, et al. *Astrophys. J.*, 656:1223–1240, 2007.

[1294] A. N. Nguyen and E. Zinner. *Science*, 303:1496–1499, 2004.

[1295] G. K. Nicolussi, A. M. Davis, M. J. Pellin, et al. *Science*, 277:1281–1283, 1997.

[1296] G. K. Nicolussi, A. M. Davis, M. J. Pellin, et al. *Meteorit. Planet. Sci. Suppl.*, 32:A99, 1997.

[1297] G. K. Nicolussi, M. J. Pellin, R. S. Lewis, et al. *Geochim. Cosmochim. Acta*, 62:1093–1104, 1998.

[1298] G. K. Nicolussi, M. J. Pellin, R. S. Lewis, et al. *Meteorit. Planet. Sci. Suppl.*, 33:A116, 1998.

[1299] G. K. Nicolussi, M. J. Pellin, R. S. Lewis, et al. *Astrophys. J.*, 504:492–499, 1998.

[1300] J. C. Niemeyer. *Astrophys. J. Lett.*, 523:L57–L60, 1999.

[1301] L. R. Nittler. In *Astrophysical Implications of the Laboratory Study of Presolar Materials*, T. J. Bernatowicz and E. Zinner (editor). American Inst. Phys., Woodbury (New York), 1997, pp 59–82.

[1302] L. R. Nittler, S. Amari, E. Zinner, S. E. Woosley, and R. S. Lewis. *Astrophys. J. Lett.*, 462:L31–L34, 1996.

[1303] L. R. Nittler and P. Hoppe. *Astrophys. J. Lett.*, 631:L89–L92, 2005.

[1304] L. R. Nittler, P. Hoppe, C. M. O'D. Alexander, et al. *Astrophys. J. Lett.*, 453:L25–L28, 1995.

[1305] L. R. Nittler and C. M. O'D. Alexander. 30th *Lunar and Planetary Science Conf.*, Abstract #2041, 1999.

[1306] L. R. Nittler, C. M. O'D. Alexander, R. Gallino, et al. *Astrophys. J.*, 682:1450–1478, 2008.

[1307] L. R. Nittler, C. M. O'D. Alexander, X. Gao, R. M. Walker, and E. Zinner. *Astrophys. J.*, 483:475–495, 1997.

[1308] L. R. Nittler, C. M. O'D. Alexander, X. Gao, R. M. Walker, and E. K. Zinner. *Nature*, 370:443–446, 1994.

[1309] L. R. Nittler, C. M. O'D. Alexander, and A. N. Nguyen. *Meteorit. Planet. Sci. Suppl.*, 41:Abstract #5316, 2006.

[1310] L. R. Nittler, C. M. O'D. Alexander, F. J. Stadermann, and E. Zinner. 36th *Lunar and Planetary Science Conf.*, Abstract #2200, 2005.

[1311] L. R. Nittler, C. M. O'D. Alexander, F. J. Stadermann, and E. K. Zinner. *Meteorit. Planet. Sci. Suppl.*, 40:Abstract #5208, 2005.

[1312] L. R. Nittler, R. M. Walker, E. Zinner, P. Hoppe, and R. S. Lewis. 24th *Lunar and Planetary Science Conf.*, Abstract #1087, 1993.

[1313] L. Nobili and R. Turolla. *Astrophys. J.*, 333:248–255, 1988.

[1314] L. Nobili, R. Turolla, and I. Lapidus. *Astrophys. J.*, 433:276–286, 1994.

[1315] C. Noël, Y. Busegnies, M. V. Papalexandris, V. Deledicque, and A. El Messoudi. *Astron. Astrophys*, 470:653–659, 2007.

[1316] I. Nofar, G. Shaviv, and S. Starrfield. *Astrophys. J.*, 369:440–450, 1991.

[1317] K. M. Nollett, M. Busso, and G. J. Wasserburg. *Astrophys. J.*, 582:1036–1058, 2003.

[1318] K. Nomoto. *Astrophys. J.*, 257:780–792, 1982.

[1319] K. Nomoto. *Astrophys. J.*, 253:798–810, 1982.

[1320] K. Nomoto and M. Hashimoto. *Phys. Rep.*, 163:13–36, 1988.

[1321] K. Nomoto and I. Iben, Jr. *Astrophys. J.*, 297:531–537, 1985.

[1322] K. Nomoto, K. Iwamoto, N. Nakasato, et al. In *Thermonuclear Supernovae*, P. Ruiz-Lapuente, R. Canal, and J. Isern (editor). Kluwer Acad. Publ., Dordrecht, The Netherlands, 1997, pp 349–378.

[1323] K. Nomoto and Y. Kondo. *Astrophys. J. Lett.*, 367:L19–L22, 1991.

[1324] K. Nomoto, T. Moriya, M. Tanaka, et al. *Proc. Science, PoS(NIC-XI)* 030 (10 pp), 2010.

[1325] K. Nomoto, K. Nariai, and D. Sugimoto. *Pub. Astron. Soc. Jpn.*, 31:287–298, 1979.

[1326] K. Nomoto, H. Saio, M. Kato, and I. Hachisu. *Astrophys. J.*, 663:1269–1276, 2007.

[1327] K. Nomoto and D. Sugimoto. *Pub. Astron. Soc. Jpn.*, 29:765–780, 1977.

[1328] K. Nomoto, D. Sugimoto, and S. Neo. *Astrophys. Space Sci.*, 39:L37–L42, 1976.

[1329] K. Nomoto, T. Suzuki, T. Shigeyama, et al. *Nature*, 364:507–509, 1993.

[1330] K. Nomoto, F.-K. Thielemann, and J. C. Wheeler. *Astrophys. J. Lett.*, 279:L23–L26 (Erratum: Astrophys. J. Lett. 283:L25), 1984.

[1331] K. Nomoto, F.-K. Thielemann, and K. Yokoi. *Astrophys. J.*, 286:644–658, 1984.

[1332] K. Nomoto, N. Tominaga, M. Tanaka, K. Maeda, and H. Umeda. In *Massive Stars as Cosmic Engines*, F. Bresolin, P. A. Crowther, and J. Puls (editor). Cambridge Univ. Press, Cambridge (UK), 2008, pp 463–470.

[1333] K. Nomoto, H. Yamaoka, O. R. Pols, et al. *Nature*, 371:227–229, 1994.

[1334] M. Notani, S. Kubono, T. Teranishi, et al. *Nucl. Phys. A*, 746:113c–117c, 2004.

[1335] T. Nozakura, S. Ikeuchi, and M. Y. Fujimoto. *Astrophys. J.*, 286:221–231, 1984.

[1336] P. Nugent, M. Sullivan, R. Ellis, et al. *Astrophys. J.*, 645:841–850, 2006.

[1337] P. E. Nugent, M. Sullivan, S. B. Cenko, et al. *Nature*, 480:344–347, 2011.

[1338] W. L. Oberkampf and C.J. Roy. *Verification and Validation in Scientific Computing*. Cambridge Univ. Press, Cambridge, UK, 2010.

[1339] A. Obertelli, T. Baugher, D. Bazin, et al. *Phys. Lett. B*, 701:417–421, 2011.

[1340] C. M. O'D. Alexander, P. Swan, and R. M. Walker. *Nature*, 348:715–717, 1990.

[1341] T. Oda, M. Hino, K. Muto, M. Takahara, and K. Sato. *Atom. Data Nucl. Data*, 56:231–403, 1994.

[1342] R. Oechslin, H.-T. Janka, and A. Marek. *Astron. Astrophys.*, 467:395–409, 2007.

[1343] E. O. Ofek, M. Muno, R. Quimby, et al. *Astrophys. J.*, 681:1464–1469, 2008.

[1344] Y. Oganessian. *J. Phys. G Nucl. Partic.*, 34:R165–R242, 2007.

[1345] K. Ogata, M. Kan, and M. Kamimura. *Prog. Theor. Phys.*, 122:1055–1064, 2009.

[1346] H. Ögelman, K. Beuermann, and J. Krautter. *Astrophys. J. Lett.*, 287:L31–L34, 1984.

[1347] J. B. Oke and J. E. Gunn. *Astrophys. J.*, 266:713–717, 1983.

[1348] F. Olivares E., M. Hamuy, G. Pignata, et al. *Astrophys. J.*, 715:833–853, 2010.

[1349] K. Oohara and T. Nakamura. *Prog. Theor. Phys.*, 82:535–554, 1989.

[1350] J. R. Oppenheimer and G. M. Volkoff. *Phys. Rev.*, 55:374–381, 1939.

[1351] E. S. Oran and A. M. Khokhlov. *Phil. Trans. R. Soc. Lond.*, 357:3539–3551, 1999.

[1352] A. M. Orishich, I. F. Shaikhislamov, and V. G. Posukh. *Laser Part. Beams*, 14:63–70, 1996.

[1353] S. Orlando and J. J. Drake. *Mon. Not. R. Astron. Soc.*, 419:2329–2337, 2012.

[1354] S. Osher and J. A. Sethian. *J. Comput. Phys.*, 79:12–49, 1988.

[1355] D. E. Osterbrock and G. J. Ferland. *Astrophysics of Gaseous Nebulae and Active Galactic Nuclei*. 2nd Ed., University Science Books, Mill Valley (California), 2006.

[1356] U. Ott and F. Begemann. *Astrophys. J. Lett.*, 353:L57–L60, 1990.

[1357] U. Ott, A. Besmehn, K. Farouqi, et al. *Pub. Astron. Soc. Aust.*, 29:90–97, 2012.

[1358] U. Ott, Q.-Z. Yin, and C.-T. Lee. *Mem. Soc. Astron. Ital.*, 77:891–896, 2006.

[1359] F. Özel, A. Gould, and T. Güver. *Astrophys.J.*, 748:5 (5pp), 2012.

[1360] B. Paczyński. *Acta Astronom.*, 15:197–210, 1965.

[1361] B. Paczyński. *Acta Astronom.*, 20:47–58, 1970.

[1362] B. Paczyński. *Annu. Rev. Astron. Astr.*, 9:183–208, 1971.

[1363] B. Paczyński. *Astrophys. Lett.*, 11:53–55, 1972.

[1364] B. Paczyński. *Astrophys. J.*, 264:282–295, 1983.

[1365] B. Paczyński. *Astrophys. J.*, 267:315–321, 1983.

[1366] B. Paczyński. *Astrophys. J. Lett.*, 308:L43–L46, 1986.

[1367] B. Paczyński. *Acta Astronom.*, 41:257–267, 1991.

[1368] B. Paczyński. *Astrophys. J. Lett.*, 494:L45–L48, 1998.

[1369] B. Paczyński and M. Proszynski. *Astrophys. J.*, 302:519–529, 1986.

[1370] B. Paczyński and A. N. Zytkow. *Astrophys. J.*, 222:604–611, 1978.

[1371] D. Page and A. Cumming. *Astrophys. J. Lett.*, 635:L157–L160, 2005.

[1372] G. Pagliara, M. Herzog, and F. K. Röpke. *Phys. Rev. D*, 87:103007 (8 pp), 2013.

[1373] R. Pakmor, S. Hachinger, F. K. Röpke, and W. Hillebrandt. *Astron. Astrophys.*, 528:A117 (9 pp), 2011.

[1374] R. Pakmor, M. Kromer, F. K. Röpke, et al. *Nature*, 463:61–64, 2010.

[1375] R. Pakmor, M. Kromer, S. Taubenberger, et al. *Astrophys. J. Lett.*, 747:L10 (5 pp), 2012.

[1376] R. Pakmor, M. Kromer, S. Taubenberger, and V. Springel. *Astrophys. J. Lett.*, 770:L8 (7 pp), 2013.

[1377] R. Pakmor, F. K. Röpke, A. Weiss, and W. Hillebrandt. *Astron. Astrophys.*, 489:943–951, 2008.

[1378] A. Palacios, M. Arnould, and G. Meynet. *Astron. Astrophys.*, 443:243–250, 2005.

[1379] N. Panagia and K. W. Weiler. *Astron. Astrophys.*, 82:389–391, 1980.

[1380] T. Pankey, Jr. *Possible Thermonuclear Activities in Natural Terrestrial Minerals.* PhD thesis, Howard Univ., Washington DC, 1962.

[1381] I. V. Panov and H.-T. Janka. *Astron. Astrophys.*, 494:829–844, 2009.

[1382] J. C. B. Papaloizou, J. E. Pringle, and J. MacDonald. *Mon. Not. R. Astron. Soc.*, 198:215–220, 1982.

[1383] F. Paresce, M. Livio, W. Hack, and K. Korista. *Astron. Astrophys.*, 299:823–834, 1995.

[1384] S. Parete-Koon, W. R. Hix, M. S. Smith, et al. *Astrophys. J.*, 598:1239–1245, 2003.

[1385] A. Parikh, J. José, C. Iliadis, F. Moreno, and T. Rauscher. *Phys. Rev. C*, 79:045802 (12 pp), 2009.

[1386] A. Parikh, J. José, F. Moreno, and C. Iliadis. *Astrophys. J. Suppl. S.*, 178:110–136, 2008.

[1387] A. Parikh, J. José, G. Sala, and C. Iliadis. *Prog. Part. Nucl. Phys.*, 69:225–253, 2013.

[1388] A. Parikh, J. José, I. R. Seitenzahl, and F. K. Röpke. *Astron. Astrophys.*, 557:A3 (11 pp), 2013.

[1389] A. Parikh, K. Wimmer, T. Faestermann, et al. *Phys. Rev. C*, 83:045806 (8 pp), 2011.

[1390] S. H. Park, S. Kubono, K. I. Hahn, et al. *Phys. Rev. C*, 59:1182–1184, 1999.

[1391] M. Parthasarathy, D. Branch, E. Baron, and D. J. Jeffery. *B. Astron. Soc. India*, 34:385–391, 2006.

[1392] M. Parthasarathy, D. Branch, D. J. Jeffery, and E. Baron. *New Astron. Rev.*, 51:524–538, 2007.

[1393] D. J. Patnaude, C. Badenes, S. Park, and J. M. Laming. *Astrophys. J.*, 756:6 (8 pp), 2012.

[1394] J. Patterson. *Astrophys. J. Suppl. S.*, 54:443–493, 1984.

[1395] A. B. C. Patzer, A. Gauger, and E. Sedlmayr. *Astron. Astrophys.*, 337:847–858, 1998.

[1396] B. Paxton, L. Bildsten, A. Dotter, et al. *Astrophys. J. Suppl. S.*, 192:3 (35 pp), 2011.

[1397] B. Paxton, M. Cantiello, P. Arras, et al. *Astrophys. J. Suppl. S.*, 208:4 (42 pp), 2013.

[1398] C. Payne-Gaposchkin. *Astrophys. J.*, 83:173–176, 1936.

[1399] C. Payne-Gaposchkin. *Astrophys. J.*, 83:245–251, 1936.

[1400] C. H. Payne-Gaposchkin. *The Galactic Novae*. North-Holland Pub., Amsterdam, The Netherlands, 1957.

[1401] M. J. Pellin, W. F. Calaway, A. M. Davis, et al. 31st *Lunar and Planetary Science Conf.*, Abstract #1917, 2000.

[1402] M. J. Pellin, M. R. Savina, E. Tripa, et al. *Meteorit. Planet. Sci. Suppl.*, 37:Abstract #5245, 2002.

[1403] M. J. Pellin, R. N. Thompson, Z. Ma, et al. 25th *Lunar and Planetary Science Conf.*, Abstract #1063, 1994.

[1404] M. J. Pellin, C. E. Young, W. F. Calaway, et al. *Phil. Trans. R. Soc. Lond.*, 333:133–143, 1990.

[1405] F. Peng, E. F. Brown, and J. W. Truran. *Astrophys. J.*, 654:1022–1035, 2007.

[1406] F. Peng and C. D. Ott. *Astrophys. J.*, 725:309–312, 2010.

[1407] A. Perego, S. Rosswog, R. M. Cabezón, et al. *Mon. Not. R. Astron. Soc.*, 443:3134–3156, 2014.

[1408] H. B. Perets, A. Gal-Yam, P. A. Mazzali, et al. *Nature*, 465:322–325, 2010.

[1409] M. A. Pérez-García, F. Daigne, and J. Silk. *EAS Publications*, 61:331–335, 2013.

[1410] S. Perlmutter, G. Aldering, G. Goldhaber, et al. *Astrophys. J.*, 517:565–586, 1999.

[1411] S. Perlmutter, S. Gabi, G. Goldhaber, et al. *Astrophys. J.*, 483:565–581, 1997.

[1412] S. Perlmutter, C. R. Pennypacker, G. Goldhaber, et al. *Astrophys. J. Lett.*, 440:L41–L44, 1995.

[1413] M. M. Phillips. *Astrophys. J. Lett.*, 413:L105–L108, 1993.

[1414] M. M. Phillips, K. Krisciunas, N. B. Suntzeff, et al. *Astron. J.*, 131:2615–2627, 2006.

[1415] M. M. Phillips, W. Li, J. A. Frieman, et al. *Pub. Astron. Soc. Pac.*, 119:360–387, 2007.

[1416] W. H. Pickering. *Observatory*, 18:436–436, 1895.

[1417] L. Piersanti, S. Cassisi, I. Iben, Jr., and A. Tornambé. *Astrophys. J. Lett.*, 521:L59–L62, 1999.

[1418] W. Pietsch and F. Haberl. *Astron. Astrophys.*, 430:L45–L48, 2005.

[1419] M. Pignatari, R. Gallino, M. Heil, et al. *Astrophys. J.*, 710:1557–1577, 2010.

Bibliography 419

[1420] M. Pignatari, M. Wiescher, F. X. Timmes, et al. *Astrophys. J. Lett.*, 767:L22 (6 pp), 2013.

[1421] M. Pignatari, E. Zinner, M. G. Bertolli, et al. *Astrophys. J. Lett.*, 771:L7 (5 pp), 2013.

[1422] P. A. Pinto, R. G. Eastman, and T. Rogers. *Astrophys. J.*, 551:231–243, 2001.

[1423] P. A. Pinto and S. E. Woosley. *Nature*, 333:534–537, 1988.

[1424] T. Piran. *Rev. Mod. Phys.*, 76:1143–1210, 2004.

[1425] T. Piran, O. Bromberg, E. Nakar, and R. Sari. *Phil. Trans. R. Soc. Lond.*, 371:20273 (10 pp), 2013.

[1426] T. Piran and R. Jimenez. *Phys. Rev. Lett.*, 113:231102 (6 pp), 2014.

[1427] A. L. Piro. *Astrophys. J.*, 759:83 (8 pp), 2012.

[1428] A. L. Piro and L. Bildsten. *Astrophys. J.*, 629:438–450, 2005.

[1429] A. L. Piro, T. A. Thompson, and C. S. Kochanek. *Mon. Not. R. Astron. Soc.*, 438:3456–3464, 2014.

[1430] M. Plavec and P. Kratochvil. *B. Astron. I. Czech.*, 15:165–170, 1964.

[1431] T. Plewa. *Astrophys. J.*, 657:942–960, 2007.

[1432] T. Plewa, A. C. Calder, and D. Q. Lamb. *Astrophys. J. Lett.*, 612:L37–L40, 2004.

[1433] S. Plüschke, R. Diehl, V. Schönfelder, et al. In *4th INTEGRAL Workshop on Exploring the Gamma-Ray Universe*, A. Gimenez, V. Reglero, and C. Winkler (editor). ESA SP-459, Noordwijk, The Netherlands, 2001, pp 55–58.

[1434] F. A. Podosek and R. S. Lewis. *Earth Planet. Sc. Lett.*, 15:101–109, 1972.

[1435] F. A. Podosek, C. A. Prombo, S. Amari, and R. S. Lewis. *Astrophys. J.*, 605:960–965, 2004.

[1436] P. Podsiadlowski. *Pub. Astron. Soc. Pac.*, 104:717–729, 1992.

[1437] P. Podsiadlowski. *New Astron. Rev.*, 54:39–44, 2010.

[1438] M. Politano, S. Starrfield, J. W. Truran, A. Weiss, and W. M. Sparks. *Astrophys. J.*, 448:807–821, 1995.

[1439] S. B. Pope. *Turbulent Flows*. Cambridge Univ. Press, Cambridge, 2000.

[1440] D. M. Popper. *Pub. Astron. Soc. Pac.*, 49:283–289, 1937.

[1441] D. Potter. *Computational Physics*. John Wiley and Sons, New York, 1973.

[1442] D. Poznanski. *Mon. Not. R. Astron. Soc.*, 436:3224–3230, 2013.

[1443] L. Pradtl. *Z. Angew. Math. Mech.*, 5:136–139, 1925.

[1444] N. Prantzos and R. Diehl. *Phys. Rep.*, 267:1–69, 1996.

[1445] N. Prantzos, C. Doom, C. De Loore, and M. Arnould. *Astrophys. J.*, 304:695–712, 1986.

[1446] N. Prantzos, M.-A. Hashimoto, and K. Nomoto. *Astron. Astrophys.*, 234:211–229, 1990.

[1447] N. Prantzos, M.-A. Hashimoto, M. Rayet, and M. Arnould. *Astron. Astrophys.*, 238:455–461, 1990.

[1448] W. H. Press, S. A. Teukolsky, W. T. Vetterling, and B. P. Flannery. *Numerical Recipes: the Art of Scientific Computing.* 3rd Ed., Cambridge Univ. Press, Cambridge, UK, 2007.

[1449] D. Prialnik and A. Kovetz. *Astrophys. J.*, 281:367–374, 1984.

[1450] D. Prialnik and A. Kovetz. *Astrophys. J.*, 445:789–810, 1995.

[1451] D. Prialnik, M. Livio, G. Shaviv, and A. Kovetz. *Astrophys. J.*, 257:312–317, 1982.

[1452] D. Prialnik and M. M. Shara. *Astrophys. J.*, 311:172–182, 1986.

[1453] D. Prialnik, M. M. Shara, and G. Shaviv. *Astron. Astrophys.*, 62:339–348, 1978.

[1454] D. J. Price and J. J. Monaghan. *Mon. Not. R. Astron. Soc.*, 374:1347–1358, 2007.

[1455] C. J. Pritchet, D. A. Howell, and M. Sullivan. *Astrophys. J. Lett.*, 683:L25–L28, 2008.

[1456] C. A. Prombo, F. A. Podosek, S. Amari, and R. S. Lewis. *Astrophys. J.*, 410:393–399, 1993.

[1457] J. Pruet, R. D. Hoffman, S. E. Woosley, H.-T. Janka, and R. Buras. *Astrophys. J.*, 644:1028–1039, 2006.

[1458] J. Pruet, R. Surman, and G. C. McLaughlin. *Astrophys. J. Lett.*, 602:L101–L104, 2004.

[1459] J. Pruet, T. A. Thompson, and R. D. Hoffman. *Astrophys. J.*, 606:1006–1018, 2004.

[1460] D. Psaltis. In *Compact Stellar X-Ray Sources*, W. H. G. Lewin and M. van der Klis (editor). Cambridge Univ. Press, Cambridge, UK, 2006, pp 1–38.

[1461] Yu. P. Pskovskii. *Sov. Astron.*, 12:750–756, 1969.

[1462] Y.-Z. Qian and G. J. Wasserburg. *Phys. Rep.*, 442:237–268, 2007.

[1463] Y.-Z. Qian and S. E. Woosley. *Astrophys. J.*, 471:331–351, 1996.

[1464] T. Quinn and B. Paczyński. *Astrophys. J.*, 289:634–643, 1985.

[1465] J. L. Racusin, S. V. Karpov, M. Sokolowski, et al. *Nature*, 455:183–188, 2008.

[1466] C. M. Raiteri, R. Gallino, M. Busso, D. Neuberger, and F. Käppeler. *Astrophys. J.*, 419:207–223, 1993.

[1467] G. Rakavy and G. Shaviv. *Astrophys. J.*, 148:803–816, 1967.

[1468] W. J. M. Rankine. *Phil. Trans. R. Soc. Lond.*, 160:277–288, 1870.

[1469] W. Rapp, J. Görres, M. Wiescher, H. Schatz, and F. Käppeler. *Astrophys. J.*, 653:474–489, 2006.

[1470] F. A. Rasio and S. L. Shapiro. *Astrophys. J.*, 432:242–261, 1994.

[1471] C. Raskin, E. Scannapieco, J. Rhoads, and M. Della Valle. *Astrophys. J.*, 707:74–78, 2009.

[1472] C. Raskin, E. Scannapieco, G. Rockefeller, et al. *Astrophys. J.*, 724:111–125, 2010.

[1473] T. Rauch, V. Suleimanov, and K. Werner. *Astron. Astrophys.*, 490:1127–1134, 2008.

[1474] T. Rauscher, N. Dauphas, I. Dillmann, et al. *Rep. Prog. Phys.*, 76:066201 (38 pp), 2013.

[1475] T. Rauscher, A. Heger, R. D. Hoffman, and S. E. Woosley. *Astrophys. J.*, 576:323–348, 2002.

[1476] T. Rauscher and F.-K. Thielemann. *Atom. Data Nucl. Data*, 75:1–351, 2000.

[1477] T. Rauscher, F.-K. Thielemann, and K.-L. Kratz. *Phys. Rev. C*, 56:1613–1625, 1997.

[1478] M. Rayet, M. Arnould, M.-A. Hashimoto, N. Prantzos, and K. Nomoto. *Astron. Astrophys.*, 298:517–527, 1995.

[1479] M. Rayet, M. Arnould, and N. Prantzos. *Astron. Astrophys.*, 227:271–281, 1990.

[1480] R. Reifarth, C. Lederer, and F. Käppeler. *J. Phys. G Nucl. Partic.*, 41:053101 (42 pp), 2014.

[1481] E. C. Reifenstein, W. D. Brundage, and D. H. Staelin. *Phys. Rev. Lett.*, 22:311, 1969.

[1482] M. Reinecke, W. Hillebrandt, and J. C. Niemeyer. *Astron. Astrophys.*, 391:1167–1172, 2002.

[1483] B. A. Remington, D. Arnett, R. P. Drake, and H. Takabe. *Science*, 284:1488–1493, 1999.

[1484] B. A. Remington, R. P. Drake, and D. D. Ryutov. *Rev. Mod. Phys.*, 78:755–807, 2006.

[1485] B. A. Remington, R. P. Drake, H. Takabe, and D. Arnett. *Phys. Plasmas*, 7:1641–1652, 2000.

[1486] B. A. Remington and H.-S. Park. In *2010 NASA Laboratory Astrophysics Workshop*, D. R. Schultz (editor). NASA, 2011, pp I23 (1–17).

[1487] M. Renaud, J. Vink, A. Decourchelle, et al. *Astrophys. J. Lett.*, 647:L41–L44, 2006.

[1488] A. Renzini. In *Supernovae and Supernova Remnants*, R. McCray and Z. Wang (editor). Cambridge Univ. Press, Cambridge, UK, 1996, pp 77–86.

[1489] A. Rest, D. Scolnic, R. J. Foley, et al. *Astrophys. J.*, 795:44 (34 pp), 2014.

[1490] A. Rest, B. Sinnott, and D. L. Welch. *Pub. Astron. Soc. Aust.*, 29:466–481, 2012.

[1491] A. Rest, D. L. Welch, N. B. Suntzeff, et al. *Astrophys. J. Lett.*, 681:L81–L84, 2008.

[1492] M. Revnivtsev, E. Churazov, M. Gilfanov, and R. Sunyaev. *Astron. Astrophys.*, 372:138–144, 2001.

[1493] S. P. Reynolds, K. J. Borkowski, U. Hwang, et al. *Astrophys. J. Lett.*, 668:L135–L138, 2007.

[1494] L. Rezzolla, B. Giacomazzo, L. Baiotti, et al. *Astrophys. J. Lett.*, 732:L6 (6 pp), 2011.

[1495] R. O. Rhodes. *The Making of the Atomic Bomb*. Simon & Schuster, New York, 1986.

[1496] G. B. Riccioli. *Almagestvm novvm astronomiam veterem novamqve complectens observationibvs aliorvm*. Bologna, Italy, 1651.

[1497] S. Richter, U. Ott, and F. Begemann. In *Nuclei in the Cosmos II*, F. Käppeler and K. Wisshak (editor). Inst. Phys. Publ., Bristol (Philadelphia), 1993, pp 127–132.

[1498] S. Richter, U. Ott, and F. Begemann. *Nature*, 391:261–263, 1998.

[1499] R. D. Richtmyer. *Commun. Pur. Appl. Math.*, 13:297–319, 1960.

[1500] R. D. Richtmyer and K. W. Morton. *Difference Methods for Initial-Value Problems*. 2nd Ed., Krieger Publishing Co., Malabar (Florida), 1994.

[1501] A. G. Riess, A. V. Filippenko, P. Challis, et al. *Astron. J.*, 116:1009–1038, 1998.

[1502] A. G. Riess, A. V. Filippenko, D. C. Leonard, et al. *Astron. J.*, 114:722–729, 1997.

[1503] K. Riles. *Prog. Part. Nucl. Phys.*, 68:1–54, 2013.

[1504] C. Ritossa, E. Garcia-Berro, and I. Iben, Jr. *Astrophys. J.*, 460:489–505, 1996.

[1505] P. J. Roache. *Fundamentals of Verification and Validation*. Hermosa Publ., Albuquerque (New Mexico), 2009.

[1506] L. F. Roberts, D. Kasen, W. H. Lee, and E. Ramirez-Ruiz. *Astrophys. J. Lett.*, 736:L21 (5 pp), 2011.

[1507] L. F. Roberts, S. E. Woosley, and R. D. Hoffman. *Astrophys. J.*, 722:954–967, 2010.

[1508] H. P. Robertson. *Astrophys. J.*, 82:284–301, 1935.

[1509] G. Rockefeller, C. L. Fryer, P. Young, et al. *Proc. Science, PoS(NIC-X)* 119 (5 pp), 2008.

[1510] I. U. Roederer, J. J. Cowan, A. I. Karakas, et al. *Astrophys. J.*, 724:975–993, 2010.

[1511] I. U. Roederer, K.-L. Kratz, A. Frebel, et al. *Astrophys. J.*, 698:1963–1980, 2009.

[1512] I. U. Roederer, J. E. Lawler, J. J. Cowan, et al. *Astrophys. J. Lett.*, 747:L8 (5 pp), 2012.

[1513] C. E. Rolfs and W. S. Rodney. *Cauldrons in the Cosmos*. Univ. Chicago Press, Chicago (Illinois), 1988.

[1514] D. Romano and F. Matteucci. In *Classical Nova Explosions*, M. Hernanz and J. José (editor). American Inst. Phys., Melville (New York), 2002, pp 144–149.

[1515] D. Romano, F. Matteucci, P. Molaro, and P. Bonifacio. *Astron. Astrophys.*, 352:117–128, 1999.

[1516] R. T. Rood, C. L. Sarazin, E. J. Zeller, and B. C. Parker. *Nature*, 282:701–703, 1979.

[1517] F. K. Röpke. *Proc. Science, PoS(SUPERNOVA)* 024 (25 pp), 2008.

[1518] F. K. Röpke and R. Bruckschen. *New J. Phys.*, 10:125009 (20 pp), 2008.

[1519] F. K. Röpke and W. Hillebrandt. *Astron. Astrophys.*, 431:635–645, 2005.

[1520] F. K. Röpke and W. Hillebrandt. *Astron. Astrophys.*, 429:L29–L32, 2005.

[1521] F. K. Röpke, W. Hillebrandt, J. C. Niemeyer, and S. E. Woosley. *Astron. Astrophys.*, 448:1–14, 2006.

[1522] F. K. Röpke, W. Hillebrandt, W. Schmidt, et al. *Astrophys. J.*, 668:1132–1139, 2007.

[1523] F. K. Röpke, M. Kromer, I. R. Seitenzahl, et al. *Astrophys. J. Lett.*, 750:L19 (7 pp), 2012.

[1524] F. K. Röpke and J. C. Niemeyer. *Astron. Astrophys.*, 464:683–686, 2007.

[1525] F. K. Röpke, J. C. Niemeyer, and W. Hillebrandt. *Astrophys. J.*, 588:952–961, 2003.

[1526] F. K. Röpke, S. E. Woosley, and W. Hillebrandt. *Astrophys. J.*, 660:1344–1356, 2007.

[1527] W. K. Rose. *Astrophys. J.*, 152:245–253, 1968.

[1528] M. N. Rosenbluth, M. Ruderman, F. Dyson, et al. *Astrophys. J.*, 184:907–910, 1973.

[1529] R. Rosner, A. Alexakis, Y.-N. Young, J. W. Truran, and W. Hillebrandt. *Astrophys. J. Lett.*, 562:L177–L179, 2001.

[1530] S. Rosswog. *Astrophys. J.*, 634:1202–1213, 2005.

[1531] S. Rosswog. *New Astron. Rev.*, 53:78–104, 2009.

[1532] S. Rosswog. *Proc. Science, PoS(NIC-XI)* 032 (22 pp), 2011.

[1533] S. Rosswog. *ArXiv e-prints*, 2014.

[1534] S. Rosswog, D. Kasen, J. Guillochon, and E. Ramirez-Ruiz. *Astrophys. J. Lett.*, 705:L128–L132, 2009.

[1535] S. Rosswog, O. Korobkin, A. Arcones, F.-K. Thielemann, and T. Piran. *Mon. Not. R. Astron. Soc.*, 439:744–756, 2014.

[1536] S. Rosswog, M. Liebendörfer, F.-K. Thielemann, et al. *Astron. Astrophys.*, 341:499–526, 1999.

[1537] S. Rosswog, R. Speith, and G. A. Wynn. *Mon. Not. R. Astron. Soc.*, 351:1121–1133, 2004.

[1538] C. Rowland, C. Iliadis, A. E. Champagne, et al. *Astrophys. J. Lett.*, 615:L37–L40, 2004.

[1539] M. A. Ruderman. *Science*, 184:1079–1081, 1974.

[1540] A. J. Ruiter, K. Belczynski, and C. Fryer. *Astrophys. J.*, 699:2026–2036, 2009.

[1541] A. J. Ruiter, K. Belczynski, S. A. Sim, I. R. Seitenzahl, and D. Kwiatkowski. *Mon. Not. R. Astron. Soc.*, 440:L101–L105, 2014.

[1542] A. J. Ruiter, S. A. Sim, R. Pakmor, et al. *Mon. Not. R. Astron. Soc.*, 429:1425–1436, 2013.

[1543] P. Ruiz-Lapuente. *Astrophys. J.*, 612:357–363, 2004.

[1544] P. Ruiz-Lapuente. *New Astron. Rev.*, 62:15–31, 2014.

[1545] P. Ruiz-Lapuente, F. Comeron, J. Méndez, et al. *Nature*, 431:1069–1072, 2004.

[1546] D. Russell and W. H. Tucker. *Nature*, 229:553–554, 1971.

[1547] H. N. Russell. *Pub. Astron. Soc. Pac.*, 31:205–211, 1919.

[1548] S. S. Russell, J. W. Arden, and C. T. Pillinger. *Meteorit. Planet. Sci.*, 31:343–355, 1996.

[1549] D. D. Sabu and O. K. Manuel. *Nature*, 262:28–32, 1976.

[1550] D. K. Sahu, G. C. Anupama, S. Srividya, and S. Muneer. *Mon. Not. R. Astron. Soc.*, 372:1315–1324, 2006.

[1551] H. Saio and C. S. Jeffery. *Mon. Not. R. Astron. Soc.*, 333:121–132, 2002.

[1552] H. Saio and K. Nomoto. *Astrophys. J.*, 500:388–397, 1998.

[1553] P. Saizar, S. Starrfield, G. J. Ferland, et al. *Astrophys. J.*, 398:651–664, 1992.

[1554] G. Sala and M. Hernanz. *Astron. Astrophys.*, 439:1061–1073, 2005.

[1555] M. Salaris and S. Cassisi. *Evolution of Stars and Stellar Populations.* John Wiley, Chichester, UK, 2005.

[1556] A. L. Sallaska, C. Wrede, A. García, et al. *Phys. Rev. Lett.*, 105:152501 (4 pp), 2010.

[1557] E. E. Salpeter. *Astrophys. J.*, 115:326–328, 1952.

[1558] E. E. Salpeter. *Aust. J. Phys.*, 7:373–388, 1954.

[1559] E. E. Salpeter. *Mem. Soc. R. Sci. Liege*, 1:116–121, 1954.

[1560] P. J. C. Salter, M. Aliotta, T. Davinson, et al. *Phys. Rev. Lett.*, 108:242701 (5 pp), 2012.

[1561] D. J. Sand, M. L. Graham, C. Bildfell, et al. *Astrophys. J.*, 746:163 (22 pp), 2012.

[1562] R. F. Sanford. *Astrophys. J.*, 109:81–91, 1949.

[1563] A. Sarangi and I. Cherchneff. *Astrophys. J.*, 776:107 (19 pp), 2013.

[1564] A. Sarangi and I. Cherchneff. *Astron. Astrophys.*, 575:A95 (20 pp), 2015.

[1565] K. Sato. *Prog. Theor. Phys.*, 51:726–744, 1974.

[1566] K. Sato. *Prog. Theor. Phys.*, 53:595–597, 1975.

[1567] M. R. Savina, A. M. Davis, C. E. Tripa, et al. *Geochim. Cosmochim. Acta*, 67:3201–3214, 2003.

[1568] M. R. Savina, A. M. Davis, C. E. Tripa, et al. *Science*, 303:649–652, 2004.

[1569] M. R. Savina, M. J. Pellin, C. E. Tripa, et al. *Geochim. Cosmochim. Acta*, 67:3215–3225, 2003.

[1570] R. A. Scalzo, G. Aldering, P. Antilogus, et al. *Mon. Not. R. Astron. Soc.*, 440:1498–1518, 2014.

[1571] R. A. Scalzo, G. Aldering, P. Antilogus, et al. *Astrophys. J.*, 713:1073–1094, 2010.

[1572] E. Scannapieco and L. Bildsten. *Astrophys. J. Lett.*, 629:L85–L88, 2005.

[1573] B. E. Schaefer. *Astrophys. J.*, 459:438–454, 1996.

[1574] B. E. Schaefer and A. Pagnotta. *Nature*, 481:164–166, 2012.

[1575] H. Schatz, A. Aprahamian, V. Barnard, et al. *Phys. Rev. Lett.*, 86:3471–3474, 2001.

[1576] H. Schatz, A. Aprahamian, J. Goerres, et al. *Phys. Rep.*, 294:167–264, 1998.

[1577] H. Schatz, L. Bildsten, and A. Cumming. *Astrophys. J. Lett.*, 583:L87–L90, 2003.

[1578] H. Schatz, L. Bildsten, A. Cumming, and M. Wiescher. *Astrophys. J.*, 524:1014–1029, 1999.

[1579] H. Schatz and K. E. Rehm. *Nucl. Phys. A*, 777:601–622, 2006.

[1580] E. Schatzman. *Ann. Astrophys.*, 12:281–286, 1949.

[1581] E. Schatzman. *Ann. Astrophys.*, 14:294–304, 1951.

[1582] E. Schatzman. *Ann. Astrophys.*, 21:1–17, 1958.

[1583] E. Schatzman. In *Star Evolution*, L. Gratton (editor). Academic Press, New York, 1963, pp 389–393.

[1584] O. H. Schindewolf. *Neues Jahrb. Geol. Paleontaol. Monatsh.*, 10:457–465, 1954.

[1585] E. M. Schlegel. *Mon. Not. R. Astron. Soc.*, 244:269–271, 1990.

[1586] B. P. Schmidt, R. P. Kirshner, R. G. Eastman, et al. *Astrophys. J.*, 432:42–48, 1994.

[1587] B. P. Schmidt, N. B. Suntzeff, M. M. Phillips, et al. *Astrophys. J.*, 507:46–63, 1998.

[1588] W. Schmidt and J. C. Niemeyer. *Astron. Astrophys.*, 446:627–633, 2006.

[1589] W. Schmidt, J. C. Niemeyer, W. Hillebrandt, and F. K. Röpke. *Astron. Astrophys.*, 450:283–294, 2006.

[1590] M. F. Schmitz and C. M. Gaskell. In *Supernova 1987A in the Large Magellanic Cloud*, M. Kafatos and A. G. Michalitsianos (editor). Cambridge Univ. Press, Cambridge, UK, 1988, pp 112–115.

[1591] D. N. Schramm. *Astrophys. J.*, 185:293–302, 1973.

[1592] D. N. Schramm and W. D. Arnett. *Astrophys. J.*, 198:629–639, 1975.

[1593] E. Schreier, R. Levinson, H. Gursky, et al. *Astrophys. J. Lett.*, 172:L79–L89, 1972.

[1594] E. Schwartzman, A. Kovetz, and D. Prialnik. *Mon. Not. R. Astron. Soc.*, 269:323–338, 1994.

[1595] G. J. Schwarz. *Astrophys. J.*, 577:940–950, 2002.

[1596] G. J. Schwarz, S. N. Shore, S. Starrfield, et al. *Mon. Not. R. Astron. Soc.*, 320:103–123, 2001.

[1597] G. J. Schwarz, S. N. Shore, S. Starrfield, and K. M. Vanlandingham. *Astrophys. J.*, 657:453–464, 2007.

[1598] P. Scott, M. Asplund, N. Grevesse, M. Bergemann, and A. J. Sauval. *Astron. Astrophys.*, 573:A26 (33 p), 2015.

[1599] P. Scott, N. Grevesse, M. Asplund, et al. *Astron. Astrophys.*, 573:A25 (19 pp), 2015.

[1600] E. R. Seaquist and M. F. Bode. In *Classical Novae*, M. F. Bode and A. Evans (editor). 2nd Ed., Cambridge Univ. Press, Cambridge, UK, 2008, pp 141–166.

[1601] E. Sedlmayr. In *Molecules in the Stellar Environment*, U. G. Jorgensen (editor). Springer-Verlag, Berlin, Germany, 1994, pp 163–185.

[1602] L. I. Sedov. *Similarity and Dimensional Methods in Mechanics*. Academic Press, New York, 1959.

[1603] P. A. Seeger, W. A. Fowler, and D. D. Clayton. *Astrophys. J. Suppl. S.*, 11:121–166, 1965.

[1604] I. R. Seitenzahl, F. Ciaraldi-Schoolmann, F. K. Röpke, et al. *Mon. Not. R. Astron. Soc.*, 429:1156–1172, 2013.

[1605] I. R. Seitenzahl, F. X. Timmes, and G. Magkotsios. *Astrophys. J.*, 792:10 (7 pp), 2014.

[1606] A. M. Serenelli. *Astrophys. Space Sci.*, 328:13–21, 2010.

[1607] A. M. Serenelli and M. Fukugita. *Astrophys. J. Lett.*, 632:L33–L36, 2005.

[1608] F. D. Seward and P. A. Charles. *Exploring the X-ray Universe*. 2nd Ed., Cambridge Univ. Press, Cambridge, UK, 2010.

[1609] A. W. Shafter. In *Classical Nova Explosions*, M. Hernanz and J. José (editor). American Inst. Phys., Melville (New York), 2002, pp 462–471.

[1610] N. M. Shakhovskoi. *Sov. Astron. Letters*, 2:107–108, 1976.

[1611] N. M. Shakhovskoi and Yu. S. Efimov. *Sov. Astron.*, 16:7–9, 1972.

[1612] A. Shankar and D. Arnett. *Astrophys. J.*, 433:216–228, 1994.

[1613] A. Shankar, D. Arnett, and B. A. Fryxell. *Astrophys. J. Lett.*, 394:L13–L15, 1992.

[1614] P. R. Shapiro and P. G. Sutherland. *Astrophys. J.*, 263:902–924, 1982.

[1615] M. M. Shara. *Astrophys. J.*, 243:926–934, 1981.

[1616] M. M. Shara. *Astrophys. J.*, 261:649–660, 1982.

[1617] M. M. Shara. *Astron. J.*, 107:1546–1550, 1994.

[1618] M. M. Shara, M. Livio, A. F. J. Moffat, and M. Orio. *Astrophys. J.*, 311:163–171, 1986.

[1619] M. M. Shara, A. F. J. Moffat, and R. F. Webbink. *Astrophys. J.*, 294:271–285, 1985.

[1620] M. M. Shara and D. Prialnik. *Astron. J.*, 107:1542–1545, 1994.

[1621] M. M. Shara, O. Yaron, D. Prialnik, A. Kovetz, and D. Zurek. *Astrophys. J.*, 725:831–841, 2010.

[1622] G. Shaviv. *The Life of Stars. The Controversial Inception and Emergence of the Theory of Stellar Structure.* Springer-Verlag, Berlin (Germany), 2009.

[1623] G. Shaviv. *The Synthesis of the Elements. The Astrophysical Quest for Nucleosynthesis and What It Can Tell Us About the Universe.* Springer-Verlag, Berlin (Germany), 2012.

[1624] W. Shea. In *1604-2004: Supernovae as Cosmological Lighthouses*, M. Turatto, S. Benetti, L. Zampieri, and W. Shea (editor). Astron. Soc. Pac. Conf. Series, San Francisco (California), 2005, pp 13–20.

[1625] K. J. Shen and L. Bildsten. *Astrophys. J.*, 660:1444–1450, 2007.

[1626] J. E. Shepherd and J. H. S. Lee. In *Major Research Topics in Combustion*, M. Y. Hussaini, A. Kumar, and R. G. Voigt (editor). Springer-Verlag, New York, 1992, pp 439–487.

[1627] O. T. Sherman. *Mon. Not. R. Astron. Soc.*, 47:14–18, 1886.

[1628] M. Shibata and K. ō. Uryū. *Phys. Rev. D*, 62:087501 (4 pp), 2000.

[1629] T. Shigeyama, K. Nomoto, and M. Hashimoto. *Astron. Astrophys.*, 196:141–151, 1988.

[1630] I. S. Shklovskii. *Astrophys. J.*, 148:L1–L4, 1967.

[1631] I. S. Shklovskii and C. Sagan. *Intelligent Life in the Universe.* Holden-Day, San Francisco (California), 1966.

[1632] S. N. Shore. In *Classical Nova Explosions*, M. Hernanz and J. José (editor). American Inst. Phys., Melville (New York), 2002, pp 175–187.

[1633] S. N. Shore. *The Tapestry of Modern Astrophysics.* Wiley-VCH, Weinheim, Germany, 2002.

[1634] S. N. Shore. *Astrophysical Hydrodynamics. An Introduction.* Wiley-VCH, Weinheim, Germany, 2007.

[1635] S. N. Shore. In *Classical Novae*, M. F. Bode and A. Evans (editor). 2nd Ed., Cambridge Univ. Press, Cambridge, UK, 2008, pp 194–231.

[1636] S. N. Shore. In *RS Ophiuchi (2006) and the Recurrent Nova Phenomenon*, A. Evans, M. F. Bode, T. J. O'Brien, and M. J Darnley (editor). Astron.Soc. Pac. Conf. Series, San Francisco (California), 2008, pp 19–30.

[1637] S. N. Shore. *B. Astron. Soc. India*, 40:185–212, 2012.

[1638] S. N. Shore. *Astron. Astrophys.*, 559:L7 (4 pp), 2013.

[1639] S. N. Shore and R. D. Gehrz. *Astron. Astrophys.*, 417:695–699, 2004.

[1640] S. N. Shore, S. J. Kenyon, S. Starrfield, and G. Sonneborn. *Astrophys. J.*, 456:717–737, 1996.

[1641] S. N. Shore, G. Schwarz, H. E. Bond, et al. *Astron. J.*, 125:1507–1518, 2003.

[1642] S. N. Shore, G. Sonneborn, S. Starrfield, R. Gonzalez-Riestra, and R. S. Polidan. *Astrophys. J.*, 421:344–349, 1994.

[1643] S. N. Shore, G. Sonneborn, S. G. Starrfield, et al. *Astrophys. J.*, 370:193–197, 1991.

[1644] S. N. Shore, S. Starrfield, R. Gonzalez-Riestra, P. H. Hauschildt, and G. Sonneborn. *Nature*, 369:539–541, 1994.

[1645] S. N. Shore, S. Starrfield, and G. Sonneborn. *Astrophys. J. Lett.*, 463:L21–L24, 1996.

[1646] F. H. Shu. *The Physical Universe. An Introduction to Astronomy.* University Science Books, Sausalito (California), 1982.

[1647] W. Sidgreaves. *Astrophys. J.*, 14:366–367, 1901.

[1648] W. Sidgreaves. *Mon. Not. R. Astron. Soc.*, 62:137–156, 1901.

[1649] T. Siegert, R. Diehl, M. G. H. Krause, and J. Greiner. *Astron. Astrophys.*, 579:A124 (7 pp), 2015.

[1650] R. Sienkiewicz. *Astron. Astrophys.*, 45:411–416, 1975.

[1651] R. Sienkiewicz. *Astron. Astrophys.*, 85:295–301, 1980.

[1652] J. M. Silverman and A. V. Filippenko. *Mon. Not. R. Astron. Soc.*, 425:1917–1933, 2012.

[1653] J. M. Silverman, R. J. Foley, A. V. Filippenko, et al. *Mon. Not. R. Astron. Soc.*, 425:1789–1818, 2012.

[1654] J. M. Silverman, M. Ganeshalingam, and A. V. Filippenko. *Mon. Not. R. Astron. Soc.*, 430:1030–1041, 2013.

[1655] J. M. Silverman, M. Ganeshalingam, W. Li, and A. V. Filippenko. *Mon. Not. R. Astron. Soc.*, 425:1889–1916, 2012.

[1656] J. M. Silverman, J. J. Kong, and A. V. Filippenko. *Mon. Not. R. Astron. Soc.*, 425:1819–1888, 2012.

[1657] S. A. Sim, F. K. Röpke, W. Hillebrandt, et al. *Astrophys. J. Lett.*, 714:L52–L57, 2010.

[1658] V. A. Simonenko, D. A. Gryaznykh, I. A. Litvinenko, V. A. Lykov, and A. N. Shushlebin. *Astron. Letters*, 38:305–320, 2012.

[1659] V. A. Simonenko, D. A. Gryaznykh, I. A. Litvinenko, V. A. Lykov, and A. N. Shushlebin. *Astron. Letters*, 38:231–237, 2012.

[1660] B. Singh, J. L. Rodriguez, S. S. M. Wong, and J. K. Tuli. *Nucl. Data Sheets*, 84:487–563, 1998.

[1661] E. M. Sion, M. J. Acierno, and S. Tomczyk. *Astrophys. J.*, 230:832–838, 1979.

[1662] E. M. Sion and S. G. Starrfield. *Astrophys. J.*, 303:130–135, 1986.

[1663] E. M. Sion and S. G. Starrfield. *Astrophys. J.*, 421:261–268, 1994.

[1664] J. Smak. *Acta Astronom.*, 32:213–224, 1982.

[1665] J. Smak. *Acta Astronom.*, 32:199–211, 1982.

[1666] L. Smarr, J. R. Wilson, R. T. Barton, and R. L. Bowers. *Astrophys. J.*, 246:515–525, 1981.

[1667] S. J. Smartt. *Annu. Rev. Astron. Astr.*, 47:63–106, 2009.

[1668] D. M. Smith. *New Astron. Rev.*, 48:87–91, 2004.

[1669] M. S. Smith, W. R. Hix, S. Parete-Koon, et al. In *Classical Nova Explosions*, M. Hernanz and J. José (editor). American Inst. Phys., Melville (New York), 2002, pp 161–166.

[1670] N. Smith, K. H. Hinkle, and N. Ryde. *Astron. J.*, 137:3558–3573, 2009.

[1671] C. Sneden, J. J. Cowan, and R. Gallino. *Annu. Rev. Astron. Astr.*, 46:241–288, 2008.

[1672] C. Sneden, J. J. Cowan, J. E. Lawler, et al. *Astrophys. J.*, 591:936–953, 2003.

[1673] M. A. J. Snijders, T. J. Batt, P. F. Roche, et al. *Mon. Not. R. Astron. Soc.*, 228:329–376, 1987.

[1674] V. V. Sobolev. *Moving Envelopes of Stars*. Harvard Univ. Press, Cambridge (Massachusetts), 1960.

[1675] G. A. Sod. *J. Comput. Phys.*, 27:1–31, 1978.

[1676] A. M. Soderberg, S. R. Kulkarni, E. Nakar, et al. *Nature*, 442:1014–1017, 2006.

[1677] N. Soker. In *Binary Paths to Type Ia Supernovae Explosions*, R. Di Stefano, M. Orio, and M. Moe (editor). Cambridge Univ. Press, Cambridge, UK, 2013, pp 72–75.

[1678] N. Soker, E. García-Berro, and L. G. Althaus. *Mon. Not. R. Astron. Soc.*, 437:L66–L70, 2014.

[1679] J. L. Sokoloski, G. J. M. Luna, K. Mukai, and S. J. Kenyon. *Nature*, 442:276–278, 2006.

[1680] J.-E. Solheim and L. R. Yungelson. In *14th European Workshop on White Dwarfs*, D. Koester and S. Moehler (editor). Astron. Soc. Pac. Conf. Series, San Francisco (California), 2005, pp 387–392.

[1681] W. M. Sparks. *Astrophys. J.*, 156:569–595, 1969.

[1682] W. M. Sparks and G. S. Kutter. *Astrophys. J.*, 321:394–403, 1987.

[1683] W. M. Sparks, S. Starrfield, and J. W. Truran. *Astrophys. J.*, 220:1063–1075, 1978.

[1684] T. Spillane, F. Raiola, C. Rolfs, et al. *Phys. Rev. Lett.*, 98:122501 (4 pp), 2007.

[1685] T. Spillane, C. E. Rolfs, and J. S. Schweitzer. *Proc. Science, PoS(NIC-X)* 016 (5 pp), 2008.

[1686] A. Spitkovsky, Y. Levin, and G. Ushomirsky. *Astrophys. J.*, 566:1018–1038, 2002.

[1687] V. Springel. *Mon. Not. R. Astron. Soc.*, 364:1105–1134, 2005.

[1688] V. Springel. *Annu. Rev. Astron. Astr.*, 48:391–430, 2010.

[1689] V. Springel. *ArXiv e-prints*, 2014.

[1690] J. Spyromilio and J. Bailey. *P. Astron. Soc. Aust.*, 10:263–264, 1993.

[1691] J. Spyromilio and B. Leibundgut. *Mon. Not. R. Astron. Soc.*, 283:L89–L93, 1996.

[1692] A. Spyrou, S.J. Quinn, A. Simon, et al. *Proc. Science, PoS(NIC-XIII)* 025 (8 pp), 2014.

[1693] G. Srinivasan. *Astron. Astrophys. Rev.*, 11:67–96, 2002.

[1694] F. J. Stadermann, T. K. Croat, T. J. Bernatowicz, et al. *Geochim. Cosmochim. Acta*, 69:177–188, 2005.

[1695] F. J. Stadermann, C. Floss, M. Bose, and A. S. Lea. *Meteorit. Planet. Sci.*, 44:1033–1049, 2009.

[1696] F. J. Stadermann, R. M. Walker, and E. Zinner. *Meteorit. Planet. Sci. Suppl.*, 34:A111, 1999.

[1697] D. H. Staelin and E. C. Reifenstein. *Science*, 162:1481–1483, 1968.

[1698] I. H. Stairs. *Science*, 304:547–552, 2004.

[1699] R. J. Stancliffe, M. Lugaro, C. Ugalde, et al. *Mon. Not. R. Astron. Soc.*, 360:375–379, 2005.

[1700] K. Z. Stanek, T. Matheson, P. M. Garnavich, et al. *Astrophys. J. Lett.*, 591:L17–L20, 2003.

[1701] S. Starrfield. *Mon. Not. R. Astron. Soc.*, 152:307–322, 1971.

[1702] S. Starrfield. *Mon. Not. R. Astron. Soc.*, 155:129–137, 1971.

[1703] S. Starrfield. In *Classical Novae*, M. F. Bode and A. Evans (editor). John Wiley and Sons, Chichester, UK, 1989, pp 39–60.

[1704] S. Starrfield. *AIP Adv.*, 4:041007 (14 pp), 2014.

[1705] S. Starrfield, R. D. Gehrz, and J. W. Truran. In *Astrophysical Implications of the Laboratory Study of Presolar Materials*, T. J. Bernatowicz and E. Zinner (editor). American Inst. Phys., Woodbury (New York), 1997, pp 203–234.

[1706] S. Starrfield, C. Iliadis, W. R. Hix, F. X. Timmes, and W. M. Sparks. *Astrophys. J.*, 692:1532–1542, 2009.

[1707] S. Starrfield, S. N. Shore, W. M. Sparks, et al. *Astrophys. J. Lett.*, 391:L71–L74, 1992.

[1708] S. Starrfield, W. M. Sparks, and G. Shaviv. *Astrophys. J. Lett.*, 325:L35–L38, 1988.

[1709] S. Starrfield, W. M. Sparks, and J. W. Truran. *Astrophys. J.*, 291:136–146, 1985.

[1710] S. Starrfield, W. M. Sparks, and J. W. Truran. *Astrophys. J. Lett.*, 303:L5–L9, 1986.

[1711] S. Starrfield, W. M. Sparks, J. W. Truran, and M. C. Wiescher. *Astrophys. J. Suppl. S.*, 127:485–495, 2000.

[1712] S. Starrfield, F. X. Timmes, W. R. Hix, et al. *Astrophys. J. Lett.*, 612:L53–L56, 2004.

[1713] S. Starrfield, J. W. Truran, M. Politano, et al. *Phys. Rep.*, 227:223–234, 1993.

[1714] S. Starrfield, J. W. Truran, and W. M. Sparks. *Astrophys. J.*, 226:186–202, 1978.

[1715] S. Starrfield, J. W. Truran, W. M. Sparks, and M. Arnould. *Astrophys. J.*, 222:600–603, 1978.

[1716] S. Starrfield, J. W. Truran, W. M. Sparks, and G. S. Kutter. *Astrophys. J.*, 176:169–176, 1972.

[1717] S. Starrfield, J. W. Truran, M. C. Wiescher, and W. M. Sparks. *Mon. Not. R. Astron. Soc.*, 296:502–522, 1998.

[1718] A. W. Steiner, J. M. Lattimer, and E. F. Brown. *Astrophys. J.*, 722:33–54, 2010.

[1719] A. W. Steiner, J. M. Lattimer, and E. F. Brown. *Astrophys. J. Lett.*, 765:L5 (5 pp), 2013.

[1720] T. Stephan, A. M. Davis, M. J. Pellin, et al. 45th *Lunar and Planetary Science Conf.*, Abstract #2242, 2014.

[1721] T. Stephan, A. M. Davis, M. J. Pellin, et al. *Meteorit. Planet. Sci. Suppl.*, 74:Abstract #5192, 2011.

[1722] F. R. Stephenson and D. A. Green. *Historical Supernovae and Their Remnants.* Clarendon Press, Oxford, UK, 2002.

[1723] J. Stevens, E. F. Brown, A. Cumming, R. Cyburt, and H. Schatz. *Astrophys. J.*, 791:106 (8 pp), 2014.

[1724] J. M. Stone and M. L. Norman. *Astrophys. J. Suppl. S.*, 80:753–790, 1992.

[1725] R. Stothers. *Isis*, 68:443–447, 1977.

[1726] G. Stratta, B. Gendre, J. L. Atteia, et al. *Astrophys. J.*, 779:66 (14 pp), 2013.

[1727] F. J. M. Stratton and W. H. Manning. *Atlas of Spectra of Nova Herculis 1934.* Solar Physics Observ., Cambridge, 1939.

[1728] M. Stritzinger and B. Leibundgut. *Astron. Astrophys.*, 431:423–431, 2005.

[1729] M. Stritzinger, B. Leibundgut, S. Walch, and G. Contardo. *Astron. Astrophys.*, 450:241–251, 2006.

[1730] M. Stritzinger, P. A. Mazzali, J. Sollerman, and S. Benetti. *Astron. Astrophys.*, 460:793–798, 2006.

[1731] T. Strohmayer and L. Bildsten. In *Compact Stellar X-Ray Sources*, W. H. G. Lewin and M. van der Klis (editor). Cambridge Univ. Press, Cambridge, UK, 2006, pp 113–156.

[1732] T. E. Strohmayer and E. F. Brown. *Astrophys. J.*, 566:1045–1059, 2002.

[1733] T. E. Strohmayer, W. Zhang, and J. H. Swank. *Astrophys. J. Lett.*, 487:L77–L80, 1997.

[1734] T. E. Strohmayer, W. Zhang, J. H. Swank, et al. *Astrophys. J. Lett.*, 469:L9–L12, 1996.

[1735] R. M. Stroud. *Workshop on Cometary Dust in Astrophysics*, Abstract #6011, 2003.

[1736] R. M. Stroud, M. F. Chisholm, P. R. Heck, C. M. O. Alexander, and L. R. Nittler. *Astrophys. J. Lett.*, 738:L27 (5 pp), 2011.

[1737] R. M. Stroud, L. R. Nittler, and C. M. O. Alexander. *Science*, 305:1455–1457, 2004.

[1738] R. M. Stroud, E. K. Zinner, and F. Gyngard. 46th *Lunar and Planetary Science Conf.*, Abstract #2576, 2015.

[1739] H. E. Suess and H. C. Urey. *Rev. Mod. Phys.*, 28:53–74, 1956.

[1740] B. E. K. Sugerman, B. Ercolano, M. J. Barlow, et al. *Science*, 313:196–200, 2006.

[1741] D. Sugimoto and M. Y. Fujimoto. *Pub. Astron. Soc. Jpn.*, 30:467–482, 1978.

[1742] D. Sugimoto and S. Miyaji. In *Fundamental Problems in the Theory of Stellar Evolution*, D. Sugimoto, D. Q. Lamb, and D. N. Schramm (editor). Reidel, Dordrecht, The Netherlands, 1981, pp 191–206.

[1743] D. Sugimoto and K. Nomoto. *Space Sci. Rev.*, 25:155–227, 1980.

[1744] V. Suleimanov, J. Poutanen, and K. Werner. *Astron. Astrophys.*, 527:A139 (12 pp), 2011.

[1745] A. Summa, A. Ulyanov, M. Kromer, et al. *Astron. Astrophys.*, 554:A67 (10 pp), 2013.

[1746] R. Surman and G. C. McLaughlin. *Astrophys. J.*, 618:397–402, 2005.

[1747] R. Surman, G. C. McLaughlin, and W. R. Hix. *Astrophys. J.*, 643:1057–1064, 2006.

[1748] R. Surman, G. C. McLaughlin, M. Ruffert, H.-T. Janka, and W. R. Hix. *Astrophys. J. Lett.*, 679:L117–L120, 2008.

[1749] P. G. Sutherland. In *Supernovae*, A. G. Petschek (editor). Springer-Verlag, Berlin, Germany, 1990, pp 111–142.

[1750] A. Suzuki and T. Shigeyama. *Astrophys. J. Lett.*, 723:L84–L88, 2010.

[1751] J. Svestka. *Astrophys. Space Sci.*, 45:21–25, 1976.

[1752] J. H. Swank, R. H. Becker, E. A. Boldt, et al. *Astrophys. J.*, 212:L73–L76, 1977.

[1753] D. A. Swartz, A. V. Filippenko, K. Nomoto, and J. C. Wheeler. *Astrophys. J.*, 411:313–322, 1993.

[1754] E. Symbalisty and D. N. Schramm. *Astrophys. Lett.*, 22:143–145, 1982.

[1755] E. M. D. Symbalisty, D. N. Schramm, and J. R. Wilson. *Astrophys. J. Lett.*, 291:L11–L14, 1985.

[1756] T. Szalai, J. Vinkó, Z. Balog, et al. *Astron. Astrophys.*, 527:A61 (14 pp), 2011.

[1757] F.M. Szasz. *The Day the Sun Rose Twice.* Cambridge Univ. Press, Cambridge, UK, 1984.

[1758] R. E. Taam. *Astrophys. J.*, 237:142–147, 1980.

[1759] R. E. Taam. *Astrophys. J.*, 242:749–755, 1980.

[1760] R. E. Taam. *Astrophys. J.*, 241:358–366, 1980.

[1761] R. E. Taam. *Astrophys. J.*, 247:257–266, 1981.

[1762] R. E. Taam. *Astrophys. Space Sci.*, 77:257–265, 1981.

[1763] R. E. Taam. *Astrophys. J.*, 258:761–769, 1982.

[1764] R. E. Taam and R. E. Picklum. *Astrophys. J.*, 224:210–216, 1978.

[1765] R. E. Taam and R. E. Picklum. *Astrophys. J.*, 233:327–333, 1979.

[1766] R. E. Taam, S. E. Woosley, and D. Q. Lamb. *Astrophys. J.*, 459:271–277, 1996.

[1767] R. E. Taam, S. E. Woosley, T. A. Weaver, and D. Q. Lamb. *Astrophys. J.*, 413:324–332, 1993.

[1768] F. Taddia, J. Sollerman, G. Leloudas, et al. *Astron. Astrophys.*, 574:A60 (31 pp), 2015.

[1769] F. Taddia, M. D. Stritzinger, M. M. Phillips, et al. *Astron. Astrophys.*, 545:L7 (8 pp), 2012.

[1770] A. Tajitsu, K. Sadakane, H. Naito, A. Arai, and W. Aoki. *Nature*, 518:381–384, 2015.

[1771] K. Takahashi, J. Witti, and H.-T. Janka. *Astron. Astrophys.*, 286:857–869, 1994.

[1772] G. A. Tammann and B. Leibundgut. *Astron. Astrophys.*, 236:9–14, 1990.

[1773] T.-H. Tan and J. E. Borovsky. *J. Plasma Phys.*, 35:239–256, 1986.

[1774] W. P. Tan, J. L. Fisker, J. Görres, M. Couder, and M. Wiescher. *Phys. Rev. Lett.*, 98:242503 (4 pp), 2007.

[1775] W. P. Tan, J. Görres, M. Beard, et al. *Phys. Rev.C*, 79:055805 (11 pp), 2009.

[1776] M. Tanaka, P. A. Mazzali, S. Benetti, et al. *Astrophys. J.*, 677:448–460, 2008.

[1777] H. Tananbaum, H. Gursky, E. M. Kellogg, et al. *Astrophys. J. Lett.*, 174:L143–L149, 1972.

[1778] M. Tang and E. Anders. *Geochim. Cosmochim. Acta*, 52:1235–1244, 1988.

[1779] V. Tatischeff and M. Hernanz. *Astrophys. J. Lett.*, 663:L101–L104, 2007.

[1780] T. M. Tauris and T. Sennels. *Astron. Astrophys.*, 355:236–244, 2000.

[1781] Y. Tawara, S. Hayakawa, and T. Kii. *Pub. Astron. Soc. Jpn.*, 36:845–853, 1984.

[1782] Y. Tawara, T. Kii, S. Hayakawa, et al. *Astrophys. J. Lett.*, 276:L41–L44, 1984.

[1783] G. Taylor. *R. Soc. London Proc.*, 201:159–174, 1950.

[1784] G. Taylor. *R. Soc. London Proc.*, 201:175–186, 1950.

[1785] H. H. Teng, Z. L. Jiang, and Z. M. Hu. *Acta Mech. Sinica*, 23:343–349, 2007.

[1786] K. D. Terry and W. H. Tucker. *Science*, 159:421–423, 1968.

[1787] L.-S. The, D. D. Clayton, R. Diehl, et al. *Astron. Astrophys.*, 450:1037–1050, 2006.

[1788] L.-S. The, D. D. Clayton, L. Jin, and B. S. Meyer. *Astrophys. J.*, 504:500–515, 1998.

[1789] L.-S. The, M. D. Leising, and D. D. Clayton. *Astrophys. J.*, 403:32–36, 1993.

[1790] The American Institute of Aeronautics and Astronautics. *AIAA Guide for the Verification and Validation of Computational Fluid Dynamics Simulations.* American Inst. Aeronautics & Astronautics, Reston (Virginia), 1998.

[1791] F.-K. Thielemann, A. Arcones, R. Käppeli, et al. *Prog. Part. Nucl. Phys.*, 66:346–353, 2011.

[1792] F. K. Thielemann and W. D. Arnett. *Astrophys. J.*, 295:604–619, 1985.

[1793] F.-K. Thielemann, F. Brachwitz, C. Freiburghaus, et al. *Prog. Part. Nucl. Phys.*, 46:5–22, 2001.

[1794] F.-K. Thielemann, F. Brachwitz, P. Höflich, G. Martinez-Pinedo, and K. Nomoto. *New Astron. Rev.*, 48:605–610, 2004.

[1795] F.-K. Thielemann, M.-A. Hashimoto, and K. Nomoto. *Astrophys. J.*, 349:222–240, 1990.

[1796] F.-K. Thielemann, K. Nomoto, and M. Hashimoto. In *Supernovae*, S. A. Bludman, R. Mochkovitch, and J. Zinn-Justin (editor). 1994, pp 629–676.

[1797] F.-K. Thielemann, K. Nomoto, K. Iwamoto, and F. Brachwitz. In *Thermonuclear Supernovae*, P. Ruiz-Lapuente, R. Canal, and J. Isern (editor). Kluwer Acad. Publ., Dordrecht, The Netherlands, 1997, pp 485–514.

[1798] F.-K. Thielemann, K. Nomoto, and K. Yokoi. *Astron. Astrophys.*, 158:17–33, 1986.

[1799] B. C. Thomas, C. H. Jackman, A. L. Melott, et al. *Astrophys. J. Lett.*, 622:L153–L156, 2005.

[1800] B. C. Thomas, A. L. Melott, B. D. Field, and B. J. Anthony-Twarog. *Astrobiol.*, 8:9–16, 2008.

[1801] B. C. Thomas, A. L. Melott, C. H. Jackman, et al. *Astrophys. J.*, 634:509–533, 2005.

[1802] T. A. Thompson, A. Burrows, and B. S. Meyer. *Astrophys. J.*, 562:887–908, 2001.

[1803] S. E. Thorsett. *Nature*, 356:690–691, 1992.

[1804] F. X. Timmes. *Astrophys. J. Suppl. S.*, 124:241–263, 1999.

[1805] F. X. Timmes, R. D. Hoffman, and S. E. Woosley. *Astrophys. J. Suppl. S.*, 129:377–398, 2000.

[1806] F. X. Timmes and J. C. Niemeyer. *Astrophys. J.*, 537:993–997, 2000.

[1807] F. X. Timmes and S. E. Woosley. *Astrophys. J.*, 396:649–667, 1992.

[1808] F. X. Timmes, S. E. Woosley, and T. A. Weaver. *Astrophys. J. Suppl. S.*, 98:617–658, 1995.

[1809] F. X. Timmes, M. Zingale, K. Olson, et al. *Astrophys. J.*, 543:938–954, 2000.

[1810] P. Todini and A. Ferrara. *Mon. Not. R. Astron. Soc.*, 325:726–736, 2001.

[1811] S. Toonen, G. Nelemans, and S. Portegies Zwart. *Astron. Astrophys.*, 546:A70 (16 pp), 2012.

[1812] E. F. Toro. *Riemann Solvers and Numerical Methods for Fluid Dynamics. A Practical Introduction.* Springer-Verlag, Berlin, Germany, 2009.

[1813] I. S. Towner and J. C. Hardy. In *Symmetries and Fundamental Interactions in Nuclei*, E. M. Henley and W. C. Haxton (editor). World Scientific, Singapore, 1995, pp 183–250.

[1814] D. M. Townsley and L. Bildsten. *Astrophys. J.*, 600:390–403, 2004.

[1815] D. M. Townsley, A. C. Calder, S. M. Asida, et al. *Astrophys. J.*, 668:1118–1131, 2007.

[1816] D. M. Townsley, A. P. Jackson, A. C. Calder, et al. *Astrophys. J.*, 701:1582–1604, 2009.

[1817] C. Travaglio, R. Gallino, S. Amari, et al. *Astrophys. J.*, 510:325–354, 1999.

[1818] C. Travaglio, R. Gallino, E. Arnone, et al. *Astrophys. J.*, 601:864–884, 2004.

[1819] C. Travaglio, R. Gallino, M. Busso, and R. Gratton. *Astrophys. J.*, 549:346–352, 2001.

[1820] C. Travaglio, R. Gallino, T. Rauscher, et al. *Astrophys. J.*, 795:141 (8 pp), 2014.

[1821] C. Travaglio, R. Gallino, T. Rauscher, F. K. Röpke, and W. Hillebrandt. *Astrophys. J.*, 799:54 (13 pp), 2015.

[1822] C. Travaglio, W. Hillebrandt, M. Reinecke, and F.-K. Thielemann. *Astron. Astrophys.*, 425:1029–1040, 2004.

[1823] C. Travaglio, F. K. Röpke, R. Gallino, and W. Hillebrandt. *Astrophys. J.*, 739:93 (19 pp), 2011.

[1824] J. W. Truran. *Prog. Part. Nucl. Phys.*, 6:177–190, 1981.

[1825] J. W. Truran. In *Essays in Nuclear Astrophysics*, C. A. Barnes, D. D. Clayton, and D. N. Schramm (editor). Cambridge Univ. Press, Cambridge, UK, 1982, pp 467–494.

[1826] J. W. Truran, W. D. Arnett, and A. G. W. Cameron. *Can. J. Phys.*, 45:2315–2332, 1967.

[1827] J. W. Truran and M. Livio. *Astrophys. J.*, 308:721–727, 1986.

[1828] J. W. Truran, M. Livio, J. Hayes, S. Starrfield, and W. M. Sparks. *Astrophys. J.*, 324:345–354, 1988.

[1829] X. L. Tu, H. S. Xu, M. Wang, et al. *Phys. Rev. Lett.*, 106:112501 (5 pp), 2011.

[1830] Y. Tuchman and J. W. Truran. *Astrophys. J.*, 503:381–386, 1998.

[1831] W. H. Tucker and K. D. Terry. *Science*, 160:1138–1139, 1968.

[1832] J. K. Tuli. *Nuclear Wallet Cards*. 8th Ed., Brookhaven National Laboratory, Upton (New York), 2011.

[1833] C. Tur, S. M. Austin, A. Wuosmaa, et al. *Proc. Science, PoS(NIC-IX)* 050 (7 pp), 2006.

[1834] C. Tur, A. Heger, and S. M. Austin. *Astrophys. J.*, 671:821–827, 2007.

[1835] C. Tur, A. Heger, and S. M. Austin. *Astrophys. J.*, 718:357–367, 2010.

[1836] M. Turatto, S. Benetti, and A. Pastorello. In *Supernova 1987A: 20 Years After: Supernovae and Gamma-Ray Bursters*, S. Immler, K. Weiler, and R. McCray (editor). American Inst. Phys., New York, 2007, pp 187–197.

[1837] M. Turatto, E. Cappellaro, I. J. Danziger, et al. *Mon. Not. R. Astron. Soc.*, 262:128–140, 1993.

[1838] R. Turolla, L. Nobili, and M. Calvani. *Astrophys. J.*, 303:573–581, 1986.

[1839] A. Tziamtzis, P. Lundqvist, P. Gröningsson, and S. Nasoudi-Shoar. *Astron. Astrophys.*, 527:A35 (14 pp), 2011.

[1840] A. Uomoto and R. P. Kirshner. *Astrophys. J.*, 308:685–690, 1986.

[1841] V. V. Usov. *Nature*, 357:472–474, 1992.

[1842] V. P. Utrobin. *Astron. Astrophys.*, 306:219–231, 1996.

[1843] S. van den Bergh. *Astrophys. J.*, 413:67–69, 1993.

[1844] S. van den Bergh. *Pub. Astron. Soc. Pac.*, 106:689–695, 1994.

[1845] S. van den Bergh and G. A. Tammann. *Annu. Rev. Astron. Astr.*, 29:363–407, 1991.

[1846] S. D. van Dyk, C. Y. Peng, A. J. Barth, et al. *Pub. Astron. Soc. Pac.*, 111:313–320, 1999.

[1847] H. M. van Horn and C. J. Hansen. *Astrophys. J.*, 191:479–482, 1974.

[1848] M. H. van Kerkwijk, C. G. Bassa, B. A. Jacoby, and P. G. Jonker. In *Binary Radio Pulsars*, F. A. Rasio and I. H. Stairs (editor). Astron. Soc. Pac. Conf. Series, San Francisco (California), 2005, pp 357–370.

[1849] J. van Paradijs. *Astron. Astrophys.*, 107:51–53, 1982.

[1850] J. van Paradijs and J. E. McClintock. In *X-Ray Binaries*, W. H. G. Lewin, J. van Paradijs, and E. P. J. van den Heuvel (editor). Cambridge Univ. Press, Cambridge, UK, 1995, pp 58–125.

[1851] L. van Wormer, J. Goerres, C. Iliadis, M. Wiescher, and F.-K. Thielemann. *Astrophys. J.*, 432:326–350, 1994.

[1852] K. M. Vanlandingham, G. J. Schwarz, S. N. Shore, S. Starrfield, and R. M. Wagner. *Astrophys. J.*, 624:914–922, 2005.

[1853] K. M. Vanlandingham, S. Starrfield, and S. N. Shore. *Mon. Not. R. Astron. Soc.*, 290:87–98, 1997.

[1854] K. M. Vanlandingham, S. Starrfield, S. N. Shore, and G. Sonneborn. *Mon. Not. R. Astron. Soc.*, 308:577–587, 1999.

[1855] H. K. Versteeg and W. Malalasekera. *An Introduction to Computational Fluid Dynamics: The Finite Volume Method.* 2nd Ed., Prentice Hall, New Jersey, 2007.

[1856] I. V. Veryovkin, W. F. Calaway, C. E. Tripa, M. J. Pellin, and D. S. Burnett. *AGU Fall Meeting Abstracts*, A1185, 2004.

[1857] I. V. Veryovkin, C. E. Tripa, A. V. Zinovev, S. V. Baryshev, and M. J. Pellin. 42nd *Lunar and Planetary Science Conf.*, Abstract #2790, 2011.

[1858] J. Vesic, A. Cvetinovic, M. Lipoglavsek, and T. Petrovic. *Eur. Phys. J. A*, 50:153 (9 pp), 2014.

[1859] J. Vink. *Adv. Space Res.*, pp 976–986, 2005.

[1860] J. Vink, J. Bleeker, K. van der Heyden, et al. *Astrophys. J. Lett.*, 648:L33–L37, 2006.

[1861] A. Virag, B. Wopenka, S. Amari, et al. *Geochim. Cosmochim. Acta*, 56:1715–1733, 1992.

[1862] F. J. Virgili, C. G. Mundell, V. Pal'shin, et al. *Astrophys. J.*, 778:54 (18 pp), 2013.

[1863] F. Vissani. *J. Phys. G Nucl. Partic.*, 42:013001 (31 pp), 2015.

[1864] D. W. Visser, J. A. Caggiano, R. Lewis, et al. *Phys. Rev.C*, 69:048801 (4 pp), 2004.

[1865] C. Vollmer, P. Hoppe, and F. E. Brenker. *Astrophys. J.*, 769:61 (8 pp), 2013.

[1866] C. Vollmer, P. Hoppe, F. J. Stadermann, C. Floss, and F. E. Brenker. *Geochim. Cosmochim. Acta*, 73:7127–7149, 2009.

[1867] M. Volmer and A. Weber. *Z. Phys. Chem.*, 119:277–301, 1925.

[1868] J. von Neumann. *Progress Report to the National Defense Research Committee Div. B,* No. 238, OSRD-549, 1942.

[1869] J. von Neumann. In *Blast Wave*, H. A. Bethe, K. Fuchs, J. O. Hirschfelder, J. L. Magee, R. E. Peierls, and J. von Neumann (editor). LA-2000, Los Alamos Scientific Laboratory (New Mexico), 1947, pp 27–55.

[1870] J. Von Neumann and R. D. Richtmyer. *J. Appl. Phys.*, 21:232–237, 1950.

[1871] C.F. von Weizsäcker. *Phys. Z.*, 39:633–646, 1938.

[1872] W. von Witsch, A. Richter, and P. von Brentano. *Phys. Rev.*, 169:923–932, 1968.

[1873] R. V. Wagoner. *Astrophys. J. Suppl. S.*, 18:247–295, 1969.

[1874] R. V. Wagoner. *Astrophys. J. Lett.*, 214:L5–L7, 1977.

[1875] R. Walder, D. Folini, and S. N. Shore. *Astron. Astrophys.*, 484:L9–L12, 2008.

[1876] A. G. Walker. *Mon. Not. R. Astron. Soc.*, 95:263–269, 1935.

[1877] M. F. Walker. *Pub. Astron. Soc. Pac.*, 66:230–232, 1954.

[1878] P. Walker. *Nature Physics*, 7:281–282, 2011.

[1879] R. K. Wallace and S. E. Woosley. *Astrophys. J. Suppl. S.*, 45:389–420, 1981.

[1880] R. K. Wallace and S. E. Woosley. In *High Energy Transients in Astrophysics*, S. E. Woosley (editor). American Inst. Phys., New York, 1984, pp 319–324.

[1881] R. K. Wallace, S. E. Woosley, and T. A. Weaver. *Astrophys. J.*, 258:696–715, 1982.

[1882] A. Wallner, Y. Ikeda, W. Kutschera, et al. *Nucl. Instrum. Meth. B*, 172:382–387, 2000.

[1883] A. Wallner, K. Melber, S. Merchel, et al. *Nucl. Instrum. Meth. B*, 294:496–502, 2013.

[1884] C. Wallner, T. Faestermann, U. Gerstmann, et al. *New Astron. Rev.*, 48:145–150, 2004.

[1885] S. Wanajo. *Astrophys. J.*, 647:1323–1340, 2006.

[1886] S. Wanajo. *Astrophys. J. Lett.*, 666:L77–L80, 2007.

[1887] S. Wanajo, M.-A. Hashimoto, and K. Nomoto. *Astrophys. J.*, 523:409–431, 1999.

[1888] S. Wanajo and H.-T. Janka. *Astrophys. J.*, 746:180 (15 pp), 2012.

[1889] S. Wanajo, Y. Sekiguchi, N. Nishimura, et al. *Astrophys. J. Lett.*, 789:L39 (6 pp), 2014.

[1890] B. Wang, X. Chen, X. Meng, and Z. Han. *Astrophys. J.*, 701:1540–1546, 2009.

[1891] B. Wang and Z. Han. *New Astron. Rev.*, 56:122–141, 2012.

[1892] L. Wang. In *Cosmic Explosions in Three Dimensions: Asymmetries in Supernovae and Gamma-Ray Bursts*, P. Höflich, P. Kumar, and J. C. Wheeler (editor). Cambridge Univ. Press, Cambridge, UK, 2004, pp 17–29.

[1893] L. Wang, D. Baade, A. Clocchiatti, et al. *Astro2010: The Astronomy and Astrophysics Decadal Survey* (7 pp), 2009.

[1894] L. Wang, D. Baade, P. Höflich, et al. *Astrophys. J.*, 591:1110–1128, 2003.

[1895] L. Wang, D. Baade, and F. Patat. *Science*, 315:212–214, 2007.

[1896] L. Wang and J. C. Wheeler. *Annu. Rev. Astron. Astr.*, 46:433–474, 2008.

[1897] L. Wang, J. C. Wheeler, and P. Höflich. *Astrophys. J. Lett.*, 476:L27–L30, 1997.

[1898] L. Wang, J. C. Wheeler, Z. Li, and A. Clocchiatti. *Astrophys. J.*, 467:435–445, 1996.

[1899] M. Wang, G. Audi, A. H. Wapstra, et al. *Chinese Phys. C*, 36:1603–2014, 2012.

[1900] W. Wang, M. J. Harris, R. Diehl, et al. *Astron. Astrophys.*, 469:1005–1012, 2007.

[1901] Z.-R. Wang. *Astrophys. Space Sci.*, 305:207–210, 2006.

[1902] Z. R. Wang, Q.-Y. Qu, and Y. Chen. *Astron. Astrophys.*, 318:L59–L61, 1997.

[1903] Z. R. Wang, Y. Zhao, M. Li, and Q. L. Zhou. In *1604-2004: Supernovae as Cosmological Lighthouses*, M. Turatto, S. Benetti, L. Zampieri, and W. Shea (editor). Astron. Soc. Pac. Conf. Series, San Francisco (California), 2005, pp 48–52.

[1904] R. A. Ward and W. A. Fowler. *Astrophys. J.*, 238:266–286, 1980.

[1905] B. Warner. *Mon. Not. R. Astron. Soc.*, 227:23–73, 1987.

[1906] B. Warner. *Cataclysmic Variable Stars*. Cambridge Univ. Press, Cambridge, 1995.

[1907] B. Warner. In *Classical Nova Explosions*, M. Hernanz and J. José (editor). American Inst. Phys., Melville (New York), 2002, pp 3–15.

[1908] G. J. Wasserburg, A. I. Boothroyd, and I.-J. Sackmann. *Astrophys. J. Lett.*, 447:L37–L40, 1995.

[1909] T. A. Weaver and S. E. Woosley. *Ann. NY Acad. Sci.*, 336:335–357, 1980.

[1910] T. A. Weaver, S. E. Woosley, and G. M. Fuller. In *Numerical Astrophysics*, J. M. Centrella, J. M. Leblanc, and R. L. Bowers (editor). Jones and Bartlett Publ., Boston (Massachusetts), 1985, pp 374–388.

[1911] T. A. Weaver, G. B. Zimmerman, and S. E. Woosley. *Astrophys. J.*, 225:1021–1029, 1978.

[1912] R. F. Webbink. *Astrophys. J.*, 277:355–360, 1984.

[1913] C. Weber, V.-V. Elomaa, R. Ferrer, et al. *Phys. Rev. C*, 78:054310 (18 pp), 2008.

[1914] N. N. Weinberg and L. Bildsten. *Astrophys. J.*, 670:1291–1300, 2007.

[1915] N. N. Weinberg, L. Bildsten, and E. F. Brown. *Astrophys. J. Lett.*, 650:L119–L122, 2006.

[1916] N. N. Weinberg, L. Bildsten, and H. Schatz. *Astrophys. J.*, 639:1018–1032, 2006.

[1917] G. Weirs, V. Dwarkadas, T. Plewa, C. Tomkins, and M. Marr-Lyon. *Astrophys. Space Sci.*, 298:341–346, 2005.

[1918] M. K. Weisberg, T. J. McCoy, and A. N. Krot. In *Meteorites and the Early Solar System II*, D. S. Lauretta and H. Y. McSween, Jr. (editor). University of Arizona Press, Tucson (Arizona), 2006, pp 19–52.

[1919] A. Weiss, W. Hillebrandt, H.-C. Thomas, and H. Ritter. *Cox and Giuli's Principles of Stellar Structure*. 2nd Ed., Cambridge Scientific Pub., Cambridge, UK, 2004.

[1920] A. Weiss and J. W. Truran. *Astron. Astrophys.*, 238:178–186, 1990.

[1921] V. F. Weisskopf and D. H. Ewing. *Phys. Rev.*, 57:472–485 (Erratum: Phys. Rev. 57:935), 1940.

[1922] A. J. Wesselink. *Bull. Astron. Inst. Neth.*, 10:91–98, 1946.

[1923] R. M. West, A. Lauberts, H.-E. Schuster, and H. E. Jorgensen. *Astron. Astrophys.*, 177:L1–L3, 1987.

[1924] J. C. Wheeler. *Astrophys. J.*, 214:560–565, 1977.

[1925] J. C. Wheeler. *Rep. Prog. Phys.*, 44:85–138, 1981.

[1926] J. C. Wheeler and R. P. Harkness. In *Galaxy Distances and Deviations from Universal Expansion*, B. F. Madore and R. B. Tully (editor). Reidel, Dordrecht, The Netherlands, 1986, pp 45–54.

[1927] J. C. Wheeler and R. P. Harkness. *Rep. Prog. Phys.*, 53:1467–1557, 1990.

[1928] J. C. Wheeler and R. Levreault. *Astrophys. J. Lett.*, 294:L17–L20, 1985.

[1929] J. C. Wheeler, I. Yi, P. Höflich, and L. Wang. *Astrophys. J.*, 537:810–823, 2000.

[1930] J. Whelan and I. Iben, Jr. *Astrophys. J.*, 186:1007–1014, 1973.

[1931] R. C. Whitten, W. J. Borucki, J. H. Wolfe, and J. Cuzzi. *Nature*, 263:398–400, 1976.

[1932] D. T. Wickramasinghe and L. Ferrario. *Pub. Astron. Soc. Pac.*, 112:873–924, 2000.

[1933] M. Wiescher, J. Gorres, F.-K. Thielemann, and H. Ritter. *Astron. Astrophys.*, 160:56–72, 1986.

[1934] E. P. Wigner. *Phys. Rev.*, 70:606–618, 1946.

[1935] E. P. Wigner. *Phys. Rev.*, 70:15–33, 1946.

[1936] E. P. Wigner and L. Eisenbud. *Phys. Rev.*, 72:29–41, 1947.

[1937] R. Wijnands. *Astrophys. J. Lett.*, 554:L59–L62, 2001.

[1938] R. Wijnands, C. O. Heinke, D. Pooley, et al. *Astrophys. J.*, 618:883–890, 2005.

[1939] D.B. Williams and C.B. Carter. *Transmission Electron Microscopy: A Textbook for Materials Science*. 2nd Ed., Springer, New York, 2009.

[1940] F.D. Williams. *Combustion Theory*. 2nd Ed., Addison-Wesley, Menlo Park (California), 1985.

[1941] S. C. Williams, M. F. Bode, M. J. Darnley, et al. *Astrophys. J. Lett.*, 777:L32 (4 pp), 2013.

[1942] J. R. Wilson. *Astrophys. J.*, 163:209–219, 1971.

[1943] J. R. Wilson. *Phys. Rev. Lett.*, 32:849–852, 1974.

[1944] J. R. Wilson. In *Numerical Astrophysics*, J. M. Centrella, J. M. Leblanc, and R. L. Bowers (editor). Jones and Bartlett Publ., Boston (Massachusetts), 1985, pp 422–434.

[1945] J. R. Wilson, R. Couch, S. Cochran, J. Le Blanc, and Z. Barkat. *Ann. NY Acad. Sci.*, 262:54–64, 1975.

[1946] J. R. Wilson, R. Mayle, S. E. Woosley, and T. Weaver. *Ann. NY Acad. Sci.*, 470:267–293, 1986.

[1947] O. C. Wilson. *Astrophys. J.*, 90:634–636, 1939.

[1948] R. E. Wilson. *Astron. Astrophys.*, 99:43–47, 1981.

[1949] R. G. Wilson. *Int. J. Mass Spectrom.*, 143:43–49, 1995.

[1950] J. A. Winger, D. P. Bazin, W. Benenson, et al. *Phys. Rev. C*, 48:3097–3105, 1993.

[1951] P. F. Winkler, G. Gupta, and K. S. Long. *Astrophys. J.*, 585:324–335, 2003.

[1952] R. D. Wolstencroft and J. C. Kemp. *Nature*, 238:452, 1972.

[1953] D. Wood. *Mon. Not. R. Astron. Soc.*, 194:201–218, 1981.

[1954] K. S. Wood, J. F. Meekins, D. J. Yentis, et al. *Astrophys. J. Suppl. S.*, 56:507–649, 1984.

[1955] P. R. Wood. *Astrophys. J.*, 190:609–630, 1974.

[1956] W. M. Wood-Vasey, A. S. Friedman, J. S. Bloom, et al. *Astrophys. J.*, 689:377–390, 2008.

[1957] C. E. Woodward, G. F. Lawrence, R. D. Gehrz, et al. *Astrophys. J. Lett.*, 408:L37–L40, 1993.

[1958] P. Woodward and P. Colella. *J. Comput. Phys.*, 54:115–173, 1984.

[1959] S. Woosley and T. Janka. *Nature Phys.*, 1:147–154, 2005.

[1960] S. E. Woosley. *Nature*, 269:42–44, 1977.

[1961] S. E. Woosley. In *Nucleosynthesis and Chemical Evolution*, J. Audouze, C. Chiosi, and S. E. Woosley (editor). Geneva Obs., Sauverny, Switzerland, 1986, pp 1–195.

[1962] S. E. Woosley. *Astrophys. J.*, 330:218–253, 1988.

[1963] S. E. Woosley. In *Supernovae*, A. G. Petschek (editor). Springer-Verlag, Berlin, Germany, 1990, pp 182–212.

[1964] S. E. Woosley. *Astrophys. J.*, 405:273–277, 1993.

[1965] S. E. Woosley. *Astrophys. J.*, 476:801–810, 1997.

[1966] S. E. Woosley. In *Thermonuclear Supernovae*, P. Ruiz-Lapuente, R. Canal, and J. Isern (editor). Kluwer Acad. Publ., Dordrecht, The Netherlands, 1997, pp 313–336.

[1967] S. E. Woosley. *Astrophys. J.*, 525:C924–C925, 1999.

[1968] S. E. Woosley, W. D. Arnett, and D. D. Clayton. *Astrophys. J.*, 175:731–749, 1972.

[1969] S. E. Woosley, W. D. Arnett, and D. D. Clayton. *Astrophys. J. Suppl. S.*, 26:231–312, 1973.

[1970] S. E. Woosley, T. S. Axelrod, and T. A. Weaver. In *Stellar Nucleosynthesis*, C. Chiosi and A. Renzini (editor). Reidel, Dordrecht, The Netherlands, 1984, pp 263–293.

[1971] S. E. Woosley and J. S. Bloom. *Annu. Rev. Astron. Astr.*, 44:507–556, 2006.

[1972] S. E. Woosley, D. Hartmann, and P. A. Pinto. *Astrophys. J.*, 346:395–404, 1989.

[1973] S. E. Woosley, D. H. Hartmann, R. D. Hoffman, and W. C. Haxton. *Astrophys. J.*, 356:272–301, 1990.

[1974] S. E. Woosley and W. C. Haxton. *Nature*, 334:45–47, 1988.

[1975] S. E. Woosley and A. Heger. *Phys. Rep.*, 442:269–283, 2007.

[1976] S. E. Woosley, A. Heger, A. Cumming, et al. *Astrophys. J. Suppl. S.*, 151:75–102, 2004.

[1977] S. E. Woosley, A. Heger, and T. A. Weaver. *Rev. Mod. Phys.*, 74:1015–1071, 2002.

[1978] S. E. Woosley and R. D. Hoffman. *Astrophys. J.*, 395:202–239, 1992.

[1979] S. E. Woosley and W. M. Howard. *Astrophys. J. Suppl. S.*, 36:285–304, 1978.

[1980] S. E. Woosley and D. Kasen. *Astrophys. J.*, 734:38 (27 pp), 2011.

[1981] S. E. Woosley, N. Langer, and T. A. Weaver. *Astrophys. J.*, 411:823–839, 1993.

[1982] S. E. Woosley, N. Langer, and T. A. Weaver. *Astrophys. J.*, 448:315–338, 1995.

[1983] S. E. Woosley and R. E. Taam. *Nature*, 263:101–103, 1976.

[1984] S. E. Woosley and T. A. Weaver. In *Essays in Nuclear Astrophysics*, C. A. Barnes, D. D. Clayton, and D. N. Schramm (editor). Cambridge Univ. Press, Cambridge, UK, 1982, pp 377–400.

[1985] S. E. Woosley and T. A. Weaver. In *High Energy Transients in Astrophysics*, S. E. Woosley (editor). American Inst. Phys., New York, 1984, pp 273–297.

[1986] S. E. Woosley and T. A. Weaver. *Annu. Rev. Astron. Astr.*, 24:205–253, 1986.

[1987] S. E. Woosley and T. A. Weaver. In *Supernovae*, S. A. Bludman, R. Mochkovitch, and J. Zinn-Justin (editor). North Holland Pub., Amsterdam, The Netherlands, 1994, pp 63–154.

[1988] S. E. Woosley and T. A. Weaver. *Astrophys. J.*, 423:371–379, 1994.

[1989] S. E. Woosley and T. A. Weaver. *Astrophys. J. Suppl. S.*, 101:181–235, 1995.

[1990] S. E. Woosley, J. R. Wilson, G. J. Mathews, R. D. Hoffman, and B. S. Meyer. *Astrophys. J.*, 433:229–246, 1994.

[1991] B. Wopenka, Y. C. Xu, E. Zinner, and S. Amari. *Geochim. Cosmochim. Acta*, 106:463–489, 2013.

[1992] H. Worpel, D. K. Galloway, and D. J. Price. *Astrophys. J.*, 772:94 (14 pp), 2013.

[1993] C. Wrede. *AIP Adv.*, 4:041004 (17 pp), 2014.

[1994] C.-C. Wu, M. Leventhal, C. L. Sarazin, and T. R. Gull. *Astrophys. J. Lett.*, 269:L5–L9, 1983.

[1995] Z.-Z. Xi and S.-R. Bo. *Acta Astronom. Sinica*, 13:1–21, 1965.

[1996] Y. Xu, K. Takahashi, S. Goriely, et al. *Nucl. Phys. A*, 918:61–169, 2013.

[1997] T. Yada, C. Floss, F. J. Stadermann, et al. *Meteorit. Planet. Sci.*, 43:1287–1298, 2008.

[1998] R. Z. Yahel, W. Brinkmann, and A. Braun. *Astron. Astrophys.*, 176:223–234, 1987.

[1999] D. G. Yakovlev, P. Haensel, G. Baym, and C. Pethick. *Phys. Usp.*, 56:289–295, 2013.

[2000] M. Yamanaka, K. S. Kawabata, K. Kinugasa, et al. *Astrophys. J. Lett.*, 707:L118–L122, 2009.

[2001] O. Yaron, D. Prialnik, M. M. Shara, and A. Kovetz. *Astrophys. J.*, 623:398–410, 2005.

[2002] Q.-Z. Yin, C.-T. A. Lee, and U. Ott. *Astrophys. J.*, 647:676–684, 2006.

[2003] S.-C. Yoon, P. Podsiadlowski, and S. Rosswog. *Mon. Not. R. Astron. Soc.*, 380:933–948, 2007.

[2004] T. Yoshida, T. Suzuki, S. Chiba, et al. *Astrophys. J*, 686:448–466, 2008.

[2005] T. Yoshida, M. Terasawa, T. Kajino, and K. Sumiyoshi. *Astrophys. J.*, 600:204–213, 2004.

[2006] P. A. Young, C. L. Fryer, A. Hungerford, et al. *Astrophys. J.*, 640:891–900, 2006.

[2007] P. A. Young, E. E. Mamajek, D. Arnett, and J. Liebert. *Astrophys. J.*, 556:230–244, 2001.

[2008] T. R. Young and D. Branch. *Astrophys. J. Lett.*, 342:L79–L82, 1989.

[2009] S. Yu and C. S. Jeffery. *Mon. Not. R. Astron. Soc.*, 417:1392–1401, 2011.

[2010] T. Yu, B. S. Meyer, and D. D. Clayton. *Astrophys. J.*, 769:38 (6 pp), 2013.

[2011] L. R. Yungelson and M. Livio. *Astrophys. J.*, 497:168–177, 1998.

[2012] L. R. Yungelson and M. Livio. *Astrophys. J.*, 528:108–117, 2000.

[2013] L. R. Yungelson, M. Livio, A. V. Tutukov, and R. A. Saffer. *Astrophys. J.*, 420:336–340, 1994.

[2014] H. Yurimoto. In *Origin of Matter and Evolution of Galaxies 2011*, S. Kubono, T. Hayakawa, T. Kajino, H. Miyatake, T. Motobayashi, and K. Nomoto (editor). American Inst. Phys., Melville (New York), 2012, pp 139–141.

[2015] M. Zamfir, A. Cumming, and C. Niquette. *Mon. Not. R. Astron. Soc.*, 445:3278–3288, 2014.

[2016] T. J. Zega, L. R. Nittler, F. Gyngard, et al. *Geochim. Cosmochim. Acta*, 124:152–169, 2014.

[2017] T. J. Zega, C. M. O'D. Alexander, L. R. Nittler, and R. M. Stroud. *Astrophys. J.*, 730:83 (10 pp), 2011.

[2018] Y. B. Zel'dovich and Y. P. Raizer. *Physics of Shock Waves and High-Temperature Hydrodynamic Phenomena*. Academic Press, New York, 1967.

[2019] Y.B. Zel'dovich. *Zh. Eksp. Teor. Fiz.*, 10:542–568, 1940.

[2020] B.-B. Zhang, B. Zhang, K. Murase, V. Connaughton, and M. S. Briggs. *Astrophys. J.*, 787:66 (9 pp), 2014.

[2021] G. Zhang, M. Méndez, D. Altamirano, T. M. Belloni, and J. Homan. *Mon. Not. R. Astron. Soc.*, 398:368–374, 2009.

[2022] F.-Y. Zhao, R. G. Strom, and S.-Y. Jiang. *Chinese J. Astron. Ast.*, 6:635–640, 2006.

[2023] X. Zhuge, J. M. Centrella, and S. L. W. McMillan. *Phys. Rev. D*, 50:6247–6261, 1994.

[2024] M. Zingale, C. M. Malone, A. Nonaka, A. S. Almgren, and J. B. Bell. *ArXiv e-prints*, 2014.

[2025] M. Zingale, F. X. Timmes, B. Fryxell, et al. *Astrophys. J. Suppl. S.*, 133:195–220, 2001.

[2026] M. Zingale, S. E. Woosley, J. B. Bell, M. S. Day, and C. A. Rendleman. *J. Phys. Conf. Ser.*, 16:405–409, 2005.

[2027] E. Zinner, M. Jadhav, F. Gyngard, and L. R. Nittler. *Meteorit. Planet. Sci. Suppl.*, 73:Abstract #5137, 2010.

[2028] E. Zinner, M. Jadhav, F. Gyngard, and L. R. Nittler. 42nd *Lunar and Planetary Science Conf.*, Abstract #1070, 2011.

[2029] E. K. Zinner. *Annu. Rev. Earth Pl. Sc.*, 26:147–188, 1998.

[2030] E. K. Zinner. *Meteorit. Planet. Sci.*, 33:549–564, 1998.

[2031] E. K. Zinner. In *Meteorites, Comets and Planets: Treatise on Geochemistry, Vol. 1*, A. M. Davis (editor). 2nd Ed., Elsevier B. V., Amsterdam, The Netherlands, 2014, pp 181–213.

[2032] E. K. Zinner, S. Amari, R. Guinness, et al. *Geochim. Cosmochim. Acta*, 71:4786–4813, 2007.

[2033] E. K. Zinner, S. Amari, R. Guinness, et al. *Geochim. Cosmochim. Acta*, 67:5083–5095, 2003.

[2034] E. K. Zinner, S. Amari, and R. S. Lewis. *Astrophys. J. Lett.*, 382:L47–L50, 1991.

[2035] E. K. Zinner, F. Gyngard, and L. R. Nittler. 41st *Lunar and Planetary Science Conf.*, Abstract #1359, 2010.

[2036] E. K. Zinner, T. Ming, and E. Anders. *Nature*, 330:730–732, 1987.

[2037] E. K. Zinner, F. Moynier, and R. M. Stroud. *P. Natl. Acad. Sci. USA*, 108:19135–19141, 2011.

[2038] E. K. Zinner and M. Tang. 19th *Lunar and Planetary Science Conf.*, Abstract #1323, 1988.

[2039] M. E. Zolensky, C. Pieters, B. Clark, and J. J. Papike. *Meteorit. Planet. Sci.*, 35:9–29, 2000.

Printed in the United States
by Baker & Taylor Publisher Services